GENETICS, PALEONTOLOGY, AND MACROEVOLUTION

Genetics, Paleontology, and Macroevolution

JEFFREY LEVINTON
State University of New York at Stony Brook

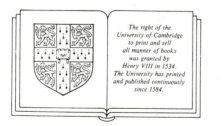

The right of the
University of Cambridge
to print and sell
all manner of books
was granted by
Henry VIII in 1534.
The University has printed
and published continuously
since 1584.

CAMBRIDGE UNIVERSITY PRESS
Cambridge
New York   New Rochelle   Melbourne   Sydney

Published by the Press Syndicate of the University of Cambridge
The Pitt Building, Trumpington Street, Cambridge CB2 1RP
32 East 57th Street, New York, NY 10022, USA
10 Stamford Road, Oakleigh, Melbourne 3166, Australia

© Cambridge University Press 1988

First published 1988

Printed in the United States of America

Library of Congress Cataloging-in-Publication Data

Levinton, Jeffrey S.

Genetics, paleontology, and macroevolution.

Bibliography: p.
1. Evolution.   2. Genetics.   3. Paleontology.
I. Title.
QH366.2.L47 1987   575   87-11730

British Library Cataloguing in Publication Data

Levinton, Jeffrey S.

Genetics, paleontology and macroevolution.

1. Evolution
I. Title
575   QH366.2

ISBN 0-521-24933-3

For Joan, as ever always

 To talk casually
About an iris flower
Is one of the pleasures
Of the wandering journey
   Matsuo Bashō

Contents

Preface

I have so many things to write about, that my head is as
full of oddly assorted ideas, as a bottle on the table is filled
with animals.
Charles Darwin, 1832, Rio de Janeiro

Evolutionary biology enjoys the peculiar dual status of being that
subject which clearly unites all biological endeavors, while occasionally
seeming to be nearly as remote from complete understanding as when
Darwin brought it within the realm of materialistic science. Somehow,
the basic precepts first proposed by Darwin have never been either fully
accepted nor disposed, to be followed by a movement towards further
progress in some other direction. The arguments of today—the questions of natural selection and adaptation, saltation versus gradualism,
and questions of relatedness among organisms—are not all that different from those discussed one hundred years ago, even if the research
materials seem that much more sophisticated.

Darwin espoused thinking in terms of populations. His approach
was open to experimentation, but this had to await the (re)discovery
of genetics a half a century later, before a major impediment to our
understanding could be thrown aside. As it turned out, the rediscovery
of genetics was initially more confusing than helpful to our understanding of evolution. The rediscovery of genetically transmissible discrete
traits revived saltationism, and it took over a decade for biologists to
realize that there was no conflict between the origin of discrete variants
and the theory of natural selection. In the twentieth century, the focus of experimentalists has moved towards processes occurring within
populations. But many of the inherently most fascinating questions lie
at higher taxonomic levels, or at greater distances of relationship than
between individuals in a population. The questions are both descriptive and mechanistic. We would like to know just how to describe the
difference between a lizard and an elephant, in terms that would make
it possible to conceive of the evolutionary links between them. We
are only now beginning to do this, principally at the molecular genetic
level. Differences in nucleotide sequences are beginning to have more
meaning at this level, especially because of the emerging knowledge of

gene regulation. But we would also like to understand the mechanisms behind the evolutionary process at higher levels of morphological organization. This inevitably involves a knowledge of history with all the limitations that that subject embraces. Just how can we be sure about biological historical facts? Surely the fossil record must come into play here, even if it is scattered in preservation.

I will try here to provide an approach to studying macroevolution, which I define to be the study of transitions between related groups of distant taxonomic rank. The formula is simple. First we must have a sound systematic base that is derived from a well-established network of genealogical relationships. Otherwise we cannot ask the appropriate questions in the first place. Second we must be able to describe the differences between organisms in molecular, developmental, morphological, and genetic terms. Third we must understand the processes of evolution at all levels from the nature of polymorphisms to the appearance and extinction of major groups. Finally, we must have a criterion by which adaptation can be judged. It may not be true that one group is inherently superior to another unrelated group. But if we cannot devise a criterion for increases in performance, even in biologically complex organisms, then we will not be able to test Darwin's claim that evolution involves improvement (not perfection) in a given context of an organism-environment relationship.

Because the problems require such a broad scope of approaches and solutions, our understanding of macroevolution is often mired in arguments that appear, then disappear, then reappear, with no real sense of progress. The saltationist-gradualist argument has had such a history, simply because of our lack of knowledge as to what saltation really means and the usual lack of a good historical record. Because evolutionary biologists tend to reason by example, it is easy to "prove a point" by citing a hopelessly obscure case or one which may turn out to be unusual. Yet it seems fruitless to settle an argument by counting up all of the examples to prove a claim, without some theoretical reason to expect the majority of cases to fit in the first place. This danger is endemic to a science which depends upon history. Most biologists would be quite disappointed if evolutionary biology were nothing much more than a form of stamp collecting. We look for theories and principles.

It is my hope that this volume will provide a framework within which to view macroevolution. I don't pretend to solve the important issues, but I do hope to redirect graduate students and colleagues toward some fruitful directions of thought. While I like to think that this is a balanced presentation, my shortcomings and prejudices will often surface. In particular, this volume will resort to advocacy when attacking the view of evolution that speciation is a fundamental level of evolutionary change in the macroevolutionary perspective, and that the neo-Darwinian movement and the Modern Synthesis somehow undermined our ability to understand the process of evolution and brought us to our present pass of misunderstanding. The recent "born again" moves toward saltationism, and the staunchly ideological adherence to related restrictive concepts, such as punctuated equilibria, are great leaps backward and have already led many toward unproductive dead ends that are more filled with rhetoric than scientific progress. Ultimately this is a pity, since some of these ideas have been interesting, and have exposed unresolved issues in evolutionary theory.

While this book is principally meant to be a blueprint for the study of macroevolution, I found it necessary to discuss certain areas at an elementary level. This is partially owing to the heterogeneous audience that I anticipate. I doubt that most paleontologists will be aware of the details of genetics, and neontologists will similarly benefit from some geological introduction.

Many colleagues were very generous with their time in reviewing this manuscript. I thank the following who reviewed one or more chapters. Richard K. Bambach (Chapters 1–8), Michael J. Bell (Chapters 3, 4, 7), Stefan Bengtson (Chapters 7, 8), John T. Bonner (Chapters 1–8), Peter W. Bretsky, Jr. (Chapters 7, 8), Brian Charlesworth (Chapters 3, 7, 8), John Cisne (Chapter 7), Richard Cowan (Chapters 6, 7), Gabriel Dover (part of Chapter 3), Walter Eanes (Chapters 3, 4), Joseph Felsenstein (Chapter 2), Karl Flessa (Chapter 8), Douglas Futuyma (Chapters 1, 3, 4), Paul Harvey (part of Chapter 6), Max Hecht (Chapters 1–8), George Lauder (Chapter 6), Jack Sepkoski (Chapter 8), David Wake (Chapter 5), and especially David Jablonski (Chapters 1–9). This sounds like extensive reviewing; but consider my extensive ignorance.

I also have been lucky to have had conversations or correspondence with many individuals who gave me useful information, their unpublished works, letters, insights, and important references. Among them, I am grateful to Bill Atchley, David Wake, Björn Kurtén, Lars Werdelin, Steve Orzack, John Maynard Smith, Brian Charlesworth, Michael Bell, Pete Bretsky, Gabriel Dover, Steve Farris, Steve Stanley, Doug Futuyma, Walter Eanes, Curt Teichert, George Oster, Richard Reyment, Jürgen Schöbel, Max Hecht, Russ Lande, Art Boucot, Ledyard Stebbins, Vjaldar Jaanusson, Ernst Mayr, George Gaylord Simpson, Jack Sepkoski, and Urjö Haila.

This manuscript was typeset using the Document Composition Facility at the Biological Sciences Computing Facility at Stony Brook. I am very grateful to Dave Van Voorhees, who, in the main, formatted the manuscript into appropriate files. Scott Ferson, Kent Fiala, and Jim Rohlf were infinitely patient with our questions, and all contributed materially to our ability to produce the final product. I am also very grateful to Mitzi Eisel and to Marie Gladwish for skillfully preparing most of the figures. I also thank Richard Ziemacki, Helen Wheeler, Jim DeMartino, Peter–John Leone, and especially Rhona Johnson, all of Cambridge University Press, for their patience and kindness. Most of all, I am grateful to my wife Joan, who made life so easy (at least for me) while I prepared the manuscript.

I am very grateful for the hospitality of Staffan Ulfstrand, Zoology Department of the University of Uppsala; Gabriel Dover of the Department of Genetics and Kings College, University of Cambridge; Catherine Thiriot, Odile Mayzaud, and Patrick Mayzaud, all of the Station Zoologique, Villefranche-Sur-Mer, France; and Jacques Soyer, Laboratoire Arago, Banyuls-Sur-Mer, France. I also am deeply grateful to the Guggenheim Foundation, which mainly supported the writing of this work.

Banyuls-Sur-Mer and Stony Brook

Chapter 1

Macroevolution: The Problem and the Field

Darwin first shewed that evolution is a problem and no vain riddle
W. Bateson

The Process and the Field of Macroevolution

The Return of Macroevolution

Macroevolution has returned to occupy center stage in evolutionary biology. We have long desired to know how best to describe and organize the diversity of life, and to know how and why this diversity came to be. No mystery is more intriguing than why we have amoebas and horses, or dandelions and palms. The child's first walk in a meadow, seeing flowers and butterflies for the first time, can inspire the same wonder in the most sophisticated biologist walking those same tracks many years later.

The return to this larger perspective comes from many quarters of biology and paleontology. The advances in molecular genetics and developmental biology in recent years have only increased our confidence that the nature of living systems can be understood mechanistically; we can now imagine the possibility of describing the difference between organisms in terms of their genes, gene products, and spatial organization. The large-scale collation of fossil data and a new understanding of the history of the earth have brought similar increases of confidence among geologists and paleontologists. But we should not overlook some significant changes in fields such as systematics, and the

1

crucial groundwork in population biology established through the advances of the neo-Darwinian movement and the Modern Synthesis. All these place us in position to answer questions that could not even be asked very seriously just a few decades ago. Our fascination with the diversity of life may yet be satisfied with intellectual progress beyond Darwin.

Definition of the Process of Macroevolution

I define macroevolution in order to free it from any dependence upon specific controversies of the past few years and to outline a field that is now beginning to emerge. I define the process of macroevolution to be (Levinton 1983) **the sum of those processes that explain the character-state transitions that diagnose evolutionary differences of major taxonomic rank.**

This definition of macroevolution focuses upon character-state differences (defined in Chapter 2) rather than on jumps, for example, from one taxon to another of great distance. My definition is noncommittal to any particular taxonomic level. I believe that one should eschew definitions of macroevolution such as (a) evolution above the species level (e.g., Eldredge and Cracraft 1980; Stebbins and Ayala 1981) or (b) evolution caused by speciation and selection among species (e.g., Stanley 1979). These definitions presume that major transitions can be analyzed properly only by examining speciation and other processes occurring at the species level and above, and they restrict our views toward alternative hypotheses.

It is not useful to distinguish sharply between microevolution and macroevolution, as I will show in this volume. The taxonomic rank marking any dichotomy between microevolution and macroevolution would depend on the kind of transition being studied. Our impression of "major" degrees of evolutionary change is inherently qualitative and not fixed at any taxonomic rank across all major taxonomic groups. This is apparent when we consider transitions whose importance may rely upon many characters, or just one. For the cichlid fishes, a synarthrosis between the lower pharyngeal jaws, a shift of insertion of the fourth levator externus muscles, and the development of synovial joints between the upper pharyngeal jaws and the basicranium may be necessary (but not sufficient) for the morphological diversification of species with differing food collection devices (Liem

1973). On the other hand, the evolution of the mammals involved a large number of integrated physiological and morphological traits, and these were acquired over a long period of time (Kemp 1982). Yet both fall well within the province of macroevolutionary change, because of the potential at least for evolutionary differences spanning large chasms of taxonomic rank.

A second reason for an unrestricted definition of the taxonomic level required to diagnose macroevolutionary change is the variation in higher level taxonomic splitting among major groups (Van Valen 1973a). There is no simple way of drawing an equivalence between families of mammals and mollusks; comparisons of rates of evolution between groups at "comparable" taxonomic levels (e.g., Stanley 1973a) are therefore usually invalid (Levinton 1983; Van Valen 1973a). This point is illustrated well by qualitative studies on hybridity and genetic and phenotypic distance within groups of species of similar taxonomic distance from different phyla. The taxonomist tends to use a qualitative threshold of phenetic difference to define significant evolutionary distance. Thus the ferret and the stoat are placed in different genera, even though they hybridize and produce fertile offspring. Crosses between congeneric species of frogs, however, do not usually produce viable, let alone fertile, offspring.

The difficulty of distilling an unambiguous definition of macroevolution is influenced by our current ignorance of the relationship between morphological and genetic divergence among distantly related taxa. By what proportion of the genome do chimpanzees and man differ? Despite our available estimates of genetic differentiation from sequenced DNA and protein amino acid sequences, allozymes, and karyotypes, we cannot draw a parallel with our knowledge of morphological differences. We are crippled by this ignorance when seeking to judge how "hard" it is for evolutionary transition to take place. What is our standard of difficulty? Genetic? Functional morphological? Developmental? Worse than that, what if interactions among these three occur?

My last justification for a somewhat loose definition is that it permits an expansion of previous evolutionary theory to embrace the larger scale hierarchical processes (see below) and higher level taxonomic variations previously ignored by the bulk of evolutionary biologists, except in passing or in gratuitous extrapolation from lower taxonomic levels of concern. It is my hope that my definition will eventually not be needed and that "macroevolution" will merge with "microevolution" to become a discipline without a needless dichotomy. The need for the return of macroevolution today, in my view, is more to sell the expansion of approaches than to necessarily dismiss any previous theory.

Definition of the Scope of Macroevolution

The discipline of macroevolution should include those fields that are needed to elucidate the processes involved in accomplishing the change from one taxonomic state to another of significant distance. These fields can be organized around several basic questions:

1. How do we establish the phylogenetic relationships among taxa?

2. What is the nature of evolutionary novelty?

3. How do genetic, developmental, and morphological constraints channel the course of morphological and genetic evolution?

4. What are the patterns of change and what processes regulate the rate of evolutionary change from one character state to another?

5. What ecological processes regulate the commonality of timing of evolutionary radiations and extinctions? What is the role of extinction in the evolutionary potential of newly evolved or surviving groups?

6. What ecological processes regulate morphological and species diversity? To what degree do these effects have evolutionary consequences for any given group?

In the following chapters I will try to support the following assertions:

1. Systematics is the linchpin of macroevolutionary studies. Without an acceptable network of phylogenetic relationships it is impossible to investigate the possible paths of major evolutionary change (Chapter 2).

2. The nature of evolutionary novelty is probably the most studied and still the most confused element of evolutionary biology. The presence of discontinuity in morphological state can be explained

readily using the available data and theory of genetics (Chapters 3 and 5). The mechanisms behind the discontinuities are more poorly understood and may relate to a complex interaction between genetic and developmental processes (Chapter 5). The epigenetic processes are also subject to genetic control and thus a spectrum of resultant morphologies can be discontinuous.

3. There is no evidence that morphological evolution is accelerated or associated with speciation, except as an effect of ecologically unique circumstances leading to directional selection. Intraspecific variation during the history of a species is the stuff of interspecific morphological differentiation (Chapter 4).

4. Developmental biology presents a pattern consistent with conservatism in evolution. We know enough to judge that the part of the genome specifying major body plans contains strongly conservative elements behind a superposed record of extensive evolutionary change (e.g., in the vertebrates). We do not understand as yet the genetic basis of the conservatism, nor can we be sure whether it represents either a developmental constraint or stabilizing selection (Chapter 5).

5. The nature of form is best understood within the framework of Adolph Seilacher's concept of **Constructional Morphology**. Constructional, Phylogenetic-Developmental, and Functional Morphological factors interact to determine form. This combination tends to make evolutionary pathways often eccentric and not conducive to predictions from "ground up" engineering approaches to optimality. Having recognized historical constraints, however, optimality approaches can be used to gauge the performance of alternative morphotypes. Indeed, without such an approach, studies of adaptation would be vacuous (Chapter 6).

6. Having understood the nature of variation, we find little evidence that the fossil record consists of anything more than the standard variation within populations that can be studied by evolutionary biologists. The process of macroevolution need not invoke paroxysmal change in genetics or morphology. The genetic basis of

morphological change, nevertheless, involves a considerable variety of mechanisms. Morphological evolution is not the necessary consequence of speciation, though it may be a cause of speciation (Chapters 4 and 7).

7. *Baupläne* are evolved piecemeal. Trends leading to complex forms consist of a large number of specific changes acquired throughout the history of the origin of the derived *bauplan* (Chapter 7). Subsequently, however, stability is common. Some trends, such as a general increase in invertebrate predator defense, and reductions in variation of morphologies, are probably due, to a degree, to the selective success and extinction of different taxa. Even though speciation rate is not related causally to the origin of the novelty, intertaxon survival, sometimes due to random extinction, has been a crucial determinant of the present and past complexion of the biotic world (Chapter 8).

8. Although earth history has had a clear impact on diversification and standing diversity, patterns of taxonomic longevity may have had a distinctly random component. Major differences in biology may have consequences for rates of morphological evolution and speciation, but patterns of distribution within these groups may reflect random appearance-extinction processes (Chapter 8).

9. Mass extinctions and radiations are a fact of the fossil record. But both are more easily recognized by changes in the biota than by any recognizable physical events. Means of distinguishing among current hypotheses of regulation of mass extinction and radiation are equivocal at best (Chapter 8).

10. The evolutionary and the fossil records support the Theory of Commitment. Functional integration, habitat selection, and decrease in flexibility of developmental mechanisms increase over time, thus reducing a population's ability for extensive evolutionary change. The interaction of these three factors contributes to stasis. It is an important task of evolutionary biology to understand how these three factors interact to specify the potential variability in populations upon which natural selection can work (Chapter 9).

Is Macroevolution Something Apart from Microevolution?

Arguments surrounding macroevolution have focused in the past upon Goldschmidt's conception of macroevolution as a singular, revolutionary step from an ancestral to a descendant species, via a systemic mutation (see below). Even more recent controversies such as the importance of speciation in morphological evolution, debate the reality of a decoupling of microevolution from macroevolution (e.g., Stanley 1975). In this context, if the microevolutionists win, then there is no such thing as macroevolution. If the macroevolutionists gain favor, then microevolution exists, but it is a minor part of a much larger set of evolutionary constructs. Macroevolutionist claims began by relegating microevolution to the ash heap of history (e.g., Gould 1980a). Subsequent arguments have softened, only emphasizing the expansion of evolutionary theory offered by macroevolutionary considerations (Gould 1982a).

Is the dichotomy very useful? For one group to "win" conveniently ensures the irrelevance of the other to major contributions in evolutionary theory. The focus of this argument is at the speciation threshold of evolution. But I hope that the reader realizes already that there is much more to paleontological and neontological macroevolutionary arguments than the nature of speciation. Macroevolution must be a field that embraces the ecological theater including the range of time scales of the ecologist, to the sweeping historical changes only available to paleontological study. It must include the peculiarities of history, which must have had singular effects on the directions that the composition of the world's biota took (e.g., the splitting of continents, the establishment of land and oceanic isthmuses). It must take the entire network of phylogenetic relationships and superpose a framework of genetic relationships. Then the nature of constraint of evolutionary directions and the qualitative transformation of ancestor to descendant over major taxonomic distances must be explained.

These are not fields that have been tackled with gusto by the founders of the Modern Synthesis in the past fifty years. Instead, the followers of the Modern Synthesis have been devising theories oriented toward explaining changes in gene frequencies or small-scale evolutionary events, leaving it to someone else to go through the trouble of working in larger time scales and considering the larger historical scale so important to the grand sweep of evolution within sight of the horizon of the paleontologist. The developmental/genetic mechanisms that generate variation (what used to be called physiological genetics) have also been neglected. Population geneticists assume variation, but do not study how it is generated nearly as much as they worry about the fate of variation as it is selected, or lost by stochastic processes.

The renaissance in the paleontological analysis of major trends in evolution (e.g., Raup 1976*a, b*; Sepkoski et al. 1981), a revival of interest in the relationship of development to evolution (Bonner 1982; Gould 1977; Raff and Kaufman 1983), a possible future understanding of the genetics of development (e.g., Davidson 1982), and major advances in the understanding of the evolutionary structure of the genome (Nei and Koehn 1983) all place us at a period in history where such macroevolutionary questions can now be asked. I wonder if any success could have been achieved if these questions had been asked thirty to forty years ago.

The field of macroevolution depends upon an important consideration, usually understood implicitly but avoided in practice by most evolutionary biologists in their research. The evolutionary process is so complex and so interactive at all levels of the evolutionary hierarchy that no study of one locus or even one trait will usually explain what we want to explain, which is how and why a given historical evolutionary event took place.

Evolutionary biology and the field of observational astronomy share the same intellectual problems. Astronomers search the heavens, accumulate logs of stars, analyze various energy spectra, and note motions of bodies in space. A set of physical laws permits interpretations of the present "snapshot of the universe" afforded by the various telescopic techniques available to us. To the degree that the physical laws permit unambiguous interpretations, conclusions can be drawn about the consistency of certain observations with hypotheses. Thus rapid and cyclical changes in light intensity led to the proof of the reality of pulsars. The large-scale structure of the universe inspires a more historical hypothesis: the big bang origin of the universe.

The evolutionary biologist confronts the same problem. A set of organisms exists today in a partially measurable state of spatial, morphological, and chemical relationships. We have a set of physical and biological laws that might be used to construct predictions about the outcome of the evolutionary process. But, as we all know, we are not very successful, except at solving seemingly trivial problems. We have plausible explanations for the reason why moths living in industrialized areas are rich in dark pigment, but we don't know whether or why life arose more than once or why some groups became extinct (e.g., the dinosaurs) while others managed to survive (e.g., horseshoe crabs). Either our laws are inadequate and we have not described the available evidence properly, or no laws can be devised to predict uniquely what should have happened in the history of life. It is the field of macroevolution that should consider such issues.

Hierarchy and Evolutionary Analysis

We need a context within which to study macroevolution. J. W. Valentine (1968, 1969) first suggested to paleontologists that large-scale evolutionary studies should use a hierarchical framework. This issue has been revived recently (e.g., Gould 1982a; Vrba and Eldredge 1984; Eldredge 1985; Salthe 1985; Vrba and Gould 1986). Allen and Starr (1982) provide many examples of the necessity of a hierarchical approach.

Hierarchies can be used either as an epistemological convenience, or as a necessary ontological framework for evolutionary thought. Both approaches have been taken in the past, sometimes with the same hierarchy. The standard taxonomic hierarchy is used commonly as a means to examine rates of appearance and extinction. Although different taxonomic levels may change differently over time, such studies do not assign special significance to these levels, as opposed to another set of levels that might also be studied (e.g., studying species, subfamilies, and families, as opposed to species, families, and orders). They are just conveniences, whose ascending order of ranking may correlate with differences of response (e.g., Valentine 1969). On the other hand, some regard certain taxonomic levels as fundamental and of ontological significance. Van Valen (1984a) sees the family level as a unit of adaptation. The species has been claimed to have great importance (Eldredge and Gould 1972). I and most neo-Darwinians see the organism as a fundamental level of the hierarchy, around which all other processes turn. If a given taxonomic level has meaning, it is because the traits of an organism can be traced to this taxonomic level.

We use hierarchy in the sense of a series of nested sets. Higher levels are therefore more inclusive. There are at least two main hier-

archies that we must consider: organismic-taxonomic and ecological.
The organismic-taxonomic hierarchy can be ordered as:

> molecules–organelle–cell–tissue–organ
> –organism–population–species– monophyletic group

A variant of this hierarchy would include the substitution of gene–
chromosome–organism at the lower end.

The ecological hierarchy would include:

> organism–population–community

There is no necessary correspondence, however, between levels of the
ecological and organismic-taxonomic hierarchies.

If all processes could be studied exclusively with the smallest units
of the hierarchy, then two conclusions would readily follow. First, it
would not be necessary to study higher levels, i.e., there would be no
macroscopic principles. Second, higher levels would be simple sums
of the lower ones, with no unique characteristics of their own. The
first principle might lead a geneticist to claim that once genes are understood, the entire evolutionary process could be visualized as gene–
environmental interactions, with no consideration of the properties of
cells, organisms, species, or monophyletic groups. The second might
lead a paleontologist to argue that patterns of ordinal standing diversity are a direct reflection of species diversity (e.g., Sepkoski 1978).

Taking the hierarchy as given, we can ask the following questions:

1. Can one learn about the higher levels from the lower?

2. Can one understand processes at a given level without resorting
   to knowledge of other levels?

3. Is there any principle of interaction among levels, such as unidirectional effects exerted by lower levels on higher levels (e.g.,
   those of genes on individual survival) but not the reverse (the
   effect of survival of individual organisms on the future presence
   of the gene)?

The first question raises the issue of reductionism, a major area
of controversy in biology (e.g., Ayala and Dobzhansky 1974; Dawkins
1983; Lewontin 1970; papers in Sober 1984a; G.C. Williams 1966, 1985;
Wimsatt 1980; Vrba and Eldredge 1984). It is a common belief that all

aspects of biological organization can be explained if the entire genome were sequenced and all the nature and sequence of all proteins were known. In parallel with this argument, several biologists have proposed the gene as the unit of selection and the primary target of understanding. A theory at the level of the gene would then be extrapolated to a theory of the entire genome. In one case (G. C. Williams 1966) the claim was a healthy antidote to the proposal that certain forms of evolution can only be explained at another level of the hierarchy, the population (e.g., Wynne-Edwards 1962).

Although reductionism is often an object of strong criticism, there seems to be much confusion about definitions. There are at least three concepts that are often freely intermixed. First, reductionism can imply the presence of a **reducing science**, which can explain all phenomena within its own framework. In this conception of reductionism, biological constructs such as species, cells, and amino acids could be described in terms of the language and laws of physics. In evolutionary biology, the language and processes of Mendelian genetics might be substituted by the language and processes of molecular biology (Schaffner 1974). Second, reductionism is often used to imply **atomism**, where all phenomena of a science can be described effectively by laws involving the smallest ontological units. Thus one might claim that the extinction of the dinosaurs could be explained with knowledge of their nucleotide sequences only. This is the type of reductionism often under attack by macroevolutionists (e.g., Gould 1983b, Vrba and Eldredge 1984). Finally, some (e.g., Wimsatt 1980) attack reductionism as an impractical attempt to explain phenomena in terms of the smallest ontological units of a science. This does not imply that it is impossible to do so, only that it is so difficult that higher constructs of a hierarchy are more practical (Nagel 1961). This argument can also be made when, in order to adequately describe another science, the use of a reduced science requires a myriad of complexities in language (e.g., translating Mendelian genetics into molecular genetics–Hull 1974).

The confusion of these types of reductionism makes debate quite difficult. For example, the geneticist Richard Goldschmidt was a reductionist of the reducing science kind (G.E. Allen 1974), even if he is remembered for immortalizing the distinct break of the species level. He believed that chromosomal effects could be reduced to physical laws. Yet, Vrba and Eldredge (1984) place him on the side of holism.

As another example, Wimsatt (1980) criticizes the reductionist program, but only because it is impractical to explain many phenomena. From this argument alone, it would not be clear that he would reject the other two types of reductionism, if his objections to workability could be addressed. On the other hand, others find that certain levels have emergent properties, which are irreducible to lower levels of a hierarchy. This opinion, presumably, would also apply if a reducing science were available. In other words, if physics could subsume all biological processes, such individuals would criticize physics if it were atomistic.

The attraction of both atomistic and reducing-science reductionism rests in their sweeping approach at explanation. If all scientific explanation could be accomplished with some minimal level constructs in a single science, then we could achieve an essentially universal language. Keats decried Newton for reducing the poetic elegance of the rainbow to its vulgar prismatic colors. If, however, such a reduction were possible, then grouping concepts such as the rainbow would be superfluous. But can we find such basic elements and a set of relationary laws in science? Do we find emergent properties in higher hierarchical levels that cannot be defined in a language derived from the lower levels?

The dream of reductionism has never been achieved, nor does it seem likely that we will learn all by resorting to explanations using only the basic elements (Popper 1974). As we study different geometries, we learn that the detail lost in switching from Euclidean geometry to topology is replaced by whole new concepts that were never previously visible (Medawar 1974). In Euclidean geometry shape is invariant, and transformations and comparisons are based upon angles, numbers of sides, and curvature about foci; topology ignores exact shape but maintains a sense of space and linear order. The transition from the former geometry to the latter involves a restriction of detail, but new concepts emerge. Thus the notion of conic sections appears in the geometry of projection.

In evolutionary biology, the gene is often employed as the smallest unit of consideration, though recent discoveries of molecular genetics muddles this a bit. Population genetics usually sees the fate of genes in terms of their contributions to fitness and stochastic processes. Complexities of genetic structure, such as epistasis and linkage, greatly complicate population genetic models. Yet it is a legitimate pursuit to ask how genes survive by virtue of their effects on the phenotype, although one might question the power of both our empirical tools and multilocus models to realistically attack population genetic problems (e.g., Lewontin 1974).

Most evolutionary biologists acknowledge a great deal of complexity in the effects of single genes on the phenotype, and emphasize the complex interactions among genes. Most adhere to the principle that the phenotype, and not the gene, is the unit of selection (e.g., Dobzhansky 1970). The integrity of the organism and its internal interactions have been emphasized by Dobzhansky (1951), Lerner (1954), and Stebbins (1974), among others. Consider Stebbins' statement (1974, p. 302) of the limited evolutionary potential of the incorporation of new alleles:

> Mutations that affect these structures and processes have an adaptive value not in direct connection with genotype-environment interactions, but through their interactions with other genes that contribute to the structures or processes involved. In higher organisms, the majority of genes contribute in one way or another to these conserved structures and processes. The adaptive value, and hence the acceptance or rejection by natural selection of most new mutations, depends not upon direct interactions between these mutations and the external environment, but upon their interaction with other genes, and their contribution to the adaptiveness of the genotype as a whole.

This is not an appeal to mysticism. Stebbins merely acknowledges that genes serve to determine a functioning phenotype in a complex manner. Genes may very well be retained by virtue of their contributions to fitness, but there is an important hierarchical level, the organism, that also shapes the fabric of genetic organization.

The organism is not the simple sum of its parts. It may well be that division of labor in some hymenoptera serves the purpose of the survival of genes, but the phenomenon of labor division cannot be explained from the genes' mere presence. The notion of levels is well entrenched within evolutionary biology, but the exact awareness of levels is not always present when evolutionary hypotheses are formulated.

The effects of individual genes on fitness can be overshadowed by other processes, which are best considered as interactions of higher levels of the hierarchy with lower levels. Consider the many studies of regional gene frequency clines discovered by students of allozyme polymorphisms over the past couple of decades. Typically, one samples over a geographic-environmental gradient and finds a spatially progressive change in allele frequency at a locus (e.g., *Adh* for *Drosophila*). The distribution and abundance of the variant alleles have been studied by those interested in the question of natural selection. There is almost universal agreement that if the functional differences among allozymes could be related to fitness, then the problem of geographic variation would be solved. But is this true?

Effects within an evolutionary hierarchical system can be transmitted downward. This property of hierarchies has been termed **downward causation** (Campbell 1974). For example, consider the following geographic situation. A step cline transects a continent, with allele *a* nearly fixed in the east while *b* is fixed in the west. Suppose that a dramatic change in structural habitat (e.g., loss of the species' requisite food plant) drives the entire western part of the species to extinction. Owing to stochastic loss, the small remaining presence of allele *b* in the east fades out. The loss of the allele has nothing to do with effects of the locus on overall fitness; it is simply a consequence of selection at a higher level of hierarchical organization, the population. In all cases where geography plays a role in genetic differences in a species, the difference between single gene selection, and group selection can be similarly ambiguous (Levins 1970).

The question of considering levels of the hierarchy without resorting to explanations at other levels is of equal importance in evolutionary investigations. This can be as much a practical issue as a philosophical one. In an empirical study of diversity in the fossil record, for example, higher taxonomic levels may be more tractable than lower ones. Valentine (1968) was a pioneer among paleontologists in considering hierarchies from a paleoecological point of view. If hierarchies are "nearly decomposable" (Simon 1962), different taxonomic levels might respond variously to the same environmental processes. But if complete reduction were possible, one might study the abundance of taxonomic families (in the taxonomic hierarchy) over geological time without resorting to lower levels (e.g., species), since families would be a convenient way to monitor changes at the lower level.

But the response of families to aspects of earth history might be entirely different from the response, for example, of species. This difference may transcend the fact that families are, of course, constructs of species and therefore may have responses that can be predicted from the aggregated species of each family. The family level might, for example, correspond ecologically to adaptive zones and therefore have its own unique response (e.g., Van Valen 1984a). It is crucial in any hierarchical analysis of a system to understand (a) to what degree it is decomposable and (b), if the hierarchy is decomposable, the nature of the differences of response of different hierarchical levels to different processes.

Consider, for example the pattern of first appearances of phyla versus those of families (Valentine 1968). Phyla show a distinct peak in the rate of first appearances early in the Phanerozoic. The pattern of first appearances of families, however, is completely different. One might argue that phyla represent major turning points in the history of life: as a response to a series of open environments, developments of major evolutionary consequence came first. By contrast, family-level divisions may represent minor evolutionary changes that came and went in response to minor changes in earth and biotic history.

This form of hierarchical analysis highlights an implicit assumption made in many taxonomic diversity studies in the fossil record (e.g., Sepkoski 1978; 1984): Species richness is proportional to richness at the (higher) taxonomic level of consideration. But as the taxonomic level increases, other considerations such as the degree of major morphological diversity must also be considered. Are Sepkoski's (1981) and Flessa and Imbrie's (1973) studies, at high taxonomic levels, analyses of species richness or of the degree of diversity of "major" morphological adaptations?

Certain measures will have entirely different meanings at different levels of the hierarchy. In the taxonomic hierarchy, the measure of individual productivity is fecundity; at the species level, however, speciation rate would be the appropriate measure. Fecundity and speciation have entirely different meanings, since speciation decouples two entities from further evolutionary connection, whereas an organism's offspring would still be part of the same interfertile population unit. Extinction also has different meanings. At the organismal level death does not necessarily entail the loss of given genes from the population; in the case of species extinction it almost invariably does. At the level of the monophyletic group entire character complexes will be lost.

Although generalizations about the interactions within hierarchies are difficult to make, certain evolutionary hypotheses are phrased most profitably in terms of a regularity of interaction within a hierarchical framework. Riedl (1978) argues, for example, that an ordering principle of evolution is "burden," which is the effect on the whole organism of a given evolutionary change. He argues that natural selection is a confrontation between the external aspects of the environment with the internal interactions of the organism. Evolution emerges from the continuing interaction between internal organismal organization and the effects of the external environment (Schmalhausen 1949). As such, the nature of internal order (we will not precisely define this for the moment) at a given time in a taxon's history is part of the measure of response to selection. This leads to the following hypothesis. With the evolution of increasing internal order the functional burden, encumbered by any given response to natural selection, increases and "with this a new lack of freedom called canalization also increases" (Riedl 1978, p. 80). In hierarchical terms, Riedl (1978) argues that as the evolution of increasing internal order (presumably of development) proceeds, any new effect of selection on any part of the system (e.g., gene) will have increasing effects on the entire system (e.g., developing embryo). Thus he predicts that the tightness of effect from the lower to the upper part of the organismal hierarchy will increase with evolution.

Jacob (1977) has proposed a related hypothesis, based upon a presumed hierarchical structure of organization within the living organism. "Highly evolved" organisms are not perfectly evolved machines at all. Rather, the process of evolution acts in the way that an engineer tinkers with an invention while "improving" it. This leads to machines and organisms that have a peculiar set of internal constraints that can be explained only by history. As Darwin (1859) recognized, the process of evolution via natural selection should build up complex and imperfect organisms with limited abilities to deal with environmental change. "Nor ought we to marvel if all the contrivances in nature be not, as far as we can judge, absolutely perfect. The wonder indeed is, on the theory of natural selection, that more cases of the want of absolute perfection have not been observed" (Darwin 1859, p. 472).

Hierarchies are thus the natural framework for the study of the evolutionary process. Having the wrong gene could conceivably extinguish a phylum. Extinguishing a phylum could, by accident, extinguish a gene. The hierarchical approach allows the organization of research programs to tackle such questions that are historical in nature.

In the context of hierarchies, the macroevolutionist critique of the Modern Synthesis rests in the belief that selection at the level of organism and levels beneath is inadequate to explain the entirety of evolution. This is predicated on the belief that processes relating to larger groups can result in evolutionary change. The principal example of such a process is the balance of speciation and extinction, which might produce biased morphological change (Eldredge and Gould 1972, Stanley 1975). This claim is not at odds with the presence of selection at lower levels of the hierarchy. Rather, it suggests an expansion of possibilities in the explanation of evolutionary trends. At the least, one can argue intuitively that extinction strongly affects the relative proportion of taxa and, therefore, the spectrum of morphologies. Since habitat destruction is often a major source of extinction, it is not very controversial to claim that extinction would not be tightly linked to individual genes in many cases. What would be controversial is to argue that such processes caused the evolution of complex morphological structures such as the cephalopod eye. Here, neo-Darwinians would stand firm in ascribing such an evolutionary process to the interactions of genes and the organism.

The Role of Type in Evolutionary Concepts

Typology and Evolution

The problem of macroevolution has always been regarded as the problem of the origin and evolution of types and the present gulf between them. A type is a class whose members share a certain set of defining traits. Such a definition implies gaps between types, or at least discrete differences in the sets of traits that define the different types. If you don't believe in types and gaps, then you don't worry much about major evolutionary jumps; but the belief in types, among species or among higher taxonomic constructs (e.g., *baupläne*) will lead you toward a deep concern about discontinuities in evolution.

I would like to distinguish among three sorts of typologies that permeate the study of biology:

1. *Essentialist Type* (or *Idealistic Type*). The type has a fixed im-

mutable essence. Minor variation is possible within the type.

2. *Modality Descriptor.* The type is of a modal form, defined by the overall properties of a population. Intermediate stages between the types are possible but uncommon, at least at present.

3. *Saltatory Type.* The type has a fixed set of properties but it is only changeable into other discrete types via a saltatory process. Intermediate stages would be claimed not to exist or to ever have existed.

The deep-seated belief in types derives from an essentialist philosophy, which views the world as a series of entities defined by their respective essences. The ordering of these entities is usually associated with a teleological view of the universe. In the biological context, species are viewed as constant and immutable. To Cuvier, for example, species were perfectly adapted to a specific environment. If the environment were eliminated or altered over time, the immutability of the species would ensure its extinction. Evolutionary hypotheses were thus inconceivable. Aristotle thought of natural selection but dismissed it in favor of a world of teleology and types. Certainly the deep-seated belief in essentialism, commonly held by as disparate a set of intellectual luminaries as Aristotle, Bacon, Mill, and Cuvier, would have tended to freeze all scientific notions of the potential mutability of species (see Hull 1973).

The problem of the biological concept of type gains modern relevance through the theory of evolution, particularly that espoused in Darwin's *On the Origin of Species by Means of Natural Selection* (1859). The pre-Darwinian notion of the perfection of design being a manifestation of the work of God accepted the types as perfectly adapted designs. It is in the post-Darwinian morass of species mutability that the essentialist notion of types takes on a nonscientific connotation. Perfection and perfect adaptedness gave way to the "law of the higgledy piggledy," as Herschel called it. Organisms were often out of step with their environment and natural selection culled out less well adapted variants. Successive forms were not necessarily perfect, according to Darwin; they only happened to be the fittest of the lot.

Aside from a decidedly nonteleological abandonment of perfection, Darwin's theory concluded that species were mutable. Darwin's conception of evolution presumed that every pair of ancestral and descendant forms comprised the endtowers of a bridge of a (not necessarily evenly) graded series of intermediates spanning the chasm. Darwin perceived that gaps between successional fossil forms could be explained by two possible shortcomings of the data of paleontology: (a) The new species arose via a string of intermediates in a small and isolated population not preserved in the fossil record and (b) the series of intermediates could not be preserved owing to frequent gaps in the fossil record. If only the gaps could be filled, then we would find our intermediates. Was Darwin right? We will discuss this issue in Chapter 7. Whether right or wrong, Darwin clearly was antitypological.

The transitional period between the dominance by typological idealists such as the English morphologist Richard Owen and the new generation of evolutionists led by Darwin and Huxley was a bit more muddled than is generally realized (see discussions in Desmond 1982; Ospovat 1981). Although Owen vigorously opposed the godless role of chance and the purposeless force of natural selection, he nevertheless came to believe in extensive gradual change from a primitive ancestor, all within a general archetype. The archetype, however, contained an essence that was to be revealed among the members by the study of homology. Thus he saw vertebrate evolution as a gradual process and even managed to find a transitional form, *Archegosaurus*, that obliterated the gap between reptiles and fish. Owen's (1859) reconstruction of the evolution of the Vertebrata even included a concept of branching and was therefore decidedly close in spirit to Darwin's (1859) hypothetical phylogeny diagram and Haeckel's later attempts at phylogenizing in the *Generelle Morphologie* (Bowler 1976).

By contrast, Thomas Huxley, "Darwin's bulldog," held at first to a typological view of species that probably derived from his adherence to Charles Lyell's concept of nonprogression in evolution (Desmond 1982, p. 90). This viewpoint led him to believe, despite evidence to the contrary, in the early Paleozoic origin of mammals, and in **persistence**, a concept that allowed no major progressive evolutionary trends. This latter belief was in conflict with Darwin, his idol, who said "I cannot help hoping that you are not quite as right as you seem to be" (quote in Desmond 1982, p. 86). In this context, Huxley's prepublication warning that Darwin's *The Origin* was too enthusiastically against saltation, seems more derived from confusion and mixed loyalties than prescience. In a way, Huxley's belief in persistence was more inimical to the establishment of evolutionary trends with empirical evidence than was Owen's idealized archetype, within which some evolutionary change was accepted.

An association of phyletic gradualism with nineteenth-century liberalism (Eldredge and Tattersall 1982; Gould and Eldredge 1977) is an oversimplification. One associates a belief in slow progress with this period in history. But Darwin eschewed the notion that evolution was to be understood as progress toward higher forms. Darwin's belief in slow evolution may indeed have derived from the Victorian belief in slow progress, but the notion of continuous gradational transformation was held in many non-Darwinian quarters in the mid-nineteenth century. Owen strongly believed in phyletic gradualism and was clearly associated with the forces of privilege and station. It apparently served his purpose to believe in evolutionary radiation, for it weakened the position of the followers of Lamarck (Desmond 1982, p. 69). His notion of transmutation had limits, and they were those that fit safely within a theistic philosophy. Darwin's conception of nature, red in tooth and claw, was, if anything, repugnant to the Victorian zeitgeist.

Huxley spoke clearly for the new emerging class of individuals whose station was to be recognized by their own efforts. Yet, until the late sixties he stood intransigently opposed to evolutionary progress while, at the same time, he fought vigorously for the working man and worked actively to help install a new generation of meritocratic professionals. As Ospovat (1981) wisely notes, the notion of phyletic evolution, with an inferred directional series of gradational forms, would have developed even if *The Origin* had never been published! The notion of gradualism came from the morphological tradition and did not originate with Darwin. Think of Lamarck, whose notion of gradual change and inevitable evolutionary directionality through acquired inheritance might have been the accepted paradigm of evolution had Darwin and Wallace not come along. As Riedl (1978) notes, even Goethe's philosophy, so clearly typological, allowed for extensive variation within the type (see also Sherrington 1949).

Essentialism Ends With The Rise of Population Thinking

The history of progress of twentieth-century biology can be broken down into four discrete periods. The terms I use to describe them are used disparately.

1. **Mutationist-Biometrician Debate.** This period covers the first two decades of this century, contemporary with the rediscovery of Mendelian variation and the early investigation of chromosomes. Two schools of thought were popular. The **biometricians**, led by such luminaries as Pearson, Galton, and Weldon, had by this time developed a battery of statistical techniques to analyze natural variation in populations. In contrast, the rediscovery of Mendelian transmission inspired another school of thought, led by deVries, Bateson, and Morgan (at first), to emphasize the discontinuous mutations found in laboratory experiments. This school saw **mutationism** as the stuff of evolution and rejected natural selection upon existing variation (Bateson 1894). The belief in quantum jumps from one type to the next by mutation versus a belief in natural selection on continuous variation was a false dichotomy. The controversy hampered the growth of population genetics for a decade (see Huxley 1942; Provine 1971). The belief in steplike differences between types (mutations) froze our outlook upon natural variation. We now appreciate that mutations occur at all levels of variation and that their presence in steplike transitions is far from being incompatible with the theory of natural selection. Mutation is understood as the source of variation upon which natural selection can act.

2. **Neo-Darwinian Period.** Covering the approximate interval 1920-37, this period was marked by the survival from the past century of a host of now defunct hypotheses such as Lamarckism and orthogenesis. But most important, Sewall Wright, J. B. S. Haldane, and R. A. Fisher laid the foundations for genetic analysis of traits and genetic changes in populations. The power of natural selection was discovered, starting from an initial report by Punnett (1915), and a debate arose about the relative importance of stochastic versus deterministic effects in population genetics. All three of the neo-Darwinian triumvirate, however, seem to have believed firmly in the preeminence of natural selection (Provine 1983; Mayr 1982a). A series of intense debates on the role of drift in small populations were extremely important in focusing attention on several empirical systems, such as *Panaxia* and *Cepaea* (Provine 1983).

3. **Modern Synthesis.** This period starts with the publication of Theodosius Dobzhansky's seminal work *Genetics and the Origins of Species* and culminates with the famous conference at Princeton in 1947 (see Jepsen, Mayr, and Simpson 1949). The theoretical advances made during the neo-Darwinian movement were incorporated into systematics, ecology, and, to a degree, paleontology. Older concepts lingering in evolutionary biology, such as orthogenesis and Lamarckism, were discarded. Along with Dobzhansky, Ernst Mayr, Bernhard Rensch, George Gaylord Simpson, and Ledyard Stebbins were crucial contributors. The period is marked by a harmony never seen before or since. Of course the neo-Darwinians were still actively contributing to evolutionary theory, and Sewall Wright contributed to the Princeton conference. Ernst Mayr (1982*a*) has argued that they did not influence the Modern Synthesis, but both Dobzhansky's (1937) and Simpson's (1944) texts show strong influence from theoretical population genetics (e.g., Provine 1983; LaPorte 1983).

From the beginning of this period, all architects of the Modern Synthesis followed their neo-Darwinian forebears in believing in the primacy of natural selection in shaping evolution. A few nagging examples of claimed random variation– for example, inversion polymorphisms in *Drosophila*–turned out to be strongly selected (e.g., Dobzhansky 1948*a, b*). This only strengthened the general feeling for the importance of natural selection. Gould (1983*a*) argues for a "hardening of the Modern Synthesis" and suggests that factors other than natural selection were actively suppressed. As the founders of the neo-Darwinian movement and its architects all believed in the primacy of natural selection from the beginning, it seems contradictory to conclude that any "hardening" could have taken place (Levinton 1984). Gould sees the thirties as a time of pluralism; if orthogenesis and Lamarckism are what he has in mind, we could have lived without this pluralism. The further move of the Modern Synthesis toward population thinking and experimental approaches was the healthiest episode in the twentieth-century history of evolutionary biology.

4. **Postsynthesis Period.** As in any historical period following a major congealing, this period is marked by disarray. At first, the Synthesis came to dominate natural history. But two movements

have directed current trends in the study of evolution. Wynne-Edwards' claim (1959) that group behaviors arise from group selection became a major concern. G.C. Williams' (1966) attack on this overall hypothesis attempted to restore the primacy of individual selection and an orientation toward the study of genic level natural selection. This response was contemporary with W. D. Hamilton's explanation of altruism in terms of benefit to the organism, and was followed by the sociobiology movement (e.g., E.O. Wilson 1975), which has been the source of intense debate and criticism.

Following the elucidation of the gene-protein specification process, a large degree of genic protein polymorphism was discovered (Lewontin and Hubby 1966; Hubby and Lewontin 1966; Harris 1966). This was surprising to the majority, who, from predictions of theory and experience with laboratory variation, saw gene loci in natural populations as relatively invariant, with rare mutants of low fitness. From this came the neutral theory of evolution, the first credible theory that incorporated stochastic processes to explain variation in nature (see Kimura 1983). Of course, many selectionist explanations for molecular variation have been tendered as well (see Chapter 3), but the issue has not yet been resolved.

The Modern Synthesis, a period during which genetics, systematics, and population genetic theory blended into a supposedly harmonious neo-Darwinian view of evolution (Mayr and Provine 1980), was also a time when typological thinking was under attack. Mayr (1942), in particular, was a great pioneer in exposing the methodologies of systematists as basically typological. He wrote,

> The taxonomist is an orderly person whose task it is to assign every specimen to a definite category (or museum drawer!). This necessary process of pigeon-holing has led to the erroneous belief among nontaxonomists that subspecies are clear-cut units that can be easily separated from one another. (Mayr 1942, p. 106)

and:

> The species has a different significance to the systematist and to the student of evolution. To the systematist it is

> a practical device designed to reduce the almost endless
> variety of living beings to a comprehensible system. The
> species is, to him, merely one member of a hierarchy of
> systematic categories. (Mayr 1942, p. 113)

Even Darwin, while believing that at least some species were in the
process of changing, and that certainly all species were mutable, held
a rather practical view of delineating species.

> In determining whether a form should be ranked a species
> or variety, the opinion of naturalists having sound judge-
> ment and wide experience seems the only guide to follow.
> (Darwin 1859, p. 47)

These quotes reflect a traditional reliance of systematists on the presence of types. But it is not always clear whether this reliance stems
from essentialism or from a practical attempt to classify the world's
creatures. It is doubtful that twentieth-century systematists adhered
to an essentialist concept of species. More likely, they incorporated
some intuitive notion of statistical recognition among modes between
more continuous morphological gradation. In the period preceding the
Modern Synthesis, most systematists saw species as distinct and definable by characteristic differences that arose by some sort of nonadaptive
process (see Provine 1983; Gould 1983a).

The Modern Synthesis substituted a new concept of species for
older concepts. The modern biological species concept (Dobzhansky
1935) defined speciation as a stage in a process "at which the once
actually or potentially interbreeding array of forms becomes segregated
into two or more separated arrays which are physiologically incapable of
interbreeding." Although this concept has been modified and redefined
in terms of the fitness of hybrids versus that of intrapopulation crosses,
the basic concept has survived and is still widely regarded as a natural
definition of species, although the suggested mechanisms of species
formation are varied (see Chapter 4).

The new definition of species has carried with it a more sophisticated concept of type, based upon a process that produces modality of
form rather than on an inherent and undefinable essence or the expectation of saltation. The biological mechanism of reproductive isolation
ensures the possibility that the forms of two daughter species can go
their separate ways. It acknowledges a materialistic basis behind the

ability of both native peoples and systematists to arrive at nearly the same species divisions. As Dobzhansky claimed:

> the living world is not a single array of organisms in which any two variants are connected by an unbroken series of intergrades, but an array of more or less distinctly separate arrays, intermediates between which are absent or at least rare. (Dobzhansky 1937, p. 4)

Dobzhansky's notion of type as modality is committed to the mechanism of speciation through reproductive isolation and certainly eschews the notion of essence. A well known critique of the reality of the biological species concept (Sokal and Crovello 1970) also avoids the issue of essentialism; it simply attempts to criticize the utility of the Dobzhansky-Mayr biological species concept to practicing systematists and claims the importance of phenetic similarity in systematic work. Typology as essentialism is properly absent from their arguments.

Both Ghiselin (1975) and Hull (1976, 1980) argue that if species are to be treated as classes (e.g., *Homo sapiens*) with a set of members (e.g., Martin Luther), then the class becomes effectively immutable and just as essentialist as pre-Darwinian notions of species or higher taxa. Hull (1976) recommends that a species be regarded as an entity with spatiotemporal and genetic continuity. As such, it effectively becomes an individual, bearing a proper name, that is, the specific name. The border between one species and another under this approach could be arbitrary, although Hull accepts that mechanisms such as Mayr's (1963) theory of speciation might tend to sharpen the borders between species. This individualistic concept is not essentialistic.

The old essentialist notions of type still pervade our thinking. The typological approach, transformed into an evolutionary guise through the late nineteenth century by great morphologists such as Gegenbaur, initiated a research program that accepted the concept of evolution yet stuck closely to an idealistic system. Coleman (1976) noted (p. 172), "Seemingly new organisms could always continue to appear [via evolution] in the world of objective reality, but the idealistically inclined morphologist claimed the power to discern the unvarying form or forms to which these appearances properly belonged." Thus, although evolution was taken to be the grand justification for the study of comparative morphology, a residual belief in typology prevented a study of variation and focused study on homology, with no consideration of process. This led the field of comparative morphology toward academic disaster in the twentieth century and prevented advancement relative to nonessentialist-dominated fields such as population genetics and molecular biology (Coleman 1976). This does not mean, however, that *baupläne* do not exist, only that a subtle essentialism has inhibited our capacity to study their possible materialistic basis.

An appropriate point of departure for the study of transitions in evolution was succinctly outlined by Dobzhansky. Two groups of organisms in two-dimensional space have a gap between them. Did one give rise to the other? If so, then why is the gap present? Is it hard to traverse? What is the pathway of the traverse? How fast was the change effected? These questions arise and can be approached objectively only when the mutability of the "types" is admitted and evolutionary relationships can be determined.

The mind-set of typology is not limited to arguments over taxonomic categories. Even the functional morphologist can be led to types, with intervening gaps where no intermediate is to be found. D'Arcy Wentworth Thompson revealed his prejudice in the following passage from his *On Growth and Form*:

> A 'principle of discontinuity,' then, is inherent in all our classifications, whether mathematical, physical, or biological, and the infinitude of possible forms, always limited, may be further reduced and discontinuity further revealed. ... The lines of the spectrum, the size families of crystals, Dalton's atomic law, the chemical elements themselves, all illustrate this principle of discontinuity. In short nature proceeds "from one type to another" among organic as well as inorganic forms; and these types vary according to their own parameters, and are defined by physico-mathematical conditions of possibility. In natural history Cuvier's "types" may not be perfectly chosen nor numerous enough, but "types" they are; and to seek for stepping stones across the gaps is to seek in vain, for ever. (Thompson 1952, p. 1094)

In the passages preceding this quotation, D'Arcy Thompson argued that the nature of growth and function had most probably erased much of the vestiges of morphology that might be used to reconstruct phylogeny. D'Arcy Thompson's views are reminiscent of those of the anti-Darwinian Mivart (1871), who also likened the differences among forms to the laws of crystallization. His typology is clearly quite different from that of the essentialists and quintessentially the opposite of Gegenbaur's. He believed, nevertheless, in some mechanism or axiomatic condition that underlies a typological system. Are the stepping stones never to be found?

Macroevolution and the Fall of Goldschmidt

Hopeful Monsters and Hopeless Mooting

Studies of macroevolution tend to either idolize or denigrate the role of the geneticist Richard Goldschmidt. I find myself in between the extremes. He is best remembered for his colorfully construed concept of **hopeful monsters** (Goldschmidt 1933, 1940), those few monstrosities that he claimed to be the stuff of major species-level saltations in evolution. He relied upon hypothetical chromosomal mutations that accumulated cryptically in populations until a threshold was breached, propelling the phenotype across an unbridgeable gap. Most of these new phenotypes were hopeless, but the rare success was the progenitor of a new species. This work has not withstood the test of time and was at variance with the fact and theory contemporary with its proposal. But Goldschmidt's work includes a more substantive thread attempting to integrate genetics, development, and evolution, which was largely ignored until the past decade, despite other standard-bearers for the approach (e.g., Waddington 1957, 1962).

After a long and successful career, Goldschmidt–a Jew–was dismissed from his academic position in Berlin. After leaving Nazi Germany Goldschmidt came to the United States and settled at the University of California at Berkeley. Among his important works in English are *Physiological Genetics* (1938) and *The Material Basis of Evolution* (1940). The latter brought him into disfavor with his contemporaries, so much so that he wrote a bitter introduction to his 1945 (*a, b*) papers on the evolution of Batesian mimicry in butterflies.

Why was Goldschmidt so isolated from the pillars of the neo-Darwinian period and the Modern Synthesis? In his book he proclaimed that "The neo-Darwinian theory of the geneticists is no longer tenable" (Goldschmidt 1940, p. 397). He argued "There is no such category as incipient species. Species and the higher categories originate in single macroevolutionary steps as completely new genetic systems" (ibid, p. 396). The first part of the book, entitled "Microevolution," described the nature of geographic and within-population variations in a species. The second part denied that this was the stuff of transspecific evolution. His adherence to this strong point of view is exemplified in his endorsement of the then recently published work of the paleontologist Otto Schindewolf (1936), who had proclaimed that the first bird had hatched from a reptile's egg.

Both of Goldschmidt's books display a strong empirical approach to the nature of variation and the varied relationship between development and genetics. But his final prescription for solving the mystery of mysteries, as Herschel described the origin of species, was dogmatic and simplistic: saltation. Goldschmidt admired simplicity– "a simplistic attitude is not a flaw but the ideal goal for a theory in science" (Goldschmidt 1940, p. 399).

Despite the apparent simplicity, Goldschmidt's views were based upon a false dichotomy between broader-scale chromosomal mutations and point mutations, which were presumed to be the neo-Darwinian basis for evolutionary change. Neo-Darwinians took variation for granted and made no strong distinction between single genes and larger genetic constructs, so long as they obeyed Mendelian rules. Goldschmidt's claims that neo-Darwinians believed solely that races were incipient species are also at variance with the many saltatory mechanisms of speciation that had been previously proposed (see Templeton 1982). In sum, Goldschmidt's characterizations of the neo-Darwinian movement were inaccurate caricatures.

Goldschmidt felt that the population geneticists of the day were too faithful to the notion that genes were independently acting entities. Some discoveries, such as the notion of position effects of genes, strengthened his suspicion that the genic theory was suspect. This feeling may have stemmed from his training which emphasized development and physiological function, as opposed to transmission genetics (G.E. Allen 1974). His interests in physical science may have also given him the standard nineteen-thirties philosophical mistrust of theories that depended upon the importance of fundamental units. His own theories of gene action required instead large scale integrated effects of chromosomes. These theories were mainly metaphorical in nature, and were shown to be untenable in subsequent decades, even if they were probably important in producing more focus in arguments about evolutionary genetics.

Aside from the problems of Goldschmidt's mechanism of the rise of novelties, his ideas of spread and speciation also were not well received. The arguments against the spread of novel and extreme variants appearing only rarely had been well understood by then and have been subsequently amplified. Rare variants tend to become extinct very rapidly. Dramatically different mutations are most likely of low fitness relative to the population mean phenotype (Fisher 1930). Relative to extreme phenotypes, mutants of less extreme form are much more common and therefore contribute in greater proportion to a population's evolutionary potential (e.g., H.J. Muller 1949).

In commenting upon Schindewolf's ideas, Mayr was confident that major saltations in evolution were nonexistent when he stated:

> No special evolutionary processes need to be postulated, even in groups where such missing links have not yet been found and where the primitive roots of the various stems always seem to be missing. (Mayr 1942, p. 297)

George Gaylord Simpson also was secure in his belief that transitional forms were common in evolution. He commented wryly:

> the argument from absence of transitional types boils down to the striking fact that such types are always lacking unless they have been found. (Simpson 1950, p. 233)

A specific controversy illustrates why Goldschmidt's ideas quickly lost favor, at least among neontologists. At the time Goldschmidt was writing his famous second book, *The Material Basis of Evolution*, Batesian mimicry in butterflies had already become the subject of genetic research. The now famous Punnett (1915), a protegé. of William Bateson, believed that the mimetic morph in species of the swallowtail butterfly genus *Papilio* could be explained by a single integrated genetic variant. Goldschmidt's (1945) interpretation of mimicry in *Papilio* followed directly from his ideas on developmental regulation, genetic integration, and saltation. He argued that major switch genes explained the evolution of a mimic from a non-mimetic ancestor. Developmental constraints would cause the same mimetic phenotype, controlled by the corresponding genotype, to appear repeatedly in different species.

Goldschmidt was far off the mark. The work of Clarke and Shepard (1960a, b, 1962, 1963) presented a detailed picture rather more consistent with a hypothesis of gradual selective buildup of the mimetic phenotype and significant differences in the genetic mechanisms behind the evolution of the mimic within populations of the same species, let alone among species (see J.R.G. Turner 1977, 1981).

To illustrate, consider the case of *Papilio memnon*, a species widely distributed in Southeast Asia (Clarke et al. 1968). Populations of *P. memnon* vary from place to place and can be non-mimetic or strikingly similar to various local models. They have, however, consistently failed to evolve red markings on the body that are characteristic of the models. Clarke and co-workers found that several closely linked loci were major contributors to the buildup of the mimetic phenotype. Mimicry was found, as in other species of *Papilio* (Clarke and Shepard 1963), to be controlled by a so-called supergene (the closely linked loci), which had most probably evolved gradually by the accumulation of closely linked allelomorphs in advantageous combinations. The selective stability of such supergenes, once evolved, has been confirmed theoretically (Charlesworth and Charlesworth 1976). Two other unlinked genes were also involved in the construction of the phenotype. Note that, in this context, gradual does **not** refer to an infinitesimally fine series of morphs ranging from non-mimetic to mimetic. Some of the contributing genes do make rather major jumps "toward" the mimetic phenotype, but these certainly do not amount to the magnitude of change required by Goldschmidt. This is only one example demonstrating that the Modern Synthesis by no means commonly accepted only insensibly small changes in evolutionary transitions.

Most important, crosses among mimetic races illustrated the differences of local evolution among races. The resemblance of progeny of interracial crosses to either of the local models was found to be better in none of the crosses, approximately as good in ten, and poorer in thirty five (Clarke et al. 1968). This refuted the notion that evolutionary outcome was constrained and involved identical genetic changes in all of the mimetic races. Also, modifiers affecting tail length, for example, were found to differ among populations.

Goldschmidt's Useful Developmental Approach

Although Goldschmidt is remembered for his saltationism, his books reveal a sophisticated notion of the nature of evolutionary novelty and potential for change. It is true that he held to an unduly strong version of the role of developmental constraint in evolutionary pathways. He also denied the value of experimental genetics in the study of transspecific evolution. These excesses should not conceal, however, his rather useful description of how to investigate directionality in evolution via developmentally mediated regularities in the determination of form.

It is an unfortunate and inaccurate caricature of the neo-Darwinian and Modern Synthesis movements that adaptive evolution is infinitely powerful and is not constrained by forces other than the natural selection of optimal forms (as claimed by Gould 1980a). A substantial part of the literature of genetics and allometry acknowledges the constraints imposed by one set of traits on the evolution of others (see H.J. Muller 1949; discussion in Charlesworth et al. 1982, pp. 476-80). Certainly Darwin was keenly aware that the evolutionary change of one trait was liable to bring about concomitant change in others, with no necessary concomitant adaptive significance (Darwin 1876, pp. 346-7), but an appreciation of this general problem falls short of Goldschmidt's profound understanding of the interrelationship of evolutionary direction and development. To recognize this understanding, the recent volume by Raff and Kaufman (1983) on development and evolution was dedicated to Goldschmidt.

Consider the evolution of snakelike form in saurian lizards through the increase in vertebral numbers and rudimentation of extremities. Goldschmidt saw typical explanations as reflecting ignorance of the origin of pattern in development. Severtzov, for example, envisaged evolutionary change toward the fossorial habit as resulting from a series of steps, each involving a transformation of one caudal into a descendant sacral vertebrum. The sacral vertebral column would be thus elongated and the position of the hind extremities shifted backward (cited in Goldschmidt 1938). Goldschmidt provided the following alternative explanation: "The evolutionary process changed primarily the basis of segmentation itself by altering its embryonic gradient and rhythm so that a larger number of segments was produced to begin with. The localization of the limb buds, and therewith the setting of the limits of thoracic and lumbar segments is a deterministic process independent of the primary segmentation" (Goldschmidt 1940, p. 339). His explanation thus suggested an alternative realistic developmental framework within which to view directionality in saurian evolution.

Physiological Genetics prescribed the general formula for the investigation of gene action. Goldschmidt not only appreciated the fact that genes often affected rate processes and that the phenotype was apparently broken up into a series of developmentally and probably genetically autonomous fields that were determined in a complex way as development proceeded, but also actively applied these principles to his studies and interpretations of genetic variation and evolutionary potential. He attempted to popularize Spemann's notion of **determination stream:**

> we see that genes actually controlling color and structure of a wing may act by controlling a determination stream of definite quantity, speed of progress, pattern of flow, and action upon different processes of morphogenesis and chemism. (pp. 195-6)

All of these notions are being actively investigated today. Garcia Bellido's (1975) important discovery of compartments in the development of *Drosophila* (see Chapter 5) is a natural extension of Goldschmidt's work on wing patterns in his favored organism, the gypsy moth, *Lymantria dispar*. Many evolutionary biologists have rediscovered the possible role of the determination stream in organizing and constraining evolutionary processes. The recent volume *Evolution and Development* (Bonner 1982) contains a summary article by most of the active proponents in this field (Maderson et al. 1982) and papers by two of the more enthusiastic supporters of the integration of development into evolutionary studies (Alberch 1982; Gould 1982b), yet Goldschmidt's *Physiological Genetics* is never cited! Such is the power of the hopeful monster concept promulgated in Goldschmidt's disastrous second volume. It effectively erased the potential positive effect of his pioneering earlier work and probably abscised his potential ability to hasten the rise of a major research field integrating evolutionary and developmental biology.

It may well be that Goldschmidt was a useful gadfly at the time. But his importance would have been far greater if he had pursued a rapprochement between his physiological approach and the results of population geneticists. These approaches were not at cross-purposes, really. They were just very different paths that apparently had little probability of crossing at the time.

The tension between the followers of the Modern Synthesis and those of Goldschmidt's heterodox views clearly gave macroevolution a bad name. The term became associated with major saltations between ancestor and descendant taxa, spanning major taxonomic distances. As discussed above, Goldschmidt's macroevolutionary concept fared so badly that his more reasonable outline of the role of developmental and genetic constraints in evolution contributed little to the arguments of the day. While Schindewolf's (1936) saltationism continued a German morphological-paleontological tradition (see Reif 1983a), it failed to elicit a following among English-speaking paleontologists and never led to a useful program of paleontological-evolutionary research in Germany. But the same period of the late 1930s and 1940s included another major movement in paleontology, pioneered largely by George Gaylord Simpson and later by Norman D. Newell (e.g., 1952).

Macroevolution and Paleontology

Simpson's Seminal Role in Rejuvenating Paleontology

Simpson's pioneering volume *Tempo and Mode in Evolution* (1944) attempted to describe, on the grand scale, variations in rate and qualitative changes in the history of life, as understood from the fossil record. With a brilliant and sweeping look at the history of life he brought home three major points:

1. The ebb and flow of fossil forms and the temporal variation seen in morphological change and distribution patterns was completely compatible with the then newly synthesized ideas of modern evolutionary biology.

2. Rates of morphological change had not been constant in the fossil record; periods of rapid change were often followed by periods of quiescence. The periods of rapid change were influenced strongly by ecological forces.

3. Rates of evolution differed widely among taxonomic groups.

This work and the later *The Major Features of Evolution* provided paleontology with the means to awaken from its long-standing isolation from the popular twentieth-century movements in evolutionary biology. It also asserted a unity of purpose among natural historians. Alas, *Tempo and Mode* did not have immediate impact among paleontologists; it was hardly reviewed by paleontological journals, though neontologists welcomed it as at once an excellent and critical discussion of the theory of the neo-Darwinians and the provider of the important additional dimension of geological time (Laporte 1983).

Simpson's works made him the father of the "taxic approach" to paleontology, a field that employs changes in taxon richness and longevity to understand major patterns in the history of life (see Chapter 8). This approach occupies a major part of the efforts of paleontologists today (e.g., Raup 1976*a, b*; Flessa and Imbrie 1973; Sepkoski 1981, 1984; Valentine 1969; Van Valen 1973*b*, 1984*a*). Simpson contrasted longevities within major taxonomic groups. The variations and general correlations of taxonomic longevity, speciation, and extinction rates have been used extensively to draw inferences about the tempo and mode of evolution (e.g., Stanley 1979; Sepkoski 1984), although many criticisms have been leveled concerning the potential of taxonomic bias generated by differences in morphological description, systematic practice, and so on, among taxonomic groups (e.g., Schopf et al. 1975; Van Valen 1973*a*; Levinton and Simon 1980).

Gould characterized Simpson's contributions:

> Simpson's synthesis unified paleontology, but at a high price indeed–at the price of admitting that no fundamental theory can arise from the study of major events and patterns in the history of life. (Gould 1980*b*, p. 170)

I find Simpson's contribution to be far more important. Simpson rightfully sought to establish a unification of basic principles between paleontology and neontology. This is only an admission that paleontological and neontological research must both acknowledge the premise that evolutionary principles operate similarly for extinct and recent biotas, though the specific circumstances might of course be radically different. Simpson believed from other evidence (e.g., Simpson 1952) that fluctuating rates of evolution in the vertebrates were due to the ecological conditions that might be observed today; but this stemmed from paleontological data, and not all paleontologists agreed (e.g., Newell 1952).

Rather than bringing the Modern Synthesis to paleontologists and thereby constraining their intellectual horizons, Simpson attempted to understand actively what data and approaches might stem uniquely from paleontological practice. He may have been wrong; only time will tell. But he certainly didn't freeze paleontology into any mold that wasn't paleontological at its root.

Paleontology had its own traditions that transcended the ability of Simpson or any other worker to alter significantly. Simpson did successfully debunk the notion of orthogenesis among North American paleontologists. But Simpson and the other founders of the Modern Synthesis failed to hasten a birth of activity on the part of paleontologists in documenting phyletic microevolutionary patterns. If this was Simpson's objective, then he failed rather miserably. As noted, his book was hardly noticed by paleontological reviewers. Indeed, until recently the "best" and most often cited quantitative paleontological studies of phyletic evolution dated no later than the 1930s (Rowe 1899, Brinkmann 1929). By contrast, Simpson's taxic approach, so quintessentially paleontological, [1] was followed and expanded later with great enthusiasm.

Paleontologists Reconsider the Fossil Record and Earth History

Perhaps the most important milestone marking the supposed integration of paleontology into the Modern Synthesis was the famous symposium for paleontologists and neontologists held at Princeton in 1947, and published as *Genetics, Paleontology, and Evolution* (Jepsen et al. 1949). A series of papers published in this volume demonstrated the excitement of the time and revealed no major discord among paleontologists and neontologists. The paleontological contributions were

---

[1]Simpson's taxonomic overturn approach probably derives from Charles Lyell (1831-3), who used the rate of taxic overturn in the mollusks as a method to estimate the amount of geological section missing in the early Tertiary. He judged the mammals to have too rapid an overturn to be useful in this regard (Rudwick 1972). It is probably no accident that Simpson used the same two phyla in his *Tempo and Mode*. These were still, in the mid-twentieth century, the best and most accessible data bases to paleontologists.

vastly larger in taxonomic level and time scale than the neontological contributions. In a similar symposium in 1980 (Levinton and Futuyma 1982), my colleagues and I were struck by a maintenance of this difference of outlook. But the general uniformity of objectives in 1947 was by now replaced by discord and a complex network of opposing camps (Lewin 1980; Schopf 1981; Futuyma et al. 1981; Templeton and Giddings 1981).

Another symposium organized by the Paleontological Society in 1949 (Woodring 1952) illustrated a major direction of paleontological research that has sustained general interest to the present day. An attempt was made to relate environmental variations in the history of the earth to fluctuations in the origins of biotic groups and temporal changes in taxonomic richness. An interesting conflict developed between those who found that temporal variation could be related to geological period boundaries, marine transgressions, and the like (Moore 1952; Newell 1952), and those who could find no such clear relationship (Cooper and Williams 1952; Simpson 1952). Simpson argued that (p. 370) "the evidence summarized in this paper is consistent with the view that most of the broad features of vertebrate history might have been much the same if the earth's crust had been static." Newell, on the other hand, was able to relate bursts of evolutionary activity with rises in sea level. At the close of the symposium (Woodring 1952, p. 386), M. K. Elias questioned the entire data analyses of some of the contributors by claiming that paleontological collections were likely to be strongly correlated with sediments available for fossil sampling. This argument was to be repeated later (Raup 1976a, b).

The attempt to establish a relationship, or lack of one, between events in earth history and major patterns of abundance of fossil groups became a primary concern in the 1960s and 1970s. Changes in taxonomic richness were related to sea level changes (Newell 1952), seafloor spreading (Valentine and Moores 1971), tectonism (Flessa and Imbrie 1973), stochastic processes (Raup et al. 1973), ocean surface area (Schopf 1974), and asteroid impacts (Alvarez et al. 1980), among other factors. A concern for the quality of paleontological data led David M. Raup and his colleagues to question the validity of taxon richness data (Raup 1976b) and to analyze temporal patterns in taxonomic diversity in a more meaningful way. Paleontologists became concerned with appearance-extinction patterns and some suggested that an equilibrium in diversity might be reached, though the various prescriptions and analyses differed considerably (Rosenzweig 1975; Webb 1969; Flessa and Levinton 1975; Levinton 1979; Sepkoski 1984).

It might be said that this movement only attempted to define the ecological theater and the grossest outlines of the evolutionary play. In some cases, however, important changes in the history of the earth were related to key evolutionary advancements. The advent of browsing organisms was, for example, related to the decline of worldwide blue- green algal mats in shallow water, a major environmental change (Stanley 1973b). Changes in the amount of oxygen in the atmosphere were related to the rise of eukaryotic organisms (Cloud 1968; Berkner and Marshall 1964), and the appearance of carbonate shells, which were effective in counteracting predation (Rhoads and Morse 1971). In this work, a great deal of distance was placed between studies of taxonomic diversity and adaptive radiation and the details of the evolutionary process at and below the species level. Most paleontological studies of evolution tended to outline the adaptive significance behind major radiations with little consideration for the detailed evolutionary processes behind morphological change. Trends at or below the species level were generally avoided and thought to be the "noise" that was to be filtered out in true evolutionary studies (e.g., Jeletzky 1955). Studies of functional morphology were usually extrapolated from studies of a single living or fossil species to broad evolutionary trends.

Paleontological Focus at Smaller Evolutionary Scales

Some important events and a few studies refocused the efforts of paleontologists on the details of the evolutionary process and led to the recognition among neontologists of the potential importance of paleontological data. The founding of the journal *Paleobiology*, the brainchild of T. J. M. Schopf, acknowledged that traditional paleontological journals had not broken away sufficiently from the old mold that allied paleontologists more to soft-rock geology than to evolutionary biology. As the issues of the journal were published (first in 1975), those paleontologists concerned with areas of interest to neontologists suddenly appeared as a discrete group. This, in many ways, set the main backdrop for a strong effort by paleontologists to enter the center stage of evolutionary biology.

A paper by Van Valen (1973a) analyzed the distribution of longevities of taxa of many disparate fossil groups and converted them into a series of log-linear survivorship curves. From this Van Valen concluded that evolution was largely a result of the tangled bank of biotic interactions conceived by Darwin. His "Red Queen" hypothesis argued that random appearances of biological challenges regulated the tempo of evolution. This claim was of great interest to neontologists and was the subject of later speculation on the effect of random change on genetic load (Maynard Smith 1976) and diversity (Stenseth and Maynard Smith 1984). It also was an important link between the many studies of taxon longevity (e.g., Levinton 1974; Stanley 1975; Simpson 1944) at higher levels (generic and above) and processes occurring during speciation.

A milestone in bringing a readjustment of paleontological focus toward the level of species and smaller-scale fossil trends was Niles Eldredge's (1971) attempt to refamiliarize paleontologists with the implications of then-current speciation theory for the study of patterns of morphological change in the fossil record. In particular, he emphasized the compatibility of Mayr's (1963) theory of peripheral isolates and genetic revolutions with observations of sudden change or gaps in the fossil record without transitional forms. Such gaps were also compatible with Darwin's (1859, p. 342) postulation of evolution in small populations in geographically restricted areas. Eldredge claimed that the neontological perspective made such gaps expectable. Ironically, the purpose of this paper was to alert paleontologists that their ignorance of current evolutionary theory, that is, the body of theory stemming from the Modern Synthesis of the 1940s, was actively hampering their ability to interpret data. A similar argument had been made by Shaw (1969), who claimed that paleontologists, in assuming that evolution occurred uniformly throughout a species range, were developing a highly inconsistent nomenclature at the species level. Eldredge's brand of the claim was later amplified and transformed into the **punctuated equilibrium** hypothesis, a full-scale attack on the relevance of the Modern Synthesis to evolution (Eldredge and Gould 1972; Gould and Eldredge 1977; Stanley 1975, 1979).

Although the empirical aspects of the punctuative theory of evolution will be discussed in detail in Chapters 4 and 7, I will now discuss briefly its relevance to the recent history of evolutionary thought. The rapid origin of species, followed by a long period of stability, was posed

(Eldredge and Gould 1972) as an alternative to so-called **phyletic gradualism**, that envisioned evolution as being even and slow and occurring by the transformation of an entire ancestral population into its modified descendants. Under the gradualism model, one could envision evolution as a stately, uniform, and slow progression from an ancestor to a descendant. Instead, the punctuated equilibria theory saw evolution as a jerky process; speciation itself was rapid and the generator of morphological change. After the speciation event, various constraints, including a centripetal force imposed by the homeostatic nature of development, tended to prevent change. The new species, deriving from a peripheral isolate, was equipped with "its own powerful homeostatic mechanism" (Eldredge and Gould 1972, p. 114).

Speciation was therefore the key event in evolution; evolutionary trends were to be envisioned as either variations in the rate of speciation (Vrba 1983) or selective deaths of daughter species in a biased morphological direction (Eldredge and Gould 1972; Gould and Eldredge 1977; Stanley 1975). This point of view led to a different perspective of evolutionary thinking. If speciation was a fundamental decoupling point in evolution, then major trends in evolution, i.e., macroevolution, should be thought of as "changes in species composition within a monophyletic group" (Eldredge and Cracraft 1980, p. 15). Phyletic evolution was not proscribed, it was simply a less important process.

These arguments about the nature of the fossil record and claims about the adequacy of the Modern Synthesis evinced a pronounced enthusiasm for the collection and analysis of paleontological lineages at successive horizons to search for gradual or punctuative patterns of morphological change. Gould and Eldredge (1977) managed to interpret most of the available evidence as favoring punctuations; others argued that the available data base was inadequate or unsupportive of sweeping generalizations (Sadler and Dingus 1982; Bookstein et al. 1978). Still others felt that the dichotomy between gradual and punctuative change obscured more than it clarified, since the pattern of morphological change does not uniquely imply a particular evolutionary process (summarized in Levinton 1983; Levinton and Simon 1980). Nevertheless, many *causes célèbres* were trotted out as paradigmatic examples of punctuation (e.g., Williamson 1981) or gradualism (Gingerich 1976; Malmgren and Kennett 1981). At the least, Eldredge and his colleagues generated a great deal of controversy and stimulated an examination of temporal change in the fossil record over smaller time scales.

In this volume, I will attempt to connect the threads of genetics, paleontology, and evolution to produce a framework for an integrated outlook on evolutionary theory and the fossil record. I will try to evaluate the evidence and see just where paleontology and neontology can meet and make productive statements about the nature of evolution and the history of life. In some cases my conclusions will be optimistic; there is cause to believe that a new understanding of evolution is at hand. In other cases, the limitations of data and theory are apparent and we are still in the dark. Like Simpson, I am convinced that paleontologists and neontologists have something to say to each other and are capable of speaking a common language.

The Main Points

1. The process of macroevolution is the sum of those processes that explain the character-state transitions that diagnose evolutionary distances of significant taxonomic rank. The field of macroevolution emphasizes those processes that contribute to our knowledge of differences among major taxa, but is not confined to evolution above the species level, or macromutations. Any process involved in the character transitions defined above is relevant to the field.

2. Biology and evolutionary thinking lends itself naturally to a hierarchical organization of the biosphere. The presence of distinct organizational levels begs the question of the reasons for their existence and the potential interactions among levels. While we should eschew the assignment of undefined (mystical) properties to these levels, it is useful to understand whether some levels are particularly important in evolution. In cases such as extinction, elimination of higher levels (e.g., populations) may cause the elimination of lower level units (e.g., genes). This is known as downward causation. In other cases, processes at lower levels (e.g., failure for a cold adapting gene to be fixed in a population) might contribute to the loss of a higher level (extinction of the species, if the environment becomes cold). In some cases such upward causation permits an interesting correspondence among levels (e.g., organismal properties determine properties of a mono-

phyletic group by functional or epigenetic constraints).

3. Typology has had a strong influence on evolutionary thinking. The great advance of Darwinism and neo-Darwinism lies in the breakdown of typology. The move towards population thinking eliminated the static view of taxa is immutable entities. This was especially true of the species concept, which now acquired a biological and materialistic basis.

4. In the twentieth century, macroevolution as a paroxysmal process was championed by the evolutionary biologist Goldschmidt and by the paleontologist Schindewolf. Both believed in sudden evolutionary change, and Goldschmidt postulated a series of "systemic mutations" that produced hopeful monsters. Speciation was believed to result from such mutations, and, therefore, intraspecific variation was meaningless in evolution. Both theory and data were incompatible with this notion, and Goldschmidt's ideas fell rightfully into disrepute. He also, however, described how developmental (physiological) genetics could be used as a tool to study directions of evolution, and this field has been revived with considerable success in recent years (see Chapter 5).

5. Twentieth-century evolutionary biology was marked by four important periods. The rediscovery of Mendelian genetic transmission provoked a debate between those who believed that mutations drove evolution, and those who saw evolution as a process dominated by natural selection operating on small degrees of variation. It was later appreciated that mutations generated variation, while natural selection, migration, and drift determined the disposition of genes in population. The neo-Darwinian period, dominated by Wright, Fisher, and Haldane, gave us the theoretical underpinnings for understanding the fate of variation in natural populations. The Modern Synthesis accomplished the elimination of unlikely notions, such as orthogenesis, and spread the neo-Darwinian ideas to systematists and ecologists. In the period since the Modern Synthesis, it has been suggested that molecular variation in natural populations is neutral, objections have arisen to the primacy of natural selection, and the discovery and development of molecular genetics has enriched our understanding of the nature of organic variation.

6. George Gaylord Simpson brought paleontology out of an obsolete era, dominated by beliefs in orthogenesis. He melded population genetics with paleontological data and concluded that there were no incompatibilities. During this period, paleontologists became interested in correlating temporal changes in diversity with changes in earth history, and several found that massive radiations and extinctions were the rule. A variety of hypotheses were proposed to explain these major changes, some involving catastrophic events. More recently it was suggested that paleontological data were incompatible with some supposed expectations of population genetics and neo-Darwinism. In particular, it was suggested that speciation was the motor behind evolutionary change, and most species were static throughout their history. This suggestion evinced a large-scale research program on evolutionary changes at smaller paleontological time scales. At present, many paleontologists still feel that the fossil record requires a major alteration of evolutionary theory, while others either see no conflict at all, or feel that the challenge itself is weak and unsubstantiated.

Chapter 2

Genealogical Research and Systematics as a Framework for the Study of Macroevolution

> Our ancestors cut off the brightness on the land from above
> and created a world of shadows...
> *Tanizaki Junichiro*

Systematics and Macroevolutionary Hypotheses

A genealogy connects the members of a set of individuals or taxa by a criterion of relationship by descent. I am not using the term in the sense of a human family tree, which purports to have all ancestors and descendants. Taxonomic genealogies usually lack many ancestors and descendants and all we can hope to do is draw the relationships of the remainder we have. The object of systematics is to produce a classification of taxa; genealogy may be one of several criteria used to construct the classification, but available classifications are often the basis for arguments in phylogeny. Specific taxonomic levels used in macroevolutionary studies are believed to represent either adaptive zones (e.g., Van Valen 1984a) or a unit whose abundance is correlated with species level diversity (e.g., Sepkoski 1984). Yet despite the obvious need to define the meaning of these taxonomic levels, genealogical reconstruction and systematics are usually avoided in macroevolutionary treatises (e.g., Gould 1977; Stanley 1979) and ignored in studies of taxonomic longevity, diversity, and rates of taxon turnover (e.g., Van Valen 1973b; Valentine 1969; Sepkoski 1981). This omission weakens

43

the clarity of macroevolutionary hypotheses, which often involve explanations of change between sets of character states in different taxa. A reliable pattern of genealogy must be established before hypotheses of process and transition can be posed (Eldredge and Cracraft 1980; Cracraft 1981). It may be preferable to keep genealogical reconstruction and classification as separate enterprises, but reality steps in. The language and thought processes of evolutionary biology are enmeshed in the language and practices of taxonomy.

Systematic Philosophy Affects the Significance of Many Evolutionary Statements

The taxonomic hierarchy lends itself to a level-specific approach of hypothesis formation. This has been commonplace in macroevolutionary studies. Two well accepted macroevolutionary hypotheses illustrate this well. First, it has been claimed that the phyla "appear" first in the fossil record, whereas lower taxonomic units follow. Second, different taxonomic levels have been said to have differing patterns of response to environmental change; the frequency of response decreases as the taxonomic level increases (e.g., Valentine 1968, 1969).

Remoteness in time could influence the assignment of two taxa to different taxonomic groups of equal rank. Raup (1983) found that the mean geologic age of first occurrences of the 27 readily preservable class-level taxa of marine invertebrates is 533 my. Twenty of the 27 taxa first occur in the Cambrian. Because high taxonomic rank is based upon genealogical relationship, overall similarity, and species richness, it is not clear whether this early origin is a function of true early morphological diversification, or just an inherent property of higher taxa, whose early origins are bound to make them subtend many subordinate taxa that arise by branching of the stem taxon.

Consider the effect of using classifications employing pure genealogy versus pure similarity in testing macroevolutionary hypotheses. With a strictly genealogical approach then, the observation that "phyla appear first" could be a tautological restatement of the systematic philosophy (Fig. 2.1). Phyla might represent the first taxonomic split in the clade's history. If overall resemblance is used in delimiting phyla (Fig. 2.1), then the hypothesis that "phyla appear first" would have a different significance. Here, we could say that major phenetic differences materialized early in the history of life. Derived taxa might be limited in

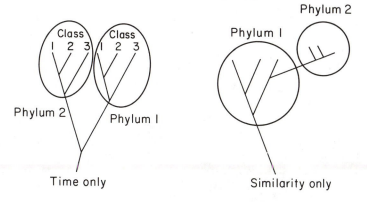

Figure 2.1: Two different classification schemes. On the left, phyla are delineated on the basis of time of branching. On the right, phyla are designated on the basis of dissimilarity, which increases on a horizontal scale.

their potential to give rise to new taxa of phylum rank.

A similar argument can be made for the hypothesis that differing taxonomic levels each have their unique responses to changes in the earth's history. To the degree that morphological similarity defines the taxa, definitions of increasingly lower taxonomic levels would correspond to increasing homogeneity of ecological response to environmental change. As the taxonomic level increases one tends to include more and more phenotypically different groups with differing ecological responses. If we group more and more phenetically different groups together into increasingly higher taxa, it stands to reason that this synthetic higher taxon will survive longer and taxa of this level will have a lower frequency of response to environmental change.

If genealogy is the only criterion used to establish classifications, then differential patterns of response by different taxonomic levels may have a different meaning. For example, more inclusive taxonomic levels may inevitably involve greater spans of geological time. The lower frequency of response of higher taxonomic levels may therefore represent a "buffering" response. Higher taxa are bound to have a lower extinction rate than lower taxa, but this may have nothing to do with morphological specialization (as in a phenetic classification); it may only be an inherent property of the taxonomic structure, where more inclusive taxa are bound to be more longevous. In the genealogical

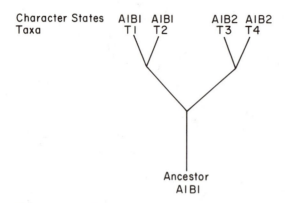

Figure 2.2: Hypothetical phylogeny of a lineage of bivalves. See text for explanation.

end member case a mixed result is also possible. With time, a taxon may become morphologically diverse and geographically widespread as cladogenesis proceeds. This would also confer on the inclusive taxon buffering against extinction. On the other hand, a strictly genealogical framework might also reveal that larger taxa in some cases have greater morphological diversity than another of comparable branching structure.

Advantages of the Genealogical Approach

A framework established from a genealogical algorithm permits a useful analysis of character variation in the context of macroevolutionary hypotheses. Many macroevolutionary hypotheses attempt to provide mechanisms to explain differential taxon longevity. Claims that taxon longevity depends upon biogeographic range (e.g., Jackson 1974; Levinton 1974; Boucot 1978) or that taxon longevity is the result of differential speciation rate or survival of species (e.g., Stanley 1975; Vrba 1983) may depend partially upon the nature of character variation within the clades under consideration. In many cases, adaptations of individuals influence the susceptibility to extinction of species and larger taxa. While speciation rate may ensure survival of a taxon, the possession of certain characters may permit an entire clade to outlast others, or might permit descendants of a given clade to invade a new habitat. The testing of such ideas requires a mapping of character transformations on genealogies.

Consider the following hypothesis: phenotypic evolution occurs because of species selection (Eldredge and Gould 1972; Stanley 1975). Levels below and above the species level are thus irrelevant to the evolutionary trend, which is a net change in character states over time. Take a hypothetical phylogeny of bivalve mollusks (Fig. 2.2). A species bears character state $A1$, representing a compressed elongate shell, and character state $B1$, representing lack of ornamentation. The clam therefore has a morphology compatible with rapid burrowing in soft substrata (Stanley 1970). Let the ancestral species split into two daughter species. A split of each daughter species results in four taxa. Extant taxa T1 and T2 bear the ancestral character states $A1$ and $B1$. Extant taxa T3 and T4 also bear state $A1$; they, however, have acquired character state $B2$, representing heavy ornamentation.

From a functional morphological point of view, the ancestral character state $A1$ interacts with the state of character B, which determines the derived state defining the genealogical groups $\{T1,T2\}$ and $\{T3,T4\}$. Let us call these two taxa "genera." In our specific example, $A1,B1$ is a functionally compatible character set, while $A1,B2$ joins two character states that would fail to be functionally harmonious under most circumstances. Squat shells with ornamentation would be preferable in stabilizing the shell on the bottom in swift currents, while elongate, compressed shells lacking ornament would be efficient in burrowing. The $A2,B1$ state is a mixed case, not much good for either function. It would therefore surprise no one if the group $\{T3,T4\}$ had a higher probability of extinction. Indeed, its evolution probably would have occurred under atypical environmental circumstances.

With this example we can make several points about the role of functional morphology in predicting relative extinction rates and the basis of extinction. First, the character A defining the $\{T1,T2,T3,T4\}$ group interacts with the state of the character defining the two included groups. The genus level of response to extinction may be defined by the special set of characters $A1,B2$, but character state $A1$ will survive in either of the two taxa: $\{T1,T2\}$, $\{T3,T4\}$. Thus taxon mortality at the genus level explains selective loss of the $A1,B2$ character complex. But this surely is not an emergent phenomenon of the genus level, as we have defined it. The inevitable retention of the $A1$ character state, moreover, is not readily identified with any taxonomic level. Indeed, it is only a matter of coincidence that selection among genera has occurred. Selective mortality can be reckoned from a simple summing of character states. Species become extinct *because of the character states they bear*; a conclusion that genus level selection occurs is, therefore, ambiguous. We can at least, however, identify the taxonomic level at which the crucial combinations of character states result in differing probabilities of extinction.

An improved degree of focus thus emerges from a genealogical approach based upon character analysis. At present, a disturbing vagueness plagues the literature. This has been reinforced by the use of taxonomic survivorship curves at many taxonomic levels, with a varied mixture of ecological and evolutionary intents. Levinton (1974), for example, employed the generic level to contrast paleoautecology with taxonomic survivorship among groups of bivalve mollusks. But the generic level was chosen as a matter of convenience, controlled by the available monographic accounts of the Bivalvia. This particular taxonomic level, which did reveal significant differences among bivalve groups, may be irrelevant for the purposes intended, simply because the character complexes involved in autecological aspects of taxon survival were concentrated at another level.

In a similar vein, variance in gastropod form has been found to decrease from the Paleozoic to the present (Cain 1977; Gilinsky 1981). This trend indicates that those taxa deviating from a modal form have tended to become extinct. Is this species selection, as claimed by Gilinsky? Of course species have become extinct. But the selection must be at the taxonomic level corresponding to the acquisition of the set of relatively poorly surviving character states. This may be at a much higher level than that of species, and can only be properly defined once a character analysis is done, set against a genealogically based systematic framework.

Genealogical and Systematic Philosophies

We must distinguish among several objectives of systematists and evolutionary biologists. We usually have at least four objectives, which are often intermixed. A **character analysis** is a study of features of individuals that may be used to construct a classification. The algorithm used to perform the character analysis may be qualitative or quantitative. A **genealogy** is a network of branchings, whose topology reflects the relationships by descent of the taxa under consideration. A **classi-**

fication is an ordering of taxa, based upon various criteria, but usually resulting in a hierarchy of successively inclusive sets (species grouped into genera, which are grouped into families, etc.). The classification may or may not be concordant with the genealogy. Finally, a **phylogeny** is an inferred genealogical history of a group, hypothesizing ancestor-descendant relationships, biogeography, etc. The genealogy is only part of the process of producing a phylogeny. Genealogies and phylogenies are hypotheses of relationships and history. To the degree that a classification is meant to reflect a genealogy, it too must be regarded as a hypothesis.

Systematics has occupied a central place in the posing and testing of macroevolutionary hypotheses. Most of the classic works in the field (Mayr 1942; Mayr 1969; Mayr et al. 1953; Simpson 1961) stressed the inherent complexity behind the traditional objectives of systematics. They agreed, however, that a useful classification should account for genealogy and morphological similarity. These two components lay behind **evolutionary systematics**, an approach that assigns taxonomic rank by means of genealogical position in a phylogenetic network and the amount of morphological divergence of a taxon from its ancestral lineage. Phylogenetic reconstruction is mixed with, or follows, classification. The two other major competing systematic philosophies take this mixed strategy to be undesirable. **Phenetics** seeks to produce classifications based upon overall similarity alone (Sokal and Sneath 1963; Sokal and Camin 1965). Genealogy is not a necessary objective of phenetic classifications, although overall similarity must have some mapping to relationship by descent (Sokal and Sneath 1963). **Phylogenetic systematics** seeks to establish a network of genealogically based relationships with no overall similarity criterion employed for classification (e.g., Hennig 1966; Camin and Sokal 1965; Kluge and Farris 1969). Phylogenetic systematists seek to cluster **monophyletic groups**, or the entire descendant subset of taxa derived from a given ancestor.

Arguments over the preferability of any of the three systems usually revolve around several desirable criteria of classifications used by evolutionary biologists:

1. **Convenience.** The system should yield a classificatory system that is not cumbersome, and should be intuitive enough for all to grasp.

2. **Congruence.** Classifications based upon different characters should yield similar results.

3. **Genealogy.** Most evolutionary biologists desire a classification that reflects phylogenetic relationships.

4. **Naturalness.** Groupings should, in some readily understood sense, reflect directly the character states used to determine the classification (Gilmour 1961).

All three approaches have their own utility but I will attempt to justify briefly why phylogenetic systematics is preferable for macroevolutionary objectives. The other two systems, however, have attractive characteristics, and all three have their respective strong disadvantages. Phenetics and phylogenetic systematics both rely upon numerical algorithms that are considerably unstable in defining groups (e.g., Rohlf et al. 1983). I believe that this accounts for the understandable controversy over which is "best," but it does not discount the larger and preferable objective of establishing genealogies, and constructing classifications based upon genealogy and character transformation.

Phylogenetic Systematics

The cladistic approach establishes networks of genealogical connections based upon uniquely shared, and evolutionarily derived, similarities—or **synapomorphies**. Any derived state is said to be **apomorphous**. Overall similarity is thought to be misleading, since it often entails groupings by shared ancestral features that may define genealogically more inclusive groups, rather than specific and closely related derived groups. Any ancestral state is said to be **plesiomorphous**.

The primary objective of cladistics is to map taxa onto a **cladogram** whose branchings signify genealogy and whose topology reflects solely the evolutionary changes in characters. Algorithms include hand calculated groupings by use of synapomorphies (e.g., Hennig 1966), computer-based algorithms that attempt to find the tree with the smallest number of evolutionary steps (the method of **parsimony**) using various assumptions that make tree calculation somewhat more tractable (e.g., Camin and Sokal 1965; Farris 1970; Goodman et al. 1982), and character compatibility algorithms that search for sets of characters whose states define the same genealogical relationships (e.g.,

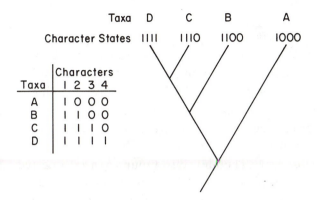

Taxa	Characters 1 2 3 4
A	1 0 0 0
B	1 1 0 0
C	1 1 1 0
D	1 1 1 1

Figure 2.3: A matrix of character states by taxa, with a cladogram, established by Hennigian principles.

Estabrook 1972, 1980; LeQuesne 1969). In some cases phenetic distances have been modified to construct genealogical relationships (e.g., Fitch and Margoliash 1967; Farris 1972; Felsenstein 1982). A second objective of cladistics is to use the genealogy to produce a classification that best represents the branching pattern.

Hennig Crystallizes the Problem   Hennig's *Phylogenetic Systematics* (1966) helped to focus most current attempts to understand genealogical relationships among taxa and to derive classifications that map logically to genealogical trees. Speciation, the splitting up of one species into two or more daughter species, is the basis for the whole system. After a number of splits, groups can be defined as a series of increasingly more inclusive nested sets (Fig. 2.3). Species are related by the branching network created by the cladogenetic process. A **monophyletic group** is thus "a group of species descended from a single (stem) species, and which includes all species descended from this stem species" (Hennig 1966, p. 73).

As cladogenesis occurs, **characters**, attributes of the organisms, change in their respective **states**. The problem is how to use character data to reckon genealogy. Many of the taxa produced have become extinct, so that we can never establish the complete network of ancestors, nor do we have an extant record of the magnitude of character transitions throughout the cladogenetic history of the group under consideration. All we have are the extant and fossil taxa, and their character states.

Groups with uniquely derived character states, **autapomorphies**, are the most closely related. Successively more distantly related groups are connected by their shared derived character states. An increasing number of synapomorphies increases the likelihood that a grouping is genealogically circumscribed. Grouping by shared ancestral characters invites the danger of producing groups that are genealogically incomplete. As shown by Fig. 2.3, it is the synapomorphies that uniquely identify genealogical closeness. The ability to group taxa into nested sets, based upon synapomorphies, would be easy if there would be compatibility of information derived from the different characters. As we shall see, incompatibility of character states is the fundamental problem in genealogical reconstruction.

Homology Criteria for homology must be established, so that features of related taxa can be identified as states of the same homologous character. Unfortunately, it is difficult to prove that a character is homologous in two species. For any evolutionary system, homology would likely imply some correspondence between the two taxa in the part of the genome controlling the character in question. This can only be inferred indirectly in most cases. Even if a direct assay of the DNA underlying the phenotypic character were possible, one would most probably find a nonidentity. The character would not be controlled by an identical set of genes, although substantial overlap is to be expected if the character states can be related at all. A second criterion of homology is that the phenotypic trait is similar in the taxa. One would not accept a genetic definition of homology alone, as genes might control several different phenotypic traits. Similarity in overall form and similarity of location within the body would be two crucial criteria.

In practice, the homology of phenotypic characters is rarely traced by systematists to the underlying genome (excepting in studies of molecular evolution). Tests for homology should involve the following criteria (Patterson 1982):

1. Two homologous character states may not exist within one organism. Different character states might exist among organisms of a single taxon or species.

2. Two character states in two organisms are of the same homolo-

gous character if they occupy the same topographic or ontogenetic position in the organism.

3. A series of character states in a series of corresponding taxa belong to one homologous character if the cladistic relationship among the taxa, defined by the character, does not contradict any genealogy defined by "truly homologous" characters.

The last two criteria require qualification and suggest other criteria. Criterion two implies that homologous characters can be "located" in different taxa. This ability is strengthened to the degree that (1) evolution produces unique phenotypic sites; and (2) evolution is slow. If evolution is very rapid and not unique (i.e., convergence is common), then the ability to identify the phenotypic expression of corresponding parts of the genome is erased. Location can also involve a temporal aspect, especially because ontogenetic data can be applied to systematic problems (e.g., DeBeer 1958; Nelson 1978; Alberch 1985).

A special set of instances may directly link homology and polarity. Developmental anomalies often reveal seemingly ancestral states. In rare instances, whales have complete limbs similar to those used in walking ancestors, despite the fact that millions of years have elapsed since the structures related to walking were presumably lost (Andrews 1921; Lande 1978). The atavistic appearance of long-lost structures in amazing detail (e.g., Andrews 1921; Kurtén 1963; Marsh 1892) seems to discredit the belief that the loss of a structure implies the loss of the genes, as a naive version of an adaptive theory of evolutionary genetics would predict (Kollar and Fisher 1980). This suggests that, for whatever reason, the genome is to a degree stable and a genetic basis for homology is possible. It also suggests another criterion for homology:

(4) Two states in two organisms can represent states of a homologous character if a developmental anomaly in one taxon produces an individual with a state largely similar to the other taxon, in the same topographic position.

This criterion must be used judiciously, since the simplicity of a character state might result in the evolutionary convergence of states of a nonhomologous character. For example, if two states represented different colors of a butterfly wing of two respective species, the appearance of taxon B's color, as a variant of taxon A, cannot guarantee

homology. Complexity of similarity is therefore an essential element
of this criterion. Unfortunately it is difficult to define a mathematical
function that relates the probability of homology to increase of similarity. One of the *bithorax* complex phenotypes (see Ouweneel 1976)
in *Drosophila melanogaster* mimics the presumed ancestral state of the
diptera, i.e., two pairs of wings. There is no strong reason to believe,
however, that this is necessarily the particular genetic route backwards
to the ancestral state. Appeals to strong similarity of detail therefore
are intuitively attractive, but no more than that.

Patterson's (1982) third test of homology, compatibility with other
homologous characters, raises both the fundamental strength and an
important weakness of the Hennigian cladistic method. This criterion
implies that homology is an hypothesis, rather than a proven statement
of genealogical connection among character states. The hypothesis of
homology for a given set of character states is therefore corroborated
if the genealogical relationships defined by synapomorphies does not
contradict others defined by other characters. Homologies of several
discrete sets of character states are thus reinforced to the degree that
the corresponding genealogical relationships defined by the discrete
sets are compatible. But what if sets of states produce incongruent
inferred genealogies? How do we decide among different trees defined
by different characters, and how do incongruities affect our hypotheses
of homology?

The first criterion of homology, that a character not be found in
two different locations on the same creature, implies that we are excluding serial homology from our discussions. Many structures–genes
for example arose in evolution by duplication. In the first descendant
taxon with a duplication, there is an ambiguity in that two structures
are homologous with one belonging to the ancestor. Subsequently,
the ontological ambiguity of homology disappears but there is still an
epistemological confusion in the status of the relationships between the
repeated structures within the same individual (e.g., two tandom genes
that arose by duplication that now serve different functions). This confusion is heightened when the duplications affect the same phenotype
in different ontological stages. Thus *Drosophila melanogaster* has larval and adult alcohol dehydrogenases that presumably serve the same
function, but the same structural gene codes for both enzymes; the difference is an upstream sequence. In this case the distinction between

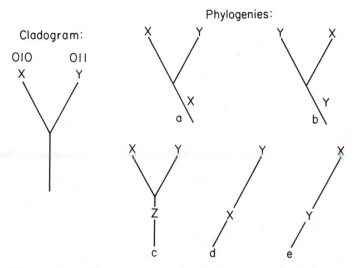

Figure 2.4: Cladograms and phylogenies. A cladogram for two taxa, with the possible phylogenies that might be implied.

serial homology and evolutionary homology breaks down.

If our objective is genealogical reconstruction alone, then it is not clear whether the genotype-phenotype distinction is all that important. Characters are characters, and homologies can be established–indeed they have been established for hundreds of years–without the benefit of knowing the genetic underpinnings. Genetic data, such as nucleotide sequences, are also sources of homology, as long as some sort of criterion of location can be employed. One must be sure, for example, that one is following the same gene through a genealogy if the sequence is to mean anything. The connection between genes and phenotype becomes important when one is interested in tracing given characters in clades, particularly with regard to evolutionary mechanism. If one believes in genetic constraints, then the DNA history is as important as a phenotypic history. This connection is most crucial if we are ever to understand the relative contributions of developmental, genetic, and functional constraints to phenotypic evolution.

Cladograms and Phylogenies   A **cladogram** is a diagram posed as a hypothesis of the genealogical relationships among a series of taxa, grouped by their synapomorphies. A **phylogeny**, by contrast, is a hypothesis depicting the exact history of the evolutionary connections

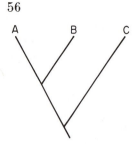

Figure 2.5: A three-taxon cladogram.

among the taxa. It may invoke a specific extinct ancestor that is not preserved as a fossil. I illustrate the distinction between a cladogram and a phylogeny in Fig. 2.4. For the two taxon case, we assume that one or the other might be an extinct species. Note that five possible phylogenetic histories can be derived from the simple cladogram connecting taxa x and y (see also Platnick 1977a, Eldredge and Cracraft 1980).

With some information on polarity of character states we can restrict the phylogenetic hypotheses somewhat. Consider three characters whose states are 0 (ancestral) or 1 (derived). Then imagine a root to the cladogram, defined by the most ancestral character states $\{0,0,0\}$. If we refer to the cladogram in Fig. 2.4, with the additional information on polarity, then the possible phylogenies are restricted to types a, c, and d. In case c an ancestor is invoked who bears the character states $\{0,0,0\}$.

When we consider three taxa, the notion of ancestral and derived character states is better defined (see Nelson and Platnick 1981 for extended discussion). Three-taxon statements devolve to the problem: are two of the taxa, A and B, more closely related to each other than to another taxon, C (Fig. 2.5)? The most distantly related taxon is defined as that one that joins in the cladogram, after uniting the first two, which share the most derived states over all characters.

A cladogram is constructed from a matrix of taxa by characters such as the one in Fig. 2.6. For this matrix we assume that 0 is the ancestral (plesiomorphic), and 1 the derived (apomorphic) state. Fig. 2.6 shows the cladogram based only upon characters 1-4. The cladogram has a **root**, which is defined by the most plesiomorphic states for all characters.

Note that characters 5 and 6 define a group that is inconsistent with

	Characters					
Taxa	1	2	3	4	5	6
A	1	1	0	0	0	0
B	0	0	1	0	1	1
C	1	1	0	0	1	1
D	0	0	1	1	0	0
E	0	1	0	0	1	1

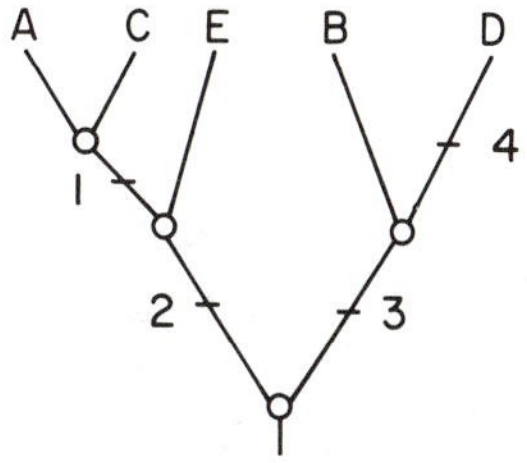

Figure 2.6: A matrix of character states by taxa for a hypothetical group. For each character, 0 is taken to be ancestral, 1 is derived. At right is a Hennigian analysis for characters 1-4, which are compatible. Character numbers on tree define groupings delineated by the next highest node (for example, group [A, C] is defined by character 1, [A, C, E] is defined by character 2).

the rooted cladogram. One might conclude that the inconsistency falsifies the hypothesis of relationship derived from the majority of the characters. We might favor the cladogram as illustrated in Fig. 2.6, as the information from four characters defines the cladogram, whereas the inconsistent hypothesis is defined by only two characters. This is often characterized as the adoption of the **least refuted hypothesis** (e.g., Lynch 1982). It is used typically by those who employ nonnumerical algorithms to generate rooted cladograms.

Rooting The Cladogram   The construction of a root for a cladogram is one of the most difficult problems of phylogenetic systematics. One needs a criterion to locate the part of the cladogram that bears the most ancestral states. This requires a criterion for identifying ancestral character states. Two main ones have been employed: **outgroups** and **ontogeny**. In a three-taxon case one taxon is more distantly related than two others that are more closely united by synapomorphies. Cladograms can be constructed by successive additions based upon three-taxon "statements" (see Nelson and Platnick 1981; Wiley 1981). The choice of an outgroup involves picking a character or set of characters which is widely agreed to have more ancestral states than is present in the first three taxa. Thus an outgroup for the bivalves might lack a shell, but have spiral cleavage and a mantle. Its attachment to the cladogram defines polarity.

The use of ontogeny derives from Haeckel's biogenetic law that ontogeny recapitulates phylogeny (Nelson 1978). The reality of the biogenetic law has long been in hot dispute (see discussion in Gould 1977). This criterion could only have utility if evolution by terminal addition occurs, and descendants either comprise a simple addition of the stages, or a form of acceleration compresses the stages into the same developmental period. Alternatively, an extension of von Baer's laws of development into evolution (von Baer didn't believe in evolution) pose hypotheses of polarity (Nelson 1978; Patterson 1982). General (ancestral) features are believed to occur early in the ontogeny of related taxa. Later ontogenetic stages represent specializations (derived states). This hypothesis supposes implicitly that early stages of ontogeny are less subject to evolutionary alteration than later stages. See Chapter 5 for further discussion of this issue.

Ontogenetic considerations show that apparent ancestry cannot be identified under certain conditions (Fink 1982):

1. When the common ancestor of two taxa evolved state *b* by adding a stage to the ontogenetic trajectory, but one of the two descendants went to state *a* by loss of the terminal state *b*;

2. Same as condition 1, only ancestor exhibits acceleration and one descendant shows slowing of development;

3. Same as condition 1, only contrasting a movement up in onset of development in an ancestor, followed by relative delay of onset in development of the descendant.

These three conditions will erase the record of character polarity. Any shuffling of stages within a sequence would destroy the directional utility of the ontogenetic order of the character states in descendants.

Alberch and Gale (1983) have recently investigated ontogeny in frogs and salamanders and demonstrate that developmental regularities might be a valid key to character state sequence. During development, digit number one is the last to appear in the frogs *Xenopus laevis*, while digit number five is the last to appear in the axolotl, *Ambystoma mexicanum*. This seems to correspond to evidence for evolutionarily-derived digit loss. The last digit produced during ontogeny is that one which is lost first. This cannot be used to determine polarity, but at least a predicted sequence defining a linear order of character states in evolution might be established from such data. Evolution could go in either direction along the sequence. Another encouraging example is McGowan's (1984) study of the development of the avian tarsus. It had been previously suggested that ratites were derived from carinates (flying birds) by an arrest of development. Carinates, however, can be shown to have an ontogenetically unique pretibial bone, while the ratites share the ascending process of the astralegus with the theropod dinosaurs. Ratites therefore are in a relatively ancestral state and their ancestral stock is thus more ancestral than that which defines the flying birds.

It is likely that many evolutionary sequences involve terminal addition, or at least, resolvable alterations of developmental sequences. Alterations of ontogenetic sequences, particularly terminal addition, have been suggested in several studies of fossil mollusks (Fisher et al. 1964; Newell 1937; Miyazaki and Mickevich 1982). The evolutionary adjustment of ontogenetic patterns in the tropical American salamander genus *Bolitoglossa* are another example. Hand and foot morphology of the species represents all stages of intermediacy between the slightly webbed, large digited structures of the upland species and the diminutive fully webbed small digited ones of the lowlands (Wake and Brame 1969). Diminutive lowland species are paedomorphic and seem to result from the retention of juvenile characters of the larger ancestors in the adults of the smaller descendants. The webbing and small size seems adaptive for the relatively more arboreal habit of the lowland forms (Wake and Brame 1969; Alberch 1981).

In his functional study, Alberch (1981) identifies the plesiomorphic state for the genus *Bolitoglossa* by an outgroup comparison. The outgroup is intermediate along the ontogenetic track, relative to the two derived species he considers carefully. Thus the root of the network (not done by Alberch in quite this manner or with this terminology) is near a point where species bear intermediate character states of the ontogenetic track. We must assume therefore that reversals are possible; one can move backward or forward along an evolutionary-ontogenetic trajectory. But suppose that we had no good outgroup for comparison. It is likely that the network might be rooted (if only ontogenetic characters were employed) near the taxon with the greatest representation of early ontogenetic character states. Given our information, this could lead to incorrect judgments about the history of the group.

We can imagine two closely related sister taxa whose difference rests on one synapomorphy. If most of the useful characters are associated only with ontogeny, and one cannot be sure as to the ancestral state of any character, only the sequence, then it may be difficult to root one group with respect to another. In other words, a conflict in rooting cladograms might arise between the use of outgroup comparisons and ontogenetic character sequences.

If there is a correlation between the order of ontogenetic stages and the stratigraphic sequence, then we would be justified in invoking fossil sequence as corroborative evidence favoring rooting of the network near the species that is both stratigraphically oldest and ontogenetically "earliest" (Miyazaki and Mickevich 1982). This conclusion is based on the sensible argument that it is more likely that evolution has proceeded forward in time, rather than backward.

The Problem of Ancestors   The root of cladograms provides information on the nature of ancestry, but does not define ancestors. Consider the cladograms and possible phylogenies in Fig. 2.4. In many cases the ancestor will never be identified, simply because no independent criteria would delimit a choice among the possible phylogenetic hypotheses. In these cases, the study of macroevolution is restricted to the study of the possible mode of transition from a taxon to its closest relative. We can only speculate what combination of character states ancestral transitional taxa might have borne.

Homoplasy: The Fundamental Problem   Homoplasy can be defined as any resemblance between two (or among more) taxa that is not due to inheritance from a common ancestor (Simpson 1961, p. 78). Parallel evolution, convergent evolution, evolutionary reversals, mimicry, and chance evolution of similarity can all produce homoplasies. Parallel evolution could imply some common evolutionary constraint (see Chapter 5) due to common ancestry of two now-distant taxa, which causes separated lineages to develop along similar phyletic paths. Convergence can cause us to mistake homologous for **analogous** structures, or features of two taxa that are similar, but not because of a common evolutionary connection.

As the degree of homoplasy increases in the evolutionary history of a group, the degree of inconsistency with the same tree among different characters must also increase. No one would doubt that homoplasy occurs; indeed character inconsistency is the rule rather than the exception (see Felsenstein 1982). As mentioned above, the concept of minimum refutation, or maximum corroboration, arose from this problem.

Some Approaches to the Homoplasy Problem   We can attempt to minimize the blurring effects of homoplasy by at least two approaches: **character compatibility**, and **parsimony**. Character compatibility involves choosing a majority set of internally consistent characters to construct the genealogy and the classification. Parsimony acknowledges conflicts but chooses the cladistic network requiring the fewest character changes. Nonnumerical approaches usually attempt to establish synapomorphies and then drop characters that imply groupings inconsistent with the majority of consistent characters (e.g., Eldredge and Cracraft 1980). This is, in effect, a compatibility analysis. Character sets may be excluded by appealing to various functional morphological considerations that might suggest convergence. Thus J.D. Smith (1976) argued that wing characters in the bats are homoplasic for functional reasons, and suggested that bats may be polyphyletic. Other nonquantitative studies compromise compatibility by sometimes accepting inconsistencies among characters (that require reversals in character state) to minimize the number of steps to make the tree. These are parsimony techniques. Of course, a compatibility analysis can also be a parsimony analysis when dropping out of inconsistent blocks of characters is the very thing required to minimize the number of steps. This is often true with small numbers of characters.

The compatibility approach was formalized by LeQuesne (1969) who recommended that monophyletic groups be defined by **cliques** of consistent character states. Fig. 2.7 shows a simple case, where classification by use of secondary compounds is employed to produce a genealogical classification. We assume that evolutionary acquisition of the compound occurs only once and that absence is the ancestral state. As can be seen two chemicals can be inconsistent with the same hypothesis of genealogical relationship. Numerical approaches have been developed, especially by Estabrook and colleagues (Estabrook 1972; Estabrook and Anderson 1978), to identify cliques of compatible characters to construct a rooted cladogram.

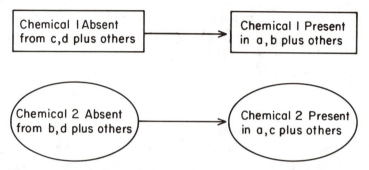

Figure 2.7: A hypothetical character compatibility analysis for secondary compounds in a set of plant taxa. The hypothesis that the occurrence pattern of chemical 1 is ideally related to the evolution of the group is incompatible with the hypothesis that chemical 2 is ideally related to the group's evolution (after Estabrook 1980).

Parsimony is employed in both nonquantitative and numerical approaches (Camin and Sokal 1965; Kluge and Farris 1969; Farris 1970). The possibility of reversal of character state is met with the notion that the hypothesis of genealogy most likely to be correct is the one that requires the minimum number of evolutionary steps to explain the tree. Of course, nature doesn't have to be parsimonious; we simply claim that we will be correct most often by assuming the minimum number of steps. If the rate of evolution is very rapid, and filled with reversals of character state, then the shortest number of steps may very well not be the correct phylogenetic solution. Should we then drop noisy characters and adopt a compatibility approach? Compatibility analysis can be criticized on the ground that it omits possibly informative data by dropping characters from the analysis. Although some characters have conflicts with the overall set, their use might contribute to some resolution, assuming that homoplasy is homogeneously scattered throughout the incompatible set. If homoplasy is known to be present in a few specific characters, it might be best to drop these from the analysis.

Camin and Sokal (1965) proposed a simple algorithm that depends upon the knowledge of polarity of evolution among the states of any character. They assume that (1) characters can be expressed in discrete states; (2) states can be ordered; (3) the ancestral state arose once; and (4) evolution is irreversible. Several algorithms have been proposed to find the tree with the smallest number of evolutionary steps (Camin and Sokal 1965; Estabrook 1968; Sneath and Sokal 1973).

Another algorithm employing parsimony was suggested by Kluge and Farris (1969), and further developed by Farris (1970). The method, like that of Camin and Sokal, depends upon the ability to code a series of states in some order, but reversals are permitted. Polarity needs to be known in order to root the cladogram.

The data recorded in the form shown above are in Fig. 2.6. The distance between any two taxa is then computed as a city block, or Manhattan metric, which computes the degree of differentiation, the **advancement index**, of taxon A from B, as:

$$d(A, B) = \sum_i \mid x(i, A) - x(i, B) \mid$$

where $d(A,B)$ is the degree of differentiation, $x(i,A)$ is the value of character state i, for taxon A.

Although taxa can be grouped in any order (see Farris 1972) the following algorithm (based upon Kluge and Farris 1969; Wiley 1981, p. 182) starts either with the most "ancestral" taxon or with a hypothetical ancestor.

1. Specify an ancestor or sister group.

2. Compute D to ancestor for each taxon, find the taxon with minimum D. Connect taxon to ancestor, creating an **interval** for that taxon.

3. Find next taxon with smallest D from ancestor/sister group.

4. Find interval that shows least difference from this selected taxon. To find this, compute the difference between the selected taxon and the interval of each taxon that is already connected to the tree. If we have a taxon, A, for example, connected to an ancestor and our next taxon is B, then we compute:

$$D[B, INT(A)] = [D(B, A) + D(B, ANC(A))]/2$$

5. Attach taxon to interval with which it differs least. To do this we construct an hypothetical common ancestor for the two taxa, such that the ancestor's character states are the median of the

character states of the first taxon, its original ancestor, and the new taxon. Using the median allows the tree to satisfy the triangle inequality (the distance between any two taxa is less than, or equal to, the distance between the two and the sum of distances between each of the two, and a third taxon).

This algorithm leads to the cladogram depicted in Fig. 2.6. In effect it works by dropping the two incompatible characters. In complex data sets, the cladogram with the smallest number of changes must be found by calculating many trees and choosing the one with the least amount of change. In these cases parsimony does not necessarily drop out incompatible characters, which may give some additional information that can minimize the length of the tree. Another algorithm is required for searching among many trees (e.g., Farris 1970). Various mainframe and microcomputer packages have become available in recent years to find the correct tree.

This algorithm has been found to be a usually most parsimonious method for establishing cladograms. It was the best method for approximating the phylogeny of the caminalcules when all characters were considered (Sokal 1983b). It performed more poorly than some phenetic approaches when smaller partitions of the the characters were employed (Sokal 1983c).

There is not space here to consider the other techniques of constructing cladograms. Of greatest interest are the techniques that estimate cladograms from molecular data (e.g., Eck and Dayhoff 1966; Fitch 1971; Fitch and Farris 1974; G.W. Moore et al. 1976). In some cases overall distance (e.g., number of amino acid differences) has been used to construct cladograms, as in the classic work of Fitch and Margoliash (1967). A fitting method is employed that minimizes the number of evolutionary steps on the constructed tree, relative to the total number of steps required to explain the original matrix of amino acid differences. Later approaches have employed parsimony to infer trees for both amino acid differences and nucleotide differences (e.g., Goodman et al. 1979, 1982). Gene duplications have been invoked to increase the consistency of characters on the tree (Goodman et al. 1982).

Efficacy of the Cladistic Methods   Felsenstein (1978, 1982, 1983) has discussed the conditions under which parsimony and character compatibility are likely to fail in producing an accurate genealogy. As might be expected, as the rate of evolution increases, the probability of reversals, convergence, etc. may increase as well. This will tend to blur the tree and diminish the ability of uniquely derived states to identify monophyletic groups. As the degree of homoplasy increases it becomes increasingly unjustifiable to have great confidence in the best solution, i.e., the shortest cladogram, although we would accept it as the best available hypothesis for the data at hand (see Sober 1983).

The worst case is that of parallel evolution. Felsenstein (1978) shows that if parallel evolution is pronounced (due to either elapsed time or increased rate of evolution), over all characters, along two isolated branches of a phylogenetic tree, relative to another intermediate branch where the evolutionary change is less, the inferred tree will be incorrect. This can be visualized as an isolated and coordinated extensive change in unrelated lineages. While this certainly is true in the formal sense, one wonders when such a problem would be encountered, given the requirement that differentiation must have a high probability in the two isolated branches.

Coordinated and independent parallel evolution can be imagined when a series of characters are correlated, as in a set of characters that respond to one change as an integrated developmental unit (e.g., Gould 1982a). In this case parallel evolution might cause the construction of the wrong tree. This may be common in studies where several characters are used, and all are essentially a single response to the same primary change. Such a case seems to apply to the neotenous evolution of salamanders. Changes in one hormone may have induced the coordinated responses of blocks of characters, with some lack of harmony among the blocks (heterochronic evolution; Etkin 1970).

A Maximum Likelihood Technique Felsenstein (1979) has suggested maximum likelihood approaches to selecting the best evolutionary tree. In order to do this, one must have the probability of evolutionary change from one state to another. In a molecular data set, for example, the transitions from one nucleotide base to another would have to be known. Given the knowledge of these probabilities, one then selects the tree with the maximum likelihood of fitting the data set. Using a few realistic assumptions, it is sometimes possible to take the maximum likelihood tree to be that which satisfies a certain criterion. For example, if the probability of evolution from an ancestral to a derived state is much higher than the probability of reversal, then the method of Camin and Sokal (1965), does very well in producing the most likely tree. This method assumes no reversals in character states.

This approach has the weakness of requiring some estimate of probabilities of evolution. In the case of nucleotide evolution, this might work well for many point mutations (see Felsenstein 1981), but significant problems would arise in the case of frame-shift mutations. In the latter case, it would not be possible to easily assign probabilities of nucleotide transitions, without knowing the specific sequence context of the mutation.

Despite this problem, the approach is attractive because it links character evolution directly to the construction of evolutionary trees. Using some simple assumptions of evolutionary change, it also permits direct comparisons of different cladistic algorithms.

Phenetics

Phenetic approaches group taxa by their overall similarity (Sokal and Sneath 1963; Sneath and Sokal 1973). It is not my purpose here to describe the techniques in detail, for they have been summarized well elsewhere (e.g., Neff and Marcus 1980; Sneath and Sokal 1973; Felsenstein 1982; Sokal 1986). A matrix of taxa by character states is used to calculate some form of correlation (or distance) matrix among the taxa. These correlations (distances) are then employed in a grouping algorithm to construct a tree, by successive pairings, according to successively decreasing correlations (e.g., the Unweighted Pair Group Method described in Sneath and Sokal 1973). The tree thus constructed has the advantage of a defined root and branching, defined by given levels of overall similarity (correlation). The classification derived from the tree is thus automatically amenable to a hierarchical organization. The groups are defined to a degree by all of the characters under consideration in producing the clustering. A set of subgroups included within a group need not have the set of character states that uniquely define the larger group. By contrast, cladistically defined groups have the important characteristic that all subgroups have the same character states as those that define the more inclusive group, as well as some unique states of their own, which define the lower taxonomic level of the subgroups.

Although phenetically based classifications are often suggested (see Sokal 1983c) for their superior stability (same classification can be derived with different data sets, addition of new data) they also have been employed expressly, and are assumed, to reflect genealogical relationships (e.g., Sokal 1983a, b, c). "It is almost a truism that an intimate relation must exist between phenetic evidence and the degree of relation by ancestry" (Sokal and Sneath 1963, p. 95). A ferociously competitive literature has lined up pheneticists against phylogenetic systematists over the respective systems' stability and success in producing accurate genealogies (e.g., discussion in Farris 1983; Rohlf and Sokal 1981; Wiley 1981; Sokal 1983c). Given the recent controversy over the apparent success in obtaining similarity of classifications based upon different character sets (congruence) (Mickevich 1978; Rohlf et. al. 1983; Schuh and Farris 1981; Sokal 1983c, 1986), one is rightfully skeptical that the final answer is at hand. None of the techniques is impressively stable, and 70 percent repeatability is about as good as you can get. Stability of all techniques seems to increase with an increasing ratio of numbers of characters to numbers of taxa (Sokal and Shao 1985; Burgman and Sokal 1986). Phenetic techniques appear to be more stable with large numbers of taxa, while cladistic techniques are superior when large numbers of characters and small numbers of taxa are analyzed (Sokal et al. 1984).

The theory behind the original phenetic approach supposed the premise of nonspecificity (Sokal and Sneath 1963). Genes had sufficiently non-specific (pleiotropic) effects across the phenotype that any large sampling of characters would reflect the genome, and, therefore, would record genealogy. Different sets of characters (e.g., cephalic and pygidial in trilobites, larval and adult in moths) would therefore lead inevitably to the same classification. Incongruencies in classifications based upon different suites of characters would therefore falsify the hypothesis.

The non-specificity hypothesis seems at variance with the results of various studies. There is considerable incongruence between classifications derived from different data sets (Mickevich 1978; Rohlf et al. 1983). Rohlf (1965) found that phenetic classifications between alternative suites of characters differed more than randomly chosen sets of characters from the entire set. Poor congruence seems a feature of both popular cladistic and phenetic algorithms.

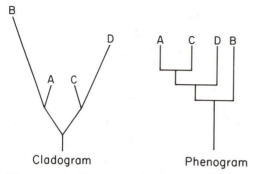

Figure 2.8: Example showing how grouping by similarity alone can be misleading in the construction of a genealogy. At left the phylogeny is indicated, with degree of morphological divergence represented by branch lengths. At right is the grouping that would be obtained using similarity alone (after Farris 1971).

Poor congruence may have little to do with a violation of the non-specificity hypothesis. Mosaic evolution–differential rates of evolution of different character sets within the same monophyletic group–would tend to produce different phenetic groupings using the different character sets (Farris 1971). Variation in relative frequencies of uniquely derived characters in taxa may be responsible for incongruities between phenetic classifications based upon different character sets (Mickevich 1978). Fig. 2.8 demonstrates how mosaic evolution could cause differential phenetic groupings based upon different character sets.

The nonspecificity hypothesis, furthermore, is not likely to be true. Although pleiotropy is common in eukaryotes localized effects of genes may be far more important. The bithorax complex in *Drosophila melanogaster*, for example, affects segmentation characters and is due to a series of tandemly arranged complementation groups (see Chapter 3). Localized chromosomal processes, such as unequal exchange, are known to affect some specific morphological characters more noticeably than others (e.g., Coen et al. 1982). Some major forms of morphological variation are known to be explained by coordinated and closely linked genes (e.g., mimicry in butterflies–Ford 1975). This major variation may be superposed on a large background of genes with diffuse effects. Finally, ontogenetic evidence (Chapter 5) suggests that the overall phenotype is probably divided into semiautonomous blocks that can evolve independently or at least respond differently to the same overall change, such as change in concentration of a morphogen. Although it is often dangerous to extrapolate from specific examples, it seems likely that the effects of genes are morphologically too localized to be expected to have the widespread effects assumed by the nonspecificity hypothesis.

Grouping by overall similarity is likely to lead to spurious conclusions from the genealogical point of view. Groups that have split off in the distant past, but have diverged little phenetically, will be grouped as close relatives. By contrast, groups that have split more recently, but have diverged phenetically to a great degree will be grouped at a lower level of overall correlation. Farris (1971) gives the simple example of classifying birds, crocodiles, mammals, and snakes. A phenetic classification would group most closely the snakes and crocodiles, on the basis of similarity of both ancestral and derived characters. The lack of divergence between these two groups would obscure the genealogy relative to the other more phenetically divergent groups. To the degree that evolutionary rates are unequal in different branches in a phylogeny, phenetic groupings will fail to link the genealogically closest branch points and place emphasis on the degree of phenetic divergence.

This point can be seen in Sokal's (1983 a, b, c) comprehensive study of the caminalcules, a group of synthetic creatures whose complete history, including phylogeny, fossils, and recent species are known by definition. The phenetic classification is superposed upon the phylogeny in Fig. 2.9. Groups A, B, C, and D, labeled by common shading patterns, are those clustered above one arbitrary level of phenetic similarity, while numbered subgroups (e.g., A1, A2) are clustered above another arbitrary, but still higher, level of similarity. B1 is the direct phyletic ancestor of B2, but the classification groups them as of equal rank. In another case, C1 gives rise to D2, which in turn gives rise to C4. The A group is of particular interest. It includes the ancestral group, A1, and a derived "radiation" (A2, A3, A4, A5, A6) as groups of equal rank. But A1 on the phenogram appears to be strongly derived, whereas it is the ancestor group of the entire clade! The groupings therefore do not present a consistent picture as to genealogy, nor is there a way to connect the groups, given the absence of information or hypotheses on polarity of evolutionary change or derived states in common.

The phenetic approach also has some properties that are specifi-

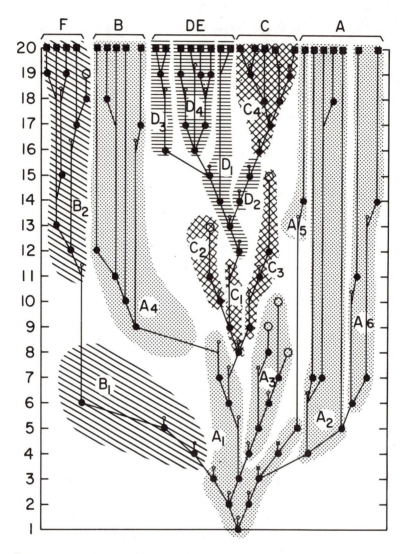

Figure 2.9: Phylogeny of the caminalcules. Shaded pattern unites group at the 0.0 phenon (similarity) level. Numbered subgroups are united at the 0.5 level. Vertical axis indicates arbitrary time units (after Sokal 1983c).

cally undesirable for macroevolutionary applications. Most important is the method of establishing similarity, which makes it difficult to explain how changes in character states define the tree generated by the clustering algorithm. The Unweighted Pair Group Method, for example, is used to establish groupings by means of general indices of similarity upon which all characters weight to some degree. This obscures our ability to define groups by a specific set of character states. As shown above, test of macroevolutionary hypotheses benefit from a taxonomic structure that permits the dissection of character states. If a phenetic approach produced a correct genealogy, then character transitions might be back-calculated and plotted on a tree of relationships. It is therefore important to understand the conditions under which a phenetic approach is liable to produce an accurate tree. The phenetic approach is especially attractive, since it automatically provides a root and a simple phenetic index of relatedness.

Under some circumstances, measures of overall similarity may lack some of the disadvantages cited above. First, if the rate of phenetic divergence for all taxa were relatively uniform, the problem of spurious groupings would disappear. In this case, time would be equivalent to character divergence. Second, if the measure of phenetic similarity were based upon a large number of complex characters, then the index of similarity would be able to resolve fine-scale differences. This would seem to be unusual in morphological characters, but might be true for complex molecular sequences. If the phenetic approach were no better than cladistic, then I would prefer the latter, which pays more attention to character evolution.

DNA Hybridization

DNA hybridization has been recently developed as a phenetic technique, which may reflect genealogy quite closely, when combined with tree-construction algorithms based upon parsimony. The technique involves hybridizing strands of DNA approximately 500 nucleotides long of an index species (whose DNA is radioactively labeled) with the DNA of a number of relatives. Heat causes the hydrogen bonds of the DNA duplexes to break down, but the sequences remain intact. The rate of dissociation of the duplexes reflects the similarity in sequence between any two test species. Repeated sequence DNA is separated before such tests are made; the dissociation is therefore related to differences in single-copy DNA.

As it turns out, the majority of the single-copy DNA consists of noncoding genes. This would suggest that evolution is random and therefore not constrained by natural selection. As suggested by discussions in Chapter 3, these sequences are just those where clock-like divergence is possible. Indeed, data of this sort collected for birds suggest a consistency with homogeneity of rate of divergence (Sibley and Ahlquist 1983).

Distances based on DNA dissociation rates are used to make a tree by means of an "average linkage" clustering method. First, the closest pair of taxa are linked; then the next step links the taxon having the smallest average distance to the previous cluster. If the rate of molecular evolution is constant, and if this is reflected in the DNA hybridization data, then this procedure should be analogous to linkages made by Wagner trees. The two closest taxa should have the most unique shared homologies, or synapomorphies. The next closest taxon should have the next most frequent shared unique genes, and so on.

In most cases, DNA hybridization data confirm the relationships that were established previously on the basis of morphological information. This approach, however has uncovered some striking differences, relative to our current understanding of the genealogical relationships of birds. For example, the Australian passerines appear to be far more closely related to each other than any are to relatives from other continents. A great deal of convergent evolution has masked the phylogenetic relationships to remarkable degree.

One might question whether this approach will be ultimately the best, especially given the rapidly expanding availability of nucleotide sequences which can be used directly in conjunction with cladistic techniques. DNA dissociation rate is related to the degree of nucleotide divergence. The very complexity of sequences makes their matching, or failure to match, something akin to matching long sequences of a poem, where different numbers of letters have been changed randomly. But there is a considerable degree of error in the measure, since 20 percent differences in sequences cannot be distinguished under experimental conditions.

The key to the efficacy of the DNA hybridization approach lies in the truth of a molecular clock. Since the genome to be considered is largely nonfunctional, the neutral theory would suggest that the rate of change is equal to the mutation rate. Sibley and Ahlquist (1983) argue that the large number of genes involved in the technique would imply some sort of average overall mutation rate. They also appeal to Van Valen's "Red Queen" (1973b) hypothesis, which would predict random change. If, however, mutation rate varies among branches of a clade, the assumption of molecular clock can produce very misleading results. If divergence is rapid along a derived branch of the clade, the presumption of a molecular clock will lead one to presume that it had originated long ago, and the taxon would be mistakenly attached toward the root of the tree. As the variance in mutation rate among subtaxa increases, the propensity for error will increase as well. As discussed in Chapter 3, our knowledge of overall mutation rates is poor.

Mitochondrial DNA

Mitochondrial DNA (mtDNA) has become an important tool in the study of divergence of closely related species and populations (see Avise and Lansman 1983; A.C. Wilson et al. 1985). The molecule's relatively short length and ease of separation makes it ideal for restriction endonuclease mapping. In vertebrates, the rate of mtDNA divergence is much higher than that of nuclear DNA, and A.C. Wilson et al. (1985) argue that this is due to the apparent inefficiency of repairing DNA damage and correcting errors of replication. Vawtor and Brown (1986) provide evidence, however, that the rate of vertebrate nuclear DNA evolution may be lower than in other groups, creating an illusion of extraordinarily high mtDNA evolution rates.

Owing to maternal inheritance, all of the mtDNA molecules in an individual are usually identical. Population bottlenecks are therefore more likely to fix rare variants than is the case for nuclear DNA. Using restriction enzyme maps, combined with sequencing of specific cloned DNA regions, it is thus possible to build up genealogies for populations and species that have diverged fairly recently. Ferris et al. (1981) estimated the age of the common mother of all chimpanzee mtDNAs at 1.9 my ago, while the common mother for the mtDNAs found in common pygmy chimps lived about 1.05 my ago. Evidence for introgression between closely related species has been found in *Drosophila* (Powell 1983). Carr et al. (1987) used restriction endonuclease cleavage maps to resolve the relationships within the genus *Xenopus*.

The Evolutionist-Phylogeneticist Conflict

Constructing the Classification and What It Means

Phylogenetic systematic practice requires the conversion of a cladogram into a classification. Hennig (1966) realized that the cladistic
system would raise conflicts with other types of classifications. First,
branch points in the cladogram are delineators of successively inclusive
(increasingly higher ranking) groups as we move toward the root. All
subgroups of a group have the same character states that define the
more inclusive group, unless recognized reversals have occurred. Any
two sister groups are defined by the corresponding two descendant
groups found "upstream" of the tree from a bifurcation. This form of
classification insists that only **monophyletic groups**, defined as all of
the descendants of a single ancestral species, be recognized. It excludes
the recognition of **paraphyletic groups**–groups that include an ancestor, but not all of the descendants. The objection to paraphyletic
groups stems from their definition by shared ancestral characters. In
many cases, this set of characters fails to define the group under consideration, as other related groups also share the ancestral character
states (e.g., Farris 1979).

  The objections to the implications of the Hennigian system have
mainly come from **evolutionary systematists** (e.g., Simpson, 1961,
1975; Mayr 1969) whose objectives only overlap partially with phylogenetic systematists. While evolutionary systematics aspires to produce
a classification based upon monophyletic groupings, the rankings of
taxa are not based exclusively upon position in a tree. Increasing degree of phenotypic difference from related taxa, and numbers of species
in a taxon both are used as criteria to raise a taxon to a higher rank.
This inevitably creates paraphyletic taxa (taxa that fail to group all
descendants derived from a respective node in a cladogram), since the
group most closely related to the divergent group is defined inevitably
by ancestral, not derived features (see Farris 1975). I stress that the
objectives of evolutionary systematists are still in the spirit of producing a classification that can be used in reconstructing the phylogeny.
The criticism that evolutionary systematists tend to erect paraphyletic
groups may be true, but such a practice need not be incompatible with
cladistic analyses. As long as the cladogram can be retrieved from the
classification one can argue that this procedure is acceptable, since no

genealogical information is lost.

Apparent progressive sequences create the most problems since evo-lutionary systematists wish to recognize grades of evolution by equal ranks, whereas the Hennigian system seems to require that more "ad-vanced" groups be of lower rank. An excellent example is the evolution of mammals. Evolutionary systematists would accept the equal rank-ing of pelycosaurs, therapsids, and mammals, because it is felt that the mammals, even though derived within the therapsids, are an important new grade of organization, which permitted an extensive evolutionary radiation. Again, as long as the cladal structure is preserved explicitly, one does not necessarily sacrifice any information. This issue lies at the heart of the analysis of fossil data, particularly that of taxonomic sur-vivorship and longevity studies (e.g., Van Valen 1973b; Levinton 1974; Raup 1978). Van Valen (1984a) remarks that such analyses would not be possible were organisms classified strictly according to Hennigian principles. What he means, I think, is that an analysis of taxonomic longevity will only have ecological meaning if ecologically equivalent groups are contrasted. To make the birds subordinate to the reptiles, for example, masks their possibly equal importance in effects within natural communities, degree of geographical coverage, similar number of species, etc. Thus, current studies of taxonomic survivorship or di-versity at, say, the family level, could benefit from the retention of the evolutionary systematists' frame of reference, as long as it does not obfuscate the genealogy.

Some have objected to the Hennigian classification system, owing to the effect of the the addition of new taxa to an existing classification, which, of course, must add still more branch points. The instability thus created becomes more worrisome as the added branch points are closer to the root, and therefore define more and more higher taxa. This is particularly true of newly discovered fossils bearing ancestral characters. As Mayr (1974) notes, the "discovery" of the birds im-mediately defines a synapomorphy with crocodiles, making the other reptiles more distantly related and increasing the overall taxonomic rank of the group [(Birds, Crocodiles)(other Reptiles)]. Using ances-tral plus derived states, birds are more divergent phenotypically from [Crocodiles, other Reptiles] than either of the two reptile groups is to the other. It therefore seems intuitively reasonable to separate the birds off in a rank equal to the crocodiles plus other reptiles (e.g.,

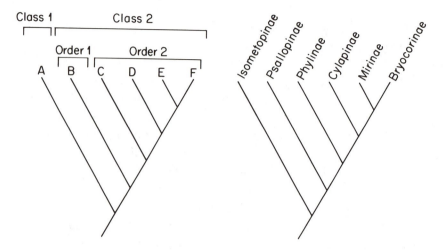

Figure 2.10: Two possible classifications that retain the genealogical information of the cladogram. On the left is a classification derived from a cladogram of six taxa. On the right is a classification of the hemipteran family Miridae (after Schuh 1976), with all taxa given the same rank.

Michener 1978).

As shown by Fig. 2.10, progressive sequences yield asymmetrical trees, which emphasize the classification problems mentioned above. Groupings of one taxon with large numbers of others are inevitable, with the sister group criterion. Thus taxon A would be of equal rank with the taxon grouping [B,C,D,E,F]. Hennig (1966) adhered to this requirement strictly, while others have found it to create taxa that are often redundant; that is, one taxon is monotypic at several rank levels (in Fig. 2.10, A is monotypic at 5 ranks, B at 4, etc.). This problem can be solved readily (e.g., Schuh 1976; Wiley 1981). Monophyly is the only essential requirement for a consistent cladistic classification. It should be possible to retrieve the cladogram from the classification; preferably, redundant taxa should be minimized. Schuh (1976) solves this problem for the hemipteran family Miridae. All of the taxa are arrayed in a linear pattern of branching, as in Fig. 2.10, which would imply a phyletic evolutionary sequence if synapomorphies along the tree are based on progressive changes of the same characters. All taxa are given equal rank, but the order of the list implies the distance along the main branch toward the taxa with the most derived states.

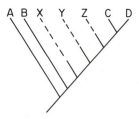

Figure 2.11: A cladogram where group [C, D] is phenetically divergent from the closest living relatives [A, B]. Taxa X, Y, and Z are
hypothetical newly discovered fossils that span the gap.

Any listing that can retrieve the cladogram is acceptable; this leads
to considerable flexibility in ranking. A proposed cladogram for the
mammals (see Chapter 7) deals similarly with such cladograms.

The potential problem posed by ranking according to degree of phenotypic divergence, or gaps, can be shown with Fig. 2.11. Here, group
[C,D] is defined on the basis of a phenotypic difference between it and
[A,B]. But what if fossil intermediates X, Y, and Z are found? The reason for the gap suddenly disappears. The use of gaps thus imposes an
instability on the definition of rank. This problem is only exacerbated
when examining the fossil record. Although groups notably divergent
from their closest relatives are common, other taxa seem to acquire
gradually that final complex of characters that gives us the total gestalt
of the taxon (Chapter 7). Yet we wish to say: "that is a mammal!"
This can be recognized implicitly in the accepted taxonomic separation of the ancestral Mesozoic mammals from the therapsids, classed
as reptiles. The use of gaps in classifications involves the use of an
ecological–evolutionary model as a classificatory criterion. The origin
of a highly divergent group is often associated with the movement into
a major new habitat and life style (Simpson 1953, Mayr 1969). The
assignment of high rank to such a divergent group is a recognition of
an ecological–morphological advancement. This criterion, of course, is
external to the genealogical structure. This would be well and good
if there was one such criterion. But what if there are others, such as
mode of development? It would be best to have a system that always
refers simply back to the genealogy. Without such a framework, the
intent behind classifications will be ambiguous, since rank is used in so
many different ways by different investigators.

One cannot overstress the loss of informativeness in employing paraphyletic groups in a classification, the necessary consequence of using gaps between a descendant group and an ancestral group as the basis of rank. One often cannot recover the cladogram. Going back to the bird-crocodile-other reptile example, we can note that the grouping [crocodiles, other reptiles] is ambiguous, as it is paraphyletic group bound by symplesiomorphies. The states uniting this group are also characteristic of the entire Amniota. The nuculid and solemyid bivalves are united on the basis of an ancestral gill, which would also define a group larger than the class Bivalvia! This hardly gives precision to the defined group Protobranchia. Yet, by describing taxon A as "strongly divergent," and referring it consistently to a cladogram, no information is lost. Wiley (1981) discusses various systems, describing degrees of divergence, to annotate classifications.

Since both cladists and evolutionary systematists seek some sort of genealogically based classification, I am sure that the common goal will tend to yield more imaginative solutions to the problem of ranks. Evolutionary systematics recognizes that monophyletic groups will be defined by certain sets of characters, but these characters will vary from group to group and can only be discovered by some sort of character analysis (e.g., Mayr 1969). This conclusion is close to the cladistic approach. We can see more fundamental issues in common between cladists and evolutionary systematists than differences, even though the two camps usually seem to attempt to accentuate the intellectual gaps. The present trend in systematics, designed to reflect phylogeny in classifications, is healthy, no matter what the particular approach. In many respects, it is rather useful that a plurality of phylogenetic approaches be maintained to help sharpen our understanding of evolutionary classification.

Before the era of evolutionary systematics, taxonomists tended to employ **restrictive monothetic** criteria, where specific characters were used to define differences at a given taxonomic level (e.g., internal characters only define ordinal level differences in brachiopods). In these idealistic classifications, restrictive monotheticism implied that key characters defined given taxonomic levels (see Mayr 1969 for discussion). This is well illustrated by the brachiopods, where certain characters were believed to define differences among genera, families, and orders (see A. Williams and Rowell 1965). The monothetic nature of brachiopod classifications of the late nineteenth century (e.g., Beecher 1891; Schuchert 1893) persisted into the middle of this century (e.g., Cooper 1944). Restrictive monothetic classifications often lead to the spurious uniting of groups with no genealogical significance. Newell (1965, 1969) has rejected such simple attempts to classify on the basis of one criterion.

Modern evolutionary systematic approaches have departed from the restrictive monothetic system used, for example, in the brachiopods. A. Williams and Rowell (1965) now recognize that many different characters may define evolutionary change in this group. They conclude that (p. 223)

> all such schemes proposed in the past are incompatible with the evolutionary history of the phylum. Previous monothetic, non-evolutionary classifications were regarded as "only catalogues . . . deliberately arranged for quick identification of stocks."

It is fair to say that all three current approaches–evolutionary systematics, phenetics, and phylogenetic systematics–have made a healthy move from classificatory and genealogical schemes that are rooted in nonevolutionary essentialism.

The Value of the Fossil Record

Hennig (1966) suggested that classifications might have to ignore fossils, in order to avoid the problem of inserting new branches into cladograms. Alternatively, coexistent fossils might be grouped at each time horizon, thus avoiding the problem of new branches. This solution would create taxonomically absurd situations, such as the placement of the same taxon at different ranks, depending upon its time horizon and the number of relatives (Wiley 1981). It has also been suggested that fossils be included in cladograms, but listed in classifications as extinct, with no rank. Alternatively, the fossil groups can be given a rank identical with their closest extant relative (see discussion in Farris 1976). I can't see these as viable approaches in a macroevolutionary context. We must have some consistent system for recovering the genealogies of the taxa under consideration. Fossils provide so much more information on morphology that it is ludicrous to exclude them when our resolution of understanding of transitions will be only reduced.

I concur with Farris (1976) that all fossil and recent taxa should be classified together.

A cloud of suspicion has risen over the utility of fossils. The value of the fossil record in systematic reconstructions has been questioned often, gently by some paleontologists (e.g., Newell 1947, 1956; Shaw 1969; Imbrie 1957), but rather strongly by others, particularly cladists (e.g., Patterson 1981, Forey 1982). As mentioned above, fossil sequence usually fails to prove ancestry, unless one has an extraordinarily complete sequence. Ancestor-descendant relationships can be safely deduced if a continuing nonrandom trend of change through a geological section can be established (see Raup and Crick 1981) with no evidence of cladogenesis (e.g., Malmgren and Kennett 1981; Ozawa 1975). There are also rare cases where cladogenesis may be traced (Grabert 1959; Prothero and Lazarus 1980).

Cases of complete ancestor–descendent records are usually those involving evolutionary transitions of small magnitude. Larger scale transitions, such as the evolution of mammals from therapsid-like ancestors, are rarely if ever recorded in such a complete and continuous sequence (Kemp 1982). While the morphological details of the transformation are fairly clear, specific statements concerning the exact phylogenetic pathway are best avoided. Because exact ancestors are difficult to identify, we are left with the unsatisfactory alternative of investigating the means by which one taxon could be transformed into the most closely related taxon. It follows from this argument that the only tenable way of eliminating the gap between hypothetical ancestral and derived sets of character states (e.g., the "mammalian condition" versus the "reptilian condition") is simply to find more transitional taxa! The data base of available fossil and living taxa, fixed upon a reliable cladistic network, is our first and foremost reference system in macroevolution.

A.B. Smith's (1984) analysis (figure 2.12) of the Echinodermata stands as an exemplar of a cladistic study that unifies the genealogical relationships of fossil and extant taxa. He notes that previous classifications suffer from a lack of focus on character transformations. Using the cladistic approach, Smith produces a genealogy of the five extant classes, and then adds extinct groups to the cladogram. Fossils shed considerable light on certain aspects of the process. First, fossils may identify character states that may have been lost in extant groups.

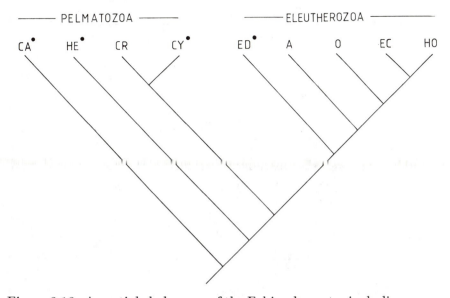

Figure 2.12: A partial cladogram of the Echinodermata, including some fossil groups. HO = Holothuroids; EC = Echinoids; O = Ophiuroids; A = Asteroids; ED = Edrioasteroids; CY = Cystoids; CR = Crinoids; HE = Helicoplacoids; CA = Carpoids. ● = extinct (after A.B. Smith 1984).

Second, extinct groups may reveal the pattern of character acquisition of traits that are now autapomorphic to the entire living group. This second point is crucial in understanding the morphological transformations that led to more derived body plans. We use this approach in Chapter 7 to examine the origin of the mammalian condition.

Some authors (Patterson 1981; Schaeffer et al. 1973; Eldredge and Cracraft 1980) have claimed that the temporal sequence of fossils should never be used as evidence for ancestor-descendant relationships. In using the fossil record, one assumes at least that the fossil record is sufficiently complete to make a determination of character-state polarity, or, at most, that younger fossils are more derived than older ones and ancestry can therefore be established. Since we know that ancient groups often survive along with their descendants, or at least descendants of the ancient groups' close relatives, the latter assumption is often false. The living fossils *Neopilina* and *Latimeria* argue against the infallibility of stratigraphic position as an indicator of ancestor–

descendant relationships.   One can even imagine hypothetical cases
where the fossil occurrence of a given descendant taxon antedates its
ancestors, because the latter group is missing from a portion of the
fossil record (Eldredge and Cracraft 1980). The closer the origins of
the two taxa, and the longer the period of coexistence, the higher
is the probability that sequence can be misinterpreted. This is not
true, however, for direct phyletic sequences. If a descendant derives
by transformation from an ancestor it is highly probable that stratigraphic sequence indicates polarity (Paul 1982). The probability of
two randomly collected specimens being preserved in the wrong order
cannot be greater than one half.

The strong limitations of temporal fossil occurrence in evolutionary reconstruction are worth emphasizing, given the large number of
paleontological studies where the inference of ancestry often turns on
stratigraphic occurrence. This is especially true when paleontologists
search for the ancestor of a large taxonomic group. Hypothethesized
ancestors to the bivalve mollusks provide convenient examples. The bivalves *Babinka* (McAlester 1965) and *Fordilla troyensis* (Pojeta et al.
1973) have both been cited as transitional forms to the mollusk class
Bivalvia. In both cases, early stratigraphic occurrence is a principal
part of the argument, though morphology also plays a role. If either
had been found in younger rocks, as in the Cretaceous, their respective
ancestral statuses would not have been as strong. The ancestral status
of the character states of both forms would also not have been a matter
of serious dispute.

In the case of *Babinka*, a series of muscle scars were linked with
the hypothetical ancestral states of the commonly cited likeness of
the hypothetical ancestral mollusk, *Neopilina* (McAlester 1965). The
anatomical claim was refuted by showing that the pedal muscle scar
pattern in *Babinka* was not homologous to the serially repeated pedal
muscle scars in *Neopilina* (see Stanley 1972, p. 166). The repeated "gill
muscle scars" in *Babinka* most probably did not represent a transitional
change between those of *Neopilina* and later bivalves, where a single
pair remains. Note that although the claim for ancestry depended
upon both stratigraphic position and character states, the refutation
was based upon an analysis of character states alone.

The case of *Fordilla troyensis* is more illuminating. This remarkable
fossil is widespread in the Lower Cambrian rocks of North America

and can also be found in the same Series in Denmark and perhaps England, Portugal, and Siberia (Pojeta and Runnegar 1974). Since all of the bivalve superclasses do not appear until Middle Ordovician time, and only the Paleaotaxodonta appear in the Early Ordovician, *Fordilla* deserved its status in 1973 as the most ancient fossil bivalve yet discovered. There is a gap of some 40 million years between it and the beginning of the known early Ordovician geographically widespread occurrences.

The unique fossil finds in New York State (Pojeta et al. 1973) permit reconstruction of internal scars, which have been used to establish its likely bivalve molluscan status. Individuals of the species appeared to have elongate, subequal adductors and a broadly inserted pallial line. The shape of the shell and position of muscle scars suggest a shallow-burrowing suspension feeder. These features are used by Pojeta and Runnegar (1974) to make a case for direct ancestry of the Bivalvia, via the group of Ordovician heteroconch families best represented by the Cycloconchidae. After speculating on the genealogical relationships between this group of families and the other bivalves, a tenuous link is even claimed between *Fordilla* and the univalved Cambrian rostroconchs.

The problem with this sort of reasoning is obvious. What if another lower Cambrian bivalve is discovered that harbors a set of character states completely different from *Fordilla*? Since the molluscan affinities of *Fordilla troyensis* were only recently appreciated (Pojeta et al. 1973), one can safely expect that some other group, now known too poorly to rise from the ranks of *incertae cedis*, will materialize soon as a competing ancestor. The preemptive claim made by Pojeta for this genus is based solely upon stratigraphic position; the ancestral states of the morphological features are thus taken simply upon faith. In this particular case, I would emphasize that there is no reason to believe, from *any* other evidence other than stratigraphic occurrence, that the character states borne by *Fordilla troyensis* are necessarily ancestral. There may have been a large and diverse bivalve fauna in Early Cambrian time that have gone unnoticed or unpreserved. This is not outlandish, given the 40 my span between *Fordilla* and later bivalve occurrences.

As it turns out, another, still older, bivalve mollusk was discovered in Early Cambrian rocks of South Australia by Peter Jell (1980), and

was reverently named *Pojetaia runnegari.* A later morphological analysis with well preserved specimens (Runnegar and Bentley 1983) established clear similarities between this form and the Palaeotaxodonta,
a group often thought to be an ancestral bivalve subclass. This discovery only emphasizes the great potential for further discoveries in Early
Cambrian rocks and the dangers of searching for ancestors. Without
questioning the specific familial affinities established by the authors
cited above, it is rather clear now that newly acquired information on
temporal occurrence will not shed further immediate light on the ancestry of the Bivalvia. Rather, additional morphological information
revealing genealogies will be decisive.

The study of evolutionary transitions between the fishes and the
tetrapods has been similarly influenced by the fossil record and by overall similarity between putative ancestors and the descendant tetrapods.
The Rhipidistia have been traditionally thought to be the tetrapod ancestors, based upon overall similarity in the skull roof, appendages, and
appropriate stratigraphic position. In particular, the presence of paired
internal nostrils (choanae) has been cited as a linking character. Rosen
et al. (1981) have claimed that the interpretation of this character is
incorrect and that the rhipidistian *Eusthenopteron* lacks choanae. By
contrast, a restudy of a Devonian lungfish from Australia suggests the
presence of choanae. Thus the restudy of characters has placed the
Dipnoi (lungfish) as the sister group of the tetrapods and completely
changed our conception of vertebrate phylogeny. Rosen et al. note
that overall similarity and the connection by stratigraphic proximity
led us astray. Whether this interpretation is correct or not, it places
the onus on paleontologists to avoid stratigraphic assessments of ancestry and classifications based upon overall similarity, which might
involve grouping with ancestral characters.

Although the case against stratigraphic position alone as a criterion for ancestry cannot be made too strongly, It is an exaggeration
to say that temporal sequence in fossil occurrence provides no useful
genealogical information (Harper 1976; Fortey and Jefferies 1982; Paul
1982). In some cases, as noted above, closely spaced samples reveal a
gradational sequence of morphological change from ancestor to descendant, with no evidence of cladogenesis. Although one can never exclude
the possibility of something happening "between the lines," such cases,
common in deep-oceanic sediments in groups such as foraminifera (e.g.,

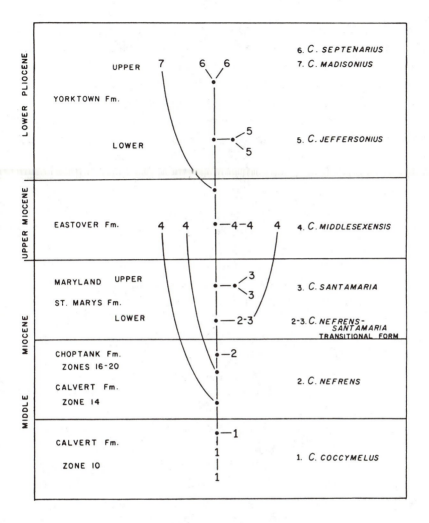

Figure 2.13: Cladogram for the Miocene-Pliocene Atlantic Coastal Plain scallop *Chesapecten*, superposed on the stratigraphic sequence (from Miyazaki and Mickevich 1982).

Grabert 1959, Malmgren and Kennett 1981, Bettenstaedt 1962), can rightfully justify temporal sequence as evidence for character polarity.

Temporal sequence may be a corroborative tool to strengthen a hypothesis of genealogy (Eldredge and Cracraft 1980, p. 58; Miyazaki and Mickevich 1982). Consider the following analysis. A systematist establishes a cladogram, which is rooted on the basis of an outgroup comparison or by an assumption of character state polarity using ontogenetic change. This analysis yields a cladogram showing an array of taxa that can be arranged from near the most ancestral state to most derived. If the stratigraphic order of the taxa occurs in the order "predicted" by the cladistic analysis, then the conclusion of character polarity based upon character analysis alone is strengthened.

An example of this sort comes from the work of Miyazaki (Miyazaki and Mickevich 1982) on the evolution of the Miocene-Pliocene scallop genus, *Chesapecten*, which appears to be have been confined to basins in the eastern coastal plain of the United States. Based upon ontogenetic change, the cladogram in Fig. 2.13 infers a genealogy and roots the tree near the taxon with the most "juvenile" features as an adult. The cladogram, as can be seen readily, is closely concordant with stratigraphic order. The cladogram and stratigraphic occurrence data can be properly considered as independent sources of evidence leading to a similar conclusion of descent.

Temporal sequence may also resolve vexing cases of convergence when other approaches fail. The two Cenozoic radiations of planktonic foraminifera demonstrate the difficulty of identifying particular morphs without good stratigraphic information (Cifelli 1969). A Paleocene radiation from globigerinid ancestors resulted in a morphologically diverse array of taxa, nearly all of which disappeared by the Oligocene. A second radiation in the Oligocene-Miocene repeated many of the species in such faithful similarity to those in the first radiation that a proper systematic assignment is impossible without the appropriate stratigraphic information.

Stratigraphic and paleogeographic position may also be used to a degree to resolve problems in character reversal and parallel evolution. In some cases, identical derived character states may be acquired independently within several monophyletic groups, when independent stratigraphic and geographic evidence isolates these groups from each other. In the ancestral trigoniacean bivalves, stratigraphic position is essential in genealogical reconstructions (Newell and Boyd 1975). The hypothesis that parallel evolution has occurred produces the most corroborated hypothesis of genealogy. Although one can only speculate on the cause of such parallelism, commonality of ground plans might result in a similar response of independent groups to an environmental change. In any case, this approach parallels the use of gene duplication events (e.g., Goodman et al. 1982) as likely devices to increase the consistency of molecular evolutionary lineages.

It might be argued that forams and mollusks are so simple that constraints will often lead to parallel or iterative evolution. One might expect repeated morphologies here but not in higher organisms such as mammals. The widespread occurrence of atavistic character states in vertebrates (e.g., Riedl 1978) makes this claim highly unlikely. In vertebrate jaws, developmental fields can be defined where correlations among characters are stronger than with characters in other putative fields (Kurtén 1953). Thus, ancestral character states often reappear in lineages as coordinated complexes. Consider the case of *Lynx lynx*, whose fossil record has been studied extensively (e.g., Werdelin 1981). In some Pleistocene and Recent specimens a coordinated appearance of the *M2* molar and a postcarnassial element derives from ancestors where the condition is completely absent (Kurtén 1963). Indeed, the *M2* molar was lost in the Felidae since the Miocene! For an unknown reason, the characters have reappeared with noticeable frequency and with sufficient morphological complexity that they can be regarded as a character reversal toward an ancestral condition.

The Main Points

1. A genealogy connects taxa by a criterion of relationship by means of descent.

2. Macroevolutionary hypotheses often depend upon available systematic structures, as they involve such factors as changes in diversity at specific taxonomic levels (e.g. family level), taxonomic survivorship curves, and phylogenetic hypotheses.

3. The approach taken to systematics can therefore affect the meaning of macroevolutionary claims. For example, the argument that "phyla appear first in the fossil record" seems tautological, if phyla are grouped purely on the order of cladogenesis. The first

splits, of necessity, would come early in biotic history. On the other hand, if phyla are organized strictly on the basis of similarity of characters, the statement might mean that some fundamental process permitted strong differences to arise early in the Phanerozoic, and subsequent phylum appearance was dampened.

4. Classifications based upon genealogical groupings are preferable, because it is then possible to map the character changes involved in transformations from one taxon to its closest relative. Genealogical connections lend themselves to hypotheses that relate phylogenetic history to adaptation. Without such a grouping, the historical constraints behind evolutionary change will be missing.

5. Phylogenetic systematics overtly attempts to construct genealogical trees and to devise classification schemes that can map simply to the genealogy. Taxa are grouped by their uniquely shared derived character states. This grouping process produces strictly monophyletic taxa. Successive groupings of taxa lead to the construction of a cladogram of genealogical relationship. The cladogram can be rooted by means of an outgroup, which is ancestral relative to the group under study.

6. Character states are used to construct the cladogram, but not all characters produce congruent trees. Homoplasy occurs when different sets of characters are in conflict; this arises chiefly from errors in identifying homologous characters, partially because of convergence, and from parallel evolution in isolated lines. Several approaches have been devised to resolve incongruency due to homoplasy. The majority set of compatible characters may be used to define the cladogram. Alternatively, the tree that requires the most parsimonious evolutionary path (minimum number of evolutionary steps and reversals) is used to construct the proper genealogy. The latter approach is more effective as it uses that information from conflicting characters that might help in resolving the tree.

7. It is crucial to identify homologous characters. Homology is a form of correspondence of characters between taxa; we hypothesize that the correspondence represents an evolutionary correspondence, that is, a space–time continuum of evolutionary

change between the two characters on the two organisms. Criteria for homology mainly involve similarity of position in space and ontogenetic appearance in the organism. Except for molecular data, we rarely know the genetic correspondence for homologous characters. Indeed, there may be none for certain morphological features that are retained by natural selection, but the underlying genes may have changed.

8. Phenetics groups taxa by overall similarity. The grouping algorithms are usually straightforward, and a root arises automatically from the grouping. Phenograms have the disadvantage that specific character transformations cannot be mapped directly, since many character changes contribute to a given grouping, often in intuitively obscure ways. Phenetic groupings will probably not correspond to genealogical grouping when the rate of evolution is highly uneven through the group. Groups that have been long divergent, but have slow rates of evolution, will be spuriously grouped as close relatives. If the rate of evolution is fairly homogeneous for many characters, then phenetic grouping should have strong genealogical significance. This lies behind the claimed efficacy of the DNA hybridization technique.

9. Evolutionary systematics groups taxa by the two criteria of overall similarity and relationship by descent. Evolutionary systematists object to cladistic classifications, as the former wish to ascribe strong significance to strong gradal changes in evolution. The overall objectives of evolutionary systematists are not all that different from those of phylogenetic systematics. Gradal changes can be accommodated in a phylogenetic systematic approach.

10. Some have claimed that fossil taxa and the fossil record are very problematic in the construction of genealogies. While it is true that the order of first appearance can be misleading as to genealogy, the order of appearance of fossils corroborates a cladogram based upon a character analysis. Fossil taxa often provide crucial links in genealogies; indeed, many character transformations could not even be imagined without fossil data. Genealogies are best constructed with the full benefit of fossil information.

Chapter 3

Genes and Polymorphism

> the organism cannot be considered as infinitely plastic and
> certainly not as being equally plastic in all directions, since
> the directions that the effects of mutations can take are, of
> course, conditioned by the entire developmental and phys-
> iological system resulting from the action of all the other
> genes already present.
> *H. J. Muller 1949*

Introduction

The estimate of genetic variation in a population and its relation to
speciation are fundamental to specific macroevolutionary hypotheses.
The discussion can best be framed as a series of questions:

1. Are species real?

2. Are the genetic differences between species of the same sort as
   intraspecific differences?

3. If the differences are qualitative, then does this make a difference
   to theories concerning the process of macroevolution?

4. Are species accidents or adaptations?

A sketch of the history of species concepts should be kept in mind
as we consider these questions. Biologists were long comfortable with
the **Linnaean concept**, which assumed the reality of species, endowed
with a platonic essence. Even if the issue of essentialism disappeared
in the twentieth century, systematists still took the species level to be

90

fundamental. In its new guise it became a **morphological species concept**. It was only with the advent of such works as Haldane (1932a), Dobzhansky (1937), and Mayr (1942) that the Darwinian notion of species mutability and the neo-Darwinian theories of genetic change were united into a general theory that saw species as arising from within-species variation, but bound by membership in a common reproductive community. Divergence between daughter species was thought to be of the same qualitative sort as geographic divergence within species, though species were believed to be under selection for prezygotic isolating mechanisms and ecological divergence from other species. Reproductive isolation is the cornerstone of the **biological species concept**. Under this concept, many of the phenotypic differences between species may not have caused, nor contribute at present to, reproductive isolation.

One conflict underlies many of the current arguments in both speciation theory and population genetics. Species are regarded either as exquisite adaptations or accidents of divergence. The first alternative was championed by Dobzhansky (1937) who regarded speciation as a process involving intense selection for balanced and integrated gene pools, and, therefore, against pairings among individuals from different gene pools. Part of this balance was achieved as two formerly isolated populations were reunited. Selection against hybridization was part of the completion of the adaptations of the two new species. The selection resulted from hybrid inferiority in either of the habitats to which the daughter species had become adapted. This view contrasts with that of H.J. Muller (1939), who believed that, after divergence, hybrid sterility arose by chance as a product of change in the genetic background, either by drift or adaptation to different biological situations. Isolated populations moved toward "ever more pronounced immiscibility as an inevitable consequence of non-mixing."

Gene Structure and Origin

Genes and DNA Sequences

Nuclear DNA is the primary genetic material. It consists of two long unbranched polymers composed of combinations of only four different deoxyribonucleotides whose bases are adenine (A), guanine (G) (A + G = the purines), cytosine (C), and thymine (T) (C + T = the pyrimidines). The nucleotides are linked like beads on a string. DNA, however, usually occurs as a helically coiled double strand. Although the order on one DNA strand is unrestricted, the second strand must have a series of nucleotides complementary to that of the first. The purine-pyrimidine pairs A-T and G-C are most stable. This pairing rule clarifies the observation that although A or G can vary substantially among different DNAs, the abundance of A equals T, and likewise for G and C.

Mutations involve a variety of mechanisms that thwart the usual exquisite regulation of DNA duplication. Individual bases may be lost or added, causing **frameshifts** of sequences. Large scale lesions in DNA may cause variable effects on the phenotype (e.g., Bender et al. 1983). As will be discussed below, a variety of stable and unstable insertions of foreign DNA sequences can occur and may exert phenotypic effects.

The action of genes is facilitated by the generation of RNA, a nucleic acid that differs from DNA in having the base uracil (U) instead of T. **Coding regions** of the DNA are "transcribed" into RNA, but other regions may be transcribed and then excised, before the final gene product is formed. Transcription is a process whereby one of the two DNA strands acts as a template for replication, based upon complementary base pairing and covalent bonding to yield the RNA strand. RNAs are often further modified before leaving the nucleus and becoming messenger RNAs (mRNA), the template for protein synthesis.

The four bases found in DNA allow for a large variety of possible orders. The message for each amino acid, a **codon**, is three bases long, so there are 4^3, or 64 possible different messages. Sixty one are used to specify the 20 amino acids used in living systems, while the remaining three are chain terminators, that is, they cause the reading of the overall message to cease. As is obvious, there are more codons than amino acids, so several triplet arrangements generate the same product. This **redundancy** is concentrated principally in the third position of the triplet codon.

A minority of the nuclear DNA consists of functioning genes. It has been estimated that the *Drosophila melanogaster* genome has about 30,000 coding genes (Spradling and Rubin 1981). Several regions (at the 5′ end) that are read before a gene are **promoters** of transcription (Fig. 3.1). These regions partly facilitate binding of RNA polymerase

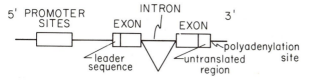

Figure 3.1: Schematic generalization of the structure of a gene.

II, the transcribing enzyme. Two are characterized by the presence of the sequences CCAT and TATA. The necessity of these regions in transcription can be tested by systematically constructing mutants with altered sequences of bases upstream of the gene (e.g., McKnight and Kingsbury 1982). Many of these sequences are conserved throughout distantly related groups of species (e.g., Falkner and Zachau 1984), though intrapopulation variation needs to be estimated. At greater cis-distances, *enhancer* sequences may be necessary to initiate or may enhance gene transcription. The remainder of the nuclear genome does not function and is often characterized as "junk."

Genes Occur in Single and Multiple Copies

Both the coding and noncoding regions of eukaryote DNA occur in **single and multiple copies**. Many genes coding for proteins such as enzymes occur as one copy, or sometimes as a series of that copy, amplified at a specific stage of development, at the same site on the chromosome. Many sequences consist of tremendous numbers of copies, as in the strongly conserved ribosomal RNA genes or the nonfunctioning human *Alu* I family, which has about 300,000 copies dispersed throughout the genome (see Arnheim 1983 for a good review). These usually involve noncoding sequences and consist of repeats of a small sequence of nucleotides. These tandemly arrayed noncoding repeats commonly account for 10 to 40 percent of the genome.

Often, a large number of multiple copies of coding sequences are concentrated on the same chromosome. The hundreds of copies of 18S and 28S ribosomal RNA genes of the clawed toad *Xenopus laevis* are all found in tandem on the same chromosome (Long and Dawid 1979). The usual similarity of sequence in these many copies is of great interest. This phenomenon has been described as **concerted evolution**, and will be discussed below. The multiple copies of both coding gene families and noncoding repeats can occur in contiguous regions on several nonhomologous chromosomes.

Introns and Exons

Nucleotide coding sequences are not necessarily continuous. Many genes are interrupted by **introns** (Fig. 3.1). These sequences are often transcribed, but later spliced out before the final mRNA leaves the nucleus. **Exons** are the regions that are translated into a gene product. In other cases the presence of certain sequences can apparently disrupt transcription, as in the introns in the 28S rRNA genes on the X chromosome of *Drosophila melanogaster* (Long and Dawid 1979). The discontinuous nature of the eukaryotic genome, therefore, requires a separate control process for splicing that must be retained in order to preserve gene function (e.g., Hamer and Leder 1979).

The significance of introns is in hot dispute. Some regard them as historical accidents whose presence requires compensating molecular adaptations such as splice junctioning. Some introns may have arisen as **transposable elements** (see below), mobile units of DNA that may affect phenotypic expression. Their spread, nevertheless, may be ensured since DNA sequences may have their own "adaptations" for dispersal and insertion throughout the genome (Orgel and Crick 1980, Doolittle and Sapienza 1980).

Evolution of New Genes

The evolution of genes involves two principal processes: the origin of genes with new functions and the evolution of genes with no major change in function of the gene product. It is improbable that a random fixation of a new nucleotide sequence might produce a new functioning protein. Most sequence changes result in gene products with reduced or unchanged function. If a change in function did occur, it would be at the expense of an already extant function, unless the need no longer exists, or the change occurred in one of a series of duplicates. This latter possibility gives rise to the following general model of gene evolution (Fig. 3.2): (1) duplication of a gene; (2) divergence of the duplicate by accumulation of mutations; and (3) changes in regulation of the new gene (Markert et al. 1975). It is possible for between-gene differentiation in gene regulation to proceed more rapidly than sequence divergence. The model presupposes that the primitive gene

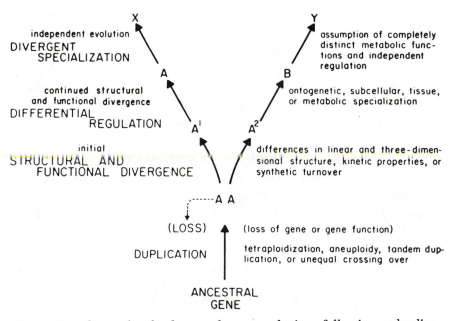

Figure 3.2: Generalized scheme of gene evolution, following a duplication (from Markert et al. 1975).

product is generalized. If it is an enzyme, then it might be able to bind to a variety of substrates. Descendant genes would be more specialized. Where the gene product is a monomer, but the functioning protein is a multimer, random association between the products of the duplicated genes would require that the genes coevolve to some extent.

Gene duplication can arise from at least two processes. Unequal exchange between homologous chromosomes can leave one chromosome with an additional copy, the other lacking any copy. Alternatively, polyploidy, common in plants but less so in animals, can result in a complete set of duplicated genes. Salmonid fishes are thought to have arisen as autotetraploids, whereas the catostomids may have arisen as allotetraploids. In the salmonids, the majority of the duplicate loci have gene products with similar tissue expressions. This is not true in the catostomids (Li 1983). The lack of divergence in the salmonids may relate to their autotetraploid origin, which presumably results in slow divergence due to tetrasomic segregation at each meiosis. In allotetraploidy, disomic segregation occurs much more rapidly and increases the opportunities for gene divergence.

Duplicated genes with little sequence divergence are common, as in malate dehydrogenase (MDH) in fishes. Near similarity in sequence, but difference in catalytic function, supports the proposed model as one plausible mechanism of gene evolution. This can be seen in the lactate dehydrogenase (LDH) genes of salmonid and cypriniform fishes, where apparent duplications have been followed by divergence in function and tissue specificity. Hemoglobin and myoglobin probably arose via divergence after a gene duplication.

A duplicate gene's function can be destroyed by random drift of regulatory elements or by changes in sequence that eliminate function. However, such nonfunctional sequences may still reside in the DNA as **pseudogenes** (see Li 1983, for summary). Pseudogenes usually have a number of nonsense or frameshift mutations that prevent the production of a functional gene product. They may also arise from the duplication of existing pseudogenes. It is unlikely that such pseudogenes are the source of novel gene function. The sequence would most likely lose the ability to specify a functional protein (Zuckerkandl 1975). Some pseudogenes bear the stamp of an mRNA origin. One of the mouse globin pseudogenes lacks introns, in comparison to its closest relatives, which have two. It may have lost its introns during the manufacture of retroviral RNA and was reinserted into the nuclear DNA as part of the DNA copy of retroviral RNA.

New genes may also have arisen from the **gene fusion** of extant functioning genes. This is most likely where the genes had arisen previously by a gene duplication event. Punctuation signals in polycistronic (multigene) mRNA normally ensure that regions corresponding to adjacent genes are translated as distinct polypeptide chains. Obliteration of these signals might cause gene fusion. If the resultant longer polypeptide is functional, then the mutation might be retained in the population. This seems to have occurred in the second and third genes *hisD* and *hisC* in the histidine operon of *Salmonella typhimurium*. The new protein aggregates to form highly active multimers (Yourno et al. 1970). Such longer proteins might evolve a complex and larger series of active substrate binding sites than their component ancestors.

Gene fusion may also have occurred with the involvement of introns. After a gene duplication event, two genes may have been brought together by recombination, and may have been separated by the intron. Subsequently the two genes plus intron may have evolved into one large transcriptional unit. This has been suggested for the origin of the pre-proinsulin gene (Douthart and Norris 1982). The accretionary nature of some genes and their protein products can be seen in the three-dimensional structure of the β-globin subunit of hemoglobin, whose exons correspond in number and amino acid sequence to the three-dimensional structural sectors of the protein (Gō 1981; Jensen et al. 1981). This situation, however, is also consistent with a model of invasion by introns, which might only be successful if invasions occurred at locations in the genes that did not disturb the individual sectors, if an intron excision mechanism didn't exist after transcription. Additionally, the invading introns might bear the genetic machinery needed for excision during the formation of mRNA (Doolittle and Sapienza 1980). All vertebrate α-globin and β-globin genes examined have the same two introns, at the same intragenic locations. This suggests long term stability in evolution, as the gene duplication event between the two gene types predates the evolution of the tetrapods; both genes are found in fishes, and other later-arising vertebrates. This overall picture will not apply universally; in the alcohol dehydrogenase (Adh) gene in *Drosophila melanogaster*, for example, functional domains are not split by introns (see Benyajati, et al. 1983).

There are approximately 1000 distinct classes of mammalian proteins, but it is not clear whether given protein classes are polyphyletic (Zuckerkandl 1975). Most genes probably evolved from genes with closely related functions. It seems unlikely that saltation from either noncoding sequences or inactive genes to a new functioning protein occurs often. Given the degree of specialization now present in proteins and their coding genes, the possibilities for the origin of proteins with wholly new functions may now be severely restricted. The stakes are now too high since enzyme activity, for example, will invariably be lowered drastically in descendants with major changes. Some evidence suggests, for example, that three-dimensional structure, which is most relevant in catalytic action, is conserved faithfully, even as amino acid sequence diverges (Phillips et al. 1983).

While functional potential may be generally restricted as evolution proceeds, discrete changes of function are still possible. This is best illustrated in the genes for lysozyme and lactalbumin. Lysozyme, widespread in vertebrates, destroys bacteria by hydrolyzing the mucopolysaccharides in cell walls. Although it has a nearly identical amino acid sequence, lactalbumin functions as one chain of the dimeric enzyme lactose synthetase in the mammary gland during lactation. Presumably a gene duplication event was followed by divergent evolution and a radical change in function.

Transposable Elements

Transposable elements (TEs) are sequences of DNA that are usually moderately repetitive and may enter and leave extant genes or other regions of nuclear DNA. Their presence was first inferred by nobel laureate Barbara McClintock, who discovered otherwise unexplainable reversions in maize phenotypes that ordinarily had Mendelian variation. The molecular nature of these elements was not understood until some thirty years later with the advent of *in situ* hybridization of mRNAs, and when a number of genes in *Drosophila melanogaster* were sequenced using the new technologies of recombinant DNA and gene cloning. TEs are found in a wide variety of eukaryotic and prokaryotic organisms.

Approximately one half of the moderately repetitive DNA of *Drosophila melanogaster* is comprised of about 30 distinct families of transposable elements (Rubin et al. 1982). At the extreme ends of each element are short inverted repeat sequences, which are, in turn, usually at the end of longer repeat sequences. TEs of *D. melanogaster* resemble the proviruses of vertebrate retroviruses.

In *Drosophila*, TEs are dispersed throughout the genome with the numbers of copies per family being approximately 20 to 50. They appear to be a varying component in the genome of different species. The families *copia* and *412* are found in all of the major drosophilid radiations but have apparently been lost repeatedly in evolution. Sequences complementary to *297* and *Tip* are restricted to species of the sophophoran radiation and are therefore of recent phylogenetic origin (G. Martin et al. 1983).

TEs usually have detrimental phenotypic effects when they are located in functioning genes. Rubin et al. (1982) investigated seven mutants at the *white* locus of *D. melanogaster* and found that five were caused by TE insertions. A spontaneous reversion to the visible wild type for the locus was accompanied by excision of the TE. The mobile elements *gypsy* and *472* are associated with mutants in the bithorax complex. Based upon limited evidence from three morphological loci,

TEs may be responsible for more than 70 percent of visible mutations (Bender et al. 1983). Given that some elements will spontaneously be excised, our measures of spontaneous mutation rate in *Drosophila* may include a small fraction that cannot be guaranteed to be transferred through the germ line.

The syndrome of hybrid dysgenesis (Kidwell et al. 1977) is due to at least two systems of TEs. In the P system, crosses between males from P and females from M strains result in the usual sterility of female progeny, but not in the reciprocal crosses. Some aspect of the female cytotype influences the outcome. P transposable elements occur in 30 to 50 copies per haploid genome in the P strain and are nearly always missing in M strains. Members of the P family transpose frequently in P-M dysgenic hybrids. Extracted chromosomes of progeny of interstrain crosses show newly acquired P elements. Dysgenesis is also characterized by a tenfold increase in the rate of phenotypic mutations, transmission ratio distortion, increased male and female recombination, and chromosomal aberrations and nondisjunction (Kidwell et al. 1977).

An apparent shift has occurred in field collections of *Drosophila melanogaster* over the past 30 years. Dominance of collections has shifted from that of M strains to that of P strains (Kidwell et al. 1981; Bingham et al. 1982). This suggests waves of invasion by successive new families of elements. The P element may have invaded *D. melanogaster* from another species.

The spread of P elements in populations is a competition between the rate of transposition and the lowering of fitness between crosses of individuals, respectively bearing and lacking the elements. If the infection rate is sufficiently great then the new element will become fixed in the population (Ginzburg et al. 1984). The story is more complicated, as there are limited numbers of copies in *Drosophila* genomes. TEs often do not occur in greater copy numbers than 50. A threshold of drastically lowered fitness may be crossed above this frequency, but autoregulation of copy number cannot be ruled out. In diploids, an autoregulatory mechanism could evolve if insertions of TEs caused a high rate of production of dominant lethals or sterile mutants (Charlesworth and Langley 1986). On the X chromosome of *Drosophila melanogaster*, the distribution of P-elements is concentrated at the distal tip relative to the rest of the euchromatic chromosome. But the distribution of *de novo* insertions is more homogeneous. The suppression of a recombination-dependent element excision process or the distal reduction of P-element elimination due to reduced unequal exchange may leave a lagging and chromosomally localized accumulation of elements (Ajioka and Eanes 1986).

Regulatory Genes and Structural Genes

Structural genes code for proteins used in the typical cellular machinery. By contrast, **regulatory genes** produce gene products that influence the action of other genes, probably by binding to sites upstream of the structural gene transcriptional unit. The latter genes act both in a *cis* and a *trans* manner; in one case, the same mutant has both *cis* and *trans* effects (Bingham and Zachar 1985). Our understanding of gene regulation is illuminated by those instances where nucleotide sequences seem to be widespread throughout a group of distantly related taxa. The homeobox sequence, for example, is found in the *engrailed* gene of *Drosophila melanogaster* and produces a transcript that binds in the 5' end of the gene (Desplan et al. 1985). This sequence is widespread in animals and mediates binding to certain DNA sequences. It is part of a larger sequence whose protein gene product has a regulatory function in the expression of genes important in development (Gehring 1987). The protein produced by the *Krüppel* gene has a finger-like structure, and can bind to DNA. This is particularly interesting as the three-dimensional structure can be modified to vary in binding (Rosenberg et al. 1986). As of this writing many enhancer and promoter sequences have been discovered to affect gene function in diverse ways. No single pattern has emerged.

The Britten-Davidson model, an early but comprehensive model of gene regulation and development, envisioned batteries of such genes, with feedback loops, as being important in development and in metabolism. While the early concern with repeated DNA as an important part of the picture is probably invalid, the possibility for regulatory messages that initiate batteries of genes is very much a live issue. A limited repertoire of regulatory genes may be involved in a wide array of initiation signals for gene action. The subject becomes complicated by those processes that may effect the rate of translation. Furthermore, the presence of some gene products in the cytosol may affect protein activity by processes such as allosteric modification. Controls of translation and final enzyme action might therefore also be properly included in a list of regulation, even if the DNA itself is not the regulated site.

Many authors (e.g., A.C. Wilson et al. 1974a; Valentine and Campbell 1975; Gould 1977 p. 405; Stanley 1979, Chapter 6) argue that regulatory and structural genes are distinct in their respective effects. Regulatory genes are seen either as the major controllers of morphological evolution (Valentine and Campbell 1975) or as possible sources of macromutations (Stanley 1979, Chapter 6).

The argument for the importance of regulatory genes in morphological evolution stems from a discordance between the pattern of nucleotide sequence difference among species for structural genes versus the degree of morphological resemblance. For example, an examination of homologous metrics among frogs shows far more phenetic resemblance than found between man and chimp (Cherry et al. 1978). Despite the phenetic resemblance, there is a major difference among frog species in protein amino acid sequences, as estimated by microcomplement fixation (Wilson et al. 1974b). This may be misleading, as frogs have strong among-taxon differences in other morphological traits, such as in the larval stage. But at the least, we can presume that raw sequence difference is not congruent to the degree of divergence among adults.

The gulf in effects between structural and regulatory genes is probably more apparent than real, and is probably based upon several misconceptions about genic variation and function. Certain gene products seem to have the exclusive function of switching on one or a battery of genes. The genes involved in hormone production, such as ecdysteroids in insects, are regulatory in this sense (Ashburner 1980). A large number of hormones travel through the vertebrate bloodstream and, through a series of intermediate processes, activate gene transcription in the nucleus. But these effects are quite variable and probably blend in effect with those of structural genes. The *lin-12* gene of *Caenorhabditis elegans* and the *Notch* gene of *Drosophila melanogaster* both encode proteins that appear to be homologous with a related protein family in mammals. The *Notch* gene seems active in the early differentiation of neuroblasts and the protein seems to be involved in intercell communication. The protein may be lodged in the cell membrane and facilitate intercell communication and gene activation (Akam 1986). If this be true, the gene is not regulatory by our definition, even if it has clear regulatory function. There is likely to be a range of such regulatory actions.

. Nucleotide sequences upstream of the coding region of a gene influence transcription, and these regions can be regulated via binding of polypeptides. Insertions of transposable elements, at a distance from the structural locus, can disrupt gene function (e.g., Zachar and Bingham 1982). These upstream sequences are also regulatory nucleotide sequences, but are not genes. We would expect that regulatory genes code for a gene product, i.e., a polypeptide, having amino acid sequences and three–dimensional structures subject to the same sort of phenomena as those encoded by, for example, enzyme loci. The regulatory proteins, however, might bind with DNA at other loci, enabling or blocking transcription. Thus these polypeptides are subject to the same amount of amino acid sequence variation as those produced by structural genes, allowing for various functional constraints. Such functional variation is obvious in the protein product of the *Krüppel* gene, as discussed above.

The RNA polymerases are gene products of regulatory loci. They bind upstream of the coding region and enable transcription to occur. Among-species differences can be found in one polymerase, while striking identity may be found for another. This variation in sequence divergence is also a feature of structural genes and may be related to constraints upon function (e.g., A.C. Wilson et al. 1977). I predict that once regulatory genes are sequenced routinely, allelic variation will be quite common, perhaps within the range of structural genes. As for some structural genes (e.g., some histones), some regulatory sequences will be faithfully conserved.

Regulatory genes probably have a large range of effects and may interact with one locus or with many others. But the same can be said for enzymes, whose interactions may be only with an external substrate, or with a large battery of other enzymes in a catalysis network. In such networks, regulation of transcription, translation, and activity of the gene product, occur partially as a function of concentrations of substrate, product, cofactors, other molecules, and the enzymes themselves. Loci near structural loci often modulate transcription (e.g., Doane 1969). Thus there is no real way to easily distinguish between many of the effects of regulatory loci and structural loci, since a structural locus can indirectly have a regulatory function via action and concentration of its gene product.

Is there thus a valid justification for believing that there are "genes for morphology," which we cannot presently understand, and a large array of insignificant genes, such as those encoding enzymes? This is uncertain. Morphology can be reckoned into cell types, spatial organization, and linear dimensions. While the presence of a given cell type is under control of regulatory genes, overall dimensions and spatial arrangement are both a partial function of cell number, and cell dimensions. These latter properties must be affected by metabolic efficiency, rates of translation of RNAs whose products are involved with cell structure, etc. The action of structural genes and feedbacks generated by their gene products must influence the emergent properties of morphological dimensions. Of course, there are probably a large number of genes typically unstudied that affect morphology as well.

A possible misconception concerning a dichotomy of effects may come from the presence of extensive polymorphism found in structural genes usually without measurable change of fitness (e.g., Eanes 1987; Harris 1966; Lewontin and Hubby 1966; Wills 1981). This leads to the notion that variation in structural genes cannot be fundamental to the phenotype (e.g., Kimura 1983). But evidence about the action of so-called regulatory genes is usually inferred from studies of inactivation of genes, such as those involved in the bithorax complex of *Drosophila melanogaster*. Such inactivation usually has profound phenotypic consequences, ergo we think that regulatory mutants have profound effects in evolution relative to structural genes. The evidence from the *white* locus (e.g., Zachar and Bingham 1982) of *Drosophila melanogaster* suggests that inactivation of loci due to TEs may be the principal cause of visible genetic variation. Insertions affecting the phenotype may be in the structural gene itself or in upstream regulatory sequences. But this very morphological variation in laboratory stocks is the source of our intuition about the unique status of morphological variation, as opposed to genetically controlled sequence variation in proteins. But animals would be no less affected by the inactivation of structural genes. Many mutations now being sequenced turn out to be explained by disruptions of coding regions (e.g., *notch* mutations in *Drosophila melanogaster*). At the *Adh* locus of *D. melanogaster*, variation in activity among alleles is probably explained by a single nucleotide difference within the structural gene (Aquadro et al., 1985). It is therefore unclear whether or not the evolution of regulatory genes does have a unique pivotal role in sudden morphological shifts. We have no large scale sampling of the average degree of sequence differentiation or polymorphism for regulatory loci, as we do for structural loci. At present we are comparing apples and oranges. At the least, the simple distinction of phenotypic effect between structural and regulatory genes seems unwarranted at present.

Phenotypes and Genetic Variation

Polygenic Determination of Traits

As we will be concentrating upon phenotypic traits, it is useful to understand the polygenic nature of most variation, including polymorphisms involving discrete morphological states. Believers in the possibility of macroevolution via hopeful monsters usually invoke mutants of major phenotypic significance (Goldschmidt 1940; Gould 1980a). This often is accompanied by the hope that so-called regulatory mutants will facilitate a switch between morphological modes. The dichotomy between major and minor genetic changes is usually hypothesized to be associated with regulatory and structural gene variation, respectively.

Workers have often mistaken discrete phenotypes for discrete genotypes, explained by a single locus of large effect. Goldschmidt (e.g., 1945) mistook the mimicry polymorphism in swallowtail butterflies to be under the control of a major switch gene. [1] Punnett, whose famous square has been used by many generations of students to diagram genetic crosses, first argued (1915) that mimetic morphs in the swallowtail genus, *Papilio*, could be explained by a single gene. His experimental evidence was impressive indeed. Crosses between mimetic and nonmimetic forms of *Papilio polytes* yielded segregating progeny of mimetic and nonmimetic morphs as if they corresponded to individual alleles at a single locus. Not only had a mutation apparently facilitated an otherwise unbridgeable vault to a new adaptive mode, it even seemingly targeted itself to an extant model species! This result squared well with the intuition that intermediate forms in evolution

---

[1] It is odd that although he believed that intrapopulation variation was not the stuff of interspecific evolution, he used a classic example of *intra*specific variation to bolster his argument.

were maladaptive (half an eye, and so on), so single mutational leaps were a necessity.

But a major discontinuity found in an experimental manipulation need not imply an evolutionary leap. Mimicry in species of *Papilio* is under polygenic control, often with several tightly linked loci controlling color, wing shape, and pattern (Turner 1977; Clarke and Shepard 1963). Modifiers contribute to determining the phenotype. Crosses between strains of mimetic and nonmimetic races of *Papilio dardanus* often result in inferior mimics, for lack of the appropriate modifiers (Clarke and Sheppard 1962).

One shouldn't leave this story with the impression that all mutations comprising the mimetic phenotype of butterflies are small in effect. This would be to take the macromutation notion too far in the opposite direction. Indeed some of the changes are quite large, particularly the effects on color differences. It is rare for a set of genes affecting a trait to all have equal and infinitesimal effects. A polygenic basis is usually found in resistance to insecticides, for example, but genes with major effect are also known (Crow 1957). The compendium of Lindsley and Grell (1968) commonly shows alleles at a locus with a wide range of expression. Most evolutionary geneticists work on the assumption of a range of effects, despite the arguments of Fisher that all changes were underlain by mutants of small effect (Turner 1983). The evidence doesn't fit this notion, nor does it permit Goldschmidtian macromutations (large scale mutations associated with a major chromosomal change). Nothing in population genetic theory, however, prohibits hopeful monsters, in the sense of major new phenotypes explained by a single allelic difference.

The evidence so far accumulated shows, however, that most morphological traits examined are under polygenic control. We need not detail here the widespread evidence. Nearly every morphological trait in plants and animals ever discovered to have a genetic basis is under polygenic control (see Wright 1978, Chapters 6, 8; Stebbins 1950; Falconer 1960). Color polymorphisms often segregate as one locus, but multilocus control is far more common. The effect of a typical polygene is on the order of 0.1 to 0.5 of the additive genetic standard deviation (Falconer 1981).

Polygenic control is not restricted to continuous traits such as bill length in birds (e.g., Boag and Grant 1981). Discontinuous traits such as digit number and vertebral number are under similar control, albeit in combination with thresholds that help determine the discrete phenotype (Wright 1934, 1935a, b; Green 1962; Lande 1978). It is, therefore, incorrect to believe that mutational jumps must of necessity occur in the evolution of discrete phenotypes. The confusion goes back to the debate toward the beginning of the century (Provine 1971) over the relative importance of biometrical (continuous) versus Mendelian (discrete) variation. This is a false distinction; yet the large mutation–small mutation false dichotomy lives on to haunt us.

The polygenic control of discrete traits can be confirmed to a degree by F1 and F2 crosses between strains that are monomorphic for two alternative discrete states. This has been done by Wright for inbred strains of guinea pigs with different numbers of digits and by Green (1962) in crosses of inbred strains of mice with different numbers of presacral vertebrae. Consider one strain selected to produce a given number of complete digits. We would expect a series of genes to have a distribution of effects, with a maximum of effect centered between thresholds determining the discrete number of digits. A cross between two such strains (Fig. 3.3) would yield an F1 progeny set with a distribution of genetic factors intermediate in genetic composition, but it may fall to one side of the threshold. Therefore, all progeny may have one number of digits. Crosses in the F2, however, should show an expansion of the distribution of genetic factors, resulting in significant numbers of individuals straddling the threshold. In guinea pigs, this results in a large number of individuals with incomplete digits (Fig. 3.3). It was the discovery of such modifiers that led Castle to agree that gradualism had an important role in evolution. Similar polygenic control was found for presacral vertebrae in mice (Green 1962). Expression in crosses among four inbred strains resulted in only seven qualitatively different presacral morphs, suggesting the presence of thresholds.

Thus the genetic architecture of both discrete and continuous traits are both polygenic and the genetic determination of the phenotype is a complex result of interactions among genes that perform different functions. The evidence suggests that control comes from much of the genome and usually cannot be restricted to switch genes of major effect. The polygenic determination of discrete traits suggests a confusion in the literature over **macromutations** and **hopeful mon-**

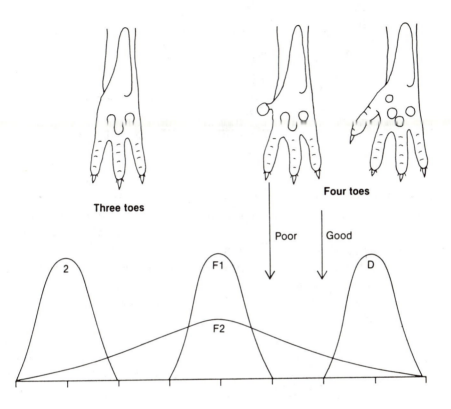

Figure 3.3: Genetic control of toe number in guinea pigs. Arrows refer to thresholds for poor and good development of four toes. A cross of strains "2" and "D" result in an F1 generation with individuals all beyond the threshold determining three digits. Segregation in the F2, however, results in an expansion of genetic variability, across the threshold required for four-digit morphs to reappear (adapted from Raff and Kaufman 1983).

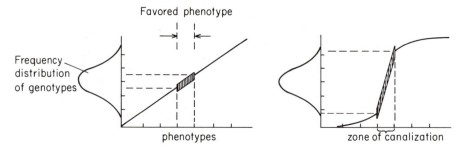

Figure 3.4: Canalization of a trait, showing the reduced phenotypic expression despite genetic variability.

sters, the terms suggested by Goldschmidt (1940) for mutations of major phenotypic effect, involving coordinated developmental changes in many characters. Discrete jumps in morphological variation should not be construed to be evidence for leaps explained solely by a single gene, though the latter certainly can occur. In Chapter 5, we discuss the hopeless nature of major developmental mutants. The negative pleiotropic effects of such mutants makes their importance in evolution unlikely.

This argument, however, does not exclude discrete **morphological** variants whose prospective functional significance may be profound. Frazzetta (1970) found a new potential site for a jaw articulation in the maxillary bone of a bolyerine snake. This change, while being of great potential functional significance, is rather easy to explain, given that joints often are controlled by effects resembling the threshold effects for vertebral articulations and toe number. A polygenically controlled thinning of the jaw at the point of the prospective joint, in combination with the crossing of a threshold for a developmental "instruction" to generate a joint, would be sufficient to result in the evolution of the new discrete character.

Canalization

Many traits have reduced variability. This can be a surprise, particularly in meristic traits, such as bristle number in *Drosophila*, where variability in populations might normally be expected. Waddington (1956) showed that many phenotypes are controlled by large numbers of genes but are nevertheless buffered against change caused by variation in the environment. He termed this buffering **canalization**. The necessity for phenotypic invariance may be attributed to potential negative effects of phenotypic deviants on fitness. In *Drosophila melanogaster* scutellar bristle number is canalized at a count of four (Rendel 1959). Strong selection can shift gene frequencies out of a zone of canalization, and thus increase the potential for expression of phenotypic variability (Fig. 3.4). Canalization may be a common explanation for phenotypes which are widespread in a species with little variation, despite the underlying polygenic nature of the trait. With a mechanism to reduce phenotypic variance, selection has no variation upon which to act, and temporal or geographic invariance in a trait may not be surprising.

Genetic Variability in Natural Populations

The measurement of genetic variability in natural populations has been discussed from different perspectives, by Lewontin (1974) and Kimura (1983). Both theoretical and empirical work in this field have centered around the following questions:

1. How much variability exists in natural populations and does it vary from one taxonomic group to the next?

2. When present, does the variability have any significance for natural selection on the phenotype?

3. Are groups of genes organized in any way that might influence their joint evolution?

4. How does intraspecific variation compare with interspecific variation?

Many believed that a measurement of the amount of genetic polymorphism in natural populations would resolve the dispute between mutually inconsistent views of genetic variation and the role of variability in natural selection. The so-called **classical** school, typified by H.J. Muller, regarded mutation as primarily a source of deleterious effects upon the phenotype. Natural populations were, therefore, predicted to have low variability. This conformed to the **wild type-mutant** dichotomy so often used in classical genetics. Wild types corresponded to the favored genotype that specified the normal phenotype. Mutants were typically deleterious in viability studies of *Drosophila* cultures. Natural selection, therefore, was directional, or purifying. A population "waited" for the rare favorable mutation that would then be fixed by selection.

By contrast, the **balance school** (Dobzhansky 1955) saw variation as common and even beneficial to natural populations. The gene combinations found in given *Drosophila* chromosome inversions were thought to be coadapted, as were heterozygous combinations within natural populations (Dobzhansky 1948*b*). This was predicated on a variety of data and beliefs. First, laboratory studies demonstrated that chromosome inversion heterozygotes often were more viable than homozygotes (Dobzhansky 1951). **Heterozygote advantage** would maintain homozygotes in the population by segregation. Second, complex environments were believed to select often for homozygotes in alternative microenvironments; this too could preserve polymorphism (e.g., Levene 1953; Haldane and Jayakar 1963; Gillespie and Langley 1974; Hedrick et al. 1976). Heterozygotes also might be more fit in environments that fluctuated regularly from one microclimate to another, given that the genes were codominant and the organism would suffer in fitness if it could not function in both microclimates (see Gillespie and Langley 1974). Finally, heterozygotes were thought to confer a superior degree of homeostasis, or superior fitness in the face of environmental change, to the organism (Haldane, 1954; Lerner 1954). Under the balance hypothesis, genetic variation was thought to be common in natural populations.

A few major types of variability have been studied extensively in natural and laboratory populations. Geneticists have long been interested in the presence of **visible morphological mutations** and their correlation with viability in natural and laboratory populations. Studies of resemblance among relatives establish a genetic component, but not usually the exact number of loci (or degree of genic polymorphism in a population) underlying the trait. Thanks to the pioneering work of Dobzhansky and the exquisite polytene salivary chromosomes of *Drosophila*, **chromosomal polymorphism** has been investigated extensively both in laboratory cage experiments and in field populations. The occurrence and distribution of **lethal genes** has been studied extensively. Finally, gel electrophoresis has been used to survey for **protein polymorphisms**. In contrast to the first three approaches, electrophoresis surveys can be made of species from diverse taxonomic origins. In recent years, recombinant DNA and direct sequencing techniques have led to estimates of **DNA sequence variation.**

Lewontin (1974) summarizes surveys morphological variation in populations, especially variants ascribable to single morphological loci. In *Drosophila mulleri* there are approximately 0.5 visible mutants per haploid genome. Other estimates of visible mutants yield a range of 0.28 to 1.19 per genome. In *D. melanogaster* there are estimated to be 0.00020 (0.5 variants/1000 loci) variant alleles per locus in a natural population. These estimates all conform to the notion that morphological genetic polymorphism for discretely recognizable phenotypes is low and consistent with the predictions of the classical school. But traits under polygenic control seem to show extensive phenotypic variance, ascribable to a good deal of genetic polymorphism. The frequency of visible mutations declines as the degree of morphological extremity of the mutation increases (Gregory 1966). There may be many more genetically based morphological changes that go undetected in experiments. Perhaps we notice only the larger-scale morphological mutants. This both justifies the relative importance of small-effect mutations as the source of most evolutionary change and emphasizes possible ubiquity of genetic polymorphism in natural populations.

The study of genetic variation raises a paradox, as the mutation rate for visible mutations (ca. 10^{-3}–10^{-2} per gamete per generation) is orders of magnitude higher than the estimated per locus mutation rate (10^{-7} – -10^{-5} per locus per generation). How can such low per locus mutation rates explain the much higher rate of appearance of morphological novelties, given that we take about 10 loci as the number explaining variation in a typical morphological trait (Falconer 1981)? Several explanations are conceivable (see discussion in Turelli 1984 and references). First, the traits may be explained by many more loci than are estimated by coventional quantitative genetic studies, which usually suggest on the order of 10 loci, representing "factors," or unlinked genetic entities. One could imagine many more linked loci explaining variation in the trait. But it is unlikely that this could usually make up for an orders of magnitude discrepancy. Next, it is possible that the loci we examine to estimate per locus mutation rate are distinct from those that control morphological change. Perhaps there are "morphology loci"–regulatory loci or changes in regulatory noncoding DNA sequences. Finally, it is possible that special changes, such as unequal crossing over, explain morphological mutants and are distinct from the mutants at individual loci. At present we cannot easily exclude any of these alternatives.

I will discuss enzyme polymorphisms in more detail below, as they have been a primary motivation for the neutral theory of molecular evolution. Most natural populations of animals, plants, and prokaryotes are quite polymorphic at soluble enzyme loci. This discovery (Hubby and Lewontin 1966; Lewontin and Hubby 1966; Harris 1966) was a great surprise to those supporters of the classical school. The occurrence of lethal genes (usually lethal when homozygous) has been studied extensively in human populations (e.g., Schull and Neel 1965) and in *Drosophila* (see summary in Lewontin 1974). Such lethals are usually explained by one or a small number of mutations, rather than by mutations of small effect scattered throughout a chromosome, whose cumulative effect is lethal. Evidence from both man and *Drosophila* suggest that the average individual carries 1 to 4 such lethals. In *Drosophila* this number is subject to a large amount of interspecific (Dobzhansky and Spassky 1954) and temporal–geographic intraspecific (Ives 1954; Band and Ives 1968) variation. It is possible that transposable elements are the major source of lethals, as they apparently are for morphological mutants.

There is no simple answer to the question of the degree of genetic variation in natural populations. This is not because answers are not available. Indeed, we are close to such an answer for enzyme polymorphisms. The lack of simplicity stems from a poor understanding of the relationship between genotype and morphological phenotype, or even functional enzyme phenotypes. This thwarts any attempt to survey the genetics of natural populations and the degree of genetic differentiation among populations of a species, and to relate these data to evolution at the phenotypic level. What does it mean, morphologically, for two populations to differ genetically by large or small amounts? We do not know.

Because we do not know the precise relationship between genotype and morphological phenotype we often use overall comparisons between genic differentiation and morphological differences, hoping that the one will, in the aggregate, tend to explain the other. The discussion above relating structural gene to morphological differences demonstrates that this can be difficult. Lewontin (1983a) shows that some obvious considerations weaken the presumption still further. First, if phenotypic effects are random with respect to genotype, and equal in magnitude over all alleles, it will be far easier statistically to distinguish differences in the mean of a morphological trait, than a difference in gene frequency. With a difference among character means at a given probability level of statistical significance, gene frequencies will differ at the same level only if the number of loci controlling the trait is less than or equal to the reciprocal of broad-sense heritability. If heritability is, as is typical, on the order of 0.5, then the equivalence is violated when the trait is explained by more than 2 loci, which must be the rule. This immediately places comparisons between the degree of morphological and genetic (allozyme data, for the most part) differentiation among populations under a statistical cloud. A random set of loci, even if they include those controlling the trait, cannot be compared validly with differences in the morphological trait. These problems are much worse when there is a nonrandom association between gene and phenotypic score. Of course, there are several examples where the magnitude of morphological and genetic differentiation are correlated (Avise 1976). But some of the discrepancies (Cherry et al. 1982) are worrisome, especially since the claim that certain forms of genetic differentiation lie behind morphological differentiation is a major issue in evolutionary biology. The struggle to measure genetic variation may have been futile, if our final objective was to produce a simple algorithm to understand emergent phenotypic variation.

Natural Selection and Phenotypic Variation

Some Definitions

Natural selection is a process that follows from the necessary conditions of: (a) genetically based phenotypic variation, and (b) differences in reproduction or survival among the variants. With these two necessary conditions it follows that: (1) if the population is out of equilibrium then the phenotypic distribution of the offspring in a population will differ from that of the parents; or (2) the phenotypic distribution will differ among age classes, beyond that expected from normal ontogenetic change (see Burian 1983, Endler 1986). The logical consequences–(1) and (2)– following conditions (a) and (b) constitute a

syllogism, not a tautology (Endler 1986). The misconception of a tautology stems from the phrase "survival of the fittest," which ignores
the logical structure of conditions and consequences.

Natural selection can also be applied to higher levels of the organismal hierarchy (see Chapter 1). At the species level, for example, species
with their traits constitute the variation, and the analogous measures
of reproduction and mortality would be speciation and extinction. The
consequence would be changes in the abundance of species. Each hierarchical level must have its own distinct heritable variation and a
measure of productivity that could differ among the units. As long as
the levels are decomposable in the hierarchical sense, the process of
natural selection and even adaptation could be applied to levels above
those of the individual.

If the change in phenotypic distribution is consonant with predictions based upon a criterion of performance (e.g., owing to natural
selection, tolerance to low salinity increases when salinity greatly decreases), then **adaptation** occurs. Adaptation is the historical process
of evolutionary change describing how natural selection interacts with
functional and developmental constraints, mutational availability, and
random processes. Natural selection is the specific part of adaptation
concerned with available variation, and the selective environment. By
contrast, **adaptedness** is a static description of the functional superiority of one or another phenotype.

Models of natural selection and adaptation often make the assumption that there are two necessary elements to predict the course of
evolution: (1) a predictable description of the relationship between
genotype and phenotype; (2) a description of the environment that
discriminates among phenotypes. There are good reasons to believe
that this is too simplistic. First, genotype-environment interactions
are such that one cannot catalogue phenotypes just by knowing the
genotype. The environment must often be specified (e.g., Gupta and
Lewontin 1982). Our definition of natural selection does not require
a separation of adaptedness from historical circumstances, as we require a specification of genotype-environment interaction. To the degree that genotype-environment interactions and environments vary,
the outcome of evolution is increasingly complicated.

The course of adaptation may be eccentric, depending upon the
genetic track taken during evolutionary change. It is well known for

example, that forward and backwards selective change of the same morphological trait does not occur at the same rate, despite the imposition of similar selection differentials. Differences in environmental history may also result in different adaptive tracks. The problem of history can be seen in a study of mutant strains of *Escherichia coli* that produce the toxin colicin (Chao and Levin 1981). In a wellmixed aquatic culture, the colicinogenic strains are at an advantage only when fairly common. When sensitive bacteria are killed, their death causes release of nutrients at random to both types of bacteria. Colicinogenic strains have lower division rates and therefore lose out unless they overwhelm the system with toxin. By contrast, in agar, colicinogenic strains come to dominate even when initially rare. The latter structured habitat permits the colicinogenic bacteria to create a barren zone, which is rich in nutrients. They then can spread at "their leisure." The fate of the gene is therefore locked up in its historical background.

Through habitat choice, behavior decisions, etc., organisms can alter their own environment. This means that in order to study natural selection properly we must be able to describe the interaction between organism and environment that determines the actual selective regime (Lewontin 1983*b*). This often means that history will thwart the prediction of clear evolutionary trajectories.

Fitness is often used interchangeably with adaptedness. Fitness should refer to the relative ability of genotypes to survive and leave offspring. One can also define fitness in terms of alleles at a locus. If we have a locus segregating for two alleles A_1 and A_2, let the respective fitnesses be W_1 and W_2. We can then define a **selection coefficient**, s, which equals $W_1/W_2 - 1$. If p is the frequency of A_1 and q is the frequency of A_2, then the change in p over one generation will be:

$$\Delta p = \frac{spq}{\bar{w}}$$

where $\bar{w}$ is the mean fitness of the entire population. Changes in allele frequencies are, therefore, associated with a fitness parameter. This expression is oversimplified and applies to haploid organisms. For a more complete discussion of selection in diploid organisms see Ewens (1969).

The assessment of relative fitness of genotypes at a locus implies a complete randomization of the background genotype (Lewontin 1974). In practice this is nearly impossible to achieve, given the great diffi-

culty of randomizing the background loci that are tightly linked to the
locus in question. This problem is not trivial and is a major source
of difficulty in interpreting the meaning of selection experiments. In a
crude experiment, selection for variants at an allozyme locus may appear to be intense, simply because the locus marks part of, or an entire
chromosome (contrast Powell 1971 with Yamazaki et al. 1983). A detailed study of fruitflies, using flanking markers at close map distance,
designed to randomize the genetic background, typically requires the
counting of tens of thousands of flies (Eanes 1984; Eanes et al. 1985).
Such studies, rarely done, show that fitness among protein phenotypes
is probably much less than usually estimated, when linked loci are
factored out (Eanes 1987).

It is difficult to measure with statistical confidence selection among
genotypes differing in fitness by as much as 1 percent. Natural selection involving such levels and less could have significant evolutionary
effects. To demonstrate that recessive lethal chromosomal mutants of
Drosophila melanogaster lowered the fitness of heterozygotes by about
one percent, Mukai and Yamaguchi (1974) had to score about one million flies. Such herculean projects have been completed only rarely.
One should remember that biologically significant selection need only
be $s > 1/2N$, and N is often 10^6.

We shall be mainly concerned with morphological characters whose
determination has both a genetic and an environmental contribution.
Let us assume further that variation in the phenotype can be arrayed
as variation along a single axis of variation. This could apply to both
continuously measurable and countable characters. Assume that the
population variation follows a normal distribution, with mean $= 0$
and standard deviation $= \sigma$. Assume also, following the methods of
quantitative genetics (see Falconer 1981), that the phenotypic scale is
determined partially by an underlying genotypic scale. We can plot
along the genotypic scale a fitness function, which depicts the relative
success of different genotypes.

Total phenotypic variance can be partitioned into a variety of genetic and environmental components. For our purposes assume that
all of the loci determining a phenotypic trait are independent, there
is no dominance, and that there are no genotype-environment interactions. We can then simply define **narrow-sense heritability**, h^2,
as the proportion of the total phenotypic variance, σ_a^2 explainable by

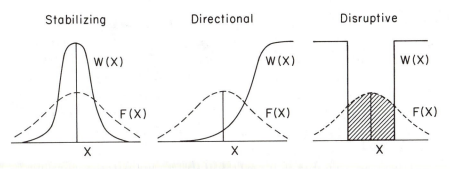

Figure 3.5: Examples of some modes of natural selection on a hypothetical normal distribution of phenotypes. Broken line is the frequency distribution of phenotypes; solid line represents fitness functions.

between-allele effects, or the additive genetic variance, G_a. Define a selection differential, S_a, which represents the difference in the means of the selected and unselected adults. If our phenotypic scale variable is z_a the change will be:

$$\Delta z_a = (G_a/\sigma_a^2)S_a = h^2 S_a$$

This formulation provides a convenient way of visualizing how selection on a continuous phenotypic character can be related to an underlying genotypic distribution.

Modes of Natural Selection

In **directional selection** (Fig. 3.5), the maximum value of the fitness function is shifted away from the mean (we use the right side as a convention). The success of a given phenotype might increase continuously with higher phenotypic value. But directional selection can involve **truncation selection** of the entire phenotypic distribution past an absolute or relative (e.g., upper ten percent of the population) threshold. This might occur when allometric considerations prohibit animals larger than a certain body size to satisfy their maintenance energy requirements, or when animals larger than a threshold size might escape the grasp of a predator.

Observations of directional selection in natural populations, where the adaptive significance is clear, are few in number (see Ford 1975; Boag and Grant 1981; Endler 1986; Seeley 1986). This should be no surprise, as selection intensities have to be quite high to see any dramatic change, and such selection will be temporary before a new equilibrium is achieved. A selection intensity among discrete morphs of only a few percent would be effectively invisible, given the swamping effect of collecting difficulties, statistical problems, spatially varying directional selection, and possible differing genotype-environment interactions in different subhabitats. Nevertheless, natural selection has been observed in a surprisingly large number of cases (see Endler 1986). Consider the following examples.

Over a hundred species of insects in British industrial regions blackened by smoke are dark in color, relative to conspecifics in unpolluted areas (Ford 1975, Chapter 14). Vegetation in industrial areas is often darkened with smokestack soot and light-colored substrates, such as lichens, are killed off by pollution. In the moth, *Biston betularia*, the black *carbonaria* variant is dominant over the recessive light-colored morph. Dark-colored morphs are preferentially killed by a variety of insectivorous birds when placed upon trees with a normal lichen cover. The light mottled color blends with the background. Birds quickly locate the light morphs against the contrasting background of the blackened vegetation (Kettlewell 1955). Kettlewell's data suggest that the selection coefficient favoring the melanic gene must be about 0.5. This would easily account for the rapid spread of the *carbonaria* morph since the middle of the past century, from negligible starting frequencies.

Intense selection also has been observed in a population of one of Darwin's finches, *Geospiza fortis* (Boag and Grant 1981; P.R. Grant 1985). Parent-offspring regressions yield an average heritability of 0.76, for a wide range of external morphological characters. In 1977 the annual rainfall on Daphne Major island (Galapagos Islands) decreased drastically, which caused a dramatic decline of plants producing small seeds. The average size and hardness of seeds increased and larger birds fed more heavily on the large hard mericarps of *Tribulus cistoides*, which had been ignored by almost all birds in earlier years. Large birds, especially those with large beaks, were able to survive because they were able to crack the large and hard seeds that predominated during the drought. The rainfall returned to normal the next year, suggesting that periods of intense selection probably occur at erratic intervals. Presumably, the selection for larger sized birds is now reversed due to a return to normal conditions and size will shift downward. A lack of intense selection will slow the reverse selection process. Some more recent evidence shows the action of stabilizing selection.

Laboratory selection experiments usually attempt to change the average value of a phenotypic trait by biased culling of phenotypes. Although deviations of several standard deviations can be achieved rapidly, this type of experiment would not reflect rates in nature. W.W. Anderson (1973) measured samples of a mixed culture of *Drosophila pseudoobscura* that had been left in high (25°C) and low temperature (16°C) incubators for 12 years. After 1.5 years, no significant divergence in body size had occurred. But after six years, and until the end of the experiment (12 years), a significant body size differentiation had developed (ca. seven percent). Body size varied inversely with temperature, and approximately 40 percent of the total phenotypic variance could be ascribed to genetic variation. This change is consistent with the commonly observed body size clines in latitudinally widespread species of *Drosophila*. Rapid evolution in body size, therefore, can be effected by the physiological effect of temperature alone.

The power of directional selection should be greatest in large populations. Genetic variability is lost more slowly in large populations and Fisher's fundamental theorem of natural selection demonstrates that the potency of natural selection is proportional to the genetic variance in a population. This is in contrast to the belief that small peripheral populations should be the site of most adaptive evolution. Experiments and theory demonstrate (Hill 1982) that large populations sustain more directional change than small ones. Thus, theory and observation refute the belief, recently popularized by punctuationalists (e.g., Eldredge and Gould 1972) that small peripheral populations are of necessity the major source of adaptive change. Random, i.e., nonadaptive, forces operate most effectively in small populations.

While large populations can sustain more directional selection, they may be spread over large areas with considerable environmental heterogeneity. If gene flow is sufficiently reduced, local populations may diverge, but a reestablishment of connection will dampen any directional shift. Large populations may therefore not accumulate much difference over long periods of time. If a broadly distributed species is divided into several geographically contiguous daughter species, then we might expect each of the daughters to go its own way. The daughter species would still have sufficiently large population size to sustain extensive directional change. Note, though, that speciation does not guarantee divergence, as evidenced by the many cases of sibling species.

Stabilizing selection operates by culling out extremes from the phenotypic distribution (Fig. 3.5). A modal phenotype is therefore favored. Stabilizing selection can occur in both continuous and discontinuous traits. In continuous traits, one would expect a range of variants, perhaps approximating a normal distribution on a linear phenotypic scale. Stabilizing selection for a modal birth weight is suggested by the survival of human infants, which is correlated with proximity to the mean birth weight of about seven pounds (Karn and Penrose 1951). In discontinuous traits such as number of segments, we must consider the model mentioned above for polygenic effects with thresholds, determining the discrete nature of the trait (e.g., n versus $n + 1$ segments). Traits such as scutellar bristle number in *Drosophila melanogaster* are often strongly canalized and extensive directional selection is necessary before the threshold is breached where variability is exposed (Rendel 1959). In less canalized traits of this sort, phenotypes differing from the mean number of (sternopleural) bristles show lowered fitness (Barnes 1968).

Stabilizing selection might be spotted in nature by sequential sampling of juvenile and adult populations. One might expect selective mortality to reduce the variance about the mean between juvenile and adult stages. Several studies have successfully recorded such a reduction, but the results are usually variable. Dunn (1942) recorded a reduction in variance of head scalation in the snake *Conopsis nasus*, but got negative results for other species. In the Gekkonid lizard *Aristelliger praesignis*, the variance of the number of toe lamellae decreases from juveniles to adults (Hecht 1952). Variance in two characters in a Cretaceous oyster diminished with increasing size (Sambol and Finks 1977). These results assume that between-generation differences in juveniles and adults correspond to within-cohort reduction of phenotypic variance.

Does any characteristic genetic structure underlie stabilizing selection? Directional selection should require a necessary reduction of genetic variance, assuming that the heterozygote does not determine the most extreme phenotype. Stabilizing selection has no such requisite. It might select for increasing heterozygosity if such genotypes determined intermediate values on the phenotypic scale. In an investigation of caudal fin ray number of the guppy *Poecilia reticulata*, in which stabilizing selection occurs, central phenotypes were found to be more heterozygous at allozyme loci than extreme phenotypes (Beardmore and Shami 1979). Although the four loci examined were unlinked, older fish had genotypic frequencies which departed significantly from the multilocus Hardy Weinberg expectation. This would suggest strong interlocus interactions on fitness. Much evidence demonstrates a correlation between genic heterozygosity and fitness, though the causal relationship is unclear (Mitton and Grant 1984).

Artificial stabilizing selection on sternopleural chaeta number (Thoday 1959) and on a wing vein trait (Scharloo 1964) in *Drosophila* both show a gradual decrease of phenotypic variance. This decrease includes a significant decrease of additive genetic variance, suggesting that the potential for evolutionary change (the response to selection) will decrease over time with stabilizing selection. These experiments however, involve intense selection and may be unrepresentative of natural populations. If enough genes contribute to determine the additive genetic variance component of the phenotypic variance, then mutation may feed variation continuously into the population (Lande 1976).

In favoring those gametes with expressed phenotypic values close to the optimum, stabilizing selection may produce negative correlations in allelic effects at closely linked loci, so that positive and negative deviations from the optimum tend to cancel. This would create a pool of hidden genetic variation that is stored in linked combinations. Recombination would decrease the correlations among loci and convert the hidden genetic variation into expressed variation (Lande 1976). Thus a mechanism for continuous generation of potential for phenotypic evolutionary change may be available. In the case without such negative correlations, where heterozygotes are intermediate in fitness, polymorphism will be lost from the population. Mutation feeds variability into the system and the polygenic nature of phenotypic determination increases the potential for mutation to increase the genetic variability affecting the trait. Stabilizing selection may, therefore, fail to reduce genetic variability sufficiently to keep a population from having a storehouse of variability capable of responding to directional selection. Schmalhausen (1949) argues that genetic variability is depleted during bouts of intense directional selection, but is gradually restored during periods of stabilizing selection.

Selection Has Not Eliminated Variability   The omnipresence of considerable heritability for morphological traits suggests that selection cannot be a potent force for removing genetic variability from a population. As heritability is a measure directly related to the potential for morphological change, we can argue that most studied populations now seem capable of change as a response to selection.

Given the commonly high heritabilities, Kimura (1983, p. 143) makes a rough calculation on the selection intensities at a locus, which I have modified slightly. Under stabilizing selection, the selection coefficient, s, equals:

$$s = -[log(1 - L_T)]h^2 / n_{nuc} h_e$$

where L_T is the segregation load, h^2 is the broad sense heritability, n_{nuc} is the number of coding nucleotide sites, and h_e is the average heterozygosity. If we take 0.1 to be an appropriate value for detectable electrophoretic heterozygosity, and the hidden electrophoretic variability to increase this value by 100 percent (Selander and Whittam 1983), we would accept $h_e = 0.2$. Kimura (1983, p. 143) argues that we should use n_{nuc} as 3.5 x 10^9 to approximate the mammalian genome, but we shall use 3.5 x 10^8 to allow for a large percentage of noncoding DNA. Finally we shall use .5 for the values of h^2 and L_T. With these assumptions, the selection coefficient is approximately 5 x 10^{-6}.

Such a low selection coefficient makes the probability of fixation close to the same order of magnitude as neutral evolution. The probability μ that a gene will be fixed by selection is:

$$\mu = \frac{S}{2N(1 - e^{-S})}$$

where N is the population size, and $S = 4N_e s$. If we assume that population size is 10^9 and N_e is 10^6, our calculated selection coefficient of 5 x 10^{-6} yields a fixation probability of approximately 10^{-8}. If $N = 1000$ and $N_e = 100$, the probability is still only 10^{-4}. These calculations suggest that genetic drift may be a major component of the change of gene frequencies of loci in control of a trait. It must be emphasized, however, that such calculations are only averages; strong selection may be operating at many loci. This calculation makes the unrealistic assumption that selection would be spread equally over all loci. Any history of strong directional selection will bias downward the degree of accumulation of genetic variability at a locus. The degree of organization of the genome may also affect such an overall calculation. There is little evidence for widespread organization, but the presence of functional multigene families complicates the picture (see below). Finally, the pleiotropic (effects on several traits) effects of genes may reduce the opportunity for fixing mutations at a locus in the population. Selection for an allele affecting one trait may be inhibited if the incorporation of the allele would cause the fixation of another deleterious trait.

Let us now consider a morphological trait. The genetic variance generated per generation for a wide variety of organisms is approximately 10^{-3} times the total environmentally controlled variance for the trait (Lande 1980a). For single characters, the genetic variance generated per generation is approximately 10^{-3} per gamete per generation per character. A few thousand generations of random drift are sufficient to cause significant changes in the genetical component of the phenotypic variance (Lande 1980a). Lande argues that, as the number of genes affecting a trait increases, the possibility for a continuous feeding of genetic variability for the trait increases similarly. Given the widespread observations of response to selection in various traits, natural populations would not be constrained to stay in the same place, except by stabilizing selection for intermediate phenotypes. Lande (1976) argues that a mutation-selection balance could maintain high levels of variability in natural populations in the face of strong stabilizing selection. Turelli (1984) argues that Lande's conclusions may depend upon unsupported assumptions of the phenotypic effects of mutation, per locus mutation rates, and the intensity of selection.

Since most morphological traits can be changed rather easily by artificial selection, stasis, the long-term constancy of an average phenotypic trait in a population, must be due mainly to a lack of net directional selection, in combination with stabilizing selection and canalization, exhaustion of required genetic variability, or genetic correlations with traits whose change would cause a compensating loss of fitness. Canalized developmental programs, favored over the generations by stabilizing selection, might constrain future variation (see Chapter 5). Aside from developmental constraints, long-term morphological stasis could involve two additional factors. First, a given morphology could be sufficiently restrictive that a loss or severe change of habitat would result in extinction. Therefore a major part of survival of living fossils would be the survival of their habitat and a probable lack of superior competitors. The similarity of sediment habitat would nevertheless maintain similarity of morphology. Stasis partially implies habitat restriction and habitat survival. Most habitat change is probably too extreme for the overall phenotype to survive. Stasis would, therefore, not be bound by any evolutionary constraint as much as it might be related to the blind alley created by habitat specialization.

The scenario here devised for stasis would involve a second crucial factor in evolution–the evolution of habitat selectivity. Of course, natural selection fashions the mechanisms of habitat choice, but habitat choice keeps the organism in the same milieu. What if the evolution of habitat specialization is suppressed? This causes selection for a generalized phenotype, capable of surviving a range of environments. The organism has some set of adaptations to respond to a wide range of challenges. By implication, this range of response is nongenetic in nature. Habitat choice permits the organism to choose its own selective regime. It is, therefore, incorrect to think of stabilizing selection simply as a culling process of phenotypes. As the degree of habitat selection changes a continuous changing interaction develops between the selective effects of the environment and the organism's ability to choose where it will live. Thus, natural selection is not a disconnected process where phenotype–genotype and environment are determined independently.

Disruptive Selection favors phenotypes (Fig. 3.5) at two or more modes in a potentially continuous distribution, and acts against phenotypes in between the modes. This form of selection is important in a set of discrete environments, each favoring a discrete phenotype. In the short run, this form of selection is an efficient means of increasing allelic variance and is a mechanism for preserving polymorphism in a population. Disruptive selection can induce reproductive incompatibility and may be a mechanism for speciation (Thoday 1959). Combined with selection for host specificity, disruptive selection may be a mechanism for sympatric speciation.

Frequency-dependent Selection occurs when the direction or intensity of selection varies with the frequencies of genotypes in the population. The rare-male effect in *Drosophila* (Ehrman 1967) is a case in point when rare male phenotypes are favored in mating, but this preference disappears as the phenotpes become more common. Frequency-dependent selection can generate complicated evolutionary trajectories when the animal is faced by conflicting selection pressures. Consider the selection imposed simultaneously by predation and mate choice. If a brightly colored male morph were present in low frequencies it might be favored as a mate over its pale-colored competitors. It might also be conspicuous enough to be taken preferentially by a visually oriented predator. If predation were relaxed, the colored morph might increase in frequency. This would dampen selective predation, as all morphs would be similar in appearance. But colored morphs might lose their advantage as conspicuously rare potential mates (see Endler 1978).

The degree to which frequency-dependent selection occurs in natural populations is unknown. An unsettled controversy still exists over whether frequency dependence is an important force acting on variation on allozyme loci (see, for example, B. Clarke and O'Donald 1964; Kimura 1983, p.263). Ellstrand and Antonovics (1985) have demonstrated frequency dependence in selection for sexual forms of a grass at low density. Lewontin (1983b) claims that ever changing selection pressures, based upon shifting interactions between genotype–phenotype and environment, and between the genotype and the phenotype, are typical in evolution.

Selection and Geographic Variation

Huxley (1939) coined the term **cline** to emphasize geographic variation in morphological and genetic traits as a common feature of natural populations. Clines may follow a linear pattern of differentiation with geographic space, or may consist of two homogeneous populations that intergrade in a spatially restricted **step cline**. All genetically based clines must be expained as a combination of the processes of drift, gene flow, and selection. **Gene flow** is the process of successful movement of genes from one subpopulation to another, and is a function of dispersal and viability success when the dispersing genotype reaches the target population. A cline may result from **primary intergradation**. Here, some spatially varying environmental property causes geographic variation in selection. Alternatively, **secondary contact** involves recent geographic mixture of two formerly differentiated populations. The cline might be maintained by mortality of both differentiated populations in the transitional zone.

While clines are easy to document in either genetic or morphological terms, the shape or width of a cline varies in response to dispersal, genetic mechanisms of character determination, and selection intensities (Slatkin 1973; Endler 1977). For example, step clines may indicate a sharp ecotone or may simply be due to a smooth selection gradient with continuous but reduced dispersal. Step clines can be produced by selection pressures that are effectively unmeasurable in natural populations (Endler 1977). Increased dispersal distance will increase the width of a cline.

Clinal variation in natural populations often suggests that gene flow is restricted, at least at the loci controlling the traits in question. For example, many studies have documented extensive latitudinal variation for physiological traits (Levinton and Monahan 1983, Lonsdale and Levinton 1985). Such differentiation may retard gene flow between geographically contiguous populations. Poor physiological performance of dispersants into adjacent populations may enforce sufficient isolation to permit gene flow restriction across the entire genome. This may permit differentiation to occur at other parts of the genome where selection pressures normally would be too small to counteract the effects of gene flow from adjacent populations. Such a process, termed **accelerating differentiation** (Christiansen and Simonsen 1978; Levinton and Lassen 1978) might be the cause of differentiation over much of the genome.

Some clines are shaped continually by natural selection. G.C. Williams et al. (1973) examined latitudinal differentiation in juveniles of the American eel, *Anguilla rostrata*. The species breeds in a restricted area of the Sargasso Sea and the newly born juveniles migrate to a series of localities along the east coast of North America. Nevertheless latitudinal clinal variation was found in juveniles indicating a period of selection at the allozyme loci examined or at linked loci. In the blue mussel *Mytilus edulis* a steep cline into the estuarine Long Island Sound from open ocean habitats is probably under the control of active selection. Immigrations of juveniles marked by open ocean genotypes soon disappear, forming a steep cline. Selective mortality probably occurs with every larval settling season. Estuarine and open ocean populations show strong differences in tolerance to salinity stress (Levinton and Lassen 1978).

Multilocus differentiation may greatly enhance the degree of gene flow restriction across a cline. Barton (1983) considered a simple model, where weak selection against heterozygotes is assumed, presuming that alternative homozygous genotypes usually would be favored with selection in a gradient, if the location in the gradient were isolated from other populations. Introgression between adjacent populations depends upon the selection intensity at each locus, the rate of recombination between adjacent loci, and the number of loci involved in the differentiation. As the degree of selection increases, relative to recombination, the barrier to gene flow is increased. If selection intensity and recombination rate are both kept constant the barrier effect increases with the square root of the number of loci involved. The role of selection in retarding gene flow is strong if the number of loci involved is large.

Do Selection Models Predict Constant Evolutionary Rates?

Proponents of the punctuated equilibrium theory (Eldredge and Gould 1972; Stanley 1979) have characterized neo-Darwinian selection models as predicting rates of phenotypic evolution to be slow and constant. Although it may be self evident from the discussion above that this is far from the case, a few points would further clarify this major misconception. Indeed, the only model that predicts such slow and constant change is the neutral theory of molecular evolution, discussed below.

Phenotypic shifts, when they occur, are liable to be rapid and of short duration, relative to longer term periods of stabilizing selection. Mimicry, for example, should involve stabilizing selection as long as the model-mimic system remains intact. If a new model is introduced and becomes much more frequent than the previous one, a bout of intense directional selection will cause a rapid shift, assuming that available genetic variation permits the natural selective shift. Climatic shifts are often equally sudden and will select for rapid evolutionary change. The rapid phenotypic shift in the Darwin's finch, *Geospiza fortis*, mentioned above, is a good example.

Both mechanisms of genetic determination of character state and some selection models suggest that rectangular (constancy, sudden change, then constancy) morphological evolution is to be expected. When traits are discrete, but determined by threshold effects, change may appear sudden, but may be underlain by a large number of genes with an elaborate aggregate control over the discrete states of the phenotype. Thus sudden changes of traits such as presence or absence of new ossification patterns (Alberch 1983), or the appearance of new ornamentation (Reyment 1982a) may be the result of rather simple genetic mechanisms (Levinton 1983).

Several models have been proposed in recent years that predict rectangular evolution of traits with standard population genetic parameters (e.g., Petry 1982; Kirkpatrick 1982). If a trait is under polygenic control, and if there is environmental/developmental variance to the trait, sudden phenotypic shifts are to be expected if an environmental change involves (1) an increase in the relative height of an adaptive peak; (2) a decrease in the depth of a valley between two fitness peaks; or (3) a shift in position of two adaptive peaks, bringing them close together. A shift may also occur when an increase in overall mutational input to the phenotyic variance occurs, or an increase in environmental–developmental variance in the character occurs. Either of these two changes may cause a chance movement of phenotypes toward another adaptive peak. A sudden shift in the location of a single peak could also cause rectangular evolution. This overall selection scheme seems plausible and would invariably predict rectangular evolution.

Let us look at one simple example of the Wrightian adaptive landscape to demonstrate how neo-Darwinian theory is consistent with rectangular phyletic evolution. Consider that instance where a landscape has two adaptive peaks. If the mean population phenotype is located at one peak, nearly all random deviations will be insufficient to move it to an adjacent valley. But a rare random shift of this magnitude would result in an extremely rapid move (Fig. 3.6), either back to the original adaptive peak or toward the second peak (Newman et al. 1985). Indeed, gradual change would only occur if the landscape itself changes and if peaks shift gradually and unidirectionally. When the two peaks are static, and when the mean phenotype is near one peak, the expected time until a random shift between phenotypic adaptive peaks increases approximately exponentially with effective population size. By contrast, the expected duration of transition between the peaks is insensitive to effective population size, and the transition time between peaks is likely to be much too short to be detected by the level of res-

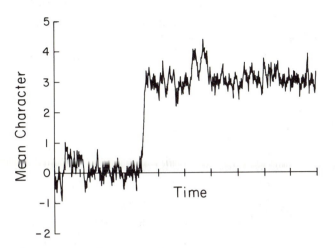

Figure 3.6: Random phenotypic evolution and the pattern of predicted evolutionary change between adaptive peaks (after Newman et al. 1985).

olution available in the geological record (Lande 1985). This would be true for either phyletic evolution or rapid phyletic evolution following or coinciding with a speciation event.

Thus there is no reason to believe that alternating periods of stasis and sudden change represent anything more than the standard expectations of typical selection models. There is nothing compelling about this pattern that points toward the punctuated equilibrium hypothesis that speciation is the driving force of morphological evolution. Douglas Futuyma (personal communication 1984) has pointed out to me a possible exception to this conclusion. Suppose that adaptation to each of two ecologically distinct environments entails a complex of genetically independent traits. Reproductive isolation can prevent recombination between them, so that populations can diverge further in gene-frequency and phenotypic space. Thus speciation can act to stabilize multilocus genotypes and permit more extensive divergence.

Stochastic Components of Evolution

Genetic Drift and the Shifting Balance Theory

Thus far, I have spoken of evolutionary change principally as a deterministic process involving natural selection. Random change in the genotypic constitution of populations–**genetic drift**–may be a strong component of genetic change. If we have a series of subdivided populations, such random forces might result in a series of populations, with alleles of a locus differing as the result of drift. The time required to fix an allele by drift is roughly the order of the effective population size. Thus, drift cannot be a very important force in even moderate-sized populations, unless there is no selection and no long-term gene flow from other populations. As mentioned above, rough calculations suggest that selection coefficients at any one locus involved in a polygenic trait may be small and comparable in effect with random forces. The aggregate effect of drift on a phenotypic character could cause random character change although this seems unlikely in natural populations (Lande 1976).

The possibility of drift is an unsolved question of population genetics. To what degree does random evolution contribute toward adaptive evolution? This is the question approached by Sewall Wright's **shifting balance theory** of evolution. Wright (1932) conceives a set of subdivided populations, among which divergence is the result of between-generation sampling biases in combination with fluctuations in selection coefficients. Because of this drifting, some populations cross fitness "valleys" and approach new peaks, which they "climb" rapidly by positive natural selection. The set of peaks and valleys comprises the **adaptive landscape**. Those populations that fortuitously reach these peaks spread at the expense of other less fortunate demes. Wright's shifting balance theory is thus a theory of adaptation, but one including random change (Provine 1983).

Although the shifting balance theory invokes random change, it is in a way more wedded to an optimalist viewpoint than would be a more standard model of natural selection. Normally, a natural selection model would predict the climbing of a peak in local phenotypic space. R.A. Fisher (1930) argued that any departure from this peak by the fixation of a major mutation would most likely move the population off the peak. By contrast, Wright envisions a series of peaks with differing heights, but with intervening valleys sufficiently shallow to permit drift to permit a movement of a population from one peak to another. But as this process occurs, it will be increasingly more difficult for populations to leave a local peak as its amplitude increases (see Templeton 1982). Thus the shifting balance theory can be seen as a

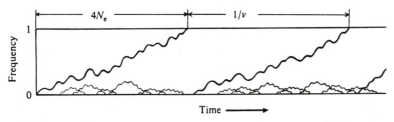

Figure 3.7: Behavior of mutant genes following their appearance in a finite population. N_e is the effective population size and v is the mutation rate. (from Kimura 1983).

mechanism maximizing the probability that a population will increasingly approach better adapted phenotypes via this process of drift and hill-climbing. Templeton (1982, p. 111) characterizes Wright's ideas: "adaptive evolution is far more efficient when natural selection is not in sole control even though, paradoxically, natural selection is the only force actually causing adaptation." Unfortunately, there is no clear evidence that Wright's model operates in natural populations.

Drift and the Neutral Theory

The neutral theory of evolution asserts that sequence changes in protein evolution result from random gains and losses of alleles in populations (see Kimura and Ohta 1971; Kimura 1983). In a finite and isolated population, as genes are transmitted from one generation to the next, random sampling effects result in the loss and gain of frequency of variants (Fig. 3.7). If a pregnant female rearing two embryos founds a new population, and has only heterozygous progeny, then a variety of random accidents can result in the loss of one or the other genes in successive generations. The same phenomenon holds for larger populations, though the probability of loss and fixation of alleles diminishes to a large degree. Random loss or fixation of an allele depends upon population size, gene flow from adjacent populations, mutation, and selection.

To determine the nature of sampling effects we must measure the **effective population size**, which is an estimate of the number of individuals contributing gametes to the next generation and the relative contributions of gametic types by different sexes. Effective population size can be far smaller than absolute population size and is especially diminished when populations fluctuate widely. Under such conditions, the effects of random sampling error are largest during those times when population size is minimal. When populations fluctuate cyclically, effective population size is calculated as the harmonic mean of population size over the number of generations considered (Wright 1938). Strong bottlenecks diminish N_e still further (Motro and Thomson 1982). Variation in offspring number can diminish N_e. Effective population size also decreases with increasing skewness in the sex ratio, as the probability of random transfer of gametes diminishes due to the disproportionate contribution of genes to the next generation by the rarer sex. If a herd is dominated by one male, for example, its gametes reach the next generation disproportionately, relative to any of the female gamete types, since each offspring has a mother and a father.

The **neutral theory of molecular evolution** states that the rate of fixation of new alleles is mainly due to mutation and random fixation-loss processes (Kimura 1983). Although a large number of mutants may appear in populations, the majority will disappear by chance. This applies also to advantageous mutants, which can be lost when they first appear and are rare. The probability of fixation of an advantageous mutant with a one percent selective advantage is approximately 0.02 (Haldane 1927). Most neutral, mildly deleterious, and probably all deleterious mutants will not be fixed within a population.

The mutation rate at a given amino acid site is believed to be on the order of 10^{-6} per site per generation. Kimura assumes an **infinite allele model**: when a new mutant appears, it occurs at a new site in which mutant alleles are not already present in the population. Let v be the mutation rate per gamete per generation. Given the presence of $2N$ chromosome sets, $2Nv$ new mutants appear in each new generation. If u is the probability of any one mutant ultimately being fixed within the population, then we can calculate the rate of substitution, k, as:

$$k = 2Nvu$$

Given an assumption of an infinite number of new alleles that can be generated with mutation, and that the low probability of mutation precludes any significant backmutation, we can estimate u, which equals $1/(2N)$. The fixation rate is, therefore,

$$k = v$$

The time to fixation for a successful neutral mutant is $4N_e$. Assuming a model of random and possibly repeated mutation at a site, k for amino acids should equal $S_{aa}/2T$, where S_{aa} is the number of amino acid site differences between homologous proteins in two species, and T is the time of divergence.

Although the overall fixation rate should be constant, the pattern of fixation on a small time scale will be erratic. Fig. 3.7 shows that most mutants will be lost. A few will gain in frequency and will be fixed in the population. Since the fixation rate is predicted to be on the order of 10^{-9} per nucleotide site per generation, we would not expect to see a constant pattern unless we examined trends covering many millions of years.

The alternative **selectionist model** would predict the following rate of fixation, where s is the selection coefficient:

$$k = 4N_e sv$$

Effective population size is an important part of the fixation process. This would suggest divergent rates of evolution among taxa with characteristically different effective population sizes. This depends upon the assumption that each new advantageous mutation is unique. Variation in selection would also suggest strong variability in rates of evolution among taxa and at different times within the history of a clade.

Molecular Polymorphism and Evolution

Amino Acid and Nucleotide Sequence Evolution

The neutral theory predicts that the fixation rate of amino acid base pair substitutions should be (1) a constant; and (2) equal to the mutation rate per gamete per generation. The second prediction would apply to neutral mutations only. Since the large majority of mutations are probably deleterious, the fixation rate should be far slower than the mutation rate. Thus we are left with the prediction of constant rates of sequence divergence occurring against a background of comparatively rapid loss of deleterious alleles appearing as mutants. This theory, therefore, has the value of a fairly restricted prediction, inconsistent with predictions based upon natural selection, which should involve large fluctuations of evolutionary rates. If we could construct a detailed phylogeny for living organisms, we should be able to plot the time of divergence versus the degree of sequence difference. If the neutral theory is valid, the slope should be approximately linear with the infinite allele model and the value of the slope (degree of sequence divergence per unit time) should be, maximally, the mutation rate.

Several other predictions are consistent with, but do not prove, the neutral theory.

1. Constraint: Molecules with fewer functional constraints should evolve at a faster rate. This would be consistent with a selectionist model.

2. Silent Substitutions: Nucleotide substitutions that fail to change the coded amino acid should occur at a rate higher than those that do change the amino acid.

3. Pseudogenes and other noncoding sequences should evolve at a rate faster than coding sequences.

4. Rates of nucleotide substitution should be independent of the nucleotide type.

Certain problems arise in testing the model. First, an accurate fossil record would be required to estimate times of splitting. It seems likely that such accuracy is usually not obtainable. Second, the neutral theory is not entirely "neutral." Kimura acknowledges that a certain proportion of the amino acid sites on a protein are highly conserved is due to constraints on function determination. Certain sites have disproportionate importance as determinants of secondary, tertiary, and quaternary structure. As long as these important sites need not be absolutely fixed in amino acid content, we would expect that the rate of molecular evolution would vary, depending upon the proportion of sites that are crucial in structure determination, and changing interactions among the sites over time as new fixation events occur. As mentioned above, the rate of mutation would be an upper limit for the rate of fixation; most likely the fixation rate would be less than the mutation rate, depending directly upon the degree of functional constraint imposed by the various amino acid sites. Kimura's theory does *not* claim that all fixations are neutral, only that the majority are, and that most mutations are deleterious and, therefore, lost.

Overall, a fixation rate of 1×10^{-9} amino acid site y^{-1} seems to be a representative intermediate value for the data that are spread over

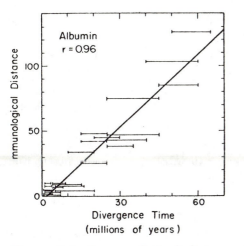

Figure 3.8: Immunological distance (determined from microcomplement fixation test) versus divergence time for albumins in mammals. The horizontal bar shows range of divergence time estimates (from A.C. Wilson et al. 1977, with permission from the Annual Review of Biochemistry, vol. 46, copyright 1977 by Annual Reviews, Inc.).

three orders of magnitude. Kimura (1983) names this unit the *pauling* for convenience of presentation. On the upper end are the fibrinopeptides, which change at a rate of 8.3 paulings; most conservative are the histones, which change at the rate of only 0.01 paulings. All rates are less than the mutation rate, as predicted by the neutral theory.

The question of constancy of molecular evolution is controversial. The notion of a molecular clock was first proposed by Zuckerkandl and Pauling (1965) and has been studied extensively by Allan Wilson and his colleagues. Microcomplement fixation has been used to measure immunological distance among proteins from different groups (e.g., Salthe and Kaplan 1966; A.C. Wilson et al. 1977). An index of dissimilarity, based upon the antibody needed to obtain a degree of fixation of complement with a test protein, relative to a reference, estimates the difference in terms of number of amino acid sites (A.C. Wilson et al. 1977). Direct amino acid and nucleotide sequencing have been used as well. Based upon these data, the summary of A.C. Wilson et al. (1977) demonstrates a correlation between elapsed time since divergence and degree of sequence divergence (Fig. 3.8).

A coup using the principle of constancy has been achieved in the study of man's ancestry. Traditional paleontological estimates placed the divergence between man and his African ape relatives back some 20 to 30 million years. Genetic studies based upon microcomplement fixation (Sarich and Wilson 1967, 1973), allozyme polymorphisms (King and Wilson 1975) and chromosome examination between man and chimp have demonstrated few differences, however. Using a clock calibrated from immunological differences among albumins, Sarich and Wilson (1973) reckoned the divergence time to be 5 ma. This estimate, first met with strong criticism among paleontologists, may prove to be closer to the mark than 20-30 million years ago.

The most commonly used model of evolution is a Poisson process, which assumes independence among sites, Markovian change, and temporal constancy in the parameters of the process (the clock assumption). On these specific points, the variance/mean ratios of divergence among lineages are usually too high to accept the clock hypothesis (Gillespie 1986). Too few points are usually available to disprove the possibility of fluctuation of rates. Using a parsimony method, Goodman et al. (1982) examined sequence differences of nearly 600 polypeptide chains of 250 species. They concluded that proteins are poor clocks and evolve at markedly nonuniform rates. Most notable are apparent accelerated rates of evolution near the times of major cladogenetic events, such as the evolution of globins during the basal radiations of the tetrapods (see Kimura 1983 for strong criticism of their methods). Rates of amino acid substitution, for cytochrome c, hemoglobins, and fibrinopeptide A among related lineages often show significantly heterogeneous rates (Langley and Fitch 1974; Ohta and Kimura 1971). If rates can be heterogeneous among lineages, there is no particular reason why they cannot vary within a lineage. Finally, Vawtor and Brown (1986) have examined mitochondrial and nuclear DNA sequence divergence in sea urchins and conclude that, while mitochondrial DNA divergence is similar to vertebrates, nuclear DNA divergence varies five to ten-fold between the two groups. Kimura (1983) suggests that changes in rates may reflect times when a constraint has been lifted, thus increasing the potential fixation rate of neutral substitutions.

Fig. 3.9 reveals a problem with constancy arguments based upon times of divergence taken from the fossil record. The top diagram shows the number of amino acid substitutions between man and various other vertebrates, as a function of divergence time. A rather good

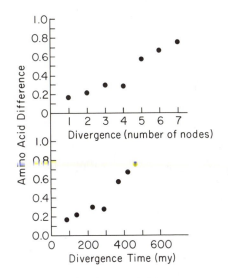

Figure 3.9: Comparison of α-hemoglobin amino acid divergence as a function of time of divergence (bottom), as compared with divergence as a function of number of nodes difference on a cladogram (top). Note that the former produces no strikingly better correlation than the latter (data from Kimura 1983, p. 67).

linear fit is obtained. The lower diagram, however, demonstrates that an equally good fit can be obtained simply by plotting divergence time versus the number of cladogenetic nodes between a given taxon and man. Absolute time, therefore, does not improve the linearity and the alternative hypothesis, degree of cladogenesis, would be as consistent with the data. Constancy is a crucial empirical issue for the neutral theory. A mere positive correlation of time with degree of divergence is not sufficient, since it only signifies that divergent evolution has occurred. Selectionist models do not preclude continual divergent evolution, in response to continuous environmental change and changing interactions within and between molecules.

Another problem with the concept of linearity is the relationship of sequence divergence among proteins to absolute time. This might make sense if mutations accumulated in long-lived germ cells at a constant rate, but would be paradoxical if mutations occurred only at the time of gamete production. If the neutral theory were correct, we might expect the appropriate positive correlation to be between degree of divergence

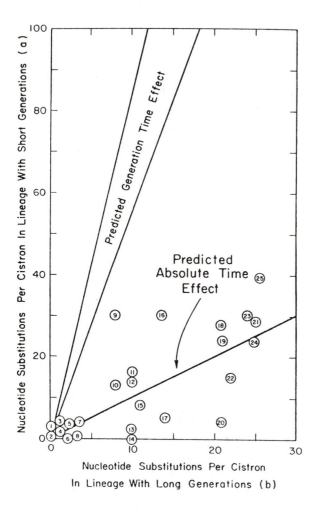

Figure 3.10: Molecular divergence, and its fit with a model based upon absolute time, versus another based upon generation time (from A.C. Wilson et al. 1977, reprinted with permission from the Annual Review of Biochemistry, vol. 46, copyright 1977 by Annual Reviews, Inc.).

and number of generations, since fixation rate can only be measured with the latter time scale. Fig. 3.10 demonstrates that absolute time is the best predictor of sequence divergence; generation time is a poor predictor. Kimura (1983) argues that the overall inverse relationship between generation time and population size might tend to equalize the absolute rate among groups, since the time to fixation is equal to $4N_e$. A species with longer generation time would likely have a lower N_e, and, therefore, a reduced time between fixations. This would balance off the difference compared with a taxon with shorter generation time, but a correlated higher N_e. Kimura thus must invoke variations in N_e to make his model fit the data, in contrast to the original, "purer" version of the neutral theory.

Some evidence may resolve the paradox of an apparent correlation of divergence with absolute time, but not with number of generations. Wu and Li (1985) have examined synonymous and nonsynonymous nucleotide substitutions in the coding regions of 11 genes from rodents and man. For synonymous substitutions, the substitution rate is about twice as high for rodents, relative to man. The most direct explanation of this difference is the shorter generation time of rodents, though there is still a paradox that a difference of two orders of magnitude in generation time yields only a two-fold difference in mutation rate. It may be that the much larger effective population size of rodents may contribute to lowering the relative rate of substitution. These results are consistent with the fact that amino acid sequence and immunological distance, metrics which reflect non-synonymous substitutions, are the types of data that have failed to show a relationship between substitution rate and generation number.

While the data do not support the simple Poisson hypothesis, more complicated models, allowing for bursts of evolution, heterogeneity among lineages, and multiple-hit mutations might leave a residual that is clock-like. Certain sequences, such as pseudogenes, noncoding regions, and third codons might conform quite well to a clock. It is also possible that overall divergence rates in single copy DNA, most of which is noncoding, might conform sufficiently to a clock hypothesis (Sibley and Ahlquist 1983).

Other aspects of variation in rate of molecular evolution are consistent with the neutral theory (see Kimura 1983, Chapter 4 and references therein). The rate of evolution at the third site of a codon, the

site of maximal code redundancy (ca. 70 percent), is twice as high as
the first two sites. In notably conservative polypeptides, such as the
histones, the rate of evolution at the third site is no different than for
rapidly evolving proteins such as fibrinopeptides. Pseudogenes, such as
a mouse pseudo α-globin gene, evolve at a much more rapid rate than
coding relatives (Miyata and Yasunaga 1981). Finally, proteins with
fewer functional constraints evolve at a much more rapid rate than
those with more constraints (A.C. Wilson et al. 1977). Certain amino
acid sites are highly conserved, as might be expected from their role
in maintaining active sites for binding with substrates, maintaining
structure, etc. (Kimura 1983, Chapter 7). The problem is that these
facts are also consistent with a selectionist model, or some compromise
model that admits some neutral change combined with selection.

Molecular Polymorphism–Selection or Drift?

Patterns of Variation   The neutral theory of evolution (Kimura and
Ohta 1971; Kimura 1983; King and Jukes 1969) also hopes to account for the surprisingly large amount of genic polymorphism found
in natural populations, first discovered by Lewontin and Hubby (1966)
and Harris (1966) for *Drosophila* and man, but subsequently found
for diploid organisms in general (Nei 1983). Kimura (1983) views this
observed variation as a reflection of the continual process of neutral
evolution. The polymorphism observed today is taken to be a steady
state of random appearance, loss, and fixation of molecular variants.

Molecular polymorphisms in natural populations present a paradox. If the balance model of genic heterozygosity is valid, then individual fitness increases with increasing heterozygosity over all loci. But
this pattern of fitness encumbers a large overdominant load as inferior
homozygotes segregate and appear in the population each generation,
along with the fitter heterozygotes.

If all of the loci are independent, then fitness is said to be **multiplicative** among the loci. This can generate seemingly intolerable
segregational loads upon a population (Lewontin and Hubby 1966).
In the case of a two-allele locus, where the selection coefficient, s, for
both homozygotes is less than unity, the segregational load at the locus,
$L_s = s/2$. Given n loci, the selective elimination per individual due to
segregational load is $\exp -sn/2$. If $s = .01$, and $n = 2000$, the selective elimination is $\exp -10$, which is roughly 1/22,000 (Kimura 1983,

p. 28). This suggests that, if selection occurs by death of the less fit prereproductive homozygotes, each individual must produce roughly 22,000 offspring to maintain a constant population size. Although this is well within the range for many plants and invertebrates that shed millions of gametes, it certainly cannot be maintained by mammals or even by crustacea such as *Hutchinsoniella*, where females lay only a few eggs. It is unlikely, moreover, that the obviously high mortality of invertebrates corresponds to anything other than the ecological dangers of successful dispersal and habitat selection.

This paradox depends upon the assumption that all of the loci contribute independently to fitness, which may not be true (King 1967; Milkman 1967; Lewontin 1974). As the number of heterozygous loci per individual increases, fitness may reach a plateau beyond which no increase in fitness would be expected. A plateau in fitness may also be reached at a minimum number of heterozygous loci, below which fitness would decrease no further. This fits our intuition, since the variance of reproductive output or mortality in most populations seems too small to accommodate the variance in fitness among genotypes predicted by the multiplicative fitness model. Even though a 2000–locus heterozygote is unlikely to be found, we would expect large variations in egg output among species with the expected range of sampled genotypes, and yet the narrowness of range is more impressive when conditions are kept constant. Lewontin (1974) argues that the upper threshold may be plausible, but cites evidence indicating that, in *Drosophila*, multiple chromosomal homozygotes at morphological loci are "less" fit than predicted by the multiplicative model. This may not apply to the gene level, however. In any event, even the presence of an upper threshold weakens the load argument (see Wills 1981, for extensive discussion).

Fig. 3.11 shows the distribution of heterozygosity for vertebrates and invertebrates. Invertebrates appear to have a greater mean and variance of heterozygosity. The neutral theory is successful in predicting the relationship between the average heterozygosity and the interlocus variance in heterozygosity (Fuerst et al. 1977). It is quite difficult to use these data to test other aspects of the neutral theory, however, since an estimate of N_e is required. The neutral theory predicts heterozygosity, H, to be (Kimura and Ohta 1971):

$$H = \frac{4N_e v}{4N_e v + 1}$$

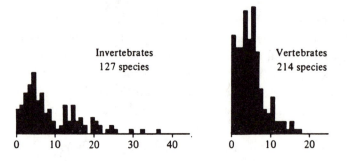

Figure 3.11: Distributions of the average heterozygosities per locus among invertebrate and vertebrate species (from Kimura 1983).

Ayala (1972) argues that if N_e equals 10^9, then H should be nearly 100 percent! Since N_e is known so imperfectly, almost any ad hoc counterclaim will weaken Ayala's criticism. Kimura (1983, p. 256) suggests a Pleistocene refuge hypothesis to invoke a bottleneck of much small N_e in the recent past. While this may be true, we are unable to take such arguments as definitive, as they are only attempts to solidify an argument with extrapolations consistent with a favored theory. A similar problem arises with Lewontin's (1974) attempt to refute the neutral theory by stating that the range of heterozygosity in natural populations implies an absurdly narrow range for N_e (factor of 4), given that we assume v to be constant among species. We usually have no idea of the magnitude of N_e. This makes for arguments but not for solutions.

In some 30 taxonomically diverse species, where estimates of population size can be made, Nei and Grauer (see Nei 1983) show that heterozygosity is less than that predicted by the neutral theory. Since N_e will be less than population size, the fit might be improved with a good estimate of N_e. Nei and Grauer argue that a model invoking overdominance demands even more heterozygosity than the neutral theory. But Ayala's criticism of a poor fit cannot be so easily dismissed, since directional selection may be the major process that reduces the degree of polymorphism, relative to the expectations of the neutral theory. Selectionists do not necessarily believe in universal overdominance at allozyme loci. Varying selection pressures in different microhabitats is a useful alternative hypothesis.

Selectionist models presume that protein polymorphisms are regulated by environmental variation. Two approaches might be taken. First, overall genetic variation might be related to aspects of environmental variability (see Hedrick et al. 1976). Thus heterozygosity would be predicted to increase in variable environments, where heterozygotes might be effectively fitter than either "specialized" homozygote. Second, functional biochemistry might be used to relate kinetic differences among allozymes to their distribution in natural populations.

A series of population cage experiments in varying and constant environments present an unclear picture of the relationship of environmental variability to heterozygosity at enzyme coding loci. Powell (1971) demonstrated that populations of *D. melanogaster*, maintained in cages variable in temperature, food, and light, had greater heterozygosity than those maintained under constant conditions. This was interpreted as support for the role of balancing selection in maintaining polymorphisms. Nei (1980) criticized these results, noting that heterozygosity had declined in *both* types of treatments, only more so in constant environments. A more recent study repeated Powell's work with *D. melanogaster* and found no difference in heterozygosity between variable and constant treatments (Yamazaki et al. 1983). This study differed in that starting populations were much larger and were maintained for six years prior to sampling. This suggests that Powell's populations were in sufficient linkage disequilibrium that selection on the allozyme loci was in response mainly to closely linked loci. Such an effect would be minimized with the much larger and longer maintained starting populations of Yamazaki et al.

Heterozygosity at allozyme loci has also been related to direct measures of fitness. A broad array of studies in recent years demonstrate a correlation between allozyme heterozygosity and somatic growth rate, developmental homeostasis as estimated by bilateral symmetry, and oxygen consumption (summarized in Mitton and Grant 1984). These results parallel a long and well known correlation between outbreeding and fitness in crops and breeding animals. In crops, the positive correlation between fitness and heterozygosity is said to be related to the establishment of a fitness-enhancing gene complement, established when two inbred strains are crossed (Jinks 1983). Mitton and Grant (1984) argue that this may be true of highly inbred breeding stocks, but is probably rarely true of more outbred natural populations. They suggest three possible hypotheses: (1) Enzymes mark blocks of chromosomes and are linked fortuitously to genes that directly control growth and development; (2) heterozygosity is correlated inversely with the degree of inbreeding–increased heterozygosity ,therefore, is correlated with reduced inbreeding depression; (3) enzyme heterozygotes confer an advantage in fitness. This might be due to the versatility of having two enzyme types instead of one in mediating metabolic pathways (Haldane 1930).

The first hypothesis would predict the presence of linkage disequilibrium which is not found extensively in protein polymorphisms in natural populations. Our measures of linkage, however, are insufficient to detect effects of closely linked loci. Eanes et al. (1985), for example, were able to show that any fitness differences in the *G6PD* locus in *Drosophila melanogaster* disappeared when the genetic background, including closely linked elements, could be subtracted. Aquadro et al. (1985) found extensive linkage disequilibrium in the *Adh* gene region of *Drosophila melanogaster*. Given the evidence for activity differences, as determined both by presence of transposable elements and single nucleotide site differences within the structural gene, linkage could well be an important component, despite the large effective population size of *D. melanogaster*. One cannot be sure that oysters or mussels are the well outbred populations they might seem, since deficiencies of allozyme heterozygotes in bivalves are rather commonplace (Levinton and Koehn 1976). The interesting successes in establishing functional differences among allozymes (see below) may point to the efficacy of the third hypothesis, but this fascinating issue remains open at present. Certainly heterozygote superiority at the enzyme level might explain the maintenance of polymorphism. But one must remember that polymorphism in the bacterium *E. coli* (Milkman 1973) argues against such inherent superiority. Since this variation comes packaged in distinct multilocus genotypes (clones), it is likely that recombination and selection are not generating a plethora of genotypes culled by natural selection (Selander and Levin 1980). One could only argue here for selection for the most heterozygous clones.

Overall environmental correlations between heterozygosity and environmental variability have been either difficult to interpret or are consistent with the neutral model. Levinton (1973, 1975) found a positive correlation between heterozygosity and environmental variability, in an among species comparison of bivalve mollusks. The correlation applied to one locus, glucose phosphate isomerase, but not consistently to the other, leucine aminopeptidase. Gooch and Schopf (1973) found no significant difference in heterozygosity between species living in variable continental shelf environments and the physically much more constant deep sea. K. Nelson and Hedgecock (1980) were able to correlate overall allozyme heterozygosity with certain environmental variables, but it is not clear how selection would be working at the enzyme loci themselves.

Geographic variation is difficult to employ in tests of the neutrality theory, mainly because geographic patterns can be explained equally well by many competing hypotheses. In some cases, however, discordant patterns of geographic variation among loci cannot be consistent both with gene flow and drift. Allele frequencies among loci may be constant, change in a step-like cline, or change in a smoother cline, over the same geographic distance (Christiansen and Frydenberg 1974). Genetic drift cannot explain such patterns. Concordant clines in different species would argue for selection (e.g., P.R. Anderson and Oakeshott 1984).

Geographic variation in protein polymorphisms can result from random fluctuations in subdivided populations (Selander and Whittam 1983). Populations of the terrestrial pulmonate snails, *Helix aspersa* (California) and *Cepaea nemoralis* (Pyrenees), show among-locus homogeneity in among-deme variance of allele frequency. Random drift among demes would predict among-locus homogeneity in differentiation among loci. In the Pyrenees, geographic differentiation at allozyme loci of *C. nemoralis* seems to best reflect a recent history of population spread, rather than natural selection. Allele frequencies show geographic variation, but there is no correlation with obvious environmental parameters such as temperature and altitude. Shell banding and color, however, are correlated with climate (J.S. Jones et al. 1980) and, in turn, seem unrelated to the allozyme differentiation.

Functional Studies The most direct test of the neutrality hypothesis would involve a systematic study of the functional biochemistry of allozyme phenotypes and a relationship of the phenotypes to overall fitness. The allozyme literature is mainly dominated by field surveys, geared either toward systematic studies, or toward elucidating the degree of geographic differentiation. But geographic patterns may be

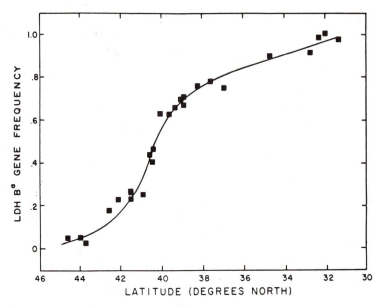

Figure 3.12: Geographical variation in an allele at the *LDH-B* locus of *Fundulus heteroclitus* (from Powers and Place 1978).

consistent with drift, migration, or selection. Unfortunately, few excellent functional studies have been done and negative evidence is unlikely to be reported with the same degree of enthusiasm as positive results. Moreover, it is nearly impossible to randomize the genetic background to the degree that viability differences among allozyme mutants can be argued as under the control of selection at that particular locus. One always worries about closely linked loci. Technical problems, relative rarity of useful geographic patterns in organisms amenable to experimentation, and unfamiliarity of evolutionary biologists with biochemistry have greatly hampered progress in this field (see Koehn et al. 1983 for a summary).

Kimura (1983) leads the reader to believe that there are no convincing examples of fitness differences for allozyme loci. To a degree this is true (Koehn et al. 1983); it is difficult indeed to randomize the genetic background, or even to expect that other loci will not interact with the ones under investigation. Nevertheless some intriguing studies ought to make even the most diehard neutralist lose a little sleep. I choose only two as examples, but Koehn et al. summarize about ten more, which are convincing to variable degrees.

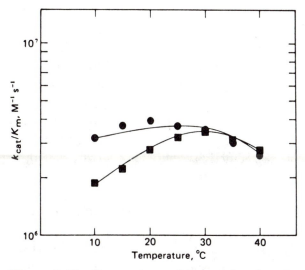

Figure 3.13: Comparison of interpolated and measured estimates for LDH k_{cat}/K_m at pH 7.50 for two genotypes of the killifish *Fundulus heteroclitus* (from Place and Powers 1978).

A north-south cline (Fig. 3.12) in coastal waters of the eastern United States has been found for a two-allele polymorphism at the heart-type lactate hydrogenase-B locus (*LDH-B*) in the killifish, *Fundulus heteroclitus* . The *LDH-B-b* allele is common in colder northern waters while the *LDH-B-a* allele dominates in warmer southern habitats (Place and Powers 1978). Reaction velocities for enzymes derived from *LDH-B-b* allele homozygotes were higher at lower temperatures (10°C) than for enzyme derived from homozygotes for the alternative allele (Fig. 3.13). ATP levels were also correlated with *LDH-B* genotype; concentrations were found to be lower in the *LDH-B-a* (southern) homozygotes.

ATP is an allosteric modifier of hemoglobin and, therefore, must affect muscular oxygen supply. At 10°C, the critical swimming speed (CTS–speed at which fish are exhausted) was found to be 3.6 body lengths sec^{-1} for homozygotes for the *LDH-B-a* allele (cold), while CTS was 4.3 for homozygotes for the *LDH-B-b* (warm) allele (DiMichelle and Powers 1982). The difference disappears at 25 degrees. The low-temperature advantage of the *LDH-B-b* allele seems related to the elevated correlated concentrations of ATP that reduce oxygen affinity and allow easier delivery of oxygen to starved tissues. This seems to be a mechanism to accelerate the activity of a cold-water fish, to compensate for its otherwise poikilothermic faithfulness to a fixed temperature-metabolism-activity relationship. Such latitudinal compensation is common for physiological characters of coastal species in latitudinal gradients (Levinton and Monahan 1983). If the northern population could not swim faster at the same low temperature, then fish would spend much of the year in a moribund state, unable to gather food or mate. But this increased activity must have a cost, leading to progressive adaptation for reduced acceleration in increasing temperatures.

A similar relationship between physiology, enzyme function, and a whole-animal manifestation of viability was found in a study of cell volume regulation in response to salinity change, in the the intertidal copepod *Tigriopus californicus.* This species occurs in temporary high tidal pools on the west coast of the United States, where evaporation often greatly increases the salinity, and, hence, the osmotic stress. To counteract water loss, cell volume regulation is accomplished partially with the aid of regulation of the concentration of free amino acids.

In *Tigriopus* a polymorphism has been investigated at a glutamate-pyruvate transaminase (*GPT*) locus (Burton and Feldman 1983). This enzyme is involved in the synthesis of alanine, an amino acid important in volume regulation when animals are first exposed to high salinities. One allele has significantly higher activity in both alanine synthesis and catabolism, resulting in a prediction of greater osmotic flexibility for copepods that are homozygous for the allele, than for the alternative allele. With increased salinity, as predicted, homozygotes for the "salinity-variation" allele accumulated alanine more rapidly and had a lower rate of mortality. As in the case of LDH-B in killifish, a lack of genetic analysis precludes a consideration of the effect of other loci.

Although these two studies are superficially convincing, one must remember that one cannot exclude the possibility of closely linked loci that regulate the activity of gene produce or produce gene products that are allosteric modifiers of the proteins under investigation. If several such genes are involved, then the importance of these studies lies much more in their value as studies of adaptation than in their pertinence to the neutral theory.

A confusing picture has developed for the alcohol dehydrogenase polymorphism, so intensively studied in *Drosophila melanogaster* (see summary in van Delden 1982). North-south clines have been observed in widely scattered regions. Two common alleles, named *Adh^f* and *Adh^s* for their fast and slow electrophoretic mobilities, differ by one amino acid. The slow allele tends to predominate at lower latitudes. The enzyme is believed to be involved in detoxification of environmental alcohols, but some uncertainty exists over this putative function. Genotype-dependent differences have been found in the michaelis constant for binding to alcohols (McDonald et al. 1980), but expected gene frequency clines near areas of alcohol stress, such as wineries, are not always found (e.g., McKenzie and McKechnie 1978). Moreover, Gibson et al. (1979) attempted selection experiments under high environmental ethanol concentrations. They found significant polygenic adaptation to alcohol resistance, but no increase as was expected for the *Adh^f* allele. Now, *in vivo* ethanol metabolism studies have shown negative results (Middleton and Kaczer 1983). The fast allele is associated with greater alcohol dehydrogenase activity than the alternative allele (B. Clarke 1975), and also may be found in increased frequency in wine cellars (Briscoe et al. 1975). Overall, the picture is confusing, despite (or because of?) much work in many laboratories. Effects on *Adh* locus enzyme activity seem to be the result of both the presence of transposable elements in the neighborhood of the gene and a single nucleotide difference within the structural locus itself. The latter seems to explain the higher activity usually found in the fast allele gene product (Aquadro et al. 1985).

Some allozyme studies have demonstrated fitness differences, but these grossly overestimate the total variation in fitness, if extrapolated to even a modest number of loci. As mentioned above, the fact that alleles do not segregate independently of others will call into question many of the studies that ignore fine scale genetic structure. Work on several loci in *E. coli* (Dykhuizen and Hartl 1983; Dykhuizen et al. 1984) and on the *G6pd* locus in *D. melanogaster* (Eanes et al. 1985) suggests that fitness differences are unmeasurable except in extreme environments, once the genetic background is subtracted. Worse than that, null alleles for enzymes commonly fail to reduce fitness (Langley et al. 1981). This result calls into question the fitness differences of as high as ten percent that are commonly estimated for allozyme loci (see Eanes 1987).

The neutral theory is thus plausible on the grand scale and seems to explain many large-scale distributions of data, such as near constancy of divergence rates, higher rates of evolution of genes and nucleotide sites with fewer functional constraints, and the predicted correlation between average heterozygosity and the variance, using an infinite allele model. Also, between-locus differences in heterozygosity are largely explained by gene size (Koehn and Eanes 1977). This is consistent with the neutral theory as larger genes constitute larger targets for mutation. There is a large degree of recorded heterogeneity of divergence rates, however. Also some intensive studies of enzyme function tend to support the hypothesis of functionally important differences among allozymes. Of course these studies lack estimates of the exact contribution of the variation to fitness, but they are suggestive. It might be argued that these are rare instances, and were only selected for further investigation because they showed suggestive geographic or temporal distributions. The large majority of protein variation may be of the neutral sort. Unfortunately the limited experimental data and broad predictive latitudes permitted by theory leave an undefinitive answer to the neutralist-selectionist controversy.

It may seem out of place to discuss genic level variation and functional studies of enzymes at some length in a treatise concerned with macroevolution. But it is at this level of concern that variation, the stuff of evolution, begins to show its effects on phenotypic evolution. At present we are ignorant of the exact role of genic level variation in evolution. Indeed, the controversies among population geneticists of the past few decades have centered around the relation of genetic polymorphism to evolution. The central battleground has shifted to protein polymorphisms, mainly because some hope of a direct mechanistic answer might be obtained. Gene level evolution may occasionally be the direct explanation for larger scale phenotypic evolution. Without such mechanistic studies, and without an awareness of this level by neontologists and paleontologists, a crucial aspect of evolution will have been missed.

There is an even more important lesson to be learned from this controversy. We are still in an embryonic stage of understanding the meaning of fitness and its relationship to defined parts of the genome. We still cannot present a comprehensive summary of fitness differences among genotypes at allozyme loci. The current estimates are too high as they fail to account for the genetic background. Without an understanding of variance in fitness, how can we predict rates of evolution with selectionist models? This problem becomes no simpler when we examine morphological traits, which are usually controlled by large numbers of loci. Complex genetic structures only complicate the task of estimating the variation of fitness in natural populations. But the most fundamental law of evolution is the relationship of variance in fitness to evolutionary rate. Can our estimates of rates of evolution, at any level, be useful when we can't quantify variation in fitness?

DNA Sequence Homogeneity: Description and Evolutionary Mechanisms

DNA sequences in multigene families within a species often vary to small degrees. Strachan et al. (1984) have examined sequence variation of the melanogaster group of *Drosophila* in the noncoding families *360* and *500*, named for the approximate usual number of nucleotide sites. The *360* family is usually restricted to the X chromosome, while the *500* family is more dispersed. In the *360* family, among-individual variation in sequence is 0.6 to 4.7 percent, while interspecific variation averages much higher, over a range of 21 to 35 percent. The same applies for the *500* family: intraspecific variation is 0.9 to 3.5 percent, while interspecific variation is 34 to 44 percent, an order of magnitude higher. Interspecific differences can be much lower, but the usual case is interspecific heterogeneity and intraspecific homogeneity. These patterns of variation in multigene families are common over many groups of plants and animals that have been investigated to date (see Arnheim 1983; Flavell 1982; Dover 1982, Dover et al. 1982; Hood et al. 1975 for summaries). Apparent homogenization has been found in globins, immunoglobins, histones, the chorion genes of silkworms, several classes of ribosomal RNA genes, and in many noncoding sequences or sequences of unknown function.

Current evidence suggests that within-species homogeneity occurs in families both of genes whose product is translated as a protein, as well as in nontranslated sequences that give rise to RNA products only, and in nontranscribed spacers between genes. Interspecific divergence, however, seems most common in sequences that apparently have no functional constraints over most of their length. In the toad *Xenopus* the 18S and 28S rRNA genes occur in hundreds of tandemly arrayed repeats. Intraspecific and interspecific homogeneity has been found. This may suggest a selective constraint upon the RNA sequence of the gene products. In the nontranscribed spacer region, however, intraspecific homogeneity in sequence is to be contrasted with extensive interspecific heterogeneity (Brown et al. 1972). A similar situation exists in a comparison of the rDNA transcribed regions of man and apes (Arnheim et al. 1980). The interspecific–intraspecific result is similar to the contrast found for the *360* and *500* families studied in the *melanogaster* subgroup of *Drosophila* mentioned above, if we remember that these latter families seem not to be functioning genes.

Gene family homogeneity presents a clear paradox to students of typical evolutionary processes. We usually presume that mutation, selection, and drift are the manipulators of variation in a population. This of course, would include sequences in noncoding regions. We would presume that noncoding regions would, if anything were, be a prime example of a target for neutral fixation of genes. But any standard model leads to immediate inconsistencies. For example, we might argue that a gene family, homogeneous in sequence over all members, arose by stochastic increases in copy number due to unequal crossing over and survival of the multiplied sequences. But, subsequently, accumulated mutations plus random fixation should lead to heterogeneity. The degree of heterogeneity should be the same in intraspecific and interspecific comparisons where the gene family is of similar size and chromosomal location. This is not true. We must believe alternatively that all families are new and have had no time for divergence, which is usually absurd, given their presence, often in large copy number, in many related species.

Natural selection would be an alternative hypothesis. All copies, even when numbering in the thousands, might be subject to such strong purifying selection that sequence homogeneity might be maintained. It is difficult to envisage the effects on the phenotype of the addition of one more mutant gene of a large gene family, all of whose members are making the same product. Could selection recognize the effect on the phenotype? Only after a degree of homogenization would the effect be large enough. A selectionist explanation is also inconsistent with intraspecific homogeneity in sequences that have no known function. Moreover, we would predict that among-gene divergence under a selective constraint would be approximately the same in intraspecific and interspecific comparisons. But relative intraspecific homogeneity and interspecific heterogeneity seem to be the rule; repeats by a variant member seem to occur.

Homogeneity in sequence among many copies suggests the action of an ongoing process maintaining homogeneity throughout gene families. Gabriel Dover (1982) has coined the term **molecular drive** to cover all processes of homogenization. Three processes lead to homogeneity: **unequal exchange, gene conversion,** and **duplicative transposition.** It may be that these processes are so different that the term molecular drive masks their diversity of process. The word "drive" is also unfortunate, as Dover intended the term to include, perhaps mainly, random processes of fixation. The term **concerted evolution** (see Arnheim 1983) has an equally unfortunate connotation of nonrandomness and directionality.

Unequal exchange, with survival of amplified chromosomal segments, would increase the number of copies within the genome. Unequal exchange is likely when there are multiple copies of a gene. Perfect register of sister chromatids at meiosis is unlikely (Kreiber and Rose 1986). Some evidence suggests that this process could lead to increased copy numbers when there is natural selection for increased amounts of the gene product (as in the amplification of metallothionene genes under metal toxicity selection), but it can also arise from random processes of gain and loss (Kimura 1983; Ohta and Dover 1983). This process is probably essential to the origin and maintenance of the large gene families and multiple copy DNA noncoding sequence families. As unequal exchange occurs, random gain and loss might result in the eventual predominance of one sequence.

Gene conversion is a process where one gene can apparently convert the sequence of another to its own sequence. The process probably involves mismatch repair of genes. It is thought to begin with the mismatching of two members of the same gene family during the imperfect pairing of two chromatids. This is followed by the replacement of the sequence in the homologous gene family member of one chromatid by that of another. This may explain similarities in sequence in gamma globins (Jeffreys 1982) and in other gene families. A biased gene conversion might maintain homogeneity among gene families, once family size has increased by unequal crossing over. The mechanisms of unequal exchange and gene conversion probably operate simultaneously to amplify and homogenize given sequences.

The mechanisms of unequal exchange and gene conversion seem most likely to occur between sister chromatids. The homogeneity among members of a gene family should therefore be greater within, than between, chromosomes. Ohta and Dover (1983), however, argue that dispersion of a multigene family onto two or more chromosomes has a quantitatively minor effect on the extent of identity among genes, unless the conversion rate between genes on nonhomologous chromosomes is low, or unless the number of nonhomologous chromosomes over which the genes are dispersed is large. Nonhomologous chromosomes do come into contact and exchange parts. This is known from classical cytogenetics to occur accidentally (see summary in M.J.D. White 1973), but such associations may also occur on a predictable basis. In germline cells of man, ribosomal DNA from two or three nonhomologous chromosomes may associate during rRNA synthesis by forming a single nucleolus. This provides the opportunity for intergene interactions. Unequal exchanges between the rDNA bearing X and Y chromosomes in *Drosophila* also underlie the cohomogenization of these two chromosomes (Coen and Dover 1983; Ohta and Dover 1983).

Transposition involves the movement of sequences from one site to another. It may involve excision of the sequence from the source site, or may involve copying and movement of the copy to the new site (duplicative transposition). The latter process increases the copy number. The TE families in Drosophila occupy 10 to 15 percent of the genome. Copy number seems restricted in maximum size, which indicates either a severe reduction of fitness or the accumulation of repressors of transposition with increasing numbers of TE's. As mentioned above, the presence of TEs within genes seems to affect fitness negatively. But there is some beneficial potential; the Ty-1 TEs in yeast can act as up and down promoters of gene transcription. This effect might favorably influence fitness.

Replacement of pre-existing families and the subsequent fixation of new variants by unequal exchange, gene conversion, and duplicative transposition can all occur by random processes. It is possible, however, that some of these processes are highly biased in direction. This is true for the P transposable element, which is highly infective. But gene conversion and unequal exchange might be biased processes as well.

In duplicative transposition, the increase of copy number will occur gradually in the population. Migration, random mating, and reassortment of chrosomosomes in diploid organisms might randomize the number of copies relative to the rate of homogenization by molecular drive. Therefore, the variance in copy number will be low, but distributed as a Poisson distribution, among individuals at any one time (Ohta and Dover 1984). This is important if there were selective differences among the nucleotide variants. If the variance is high, then fitness differences as a result of differential copy number in the population would impose a large genetic load. The load problem of course is nonexistent if the homogenization event involves a neutral substitution. It is also unimportant if a superior new type is gaining in frequency over an extant type that already serves the individual well, unless some interaction (e.g., overdominance load) among loci occurs.

Certain genes may be more vulnerable to homogenization. We might expect that those whose products interact with many other genes and their respective products might have conserved nucleotide sequences. This may be supported by differences in activity among taxa of RNA polymerase I, which transcribes only the rRNA genes, and RNA polymerase II, which transcribes all of the messenger RNA coding genes. The latter, with extensive interactions with many important genes, has a series of highly conserved sequences, to the degree that an insect gene can be transcribed with a human transcription system. In contrast, extensive among-species differentiation is known among RNA polymerase I and interspecific cell hybrids lack the interspecific compatibility of gene and transcription systems shown for RNA polymerase II. The differentiation may represent the extent to which homogenization in a multigene family requires other interactive molecules like polymerase I to coadapt (Coen and Dover 1983; Arnheim 1983). Certain sequences, such as the H sequence, found in both *antennapedia* and *bithorax* complexes of *Drosophila* (McGinnis, Levine, Haten, Kuroiwa, and Gehring 1984), may also have such an explanation behind their conservation.

An interesting case arises when molecular drive is involved in a trait that is expressed in the external morphology. It is clear that this has happened, judging from some preliminary evidence on intraspecific and interspecific comparisons. We have noted above the effect of X-Y chromosome exchange on bristles of *Drosophila melanogaster*

(Coen and Dover 1983). Another example comes from the study of the chorion genes, a multigene family found in silkworm moth species (Kafatos 1983; Rodakis and Kafatos 1982). The chorion is a complex egg covering consisting of four distinct protein classes, laid down in a precise developmental sequence. In the silkmoth, *Bombyx mori*, new protein is found, corresponding to the acquisition and increase of a new gene probably by unequal exchanges. This involves a complex amplification of variant repeats (Kafatos 1983; Goldsmith and Kafatos 1984; Jones and Kafatos 1982). The new protein provides a final outer coat to the egg case, is involved in diapause, and decreases the oxygen supply to the egg.

Consider the case where gene number for a new protein increases by random or directional processes. In the silkworm, processes of amplification would at first be invisible from the culling processes of natural selection. A few genes producing a novel chorion protein would be unimportant against an overwhelming background of "wild-type." But eventually the novel chorion genes would become sufficiently abundant that the egg would be covered by a protein that might be useful in diapause and detrimental under other circumstances. At this point natural selection could take over and select for or against the new variant. The initial process of amplification is therefore a mechanism of generation of variability, albeit a novel one, and can therefore be reconciled with natural selection.

In summary, the notion of the homogenization process as a means of stochastic spread of a variant is likely to be a major part of the overall pattern of evolution. At the least, it is a mechanism that explains patterns of homogeneity and copy number change in multigene families. Its dynamics entail slow change throughout the population, involving the processes of unequal exchange, gene conversion, and duplicative transposition. Finally, it is a source of change that must be accommodated by other interacting genes, and could be a source of variants which might be important in phenotypic evolution.

The Main Points

1. Genes typically consist of a structural coding region, and a series of leader sequences that modulate transcription. Some of these leader sequences code for polypeptides that can influence gene action.

2. Nucleotide sequences–introns–are often found within the structural gene. These sequences are eventually spliced out, so that a functioning protein can be synthesized. Other sequences, such as transposable elements, can be found in the neighborhood of the gene and may affect gene expression. In some cases the movement of transposable elements may occur at high frequency, and the mutation rate is correspondingly enhanced.

3. New genes can arise by the gradual alteration of extant genes, gene duplication followed by evolutionary change of one of the daughters, and by fusion of genes. Some genes lose function and nucleotide sequences change as a result of genetic drift; these are known as pseudogenes.

4. Previously, it was believed that distinct regulatory genes were important in evolutionary change. Recent progress greatly complicates the picture. We now know that all genes have regulatory regions, and that phenomena in development may be controlled by increasing titer of substances that could be modulated by a wide variety of mechanisms.

5. Nearly all morphological traits examined thus far are controlled by several genes, even if a few may dominate in effect. Some traits may be under the control of a large number of genes, but phenotypic states are discrete as a result of concomitantly evolved threshold effects. Such traits are said to be canalized. Despite the polygenic determination of these traits, the effective absence of phenotypic variation makes a response to normal selection pressures unlikely. Strong selection, however, can break past the threshold.

6. Natural selection can be stabilizing, directional, disruptive, or frequency-dependent. Despite the evidence for widespread selection in nature, genetic variability is rampant; hence the common strong response to selection on traits in laboratory stocks.

7. There has been a renewed focus on patterns of evolution of morphological traits. The variable effects of trait determination, patterns of selection, and the effects of stochastic variation allow for a wide variety of responses. Despite the claims of the founders

of the punctuated equilibrium theory, population genetic theory does not predict slow and even rates of evolution.

8. The neutral theory explains the large degree of molecular polymorphism in nature as a result of a balance of mutation and genetic drift. An important ancillary prediction is the molecular clock, which is run ultimately by the mutation rate. There is clear evidence of random change in pseudogenes, third codons, and a variety of nonfunctioning nucleotide sequences. The pattern is less clear for functioning genes. Some studies show strong evidence for the maintenance of polymorphism by natural selection. The rate of molecular evolution does seem to increase as the "importance" of the amino acid sequence is of lower consequence. Where the rate of nucleotide substitution is slow, it may well be that sequence is more closely associated with protein function.

9. Molecular polymorphisms present a confusing picture. Some fitness differences seem associated with protein function. The estimates of fitness differences, however, seem too high when extrapolated over the genome, or when compared to studies where the genetic background can be controlled. We probably are too ignorant of fitness differences at allozyme loci, or any others, for that matter, to be able to make predictions on the patterns and rates of evolution at this level.

10. Many sequences come in multiple copies and derive from single ancestors. These families may range from a few to hundreds of thousands of copies, and can include functioning genes (e.g., the globins) and nonfunctioning sequences. The combined processes of unequal crossing over and gene conversion may explain the duplication and homogenization of sequences observed in many families. Duplicative transposition of transposable elements may also explain the widespread occurrence of some sequences.

Chapter 4

Speciation and Trans-Specific Evolution

> To sum up, interspecific differences are of the same nature as intervarietal.
> *J.B.S. Haldane*

Speciation: A Mystery of Product and Process

Over a century after the publication of Darwin's *The Origin of Species*, we can still rightfully plead much ignorance about the nature of the speciation process. I define speciation in diploid organisms as the establishment of two groups of organisms that subsequently fail to produce fertile hybrids, or produce hybrids sufficiently infertile that natural selection against mating between the populations is likely. In his classic *Evolution: The Modern Synthesis* (1940), Julian Huxley defined two classes of speciation mechanisms: one associated with divergent natural selection operating on populations in separate and different habitats, and the other involving genetic mechanisms (e.g., polyploidy) largely independent of adaptation, and occurring more rapidly than the first. This dichotomy is not so different from Templeton's (1981) divergence-transilience distinction. The intervening period has been marked by considerable controversy over the relative frequency, and often the efficacy, of the various mechanisms.

The difficulty in understanding the speciation process is understandable as no one has ever observed it in nature, nor has anyone identified definitively what exactly happens as one species gives rise to two descendants. The time scale is part of the problem. The process must usually take longer than a lifetime of human observation. Several cases of active speciation in fishes have been noticed, but the time scale could be several thousand years (e.g., Brooks 1950). Although new species of facultatively autogamous plants may arise in only a decade or two, regularly outcrossing annuals probably take hundreds of years and woody plants probably take thousands to speciate (Stebbins 1983a). Apart from the autogamous case, these time spans are too long for a biologist, but they are also far too short for a paleontologist. For example, Williamson (1981) described a "sudden" species-level change in a snail lineage that occurred over a time span of 10–40,000 years. Cisne et al. (1980) described a "punctuative change" that could be bracketed within a 200,000-year interval. While some cases of annual varves might make the study of sudden change accessible (e.g., Bell and Haglund 1982), most geological sections preclude detailed sampling even of 10,000-year units (see Chapter 7). Speciation is thus in between the two practical levels of resolution.

If we can't easily study speciation in action, we should at least be able to describe why any two species differ. What is the reason that two populations are reproductively incompatible? Even if there is an obvious primary explanation of current incompatibility, one cannot be sure that this was the original incompatibility that caused speciation. Indeed, the theory of reproductive character displacement (Brown and Wilson 1954) argues that premating isolation mechanisms (see page 183) result from selection against hybridization, due to the presence of yet other incompatibilities, ecological or genetic. Present-day hybrid zones between species are usually quite ancient, and it may no longer be possible to learn of the mechanisms that caused the interspecies divergence in the first place.

We are usually left with the unsatisfactory alternative of describing current differences. This survey can, however, approach some interesting questions:

1. Can interspecific differences be extrapolated from intraspecific differences? It is useful to distinguish between **polymorphism**, where genetic variants are freely intermixed within a population, and **polytypism**, where certain variants are fixed in local populations. It is the extrapolation from polymorphism, to polytypism, to distinct species, that would constitute an intra-interspecific extrapolation hypothesis.

2. Is the pattern of intra-interspecific difference similar for different types of comparisons?

The first question stems from the hypothesis that a speciation event involves a unique change, never found in intraspecific variation (e.g., Goldschmidt 1940). If the extrapolation is always possible, then speciation may not be the principal cause of morphological evolution, as claimed by the advocates of the punctuated equilibrium hypothesis, but an effect of overall genetic differentiation within populations of a single species. The second question has meaning when we use one form of change (e.g., genic divergence) as a scale against which another form of change (e.g., morphological) can be gauged.

Intraspecific and Interspecific Variation

Intraspecific and Interspecific Chromosomal Variation

Modes of Chromosomal Differentiation Chromosomal polymorphisms are generated by a large variety of chromosomal interactions. Rearrangments result from exchanges when two chromatids cross each other and exchange segments, or from breaks that develop while the chromosome is being stretched on the meiotic spindle. Breakage is usually followed by reunion with the same or another chromosome (M.J.D. White 1973, Chapter 7). Rearrangements involve a variety of translocations, fusions, and reciprocal exchanges. Many such changes cause strong reductions in gametic output. In the case where a reciprocal translocation between two chromosomes results in one daughter with two centromeres and another lacking centromeres, the latter will be lost for lack of a spindle attachment. Chromosomal translocation polymorphisms can impose a high price on gamete viability. In the case of reciprocal translocation heterozygotes, gametes could have a duplication or a deficiency which, depending upon the importance and interactions of the segment, may result in reduced viability or complete nonfunction of the gamete. In some dipterans, mechanisms exist for shunting aneuploid gametes into polar bodies. On the other hand, paracentric inversions– resulting from two breaks on the same side of a centromere and inverting of the sequence–seem to have little effect on viability in dipterans. Robertsonian fusions result when an acrocentric chromosome loses its short arm and fuses with another chromosome, usually another acrocentric. This seems to occur often with little negative effect, indicating that the short arms may contain mainly heterochromatin (noncoding repeated DNA).

Owing to the heterozygote disadvantage of chromosomal polymorphisms we do not usually expect much chromosomal polymorphism; the spread of a new chromosomal variant will thus be difficult. Fixation of such a new variant will likely occur only in quite small populations by genetic drift (e.g., Wright 1940; Lande 1979a). This may not apply to some rearrangments such as: paracentric inversions in *Drosophila*, rearrangments where crossing over is suppressed, situations where aneuploid gametes are lost in polar bodies, and cases where the effects on gametes are rather modest.

Even where the spread of a new variant is suppressed by heterozygote disadvantage, gene flow may occur between adjacent populations through combination with other chromosomes where no such heterozygote problem exists. As newly fixed rearrangements accrue in the genome, the possibility for complete incompatibility with other populations may increase. In *Mus*, the effects of multiple chromosomal heterozygotes are cumulative. Thus a species with a transient polymorphism might differentiate into a series of chromosomally distinct populations, i.e., a polytypic species. As each separated population differentiates further, a series of new species might arise.

The chromosomal inversion polymorphisms of *Drosophila pseudoobscura* have been described in great detail in the famous *Genetics of Natural Populations* series, written by Theodosius Dobzhansky and his colleagues. Most populations are polymorphic for paracentric inversions on the third chromosome, and geographic variation, ascribable to selection, is common (Fig. 4.1). Although the mechanism of speciation is unknown in this group, inversions seem to have been passed on to new species. *D. persimilis*, for example, shares the *Standard* inversion with *D. pseuodoobscura* (Olvera et al. 1979). The Treeline inversion is widespread throughout the range of *D. pseudoobscura* and may have given rise to other variants. Genealogies can be established using inversions, in much the same way that other characters can be used (e.g., M.J.D. White 1973 Chapter 11).

Inversion polymorphisms of *Drosophila pseudoobscura* vary in frequency spatially and temporally as the result of natural selection. Although spatial variations in inversions were originally thought to vary only according to stochastic processes (Dobzhansky and Wright 1941),

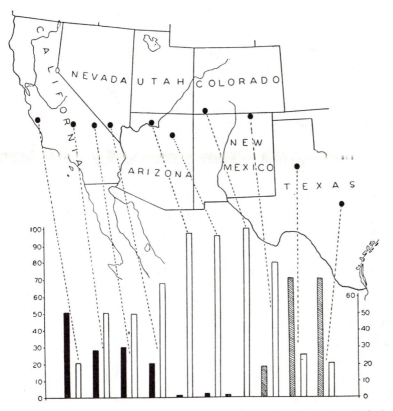

Figure 4.1: Frequencies of Standard (black), Arrowhead (white), and Pikes Peak chromosomes in populations of *Drosophila pseudoobscura* in the western United States (from Dobzhansky 1947).

later studies demonstrated correlations with altitude, season, and climate (Dobzhansky 1943, 1947, 1948*a, b*). Some viability differences among individual organisms bearing different inversions may relate to the effects of temperature and crowding (Birch 1955). In some cases strong regional differences in seasonal variation are related to climatic differences (Crumpacker and Williams 1974). In the cactus loving species, *D. pachea*, a latitudinal cline of inversion frequency is strongly correlated with climate and a change of host plant (B.L. Ward et al. 1974). No evidence proves conclusively that these differences result from characters of the chromosomes *per se*, as opposed to alleles for specific genes or groups of genes, carried by coincidence.

Dobzhansky (1947) argued that chromosomal polymorphisms were the result of heterozygote (heterokaryotype) superiority. This was based upon excesses of heterozygotes relative to Hardy-Weinberg expectations, and the convergence to intermediate frequencies of two-inversion populations in laboratory cages. Frequency-dependent selection may also explain this convergence. Studies of overall performance tend to show that heterozygotes are superior, or at least equal in performance, to homozygotes. For example, Moos (1955) showed that homozygotes for the *Chiricahua* (CH) inversion were inferior in general physiological performance to *Standard* (ST) homozygotes, which were subequal in performance with CH/ST heterozygotes.

Heterozygote superiority seems to occur only when the inversions come from the same locality (Dobzhansky 1948*b*). This may indicate that superiority is conferred by favored gene combinations and not any innate superiority of chromosomal heterozygosity.

Cases of chromosomal polytypism have been found in a wide variety of species with limited dispersal ability and usually small population size, such as in small (particularly subterranean) mammals, some Diptera, and in flightless grasshoppers (M.J.D. White 1973; Key 1974; Bush 1975; Nevo 1982). Hybrid zones between chromosomal races, when present, are usually extremely narrow (e.g., Key 1974; Nevo 1982). This may testify to reduced gene flow between adjacent populations due to poor viability of chromosomal heterozygotes.

The distinction between subspecies and species status is quite difficult in situations of chromosomal polytypism. In the Central American Peters tentmaking bat, *Uroderma bilobatum*, two cytotypically characterized populations overlap in a small embayment in Honduras (Baker 1981). The two cytotypes differ by one terminal translocation and two fusions (2N = 44, 38). Gene flow is restricted, although the hybrid zone is claimed to be quite wide. Backcross cytotypes between the two populations occur over a band of 400 km. Nevertheless, of 11 known polymorphic allozyme loci, nine have markers unique to each of the populations (Greenbaum 1981). This suggests very restricted gene flow between the two populations, despite the claimed widespread "leakage" of chromosomes across the barrier. Barton (1982) suggests that the pattern of variation seen in the differentiation zone is consistent with a hybrid zone being maintained by hybrid unfitness. The apparent great width of the zone can be explained by the obviously large dispersal distance of bats.

Chromosomally distinct races (species?) are especially common in the house mouse, *Mus musculus*. The differentiation probably occurred in the past few thousand years and was strongly influenced by migrations of *M. musculus* with its commensal, man. Hybrid zones are narrow and hybrids often show strongly reduced fertility or malformations. The typical karyotype in most localities consists of 2N = 40 acrocentrics. An isolated population in eastern Switzerland, however, is fixed for 2N = 26, suggesting the fixation of 7 fusions. This sort of isolation is common in Europe (see Capanna 1982). Populations with various degrees of fixed Robertsonian fusions have been found in the Rhaetian Alps, the Appenines, and in Sicily. Each region can be divided into populations with fixations of different chromosome numbers. In the Rhaetian populations, four metacentrics characterize all subpopulations. Other fusion types are found in successively more restricted subpopulations, until a given unit population is characterized by unique metacentrics. This suggests a process of isolation of a primordial population, followed by further substructuring into unique groups. Chromosomally distinct regions occur as islands in a sea of all acrocentric mice.

Populations of the mole rat *Spalax ehrenbergi* in Israel are also strongly polytypic, or show recent speciation (Nevo and Shaw 1972). Robertsonian fusions result in four major cytotypes (2N= 52, 54, 58, 60), which come into contact along narrow (2.8 to 0.7 km wide) zones. Paleobotanical evidence suggests that migration and subsequent differentiation in Israel occurred from 250,000 to 10,000 years ago. The four cytotypes are distributed along a north-south aridity gradient and are morphologically similar except for an inverse relationship between body size and environmental temperature. The subterranean habits and small population size of the mole rat are very conducive to such differentiation. Although premating isolation mechanisms usually exist between the cytotypes, one case still shows only postmating isolation (see p. 183). This may suggest that postmating incompatibility preceded the evolution of premating recognition mechanisms (Nevo 1982).

Fixation of Chromosomal Variants in Populations   Lande (1979a) has analyzed the dynamics of chromosome fixation and the likely deme sizes during fixation events. In a randomly mating deme, the rate of fixation of rearrangements with large heterozygous disadvantage is minuscule unless effective population size is very small. Given known rates of fixation of chromosomal variants and spontaneous rearrangement rates, Lande estimates that long-term effective deme sizes must be on the order of 30 to 800 individuals for a wide variety of mammals, lower vertebrates and insects. This suggests the ubiquitous occurrence of genetic drift in animal populations. The spread of chromosomal variants has probably occurred usually by random local extinction and colonization. Therefore, most locally fixed variants disappear when their poulation disappears via random extinction. Due to their relatively minimal effect on heterozygotes, inversions and Robertsonian fusions seem to predominate over reciprocal translocations.

There is an apparent inconsistency between the rapid speciation of mammal populations, which seems to have occurred over time spans of thousands of years, and the average fixation rate of new rearrangements in known species of mammals, which is maximally on the order of one per lineage per million years (Bush et al. 1977). It may be that most chromosomally differentiated populations are geographically restricted and thus have a high probability of extinction. The more extensive populations, such as the all-acrocentric "standard" populations of *Mus musculus*, may survive owing to their abundance. In the random case, the probability of spread of a given variant over the entire species should be $1/N$, where N is the number of demes. This probability, multiplied by the probability of fixation of a variant within a deme, would give the probability of spread through the entire species. To take the house mouse as an example, we can imagine that the number of demes must be sufficiently high that the probability of survival of the all acrocentric karyotype is assured. New populations with novel genotypes will, for the most part, go extinct.

Chromosomal incompatibility may not prove to be the primary mechanism of isolation. Genic mechanisms of postmating isolation may be important even in chromosomally polytypic populations. In the rampantly speciating Hawaiian drosophilids, many groups of species are chromosomally monomorphic. On the other hand, extensive regional differentiation in chromosomal variants can occur with minimal reproductive isolation, as in the pocket gopher *Thomomys bottae* (Patton 1972). The phenomenon of hybrid dysgenesis in *Drosophila*, discussed above, is now known to result from genic disruption by transposable elements. Crosses between different strains result in accelerated mutation rates and chromosomal disruptions (Kidwell et al. 1977). While the phenomenon is unrecognized outside of *Drosophila*, otherwise unknown alleles may be found in hybrid zones in mammals (Hafner et al. 1983), and in other groups (Whitt et al. 1973; Sage and Selander 1979; Woodruff and Gould 1980; Woodruff and Thompson 1980). This may also reflect intragenic recombination in hybrid zones that may generate novel alleles.

A strong case can be made for gene-based sterility as a mechanism of postzygotic incompatibility. In *Drosophila melanogaster* fixation of new genes in a population has a considerable probability of ensuring reduction of viability of offspring produced from crosses with other populations. At the X chromosome, fixations at about 9 percent of the genes would result in major sterility problems in females (Gans et al. 1975). In considering genes that affect more subtle aspects of mating behavior and reproduction, the potential for incompatibility is greater. Male hybrid sterility is explained similarly by several to many genes, particularly at the X chromosome (Coyne 1984). Cases of extensive chromosomal differentiation are known where reproductive compatibility is high. In the goodeid fish *Ilyodon furcidens* extensive variation in the number of metacentric chromosomes occurs within a single river basin, despite minimal allozyme divergence and full viability of laboratory crosses and backcrosses (B.J. Turner et al. 1985).

The extrapolationist hypothesis is consistent with the chromosomal differences observed among species. The mechanisms of geographic differentiation are easily related to speciation mechanisms involving the establishment of postmating isolation. Variants can be traced across subsequently differentiated populations and species (Olvera et al. 1979; M.J.D. White 1973). There is no intraspecific-interspecific dichotomy. It is not clear, however, that chromosomal incompatibility is a major genetic mechanism of speciation.

Comparisons with Morphological and Allozyme Divergence   Chromosomal evolution may be gauged against other measures of differentiation. Extensive chromosomal race formation can occur with little concomitant allozymic differentiation. In the peripheral relict pocket gopher species *Geomys tropicalis* major changes in chromosome number are not accompanied by an unusual degree of allelic substitution (Selander et al. 1974). This seems to be common in cases of extensive chromosomal differentiation (e.g., Nevo 1982; Greenbaum 1981). An interesting exception can be found in Rocky Mountain populations of *Geomys*, where extensive among-population chromosomal differentiation is accompanied by strong allozymic differentiation (Penney and Zimmerman 1976). Such major differences within one genus suggest a degree of unpredictability of an allozyme-chromosome correlation.

The rate of chromosomal variant fixation is inversely proportional to body size (Bengtsson 1980). This may relate to the longer generation time or to greater vagility of larger mammals. With relatively infrequent reproduction and few young per brood, a chromosomal abnormality would cause a significant loss of offspring. In fecund animals, loss of a few young might be matched by increased health of survivors, or accelerated production of a successive brood. Thus change might be accommodated more easily in small-bodied species. Gene flow among larger and possible geographically wide-ranging mammals is probably not an explanation for reduced divergence, since behavioral deme structuring is common among larger-bodied species (Bengtsson 1980). Although deme structuring does not guarantee reduced gene flow, it can permit such a restriction.

Relation to Morphological Evolution

Chromosomal polytypism can be correlated with geographically related reproductive isolation. The chromosomal differentiation itself probably represents random fixation in relatively small populations with low vagility. The accumulation of such fixations in isolated populations may contribute to reproductive isolation. Despite the widespread occurrence of chromosomal races, the evidence does not support any extensive concomitant morphological differentiation. For example, the three classic morphologically recognizable species of the mole rat genus *Spalax* ranging from Russia to north Africa represent at least 30 karyotypes, most of which seem to be distinct species (Nevo 1982). Although the chromosomally distinct eastern Switzerland population of *Mus musculus* was once recognized as a different species on traditional grounds as *M. poschiavanus* (see Capanna 1982 and references), numerous other isolated races are morphologically indistinguishable except by karyotype. Great karyotypic disparity among species, with little morphological differences can be observed in some rodents, foxes, insectivores, horses, and gibbons (see cited literature in Marks 1983).

I should emphasize that morphological correlations *can* be found with karyotypic differences. As an example, body size in grasshoppers seems related to the presence or absence of given inversions (M.J.D. White et al. 1963). This could be due to the presence of a few contributing genes, however. Most rearrangements seem unrelated to morphological differentiation. In *Drosophila*, rearrangements found in natural populations do not show any relationship to characters of taxonomic significance (Spieth and Heed 1972).

Chromosomal evolution has been claimed to be a cause of morphological evolution (Bush et al. 1977; A.C. Wilson et al. 1974b; A.C. Wilson et al. 1977). Under this hypothesis, chromosomal variants are regarded as having gene regulatory and morphological significance. If speciation is a cause of, or a concomitant process with, chromosomal evolution, then we would expect a correlation between speciation rate, karyotypic diversity, and morphological evolution. Rate of chromosomal evolution is assumed to be related to a measure of karyotypic diversity among extant species. Following the method of Stanley (1979) Bush et al. estimated speciation rate by taking the number of extant species and the time of origin in the fossil record for the group, and calculating a splitting rate assuming constant dichotomous splitting. There was a positive correlation between the rates of chromosomal evolution and of speciation in a study of extant species of various reptilian groups and orders of mammals. From the correlation, they inferred that karyotypic evolution is a source of morphological evolution.

Although this is possible, a casual inspection reveals some surprises. Horses have the highest speciation rate and corresponding rate of chromosomal evolution. But the living forms whose karyotypic differences are extensive, constitute a rather morphologically homogeneous group of mammals. A consideration of the fossil record of closely related horses does not increase the morphological diversity very much. Although correlated changes in body size, relative length of limbs and hypsodonty characterize the grazing equine genera, morphological similarity is very strong, to the degree that minor reinterpretations of features in fossils cause the systematic position of various groups to change radically (Woodburne and MacFadden 1982).

In contrast, the morphologically diverse Cetacea are lowest among the mammals in karyotypic diversity and speciation rate. The sei whale, *Balaenoptera borealis*, and the common dolphin, *Delphinus del-*

phis, have nearly identical karyotypes ($2N = 44$), yet they must have diverged 40 to 50 million years ago (Arnason 1972). Although the fossil record is too sparse to make an estimate of speciation rate, the Pinnipedia are similarly chromosomally homogeneous (Arnason 1972). As Bengtsson (1980, p. 38) notes

> a relationship between karyotype evolution and the evolution of regulatory genes is, at most, of highly indirect and weak nature.

The correlation observed by Bush et al. probably relates to the expected population genetic processes at reduced population size that occur during the speciation process. Karyotypic divergence is thus probably an effect of speciation, or even an occasional cause of reproductive isolation. It is not likely to be a major cause of morphological evolution.

Correlations between morphological and karyotypic evolution may occur, but only coincidentally to the speciation process. As an example, cladistic analyses of chromosomal banding patterns from 48 species of cryptodiran turtles, combined with fossil-based methods for estimating rates of karyotypic change, demonstrate that karyotypic evolution was twice as fast in turtle groups arising in the Mesozoic as in more recent splits and involved different forms of rearrangements. The deceleration in rate of change is correlated with decelerated morphological change. Some chromosomes have remained unchanged for at least 200 my (Bickham 1981). While chromosomal changes might have been involved in adaptive changes, it is likely that the initial rapid radiation of turtles was accompanied by divergent morphological evolution, which must have involved speciation among geographically separated populations. In other words, speciation could have been an effect of divergent adaptation; the tempo of karyotypic evolution would probably have tracked speciation. Chromosomal change, therefore, may not have been the cause, but was more likely the effect of evolutionary radiation and speciation.

Cherry et al. (1982) used a metric, D, to estimate proportional differences in homologous skeletal measurements, and found no substantial differences among species within genera of frogs, lizards, and mammals. Generic longevity of mammals is substantially less than for the others, and this might suggest that speciation rate accelerates mammalian morphological evolution. Alternatively, phyletic morphological evolution might be greater for mammals. A given speciation event might therefore entail more divergence. Using Van Valen's (1973a) compilation, we can calculate the ratio, R, of D within a genus to the number of species per genus. If one assumes that the average number of extant species in a genus is proportional to the number of speciation events required to generate the species richness, then the divergence/species richness ratio gives a rough estimate of the relative amount of change realized per speciation event. One gets the following: Mammals: $R = 1.90$; Lizards: $R = 1.08$; Amphibia: $R = 0.76$. A given speciation event in mammals may, therefore, entail more morphological change than in reptiles or amphibia. The relationship between the rates of morphological divergence and of speciation may therefore be coincidental, or morphological evolution may accelerate speciation.

This would solve a paradox well known to evolutionary biologists: the greatest amount of divergent evolution of morphology occurs near the beginning of the fossil record of a group. But this cannot be a time of maximum absolute number of speciation events, if any sort of exponential model of species increase applies. Thus it is not the shear number of speciation events, but a qualitative difference in evolutionary mode that increases the degree of divergence per speciation event near the beginning of the history of a radiation. Usually this seems correlated with the prior elimination of a competing group by a mass extinction (see Chapter 8). This qualitative difference undercuts the idea that speciation per se accelerates morphological evolution.

Allozymes and Interspecies Divergence

We have discussed in Chapter 3 the presence of extensive allozyme polymorphisms in natural populations. Allozyme polytypism is also common among many species (e.g., Koehn et al. 1976; Christiansen and Frydenberg 1974; Schopf and Gooch 1971), though geographic homogeneity in allele frequency is also common (Prakash et al. 1969; Ayala et al. 1972). A large number of studies, mainly in *Drosophila* species, permit estimates of the degree of intraspecific and interspecific differentiation.

Nei's (1972) index of genetic distance is commonly used to estimate allelic divergence at allozyme loci. If I_x is the average sum of the squares of the allelic frequencies over all loci for species x, I_y is the corresponding sum for species y, and I_{xy} is the average over all loci of the sum of the crossproducts of allelic frequencies for a locus, then distance D is:

$$D = -log_e I$$

where

$$I = I_{xy}/(I_x I_y)^{1/2}$$

I_x and I_y measure the average probabilities of identity over all loci of two randomly chosen homologous genes from species x or y, while I_{xy} is a measure of the average probability of identity of two randomly chosen homologous genes from the two species.

The data on the willistoni group of *Drosophila* (Ayala et al. 1974) suggests a smooth transitional increase in D from geographic populations to morphologically different species. This seems to hold generally for animals (Nei 1975): Average $D = 0.00$-0.06 between races; 0.00-0.20 between subspecies; 0.1-1.5 between sibling species; and 0.1-2.5 between nonsibling species. The species barrier does not seem to be a special level of rectangular divergence in genic identity. If these levels of organization can be regarded as stages in speciation, divergence seems to continue smoothly after speciation has progressed from the sibling species stage to a later stage of morphological divergence.

The smooth transition within a group of *Drosophila* might suggest an overall correlation between speciation rate and allozyme divergence among related groups. The fish groups Cyprinidae (minnows) and Centrarchidae (sunfishes) demonstrate that this need not be true. The North American centrarchids are depauperate in species, while rapid speciation has been the rule for the cyprinids (Avise 1977; Avise and Ayala 1976). In a study of 24 gene loci average $D = 0.63$ for centrarchids, while average $D = 0.65$ for the cyprinids. This suggests a lack of relationship between speciation rate and divergence rate. The neutral theory would predict similar divergence among species if time scales since divergence were similar. Smith (1981) has criticized this work, as the fossil record suggests a higher speciation rate for centrarchids than previously believed.

There is no necessary relationship between morphological divergence and allozymic divergence. It is true that a good correlation exists in many species of mammals and fishes (Avise 1976). But many exceptions suggest that this may be due to rather constant correlated rates of morphological and allozymic divergence with time, with no causal relationship between the two sets of traits. In the desert pupfish *Cyprinidon macularius* significant among-river morphological differentiation is not accompanied by allozymic differentiation much greater than is usually found in intraspecific comparisons of other teleosts (B. J. Turner 1983). This suggests that morphological differentiation can be rapid and independent of an allozymic scale. A similar discordance exists between patterns of color and banding and allozymic differentiation in Pyrenees and Welsh populations of the land snail *Cepaea nemoralis* (Jones et al. 1980).

Morphology

It is difficult to summarize adequately the evidence for intraspecific versus interspecific divergence in morphology. In the case of chromosomes and allozymes, one has at least the confident feeling that identifiable markers can be traced across intraspecific and interspecific barriers. In the case of morphology, different parts of the genome can exert significant control upon a given trait. There is also no uniform criterion by which one can draw equivalence between any two external morphological traits. If intraspecific variation of color morphs in butterflies can be extrapolated to interspecific comparisons, what relationship does this have to wing shape, or to time of pupation?

Consider a character that has two different states in two different species. Is there a leap in character state that can only be explained by a speciation event, or can intrapopulation polymorphism be used in a simple extrapolationist model to explain polytypism and interspecies differences? Two possible approaches can be taken. First, if hybrids and F2 generations can be obtained between the two species, quantitative genetics can be employed to learn whether the difference between alternative character states is saltatory, or polygenically controlled with extrapolation possible from within-population variation. (Remember from Chapter 3 that discrete morphs can also be polygenically controlled.) Even without a genetic approach, much can be learned from a comparative biometrical study of morphological variation at the intraspecific versus the interspecific level. But if intraspecific morphological variance is much smaller than interspecific morphological differences, one can always argue that strong directional selection occurs during the speciation event. On the other hand, if the levels of variance are about the same, one can argue that yet other charac-

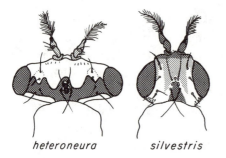

heteroneura   *silvestris*

Figure 4.2: Differences in head shape between two closely related species of Hawaiian *Drosophila* (after Val 1977).

ters are saltatory and one has not come across the "species-specific" characters.

Crosses between species and populations have been done extensively, particularly in plants where interspecific developmental incompatibilities are smaller than in animals. The minimum number of genetic factors controlling the trait is estimated by comparing the phenotypic means and variances in the two parental populations and the F1 and F2 hybrids and backcrosses. Polygenic control and the consequent possibility of extrapolation is often demonstrated by the intermediate phenotypic scores in hybrids and the expansion of phenotypic variability in the F2. In cases where intermediacy in the F1 is not found, threshold effects and polygenic inheritance usually turn out to be the rule (e.g., Wright 1934, 1935a, b; Green 1962). In most cases, the minimum number of genes for morphological traits is typically estimated as 5 to 10 with occasional values up to 20 (Lande 1983). Ten independent genetic factors were estimated to be operative in an analysis of tomato strains where two varieties differed about 56-fold in mean weight. As the haploid number of chromosomes is 12, the actual number of factors is probably larger, with some chromosome-level linkage.

Though we cannot make any universal statements, some interesting cases of interspecific variation demonstrate that the extrapolationist hypothesis is probably supportable for most trans-specific evolutionary changes. A remarkable case of extreme morphological differentiation has been discovered between two species of Hawaiian *Drosophila*, *D. heteroneura* and *D. sylvestris* (Templeton 1977; Val 1977). The pair are very similar by allozyme and cytogenetic standards, but are strikingly different in head shape (Fig. 4.2). A genetic analysis shows that at least 6 to 8 independent genetic factors control the phenotypic difference. The effects of the factors are predominantly additive, upon which is overlain a sexual dimorphism that is most likely connected by a sex-linked locus or loci whose expression is limited to males. The interspecific phenotypic difference may be quite important in premating isolation, but it could have easily evolved from intraspecific variability.

For the sibling species of *Drosophila*, genital morphology is the sine qua non of species-specific morphological characters. They often are the only means of diagnosis. Differences between species are discrete; otherwise they would not be good species characters! Coyne (1983) has recently performed a genetic analysis of genitalia differences among the siblings *D. melanogaster*, *D. simulans*, and *D. mauritania*, by substitution of different chromosomes in hybrids. Variation in genitalia is under the control of at least 4 to 5 genetic factors. There is no need to invoke any unique process in the morphological differentiation accompanying speciation.

Though species are morphologically distinct, one can find extensive regional differentiation, often equal in magnitude to interspecific differences. This can be shown, for example, in some species of the land snail genus *Cerion* on Caribbean islands (Gould 1969a; Woodruff and Gould 1980). Over distances of 100 meters, large changes in sculpture, size, and whorl number per unit size occur. Discrete, often major, variation is found commonly in species of marine snails, such as the genus *Thais* (Palmer 1985). In the three spine stickleback *Gasterosteus aculeatus* intraspecific differentiation is pronounced and of the same order as interspecific differences (e.g., Bell 1976, 1981). Within a lake in British Columbia, two probable species coexist that reflect extensions of intraspecific differences (G.L. Larson 1976; McPhail 1984). The benthic form has a heavier body, wider mouth, reduced dorsal spines and reduced lateral plates, relative to a limnetically specialized form. It is not clear that speciation occurred within this particular lake, but the body size, spine and plate polymorphisms are well known within other populations. The recent researches by Michael Bell (see Chapter 7) on a Miocene lineage demonstrate that typical intraspecific variation observed in living populations can be used as a framework to follow morphological change in fossil lineages effectively.

One example is of particular interest as it falls within the home territory of the macromutationist-speciation school. Mimicry in butterflies has been discussed above and shown to represent a polygenic system that evolved by accumulation of several new genes of varying relative effect on the phenotype. Can intraspecific variation be extrapolated to interspecific differences? Remember, this case was one of Goldschmidt's (1945) prime examples of the uniqueness of saltatory jumps.

Mimicry probably has the longest pedigree of any work integrating variation in natural populations with speciation. In 1862, Henry Walter Bates published a theory of mimetic resemblance among butterflies, stemming from his observations of intraspecific and interspecific variation in the color patterns of South American butterflies. Using the fabulous diversity of form found in Brazilian faunas, he was able to demonstrate a continuity between geographic varietal variation within a species and the common occurrence of small-ranging groups of species whose ranges were contiguous. To Bateson, this indicated that polytypism preceded speciation.

J.R.G. Turner (1981 and cited references) has investigated patterns of mimicry in the genus *Heliconius*, where Mullerian mimicry (model and mimic are poisonous) is the rule in both larvae and adults. The butterflies feed on passion flowers (Passifloraceae), which live in shaded forests. The genes involved in mimicry consist of a combination of genes of large and small effect. A large gene bridges a gap that permits the further evolution of stronger resemblance. Racial divergence within species is strong and is easily extrapolated to interspecific differences. The species pair *H. melpomene* and *H. erato* co-occur in a range of localities, each with its characteristic and quite different mutually mimetic color pattern. Laboratory crosses demonstrate complete interfertility among populations of the same species, taken from different locales. Nonmimetic relatives of both species have similar yellow and black patterns.

Brande (1979) investigated intraspecific versus interspecific variation within the genus *Mulinia* (Mactracea). *Mulinia lateralis*, for example, has a broad geographic extent from New Brunswick to Yucatan. It has given rise to one daughter species in Lake Pontchartrain, Louisiana (*M. pontchartrainensis*) and several related species also occur in the western hemisphere. Using discriminant function analysis, Brande found that the characters contributing to most of the among-locality variance within a species were also those important in among-species variation. This suggests that the features of the shell involved in intraspecific evolution are also those involved in the evolution of interspecies differences. Since shell characteristics are those expected to be crucial in bivalve adaptation (Stanley 1970) we can conclude that the speciation process is not particularly important here as a threshold in bivalve evolution. Similar results were obtained in an examination of the Miocene scallop *Chesapecten* (Kelley 1983*a*). Kelley (1983*b*) found that in some cases, characters most important in describing the variance within species were not those diagnosing differences between species. This proves little, since the "species" consisted of an ancestor-descendant series with no cladogenesis. How does one tell species apart, in this case, except by morphological change? Even in cases where true species are examined, finding such a discordance between intraspecific and interspecific variance could also indicate that times of unique ecological change induce changes in characters of otherwise low variation. Unfortunately, Brande's test applies only to confirming the continuity of intraspecific to interspecific variation. A lack of continuity yields an ambiguous result.

Brande's results follow those of other studies. B. Clarke and Murray (1969) studied variation in *Partula*, a genus of terrestrial snails found in the Society Islands of the Pacific. Though it was formerly believed that many species occupied Tahiti and Moorea, Clarke and Murray showed that only two species were present, with many individual races occupying a series of isolated valleys. Strong morphological differences may occur in direction of coiling, size, shape, and color–yet, many of the identified subspecies interbreed freely in the laboratory. There is good reason to believe that the differences among subspecies are due to genetic drift. In any event, the characters that may be used to distinguish among subspecies are the same that have been used to diagnose different species.

Brande's data on *Mulinia* allow a comparison of intraspecific versus interspecific variation. In general the degree of intraspecific variation among populations was less than that among species. This might be explained by either: (1) The power of the speciation process in morphological differentiation or (2) the passing of a sufficient amount of time to permit interspecific divergence to transpire via phyletic evolution. Apparently, the latter is the best explanation (Fig. 4.3) The recently

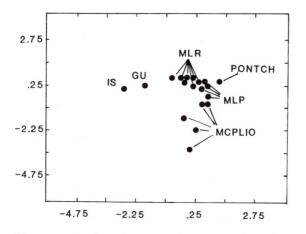

Figure 4.3: Discriminant function plot of first two axes for morphological variation within Recent populations of the bivalve *Mulinia lateralis* (**MLR**), some Pleistocene populations (**MLP**), its ancestor the Pliocene *M. congesta* (**MCPLIO**), its descendant *M. pontchartrainensis* (**PONTCH**), and other Recent species (**IS and GU**) (after Brande 1979).

derived *M. pontchartrainensis* is barely on the morphological fringes of its progenitor, *M. lateralis*. The Pliocene *M. congesta* seems to evolve gradually into its descendant *M. lateralis*. In contrast, seemingly more distantly related species are morphologically more distant as well.

Stasis is used by proponents of the punctuated equilibrium model as evidence for a centripetal force in evolution. Stasis is said to imply a set of

> genetic and developmental coherences that resist selective pressures of the moment and impose a higher level, or macroevolutionary, constraint upon changes within local populations

(Gould 1983*b*, p. 362).

This argument requires that (a) there be developmental and genetic sources of discontinuity; and (b) that these sources are mobilized mainly at speciation. As we have discussed here and will in Chapter 5, sources of discontinuity certainly exist. Our evidence, however, suggests that the sources are not associated with speciation. Yet species often are rather constant in morphology. E.E. Williams (1950), for

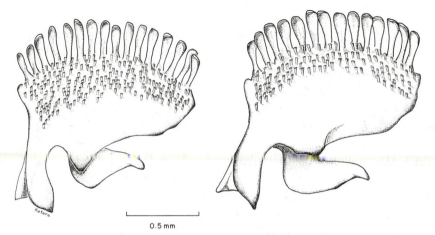

0.5 mm

Figure 4.4: Two divergent dentary morphs, found in different clones of *Poeciliopsis*. These morphologies are associated with different algal grazing behaviors (courtesy of Robert Vrijenhoek).

example, found little intraspecific variation in cervical articulations in turtles. But some interesting, and quantum, variation could be detected in comparisons among species; is this due to the sort of "resistance" suggested by Gould?

Consider pharyngeal tooth morphology in fishes. Although strong interspecific differentiation is present, intraspecific variation in pharyngeal tooth morphology is relatively slight. This might argue for a centripetal force within the history of the species. A major ecological or genetic crisis might be required to cause the evolution of new forms. This hypothesis can be falsified by examining morphological variation in tooth morphology among clones of unisexual fishes of the genus *Poeciliopsis*. Vrijenhoek (1978) has found extensive differentiation among clones for trophically significant differences in dentary morphology, involving differences in number and arrangement of teeth (Fig. 4.4). These differences coincide with interclonal niche differentiation in feeding behavior. Thus, when clones are formed, many specific and highly differentiated morphologies can be fixed within the geographic range of a species. Major variation typical of interspecific differentiaton is thus present, ready to be tapped within any species population. This seems to be common among species of fishes (Vrijenhoek 1978). Stabilizing selection must prevent these combinations from usually appearing.

Therefore, a genetic revolution is not at all necessary to break a pattern of genetic homeostasis. Very likely, natural selection and gene flow prevent the fixation of radically new morphologies. Speciation might be correlated with morphological differentiation, but this is only coincidental with spatial variation in selection pressures. Of course, there are examples where the introduction of a new allele can destabilize an otherwise regulated (canalized) trait, as in studies of the *scute* locus of *Drosophila melanogaster* and at the *tabby* locus in the mouse. But this does not have any necessary connection with speciation; it can occur as easily within a panmictic population.

Stanley's (1979) monograph on macroevolution, basically a plea for the importance of speciation in morphological evolution, unknowingly reveals the blurred distinction between his conception of speciation and divergent evolution, based upon differing selection pressures in ecologically distinctive zones. He asks (p. 72):

> Why should all populations of any established species abandon their original niche because adjacent ecological space is free for occupancy? Certainly, expansion of the original niche might be expected, but it is difficult to imagine that this could produce major adaptive shifts without fragmentation into new species. Far more likely would be the rapid invasion of adjacent ecologic space in association with divergent speciation.

This association of speciation with occupancy of divergent habitats is precisely the same as a model of divergence based upon differential selection of a polytypic species. In other words, it is an ordinary neo-Darwinian model. The appearance of a new resource or habitat would exert strong directional selection, particularly if an old resource or habitat is less abundant.

In summary, patterns of geographic variation and genetic analyses of interpopulation and interspecific crosses and backcrosses fail to substantiate the idea that speciation is a special process with regard to morphological differentiation. While one cannot say much for those characters that cannot be studied effectively, those amenable to genetic analysis only provide support for intra-interspecific extrapolation hypothesis.

Trans-Specific Stasis

As allozyme and chromosomal techniques have been applied to natural populations, it has become apparent that extensive speciation occurs with little morphological change. This is ironic in the case of the mammals, since morphological evolution in this group is believed to be rapid, relative to others (e.g., Stanley 1979; but see Gingerich 1983, and Chapter 7). The rapidity may be coincidental and may be related to adaptive shifts, rather than genetic mechanisms. This is best illustrated by the hundreds of genetically distinct races of various species of mice, accompanied by little morphological change. Consider further the rampant speciation of the genus *Peromyscus* with little morphological divergence (Kurtén 1981). If frogs change little in adult morphology, it may be that natural selection promotes little divergence.

It might be argued that cases of evolution of reproductive isolation in the absence of morphological divergence might be restricted to a small group of taxa with restricted dispersal and chronically small population sizes. Common cases of speciation without significant morphological divergence, however, are far more widespread. Morphologically nearly identical species, or **sibling species**, have long been known in groups such as *Drosophila*. Careful examination of many other groups presents a similar overall picture (see Mayr 1963, Chapter 3).

Investigations of marine species in recent years have greatly altered the traditional picture of widespread single species. Networks of nearly identical sibling species have been found in most groups of marine invertebrates. The fiddler crabs of eastern North America are a useful example (Salmon et al. 1979). Speciation in most of the temperate and in some of the subtropical North American species has occurred without significant morphological divergence. Species recognition is usually futile without a careful study of male mating behavior, which often differs among species (Crane 1975). In the species pair *Uca speciosa* and *U. spinicarpa*, morphology is identical despite allozymic difference indicative of an approximate 12 my divergence time ($D = 0.7$). Overall, the sixty species of *Uca*, spread throughout the temperate and tropical world, are similar in form.

In marine polychaetes, previous notions of broad-ranging single species have been supplanted by demonstrations of large numbers of morphologically identical species. The genus *Ophryotrocha* (Dorvilleidae) has been shown to consist of a very large number of sibling species of varying life history but strong morphological similarity (Akesson 1973). The polychaete *Capitella capitata*, previously thought to be a cosmopolitan opportunistic species, is now known to consist of a large number of sibling species (Grassle and Grassle 1974). Similar cases have also been discovered in marine mussels (Seed 1978), and the blue mussel, *Mytilus edulis*, has been recently shown to be a complex of morphologically indistinguishable sibling species (e.g., Koehn et al. 1984).

While many sibling species complexes are now being uncovered, a larger pattern of general constancy of form is apparent in many groups of organisms. Many genera are quite ancient and consist of groups of species whose overall form has deviated little from a common morphology (Wake et al. 1983). Morphological constancy goes beyond sibling species; it may be a mistake to think of speciation as bimodal in its generation of morphological divergence. The vast majority of speciation events probably beget no significant change.

If, as in most other examined cases, many morphological traits have sufficient heritability to allow extensive change during phyletic evolution or speciation, it follows that some conservative, and speciation-transcending, force preserves **trans-specific stasis**. Stabilizing selection is the most likely candidate for such stasis. The transcendance of form beyond the species level has long been recognized. Rensch (1959, p. 93), for example, notes:

> It is conspicuous that most such persisting types are not im-
> mutable species, but persistent genera, the changing species
> of which preserve a certain type of adaptation.

One might add that the generic level distinction is probably made on the basis of a preconceived notion that the genus level is an adaptively significant threshold of morphological difference.

Stanley (1985) misunderstands the significance of sibling species, noting that even if sibling species cannot be told apart in the fossil record, their lack of strong temporal change implies species stasis. But this misses the point that sibling species arise by speciation. Rampant production of sibling species is continually being recognized in more and more taxa, once nonmorphological criteria are explored (e.g., allozymes). Stanley and his allies have failed to recognize that speciation in nearly all cases usually fails to cause morphological change. Times of extensive directional evolution are blind to the number of speciation events, even if speciation may be a consequence of local adaptive divergence. This is an important distinction, as it means that speciation is an effect, not a cause, of morphological evolution.

Speciation Mechanisms

The Models

The possible mechanisms of speciation revolve around two major issues: ecology and genetics. From the ecological standpoint, the geography of speciation has been a major point of disagreement. **Allopatric** models require geographic isolation between newly forming daughter species. **Parapatric** models allow for some contiguity between diverging populations. **Sympatric** assume the possibility of random contact and mating within a population; ultimately a number of mechanisms separates the population into two or more new species. It is not often clear that these distinctions are appropriate; a continuum might be more accurate.

One major school of thought maintains that all speciation is associated with geographic separation. The **allopatric-dumbbell model** asserts that populations become separated by a geographic barrier. Populations on either side of the barrier are large and are genetically representative of the starting population, with the possible exception of some geographic variation. Subsequent divergent evolution on either side of the barrier leads to accumulated genetic differences, related to the ecological differences between the isolated habitats (Dobzhansky 1937). Mating behavior, if it is unrelated to the ecological distinctness of the isolated regions, may not diverge, and individuals of the incipient species may mate freely. The genetic differences acquired during isolation may, however, result in reduced viability of hybrids– **postmating incompatibility**– should the populations be reunited. After secondary contact between the diverging populations, postmating incompatibility between hybrids may result in selection against hybridization, resulting in one or more behavioral or morphological isolating mechanisms. From these, **premating isolation** may arise as the result of selection against hybrids. Premating isolation could also arise coincidentally when the populations are separated.

The **peripatric** model asserts that speciation occurs not by sep-

aration of large populations, but by budding at the periphery of a
species' range (Mayr 1963, 1982*b*). Gene flow and strong coadaptation
within the genome are presumed to maintain geographic homogeneity.
According to this hypothesis, small isolated populations are genetically unrepresentative of the parent population (founder effect). Mayr
(1982*b*) notes that the

> gene pool of a small either founder or relict population is
> rapidly, and more or less drastically, reorganized, resulting
> in the quick acquisition of isolating mechanisms and usually
> in drastic morphological modifications and ecological shifts.
> It involves populations that pass through a bottleneck in
> population size.

Intense selection combines with the founder effect to cause a "genetic
revolution" resulting in a rapid genetic shift. Species divergence is thus
accomplished. Alternating periods of population flushes and crashes
may rapidly accelerate the breakup of coadapted gene complexes and
cause divergence of sufficient magnitude to result in reproductive isolation from the parent (Carson 1975). The peripatric model is in the
allopatric class since it requires geographic isolation of the budded
daughter population.

The **parapatric-ecological model** argues that clinal genetic differentiation can lead to isolation among spatially contiguous populations and eventual species-level divergence in the separated populations. As discussed in Chapter 3, multilocus regional differentiation
can be an effective vehicle for isolation. This model, therefore, could
be similar to the dumbbell model.

A variety of **sympatric speciation** models do not require geographic separation. Sympatry implies that individuals are (at least at
first) physically capable of encountering one another for a sufficiently
long time for mating. Ecological models of sympatric speciation usually
assume strong disruptive selection, combined with either the evolution
of habitat loyalty (Bush 1975), or upon competitive exclusion with
a subpopulation occupying a habitat patch type (Rosenzweig 1978).
Like any model of speciation, breeding incompatibility must be established, but sympatric speciation requires ecological differentiation
as well. As a consequence, it is necessary for the newly established
daughter species to initially have genetic linkage between genes that

confer breeding incompatibility and genes that regulate ecological differentiation. Otherwise, recombination will rapidly dissociate the two and speciation will not occur.

A common theme of speciation models (there are certainly others beyond the ones cited above) is the wedding of ecological to genetic mechanisms. The peripatric model, for example, requires a specific geographic configuration–budding peripheral to the range or in an isolated area within the range–and argues for a genetic revolution within the budded population. Mayr (1963) argues that a species cohesion is maintained by gene flow, thus usually preventing divergence within the range of a species. It has been claimed that gene flow is not the mechanism of broadscale homogeneity (Ehrlich and Raven 1969; Larson et al. 1984). Uniform selection conditions throughout the range, accompanied by canalization of traits, probably combine to promote genetic homogeneity. Even if the genetic claims about peripatry may prove to be incorrect, peripheral populations do probably exist under ecologically divergent conditions. One must, therefore, make the distinction between the ecological context of speciation and the change in genetic architecture of populations during the speciation process (Templeton 1981).

Gene flow is one of the most difficult things to measure in natural populations. A direct approach requires measurements of dispersal and successful breeding of the successful immigrants. Moreover, these measurements must be done over time scales relevant to potential genetic change. Gene flow would prevent local differentiation, but the significance of gene flow depends on the strength of selection or drift. At a neutral polymorphic locus, an exchange of even one individual per generation on average would be sufficient to counter local differentiation by means of genetic drift. To counter natural selection, gene flow must be stronger as the strength of local selection increases. For gene flow to offset a fitness difference of one percent, then 100 generations of one percent replacement by immigration would have to occur (Slatkin 1987).

The short-term connections between populations that we see at present may underestimate gene flow. Ehrlich and Raven (1969) summarize evidence of dispersal and breeding phenology that indicate that populations of the checkerspot butterfly *Euphydrias* are strongly localized, with little dispersal among them. Estimates of genetic differentiation, however, suggest much greater exchange (Slatkin 1987). This may have been accomplished by rare occurrences of regional extinction, with spread by individuals over many recolonized habitats. Such bouts of regional extinction and recolonization may explain the widespread homogeneity of marine invertebrate populations (e.g., Levinton and Koehn 1976), although continuing dispersal may also be a factor.

Evidence from Current Geographic Distributions

The ubiquitous presence of geographic variation and the common ability of workers to extrapolate intraspecific variation within polytypic species to interspecific divergence is the usual justification for the importance of allopatry (or parapatry) in speciation (Mayr 1963). The evidence garnered in recent years on chromosomal races of mammals suggests that allopatry followed by divergence of isolates is a common mode of speciation in this group. The extrapolation of geographic races to species with contiguous distributions is powerful evidence, as it is difficult to imagine another explanation for a present day distribution of a series of closely related species with contiguous ranges. But it would be difficult to use such evidence to distinguish between strictly allopatric and parapatric models. Divergence will emerge in a steep selection gradient with reduced dispersal (e.g., Endler 1977; Barton 1983). Most hybridizing zones are probably quite ancient, as they connect genetically divergent species. As a result, current distributions may tell us little about the origin of genetic divergence (Barton and Hewitt 1985).

The predominance of allopatric or parapatric speciation leaves open the geographic aspect or the dumbbbell-peripatric distinction drawn above (omitting the problem of genetic revolutions). Is most speciation a budding process, or does it occur by geographic separation well within the range of an extant species? It is probable that both modes are common. The evidence mentioned above for the butterfly genus *Heliconius* suggests that polytypism can develop into speciation, a process consistent with the dumbbell model. Many newly derived species, however, are buds on the edge of the range of extant species. This has been documented in some plants (H. Lewis 1973; Gottlieb 1976; Stebbins 1983a). Diploid newly derived species of *Clarkia* in California occupy ecologically marginal habitats and are characterized by numerous chromosomal structural differences that substantially reduce the fertility of hybrids. The daughters are competitively inferior to the parent species, even on the sites where the newly budded species live. Speciation by budding in plants is facilitated by the possibility of selfing in the neospecies (see Gottlieb 1976).

The argument for sympatric speciation also derives from evidence on current distributions. In many cases, species are found to be sympatric over wide regions. Indeed, this is the common case in sibling species of *Drosophila* that can be collected easily in the same bait traps in many localities over wide geographic ranges. Can all of these cases of sympatry have arisen originally from a process of allopatric isolation and genetic divergence?

Examples of present-day broad sympatry can be quite misleading. Co-occurrence may simply reflect a recent dispersal event of a formerly allopatrically distributed species. The apparent ecological "fit" of such a species in the present community may be misinterpreted as long term-occurrence. Co-occurrence patterns of marine snails can be misleading in this way. At present, on the east coast of North America, the periwinkle *Littorina littorea* is the dominant rocky form. *L. saxatilis* tends to occur higher in the intertidal zone. *Littorina littorea* first appeared in noticeable numbers in Nova Scotia in the mid-nineteenth century. It then spread southward and may still be doing so today (Kreauter 1974). But the two species of *Littorina* (there are others) are now broadly sympatric and occupy contiguous intertidal zones as if they had been sympatric for a much greater length of time.

Some cases of sympatry would be quite difficult to dismiss in this way. Carson and Okada (cited in M.J.D. White 1982) describe pairs of species of *Drosophiella* in New Guinea and Taiwan- Okinawa, one of which breeds in male flowers, the other in the female flowers of a single plant species. Bush (1969) described the development of a series of races of the tephritid fruit fly *Rhagoletis*, based upon different host occurrence, and differentiation in some morphological characters. If the differentiation were genetic, and if host fidelity could be demonstrated, sympatric speciation would be a likely conclusion.

Futuyma and Mayer (1980) reviewed critically the requisites for sympatric speciation by means of host race formation in phytophagous insects. Strong disruptive selection and a host preference mechanism would be required. The Hopkins host selection principle–that adult host choice is influenced by larval conditioning–would provide a means to enhance the fidelity of incipient host races. Futuyma and Mayer state that the evidence for such host selection is very weak. They also show that no strong evidence exists to prove a genetic explanation for a host-plant association in *Rhagoletis*. Other cases cited in the literature can be explained equally well by an allopatric model.

A more strictly genetic mechanism of parapatric speciation has been suggested by M.J.D. White (1968). The **stasipatric** model surmises that fixation of a chromosomal rearrangement in a small population can occur within the range of a larger species population. The rearrangement thus spreads from the center and partially displaces the surrounding ancestral population. The parapatric contact between the spreading nascent species and the surrounding population is, therefore, maintained by strong heterozygote inferiority; otherwise extensive gene flow and introgression with the surrounding population would ensue. Meiotic drive, or biased survival of gametic types, would enhance the spreading potential of a novel chromosomal variant.

Genetic Architecture of Speciation

Templeton (1981, 1982) has classified speciation mechanisms within a genetic context. He defines two overall processes. **Divergence** mechanisms involve the gradual evolution of differences between populations living in ecologically distinct habitats. Divergence can occur among completely isolated populations (adaptive divergence) or along a cline if gene flow is restricted. Isolating barriers between incipient species evolve in a continuous fashion and may reflect only divergence in response to selection in different environments, or to genetic drift. The accumulation of genetic differences increases the probability of fixing differences that would produce sterile hybrids if contact between the isolated populations was reestablished. Templeton argues that adaptive divergence is the mode of speciation. By contrast, **transilience** models drive speciation from a discontinuity in which a selective barrier is overcome by other evolutionary forces such as drift. The founder principle of the peripatric hypothesis of Mayr would apply here, as would the origin of chromosomal incompatibilities among populations via inbreeding and drift. Since natural selection could operate even in cases where transilience mechanisms are in operation, a sharp distinction between the two overall modes is not possible.

Templeton also distinguishes among different genetic architectures that might differentially affect the evolutionary fate of populations placed under similar selective and stochastic forces. *Type I* consists of traits that are controlled by many genes of subequal effect. *Type II* architectures control traits by only a few genes with large effects, modified by genes with small effects.

The genetic architecture of a trait may also reflect the selective regime during the evolution of the trait. Rapid evolution owing to intense directional selection might favor the predominance of *type II* architectures in trait determination. Under strong selection, alleles with relatively large phenotypic effects would be shifted strongly in frequency. By contrast, slow divergence might favor *type I* architectures.

The genetic revolution hypothesis of Mayr is based upon several assumptions about the genetics of colonizing populations and the degree of adaptive interaction among genes. The founder effect hypothesis asserts that founding populations are extremely biased relative to the genetic content of the source population. Intense selection is thought to break up coadapted gene complexes and to strongly alter the genetic architecture of the newly derived species. These points can be examined from both theoretical and practical points of view.

Founder effects are unlikely to be very important in colonizing populations. The reduction in overall genetic variability relative to a source population is likely to be modest even in founding groups of ten or less (Lande 1979a). Even with continuous brother-sister mating, heterozygosity decreases only 19 percent per generation. Intense inbreeding of this sort would be unlikely in most animal populations but could occur in facultatively autogamous plant populations.

For morphological traits, our conclusions about the reduction of variance presume that additive (among-allele) variation is the only important factor in the determination of a trait. Under this assumption, even a few individuals still carry most of the allelic variation of the population. Some experiments on bottlenecked populations require some reconsideration of the assumption. Bryant et al. (1986) passed populations of the housefly *Musca domestica* through bottlenecks as small as one pair. Surprisingly, the additive variance for a variety of traits increased in the bottlenecked lines. These results cannot be explained by a simple additive model, which would predict a loss of additive variance. It is possible that the traits are explained by more complex genetic relationships, such as dominance of certain alleles, that are lost through the bottleneck. This would telescope the variance, but Bryant et al. argue for other effects as well. In any event, these results reduce any hope for the founder effect, except with regard to a crucial point made by Mayr: the possibility that intense selection might be operating on a novel rearrangement of genetic relationships. The shear loss of alleles still seems unlikely.

Studies of Hawaiian Drosophilidae fail to demonstrate strong genetic changes despite rampant speciation. If founder effects were important, they would be evident throughout the genome. While gene substitutions are common, there is no evidence of reductions in heterozygosity in more derived species, nor are there even significant differences between most closely related species (Rockwood et al. 1971; Carson et al. 1970; Carson and Kaneshiro 1976).

For several decades, laboratory strains of *Drosophila* have been established from a single gravid female with only rare instances of the subsequent development of inter-strain incompatibility. Cases of the evolution of laboratory reproductive incompatibility generally involve intensive disruptive selection (Thoday and Gibson 1962). Carson and Templeton (1984) argue that the failure to see speciation in bottles may relate to our consistent use of species with particular genetic architectures. They argue that species in nature commonly have gene complexes (major genes with modifiers) that would be more easily subject to rapid change in a laboratory culture. But there is no clear evidence that such differences in genetic architectures exist.

Even if founding populations are not likely to be strongly unrepresentative of source populations, the question of breakup of coadapted gene pools is a problem still equally applicable to more central populations and peripheral colonizing populations. This would be especially true if even central parts of a species range are broken up into small isolated demes. Despite many strong claims pro and con, current evidence suggests that we know little of the degree to which coadapted gene pools exist.

Maynard Smith (1982) focuses the problem by defining two alternative hypotheses about genetic organization. It is useful to speak of such organization in terms of its conceivable effects on developmental processes. The null hypothesis argues that no genetic organization is developmentally relevant on a scale larger than the gene family (a group of closely linked genes with controlling sequences). The alternative hypothesis would argue that large scale organized structures of the genome are significant in development.

On the grossest scale, the null hypothesis is well supported by available data. The absence of linkage among loci of related functions is quite common (discussion in Kimura 1983). In *Drosophila subobscura*, where paracentric inversion polymorphisms occur on all autosomal arms, no effects of polymorphism on morphology can be observed. In general, the degree of absence of such organization is surprising (J.R.G. Turner 1967). Despite some theoretical expectations (Franklin and Lewontin 1970), the genome does not usually congeal into closely linked complexes of coadapted alleles. Recombination apparently is a more potent force than selection for such linkage disequilibrium, perhaps as the result of reversing selection pressures.

While gross chromosomal organization seems unlikely, smaller scale organization nevertheless exists. First, as mentioned above, nonhomologous chromosomes may join at sites of common gene families during transcription. This is a form of large scale organization that might be strongly disrupted with changes due to recombination. Second, many sets of genes with related function are spatially contiguous. This is especially true of many multigene families, where the repeats occur in tandem arrays (e.g., the chorion genes of silkworms, Kafatos 1983). One cannot be sure whether this is just a function of history (e.g., unequal crossing over) or preservation by natural selection, as in a supergene. If homogenization processes such as gene conversion have a positive effect by "correcting" mutations, the maintenance of proximity of genes of related function might be adaptive. The same can be said for many linkage relationships among allozyme loci in mammals that are quite ancient (O'Brien and Nash 1982). The bithorax series is another example of a long-lived complex of closely linked genes. It is possible that this may relate to transcriptional order in development, but the linkage between gene order and developmental order is by no means universal. Finally, some cases of close linkage of enzyme loci with similar function or with regulatory loci are known. At the scale of the gene neighborhood, linkage disequilibrium seems to be quite important (see Aquadro et al. 1985).

Genetic drift can be shown to promote among-population divergence. Experimental studies with *Drosophila* demonstrate that divergence in chromosome inversion frequencies and mating success is more substantial when populations are started in low numbers, or when they are passed through bottlenecks, relative to those started and maintained at high numbers (Dobzhansky and Pavlovsky 1957; Dobzhansky and Spassky 1962; Santibanez and Waddington 1958). The increased phenotypic variance must increase the probability that natural selection will move the average phenotype of a population to a new adaptive peak, as variation available for selection is increased. Charlesworth et al. (1982) note that these experiments usually involve starting populations derived from interpopulation crosses, and there is good evidence that the stocks differed in background genes that interacted in fitness with genes contained in inversions. The relevance to processes in founder populations, which ought to be derived from a single locality, is therefore unclear. Divergence in the subpopulations might involve a resorting of genes from the respective local stocks to reestablish favorable combinations of modifiers.

The requisite of genetic revolutions for speciation is based on the premise of species integrity. Because he believes in the "unity of the genotype," Mayr (1963, Chapter 5) requires a mechanism to break up the coadapted gene complexes that supposedly constrain evolutionary change within the normal geographic range of a species. While there is ample evidence of intergene interactions, particularly epistatic effects on fitness of genes segregating in populations (see Templeton 1979; Charlesworth and Charlesworth 1975), there is no reason to believe that these phenomena would prevent selection from operating to change a trait, though we would certainly not expect wings to sprout on the backs of mice. Single founder effects are much less potent in enabling extensive evolutionary change by drift (Lande 1980b) or natural selection (Hill 1982). It is true, however, that a founding population could more easily be dominated by genetic drift, relative to selection. Therefore, changes that might be accomplished in just a few generations by a combination of strong drift and natural selection, would be focused in the founding populations discussed by Mayr, and Carson and Templeton (1984). But one would also expect drift at some neutral loci (e.g., allozymes) and there is no evidence for this in Hawaiian drosophilids, the prime case used by Carson and Templeton. The longevity and variability of large populations would promote both divergence and reproductive isolation.

In summary, the geographic aspect of the peripatric model is plausible. Populations may indeed usually bud off from the main range of a species. Intense selection and genetic drift may combine to cause divergence, but longer-lived large peripheral populations would be more potent than small ones in this regard. The concept of genetic revolution seems without foundation. Moreover, many of the changes occurring as the result of selection and drift at the periphery of a species range could just as easily occur in isolated demes well within the range of a species. Most important, the probability of a fitness peak shift across a deep valley is very low in a small peripheral population. Given its likely short lifespan, successful speciation would have to be very rapid. The peripatric model is, therefore, probably applicable to the degree that peripheral areas are ecologically distinct or can guarantee isolation.

The stasipatric model of speciation also requires a specific population genetic process. While the model is reputed to be a sympatric model, qualitatively different from others, it can be readily characterized as a parapatric or allopatric model. Chromosomal variants are fixed, probably by drift, in small populations, which may be peripheral or central. The spread of this subpopulation may then occur by random population fluctuation of many demes, or by ecological superiority of the phenotypes with the newly fixed rearrangements. The model requires geographic contiguity of differentiated subpopulations (it would, therefore, be a parapatric model), with some degree of isolation to allow the initial fixation within a very small deme.

Some important theoretical problems have been suggested (Lande 1979a; Futuyma and Mayer 1980). If the rearrangement spreads by hybridization with adjacent demes, as envisioned by M.J.D. White (1974), it is immediately subject to a probably strong reduction in fitness of heterokaryotypes. M.J.D. White invokes meiotic drive to eliminate the disfunctional chromosomes. This may work in those groups where they can be preferentially sequestered in polar bodies (M.J.D. White 1973). Futuyma and Mayer (1980) point out that cases cited by M.J.D. White (1968) are consistent with an allopatric model of local fixation and deme spread. Worse than that, they note that all cited cases of stasipatric speciation fail to demonstrate extensive reproductive isolation. Such isolation would probably not be established by a single fixation of a chromosome rearrangement as significant gene flow could still occur. This hardly suggests the expanding fronts of incompatibility suggested by M.J.D. White (1974).

Some models involving transilience mechanisms are supported by distributional evidence.   Polyploidy is widespread in plants and is known to occur in animals (Stebbins 1971; Ohno 1970). The occurrence of polyploidy is usually followed by the evolution of gene regulation mechanisms that effectively diploidize the population. Doubled occurrences of genes are regulated in various ways to resemble species of diploid origin (Leopoldt and Schmidtke 1982).

Since chromosome doubling usually is associated with hybridization, the effects of doubling are difficult to uncouple from those of hybridity and recombination.   It is the latter that are thought to be of importance in evolutionary potential (Levin 1982).   Chromosome doubling has some immediate ecological consequences. DNA content in plants is positively correlated with mitotic cycle time and cell size.   Polyploids usually have slower development, delayed reproduction, longer life time, larger seeds, and lower reproductive effort (Levin 1982).   Polyploidy can conceivably be directly related to the rate of morphological evolution and propulsion into new ecological milieus. Polyploidy is more common among annuals than perennials.

While polyploidy may be a source of new morphological variance, it is important to remember the common dependence of flowering plants upon apparently closely matched insect pollinators. Hybrids intermediate in form between parent species, or of highly changed form, even when autotetraploid, may be thrust out of the range of pollinator service. Stebbins and Ferlan (1956) have noted this for *Ophrys murbeckii*, a hybrid derivative from *O. fusca* and *O. lutea*. In this genus, bees pollinate by pseudocopulation, based upon the hairs, color, and form of the labellar surface. The two parent species are distinct, while the hybrid is intermediate in all respects, except in certain characters such as the small size of the labellum. Stebbins and Ferlan conclude that hybrids are being formed all of the time but most do not reproduce for lack of a pollinator. The ones that do reproduce must have some pollinator capable of servicing the flowers. When a bee pollinates a hybrid, segregating progeny will be produced and strong directional selection will ensue to adapt the plant to the new pollinator, otherwise extinction will follow.

We have discussed in Chapter 3 the common occurrence of intraspecific homogeneity in nucleotide sequence in members of multigene families, ranging from a few (e.g., some transposable element families in *Drosophila*) to a hundred thousand in number (e.g., the *alu1* family in man). This homogeneity stands in contrast to the common situation of interspecific heterogeneity. This pattern reflects the processes of unequal crossing over, gene conversion, and duplicative transposition (Dover 1982; Arnheim 1983).

The phenomenon of gene family homogeneity is spread over many taxonomic levels. In some cases, homogeneity is only found in members of a single species. This is true for families of noncoding sequences, introns, and other cases where the DNA sequence has no apparent function. Homogenization must proceed rapidly with respect to the rate of speciation. Where functional constraints appear to be important, homogeneity can be interspecific and may unite several related species. The varying degrees of interspecific differentiation can be mapped onto a cladogram like any other set of characters.

Divergence at different taxonomic levels can be seen in sequence divergence in species of cereals (Gramineae, see Flavell 1982). The haploid sizes of species of Gramineae fall in the range of 3.6 to 8.8 pg DNA and are over an order of magnitude larger than that of *Drosophila*. The majority of the "extra" DNA probably does not code for functioning genes. Over 75 percent of the DNA consists of repeated sequences, which can be divided into families, either clustered in tandem arrays or spread on several nonhomologous chromosomes.

In the species of wheat, *Aegilops squarrosa*, *A. speltoides*, and *Triticum monoccum*, nearly all of the hi repeated families are identical between the two species of *Aegilops*. About 2 to 3 percent of the genome of *A. speltoides* consists of families not found in the other two species, but most of the repeated DNA families are common to all three species. The use of restriction endonucleases reveals the presence of species-specific subfamilies that have evolved since the species diverged. Thermal stabilities of reannealed sequences between wheat and other species show a progressive increase in degree of dissimilarity as taxonomic distance increases (Smith and Flavell 1974).

The degree of meiotic chromosome pairing reflects the increasing sequence divergence in the multigene families of these species. As taxonomic divergence increases the frequency of chromosome pairing and chiasmata formation decreases. This may be related to increased divergence due to differential gene family homogenization in geographically separated groups. This might result in sterility of hybrids if the populations reestablish contact.

Processes of gene homogenization and duplication may be sufficiently different in geographically isolated species to inhibit chromosome pairing between hybridizing genomes and thus cause reproductive isolation and speciation. The correlation between multigene family divergence and decrease of chromosome pairing may lack a causal connection. It is possible that divergence in single copy genes also occurs and is the primary source of chromosome pairing failure during meiosis. By contrast to Flavell's (1982) results for *Aegilops* and *Triticum*, Rees et al. (1982) report no differences in chiasmata formation in intraspecific versus interspecific crosses among species of the cereal genus *Lolium*. Fertile hybrids can be generated despite a difference of 40 percent in nuclear DNA. In crosses between the more distantly related *Festuca drymeja* and *F. scaricea* ($2N = 14$ in both species, but 50 percent difference in nuclear DNA) chiasma formation occurs at low frequency. Nevertheless a surprising amount of pairing occurs. Pairing at pachytene is nearly complete despite DNA differences among chromosomes ranging from 25 to 42 percent. Pairing among larger chromosomes can be accomplished by means of loose ends or loops. Failure of pairing is more prominent on shorter chromosomes. Thus extensive divergence in DNA content may exert only modest effects on compatibility, but the effect may be sufficient to prevent crossing among populations with divergence in multigene or sequence family size. The relative contribution of genic and overall chromosomal compatibility to divergence is an open question, except in the many cases where chromosome divergence is apparently absent.

A similar possibility of divergence and incompatibility has been mentioned above for transposable elements involved in hybrid dysgenesis (M.G. Kidwell et al. 1977; G.M. Rubin et al. 1982). One or more unique TE's might spread in an isolated population. The probability of spread will depend upon the reduction of fitness in crosses between infected and noninfected animals, relative to the infection rate. In the P system, takeover may have taken only a few decades (M.G. Kidwell et al. 1981). This might result in the eventual accumulation of sufficient transposon family differences to affect postmating isolation, should the population be united with a conspecific but previously separated population. Interstrain incompatibility similar to hybrid dysgenesis has been observed in other species, but it is not due necessarily to the same underlying mechanism.

Current evidence casts some doubt on the role of transposable elements in speciation. This hypothesis has been proposed on both empirical (Bingham et al. 1982) and theoretical (Ginzburg et al. 1984) grounds. Hey (1987) has examined differences among semispecies of the *Drosophila athabasca* group, and between this group and *D. algonquin*. In contrast to known divergence in allozymes, inversions and morphology, there were no detectable differences in the presence of dysgenic transposable elements.

The theoretical treatment of Ginzburg et al. (1984) suggests that the drop in fitness imposed by crosses with dysgenic strains would select for avoidance of hybridization and would lead to speciation. The details of at least the known dysgenic P-M and I-R strains suggest a different interpretation (Hey 1987). First, dysgenic crosses are not reciprocal and hybrid females with limited fertility may backcross to the maternal population; the progeny will not be dysgenic, but will carry transposable elements and would inherit the cytotype associated with the elements. The transmission of elements via hybrid and partially hybrid females would destabilize the hybridization barrier. The barrier would only be stable if complete sterility were the result of crosses with strains carrying the elements. Even in this case, Hey (1987) notes that functional elements are lost rapidly, making them important as a barrier for very brief periods. As soon as individuals with dysfunctional elements appear, then hybridization can occur. The time period would appear to be less than thirty years, in the case of the P-M system of *D. melanogaster*. This brief window of opportunity may explain why Hey failed to find evidence for the action of transposable elements in speciation in *Drosophila*. These processes need not inevitably result in interpopulation incompatibility after long periods of isolation. Stebbins (1983a) cites cases of isolation among plant species populations for several million years with no apparent loss of reproductive compatibility.

Are New Species Accidents or Adaptations?

We return to the difference between Dobzhansky, who thought newly derived species were exquisite adaptations, and Muller, who thought they were accidents of divergence. The adaptive aspect of species has three potential components:

1. The species is adapted to its environment, and differentially so, from other species. The process of adaptation is primarity responsible for speciation.

2. The species has a set of adaptations for avoiding hybridization with other species. These were acquired as a result of secondary contact and selection against crossbreeding.

3. The species has a gene pool that is intimately coadapted and uniquely different from other species.

This last factor could be ambiguous in that coadaptation within the gene pool might follow speciation. One can easily see why the dispute exists, given the data presented above. There is enough diversity of genetic and ecological contexts of speciation to allow strong arguments for both points of view. Indeed, Mayr's (1963) theory of genetic revolutions is somewhat intermediate. Though strong selection is important in his hypothetical revolutions, the exact trajectory of genetical properties of the new species is unpredictable.

There are some cases in which species are recognizable as ecological adaptations. This would be expected in the adaptive divergence model of speciation, but would also be expected with a sympatric model leading to association with coexisting species such as different host plants. The fig wasps (Agaonidae) may be an example of adaptively associated speciation (Ramirez 1970). The wasps reach maturity in a male phase. Copulation takes place before the females escape from the galls inside the fig. After copulation the females emerge from the galls, and go to the anthers, which ripen synchronously with the softening of the fig and the emergence of the wasps. The female agaonids carrying pollen enter the young receptacles at the time the female phase flowers are ready for pollination. Those pollinated flowers that receive eggs become gall flowers, each nourishing a single wasp larva.

There are several hundred species of fig wasps. With a couple of exceptions, each species of New World fig (about 40 total) has its own pollinator. The same holds for the Old World figs. This can be proven by the inability of species of figs introduced into new areas to set viable seeds when their symbiont wasp species is absent. The one fig–one wasp relationship must have involved divergent evolution occurring intimately with the speciation process. Adaptations include the conformation of the ostiole and size of styles in figs, the morphology of mandibles and mandibular appendages, and the size of the ovipositor of wasps.

It is unlikely that the case of the fig wasps is typical. The examples of isolation correlated with chromosomal differentiation and genic sterility cited above cannot be associated with adaptive divergence. They seem to involve episodes of coincidental natural selection and genetic drift and subsequent incompatibility among geographically structured populations. The preponderance of evidence demonstrates that postmating sterility barriers are attributable to the effects of many loci (see Barton and Charlesworth 1984). It is unlikely that this aggregate cause of sterility can be related to any coadaptive interaction among the contributing genes. We are looking merely at accidents of divergence. The peripheral neospecies of annual plants also seem to fall in the accidental category. H. Lewis's classic study (1973) of *Clarkia* demonstrated the presence of a peripherally derived "species," differing only in the shape of the petal. The parent species is superior competitively even in the sites occupied by the new species. While the lack of superiority of the new species does not bode well for its future, the process of speciation has occurred nevertheless, and seems independent of adaptive divergence. A similar case can be made for a newly derived, reproductively isolated, population derived from the composite *Stephanomeria exigua* in sagebrush deserts of eastern Oregon. Individuals lack adaptations for freezing, germinate in the fall, and die in the winter (Gottlieb 1976). In this case speciation does not reorganize the genome, but a few changes are fixed, probably by genetic drift, that are maladaptive.

Localized adaptation may enhance interpopulation incompatibility. Adaptation to temperature in a latitudinal gradient might produce local populations that suffer strongly in fitness when individuals disperse into the "wrong" latitude. This might enhance interpopulation incompatibility, as other genetic differences build up during the physiologically enforced isolation (Levinton and Lassen 1978; Lonsdale and Levinton 1985). Thus although adaptation might enforce isolation, speciation may result from a fixation of an allele at a locus that greatly increases interpopulation sterility. The interaction between adaptation and accidental fixations, therefore, might make it difficult often to choose unambiguously between the Muller and Dobzhansky points of view.

The Punctuated Equilibria Model and Speciation

The punctuated equilibrium model purports to be an alternative to more traditional models of evolutionary change. Hopefully the discussion presented above will establish just how difficult it is to present an "alternative" to the current diversity of theories and characterizations of evolution! The model (Eldredge and Gould 1972) established a dichotomy between phyletic gradualism and punctuated equilibrium as follows:

Phyletic gradualism: Most evolutionary change arises by gradual transformation of entire species populations. The process is mediated by natural selection and is even and slow. Morphological evolution is thus a product of gradual change within populations.

Punctuated equilibrium: Speciation is the time when most evolutionary change occurs. During the rest of a species' history change is minimal or at least without a trend. Speciation, therefore, is the main cause of morphological evolution. Though speciation may not involve morphological change, the punctuated equilibrium hypothesis implies that speciation at least sets the tempo of morphological evolution.

A difficulty with the punctuated equilibrium hypothesis is its generation of the false premise that phyletic gradualism emerges as a natural property of models proposed during the Modern Synthesis (Levinton and Simon 1980; Stebbins and Ayala 1981). As mentioned above, the expectation of models of natural selection is for varying rates of evolutionary change, or usually short periods of directional change, interspersed with longer periods of stabilizing selection. Once the dichotomy is no longer valid, the punctuated equilibrium theory appears to be a solution in search of a problem.

Although I doubt that workers thought in terms of the stasis-gradualism dichotomy, even a casual inspection of the principal works of the Modern Synthesis would fall on the side of stasis. The heart and soul of Mayr's (1963) classic work is the integrity of the species. This theme reverberates Dobzhanky's (1937) seminal volume, which viewed species as stable adaptations. It is this very belief in stability that places the punctuated equilibria model within an historical context, at least for neontology (Levinton 1983).

Many proponents of punctuated equilibrium (e.g., Gould 1985a; Williamson 1981) have argued that stasis is a fundamental paradox, which the Modern Synthesis avoided and the punctuated equilibrium theory can solve. If species are inherently stable then stasis is expectable. But there is no compelling reason to expect anything other than stasis, given the probable ubiquity of stabilizing selection, and the probable rarity of the combination of novel environments and mutations required for new adaptations. As mentioned above, stabilizing selection has to be viewed as a more complicated process than mere culling of extreme phenotypes. Certainly one does not expect anything but stasis in mimetic butterflies, so long as the model does not disappear. One is strikingly impressed with the ubiquity of additive genetic variation for morphological traits. Speciation nevertheless rarely changes them. While some genetic changes may have correlated negative fitness effects on other traits, one can hardly imagine this to be universally so. Again, stabilizing selection, as qualified in Chapter 3, is probably important in stasis that transcends the speciation process. An important part of stabilizing selection involves a sort of evolutionary paralysis caused by strong habitat selection, continuous presence of similar competitors and predators. Canalization of traits as part of a devleopmental homeostatic mechanism may promote constancy. But, as mentioned above, there is no strong evidence that this promoter of constancy, or its breakdown, is linked to speciation.

The evidence cited above suggests that even a restricted punctuated equilibrium model relating speciation to phenotypic evolution is similarly unfounded. Speciation may or may not involve concomitant morphological change, depending upon the nature of the environments of the daughter species and the existence of variability. It is likely that phenotypic divergence in different environments would result in phenotypic divergence soon after isolation. Thus, in a restricted sense, the punctuated equilibrium claim of stasis throughout most of a species' history could be correct. But this is no vindication of the punctuated equilibrium model, which views speciation as the prime factor.

The ease with which it is possible to extrapolate within population variability to between-species differences, as discussed in detail above, is the best evidence that the speciation process is not a major breakpoint in the evolutionary process. No set of data points to speciation as having such a role. This effectively contradicts perhaps the most important implicit prediction of the punctuative hypothesis: that within-species variation is not the stuff of between-species variation.

The Species Selection Model

Species selection (Stanley 1975) involves selection among species whose long term result may be a morphological trend in the fossil record. Speciation events generate among-species morphological variation, while selective mortality, or differential speciation rates, would bias survival in one morphological direction (Eldredge and Gould 1972; Stanley 1975).

As mentioned in Chapter 3, a process at the species level could be analyzed from the point of view of natural selection, in the same terms as might be done at the organismal level. We presume a source of heritable variability, a reproductive mechanism, and differential "fitness" among units–in this case–species. The variability is the morphological diversity among species speciation. Differential fitness would include differential levels of speciation of different species-morphological types or differential extinction. In this framework, evolutionary trends cannot occur unless morphological variability is accumulated in speciation events, or in phyletic evolution between successive speciations. If the latter is sufficiently common, then the punctuated equilibrium hypothesis is *not* necessary for a species selection-based mechanism for directional morphological trends. In other words, we do not need to accept the punctuated equilibrium hypothesis in order to accept the notion of higher level processes. Phyletic evolution and differential speciation rates could work in the same direction to produce an overall trend in a clade (Sober 1984*b*). This should be emphasized, given our conclusion that speciation usually does not beget morphological change. Times of intense directional change are times when intrapopulation phyletic evolution is extensive and coincident with speciation owing to locally different selection pressures.

We must distinguish between species drift, species selection, and species hitch-hiking (Levinton et al. 1986). In **species drift**, morphological trends are generated due to random speciation and extinction processes, but the properties that change are species-level properties–those that are fixed throughout the species (e.g., reproductive mode, dispersal mode, pan-specific morphological characters). **Species selection** involves cases where species-level properties bias speciation or extinction rates (e.g., reduced dispersal is fixed and tends to increase the probability of speciation). This would be in contrast to the case where organismal performance results in an increase of survival of the entire species. The sum of species selection, species drift, and success stemming from individual organismal performance amounts to *sorting* (Vrba and Gould 1986). Finally, **species hitch-hiking** is the process where a given trait is proliferated in a clade, due to its accidental association with a rapidly speciating group, or with a clade having relatively low extinction rates. Hitch-hiking is analagous to genes that change in frequency due to their close linkage with other genes upon which selection or drift is acting (Maynard Smith and Haigh 1974). For example, a morph fixed in a clade of mimetic butterflies might be common due to the high speciation rate of the clade, and not because of the adaptive superiority or any fitness characteristic of the color pattern itself. By contrast, certain traits (e.g., reproductive structures) may enhance speciation or extinction, and thus spread or diminish in frequency. This has been termed the *Effect Hypothesis* (Vrba 1980).

As Maynard Smith (1983) notes, the hypothesis of species selection is usually employed to cover several quite distinct claims, which are freely confused with the punctuated equilibrium theory. Species selection can mean one of the following:

1. **Tempo Hypothesis.** Speciation is envisioned as the source of significant evolutionary change; evolution within the normal history of a species, i.e., phyletic evolution, is of insufficient magnitude to explain the scope of evolution (Eldredge and Gould 1972).

2. **Inherent Properties Hypothesis.** Selection acts upon emergent properties, such as innate capacity to speciate rapidly (Vrba 1983). If 1 and 2 apply to a given clade, then rapid evolution is assured. Here, too, the speciation rate or measure of productivity is at issue.

3. **Competitive/Adaptive Superiority.** Some trait possessed by species of one clade confers competitive superiority over others, or reduces their extinction rate, relative to others. As a result the group outsurvives the other clades (Stanley and Newman 1980).

4. **Radiations Based Upon Keystone Innovations.** The ac-

quisition of a keystone innovation permits a clade to diversify throughout a series of novel habitats. The increase in taxon richness insures its survival, relative to a sister group that fails to acquire the innovation (Liem 1973; Lauder 1981).

The first two of these fall under the punctuated equilibrium hypothesis. Both assume that speciation is important in evolution and that speciation tends to generate the sort of evolutionary change that "ordinary" phyletic evolution cannot. Point 1 certainly follows from Stanley's (e.g., 1979) claim that phyletic evolution is too sluggish to have generated the diversity of life. Point 2 depends upon point 1, and implies that those groups where speciation is inherently rapid are bound to evolve more rapidly as they have experienced more novelty-generating speciation events. If more rapidly speciating groups are more likely to spread, then selection occurs for morphological change at an hierarchical level higher than that of individual selection.

Both Vrba (1983) and Sober (1984b) point out that the tempo and inherent properties hypotheses do not require the punctuated equilibrium hypothesis. Evolution can be continuous throughout the history of the species, but species-level properties might still determine the pattern of gain and loss of certain traits and species groups. Points 3 and 4, however, do not follow uniquely from a species-level process. It does not confound intuition to believe that some groups will survive others as a result of their superior adaptations, or competitive ability. But such survival does not indicate that the origin of the superior novelties stems mainly from extinction or speciation in the first place.

At present, it is difficult to ascribe any known trends to species-level processes. Gilinsky (1981), for example, refers to the long-term decrease in the spectrum of gastropod form through geological time as **stabilizing species selection**. While it is of great interest that overall morph variability has declined over time, there is no evidence that this has much to do with either reduced speciation of extreme morphs, or accelerated speciation of common forms. Similarly, Stanley and Newman (1980) see the ascendancy of the balanomorph barnacles over the chthamalid barnacles as an example of species selection. Again, this process says nothing about the origin of the novelties characterizing the two clades in the first place, only their current relative abundance. This example accentuates the difference in process among hierarchical levels. Within-population forces were probably important in the assembly of the two body plans. The relative extinction/speciation rates, however, may have been a case of species-level sorting (Vrba and Gould 1986; Sober 1984*b*), where losers disappear as taxa, perhaps by virtue of individual inferiority. This is not species selection, at least in terms of the tempo and inherent properties hypotheses. Every balanoid organism would be expected to be superior to every chthamalid organism. Thus individual performance can be extrapolated to group-level relative survival. In any event, Dungan (1985) has disproven Stanley and Newman's hypothesis for the competitive superiority of balanoids (their supposed higher growth rate due to their low density skeleton).

The fate of separated species in a clade can be likened to the survival of demes in Wright's shifting balance theory of evolution. Gould and Eldredge (1977) call this Wright's Rule. They argue for the random generation of morphologically divergent species, that is random with respect to a trend. Sorting might create the trend. A series of related species might arise with a set of morphological features acquired through local adaptation. Owing to stochastic forces, one species might be more successful and spread at the expense of the others. This successful species might give rise to a new species flock and the process might then again be repeated. Through this complex web of speciation and extinction, an overall evolutionary trend might transcend the history of any one species. This process is what I mean by species drift, as defined above. Characters associated with the successful species would be proliferated and an evolutionary trend would be generated by a process above the species level. One should remember that all of this can be explained with either punctuated equilibrium, or with standard phyletic evolution within the history of a species.

Extinction events tied to major habitat alterations may be the best potential example of hitchhiking, where a trait disappears, but not because of its decrease of individual fitness. If a clade arises with a given fixed trait, and if that clade is restricted geographically, then a regional climatic change (e.g., a local marine regression) might eliminate the clade. Fürsich and Jablonski (1984) have suggested such a case where a clade of boring gastropods arose in the Mesozoic, but ultimately did not survive. It would seem unlikely that the extinction of this clade was due to the presence or absence of the ability to bore into hard skeletons, given the likely continual abundance of skeletalized prey. On a larger scale, mass extinctions probably have eliminated many groups, not because of their species-level properties, but because of their bad luck in being in the wrong place at the wrong time. On the other hand, some marine groups may have survived mass extinctions as a result of their relative insensitivity to pronounced changes in primary production (e.g., Levinton 1974). This would be a case of species drift.

As opposed to species drift, species selection requires a species-level property that confers relative survival or increased speciation. At the hierarchical level of species, this is the equivalent of a relatively fit genotype. Dispersal type might strongly influence species-level origins or extinctions (Jablonski 1979). In marine invertebrates, reduced dispersal seems correlated with rapid speciation and extinction (Spiller 1977; Hansen 1980). This precarious balance might shift in different cases, favoring one clade over another. If a favored clade had species with some fixed trait (e.g., pronounced ornamentation), then the trait would become more common by species hitch-hiking.

Eldredge and Gould (1972) have suggested that speciation might generate a set of new morphotypes that are randomly arrayed, but species selection would cull out a biased fraction, thus producing an evolutionary trend. It might seem absurd to expect the random generation of variant species suggested by such an undirected speciation hypothesis. A variety of functional, genetic, and developmental constraints will restrict descendant morphologies to a reduced range of morphological possibilities (Muller 1949; Maderson et al. 1982; Levinton and Simon 1980). To account for this possibility, Stanley (1979) has proposed a concept of directed speciation, where speciation events produce a strongly biased array of descendant morphologies. The concept of directed speciation is only as strong as the concept that speciation is the primary generator of morphological change in the first place. Directed speciation does little more than claim that speciation accelerates a phyletic trend. In the worst case, a single string of ancestor-descendant morphologies, unbroken by any cladogenesis would be thought to be a series of speciation events (see Chapter 7). The model here would depend upon relating speciation to morphological change, which in the case of fossil lineages would result in a tautologous relationship between the two (Levinton and Simon 1980). In many fossil cases, directed speciation amounts to uneven rates of phyletic evolution (Levinton 1983; see Chapter 7).

It might be argued that gene mutation is no less subject to biased change. Therefore, gene-level mutation is completely analogous to Wright's Rule. This might be true, with the important exception that surviving species probably consist of individuals that are in some sense functionally harmonious with their environment. Newly arising species, subject to strong direction selection, may only adopt a limited set of morphologies. This is in contrast to genic level morphs that assort within populations. Those with low fitness will eventually disappear, but will appear no less randomly than those with increased fitness. In contrast, species are not likely to have this random aspect. The analogy between gene mutation and species level divergence only holds when all possible daughter species morphologies are functionally viable. I contend that this is uncommon.

Given the undirected model of speciation, and success of a small fraction of a clade, can we distinguish between the punctuated equilibrium theory and phyletic gradualism possibilities? Slatkin (1981) uses a diffusion model of evolution to assess the possible relative contributions of speciation and phyletic evolution to such an evolutionary trend. In a random model, the relative importance of speciation and phyletic evolution in creating among-species variance in morphology depends upon: (a) the within-species phenotypic variance, V_p, generated by unpredictable changes in phyletic evolution due to either genetic drift or randomly fluctuating selection forces; and (b) the variance of phenotypic change per speciation event, V_s, generated by random changes in phenotype occurring at speciation.

For speciation to be the principal cause of between-species differences in a clade, the product of the speciation (per generation) rate, s, and V_s would have to be greater than V_p. V_p is the product of: heritability, effective population size, and the variance in the character. For mammals, s is estimated to be 0.4 per my. If heritability is 0.5, with a generation time of 2 years, and an effective population size of 1000, the contributions of speciation and phyletic evolution would be equal if the morphological variance generated by speciation were 625 times the phenotypic variance (Slatkin 1981). For a completely neutral character, the standard deviation of the change in the average value during speciation would have to be 25 times the within-species standard deviation for the speciational and phyletic components to be equal. This does not bode well for the species selection or drift processes as a generator of morphological trends, since such cases are likely to be rather rare. One would require a combination of low heritability with high effective population size to change this conclusion substantially. With the undirected speciation model any instance of directional selection will inevitably cause the phyletic component to be yet more important than the speciational component, as a directional change will occur at every generation in phyletic evolution, but not at speciation events.

These arguments suggest that it is extremely unlikely that complex structures ever arose by species drift. The potential for phyletic evolution is immense and is certainly far more potent than a mechanism which effectively depends upon a process analogous to neutral evolution. Analogous to the arguments of Chapter 3, selection is always more powerful than drift in producing trends. Given that speciation generates less variance than phyletic evolution (Slatkin 1981), this should apply even when we are comparing selection within populations to drift among species. Even in the case of species selection, Slatkin's arguments tip in favor of the power of phyletic evolution. Even without this argument, it is inconceivable how selection among species can produce the evolution of detailed morphological structures. The elaboration of some of these structures has of course taken more than the life of any one species; cladogenesis is coincidental to any major evolutionary trend, but it does not follow that it is a causal mechanism. If anything, cladogenesis may slow down the evolution of complex structures, simply because species are continually winding up in new and complex environments that might constrain the further improvement of a structure down a main evolutionary path. In contrast to the arguments of Stanley (1979) phyletic evolution is the likely source of complex adaptations, while species drift or selection is likely to bring about evolutionary trends such as changes in overall body size or degrees of ornamentation. Species selection did not form an eye or a secondary palate.

If morphological change per speciation event showed no temporal trend through the history of a clade, then morphological evolution should accelerate with time. The number of speciation events per unit time should increase as the clade becomes speciose. Species richness usually increases during the early history of a clade. But it is well known that the rate of morphological evolution usually decelerates with time. This is best recorded by the ratio of taxonomic species to families, which tends to increase with geologic time (e.g., Valentine 1969). Given the correlation of systematic rank with phenetic diversity, this trend is an indication of a decelerating rate of generation of morphological diversity. Most morphological novelties appear early in the history of a clade, and directional change tends to decelerate over time (Westoll 1949; Cisne 1975). These considerations strongly weaken the model of species selection, as it relates to accumulating change with speciation events. Some other process would have to be invoked that explains why morphological change per speciation event has declined with time. We will discuss the question of the effective spread of groups over time by their high speciation rates in Chapter 8.

While speciation is not likely the typical source of adaptive evolution, the peripheral nature of new species (Mayr 1963) often places them in ecologically marginal environments. These marginal environments have a much lower probability of survival than the larger parental species and the newly budded species would be expected to disappear. This can be seen in the broad ranging species of marine invertebrates that occasionally give rise to peripheral species. The mussel *Mytilus edulis* most probably gave rise to the Mediterranean *M. galloprovincialis*. It seems unlikely that the latter will survive, so long as its future is tied to so unstable a peripheral basin (which it is not, owing to human introductions elsewhere). Similarly, the eastern North American bivalve *Mulina lateralis* gave rise to the morphologically divergent *Mulinia pontchartrainensis*. The latter is restricted to a geologically unstable lake, and will likely go extinct. These two cases suggest that peripheral species may be morphologically divergent but will not stand the test of time. The survivors will be the large widespread species that are liable to remain morphologically static. This expectation is realized in the data of Stanley and Yang (1987), who find extensive constancy in many Atlantic coast bivalve mollusk species over many years.

In conclusion, there is little evidence to support the notion that speciation is an accelerator of evolution or even an arbiter of directional morphological trends; the potency of phyletic evolution suggests its likely primacy in the evolution of functionally integrated forms. It does not follow, however, that the present relative abundance of taxa with different morphological character complexes relates strictly to phyletic evolution. Relative extinction and origination may explain current distributions. Although relative extinction may be related to superior adaptedness conferred by given characters, it is also possible that success may relate to a given group's sheer number of species. During a crisis, a severe climatic change might favor the group that produces the most species, and spreads them over the most habitats. In this important sense, species drift and species selection may have explanatory power in understanding dominance patterns among taxa.

The Main Points

1. The process of speciation is incompletely understood because it occurs on a time scale that is inaccessible to biologists and paleontologists. Most of our information comes from static descriptions of variation within as opposed to between species.

2. Variability in chromosomes, allozymes, and morphology has been examined extensively in natural populations and species. The bulk of the evidence suggests that the sort of variation found within species is qualitively similar to that found between closely related species. In the case of allozymes, there is a smooth increase in degree of differentiation from intrapopulation, to closely related interspecific, to more distantly related interspecific comparisons.

3. While sterility is the natural criterion for recognition of different species, it is not always clear how the speciation event took place. Substitutions of alleles at single loci and fixation of chromosomal variants may both be important. Hybrid zones have often been investigated to surmise the processes that maintain isolation. Many of these, however, are probably quite ancient and species are probably long divergent.

4. Speciation has been explained as an accident of geographic divergence, resulting in fixation of alleles producing sterility, or as an adaptation that serves to avoid hybridization of separated populations that have recently come into contact. While it is clear that speciation seems associated with geographic separation of populations, divergence can occur even with modest dispersal between the populations. Most speciation events probably are due to fixation of new alleles in complete or modest isolation, resulting in sterility with the most closely related population. In

some cases, it is clear that this fixation in isolation is associated with natural selection for new environments. In a few cases, the process of adaptation is the very same one that results in the formation of new species, as when a tight relationship evolves between a population and a new host.

5. Many have suggested that dramatic genetic reorganizations occur during species formation, often in peripheral and very small populations. While this is theoretically possible, the genetics of closely related species usually fails to support such revolutions. Important exceptions include the polyploid origins of many plant and some animal species.

6. The punctuated equilibrium model argues that morphological change is associated with speciation and that species are static during their history due to some internal stabilizing mechanism. There is no evidence coming from living species to support this. If anything, recent research has demonstrated that speciation occurs typically with little or no morphological change; hence the large-scale occurrence of sibling species. As Haldane pointed out in 1932(a), within-species variability is readily extrapolated to between-species variation.

7. Trends in morphological change might emerge from patterns of speciation and extinction. These need not be associated with punctuated equilibrium. Certain species-level properties, such as geographic range, might influence speciation or extinction rates. Alternatively, a random process (species drift) might cause speciation and extinction rates to permit some taxa to expand (or contract) relative to others. Associated morphological properties might hitch-hike and expand or contract in frequency, as the result of either species selection or the species drift process. While these processes are not likely to be the reason for the evolution of complex morphological features, they might explain the predominance of a given morphology at any one time. This might, in turn, influence the variation available for adaptation. The punctuated equilibrium model is not required for species selection, since species-level processes could work in tandem with phyletic evolution to produce trends.

Chapter 5

Development and Evolution

> Fashion me, therefore, one form of a many-coloured and
> many-headed beast. There is a ring of heads both of tame
> and wild beasts, and it can change and produce them out
> of itself at will. "That is clever moulder's work," he said.
> *The Republic of Plato*

Constraint and Saltation

Developmental biology has long been a focus for evolutionary theory (Haeckel 1866; Garstang 1922; de Beer 1958; Goldschmidt 1938;
Waddington 1940; Riedl 1978; Gould 1977; Bonner 1982; Raff and
Kaufman 1983). Time and again, the concepts of **constraint** and
saltation have been formulated in terms of development. **Developmental constraints** are nonrandom channelizations of evolutionary
direction due to limitations imposed by epigenetic interactions in the
developing organism. Such interactions may strongly influence fitness,
and therefore restrict evolutionary direction. In the context of development, **saltations** are rapid evolutionary fixations of phenotypic
discontinuities governed by developmental constraints, which do not
permit continuity of form in polymorphic populations. It is the purpose of this Chapter to discuss these two concepts critically, and the
role of development in evolutionary change.

The importance of developmental integration was recognized by
the founders of the neoDarwinian movement (Orzack 1981; Ford and
Huxley 1927, 1929; Haldane 1932b; Lande 1980b), but, in the past fifty
years, embryology has tended to have more influence on the study of
phylogeny than on genetical mechanisms of evolution (de Beer 1958;

212

Nelson 1978). Embryology parted company with genetics long ago. The influence of Haeckel's biogenetic law upon embryologists tended to eliminate the expectation of much evolutionary change in embryos (Raff and Kaufman 1983). The alliance with Haeckel led to a schism among embryologists and the rise of the field of developmental biology, which adopted a more mechanistic approach and abandoned evolutionary thinking (Gross 1985).

Even though general embryology was far removed intellectually from genetics, many developmental systems, such as mutants in vertebrate skeletons, and developmental mutants in *Drosophila* and other insects were actively studied by geneticists. Sewell Wright's first research on toe number determination in guinea pigs shows a strong bent toward understanding developmental-genetic interactions. I expect that the difficult nature of quantitative genetic studies on traits with strongly determined (canalized) alternative states and the poor communication among fields led to the diminutive role of development in population genetics. It is also likely that development was simply too poorly known to be easily studied from a mechanistic point of view, let alone integrated with evolutionary genetics. Britten (1982) and Davidson (1976) both discuss the extent of our ignorance of the most basic aspects of gene action in early development. This could not have been encouraging to geneticists who were anxious to get "closer and closer to the genes."

Phylogeneticists and Developmentalists

The significance of ontogeny in evolutionary theory has been emphasized from what I shall call the **phylogeneticist** and **developmentalist** viewpoints. The phylogeneticists view ontogeny as the source of information of an organism's evolutionary history. Ernst Haeckel, that most devoted German disciple of Charles Darwin, combined a materialistic view of evolution with a belief that man's ancestry was recorded in the sequence of embryonic development. His association between the order of ontogeny of derived taxa and the order of acquisition of ancestral character states was unduly rigid and invited later strong criticism (Garstang 1922; de Beer 1958; Gould 1977). But why should development be indicative of ancestry at all? Why do primitive structures that are degenerate or absent in the adult often make an appearance albeit brief in the embryo only to later appear as vestigial structures or even to disappear? How can we explain the occasional reappearance of ancestral structures, as developmental abnormalities, that supposedly disappeared millions of years before in evolution? Why, for example, do some whales have hindlegs (Andrews 1921), do some horses develop ancestral toes instead of splints (Marsh 1892)? Some level of order in development survives the evolutionary process.

The developmentalists claim that "the diversity of structures that have been formed through the process of evolution is constrained by the rules which govern pattern formation during development" (Stock and Bryant 1981, p. 432). As such, evolutionary change is involved intimately with developmental sequences. The individual, therefore, is treated in terms of its entire ontogeny, and development is therefore both the constraint and target of selection. The developmentalist claim should be the key to the phylogeneticist claim. If we can understand whether development really does provide constraints, and if we can understand how developmental programs are shaped by evolutionary processes, then we might come to understand the limits to which phylogeny can be expressed in ontogeny. Unfortunately, nearly all evolutionary debate in the past has centered on the reverse of this thought process. Theorists have used a belief in the fixation of embryological patterns as the phylogenetic key to explain developmental sequences in extant organisms. This approach fails to take notice of many of the advances of developmental biology in the twentieth century, which point to a lability in the mechanisms of development.

Development: Constraints and Discontinuity of Form

The importance of constraint and discontinuity of form in populations has been long appreciated on both genetic and morphological grounds. Wright's (1934, 1935a) classic analysis of digit number in guinea pigs recognized the constrained nature of development, despite complex polygenic effects on traits. Quantitative geneticists working on mutations affecting vertebrate skeletons and teeth have long been aware of the presence of limited and discrete variation (e.g., Hadorn 1961; Green 1962; Grewal 1962; Gruneberg 1965; Garn and Lewis 1963). Sinnott and Dunn (1935) showed that plant development is strongly regulated by genes that affect rates, and by genetically controlled quantum changes in morphology (see also Haldane 1932b). Extreme fruit-shape differences are already established in the very earliest primordia of squash. Many of the strong differences among genotypes can be attributed to phenotypic changes early in the developmental process (e.g., Atchley et al. 1984). In plants, mutations affecting early stages of development often have the most significant final effects on overall form. Form differences are usually not the result of "mere continuations or extensions of the later stages of a growth period" (Sinnott and Dunn 1935, p. 140).

Why might there be constraints? The developmentalists provide at least some keys to the solution. Consider the quantum nature of development. It is not continuous but discontinuous. Despite the potential continuous nature of additive gene effects and the continuous possible range of concentrations of morphogens, hormones, and other developmental messages, morphological structures often come as complete structures or not at all. Of equal interest is the importance of localization in development. Embryos develop only as the result of a complex series of timing events that bring different cells into contact, or place cells or even molecules of restricted developmental potency in a proper environment for induction. The spatial position of cell groups seems crucial in the generation of morphological patterns due to postulated localized intercellular movement of dissolved substances (Turing 1952; Wolpert 1969; Garcia-Bellido 1975; Summerbell 1981), transcellular electric fields (Jaffe and Stern 1979; Nuccitelli 1983), mechanochemical interactions (e.g., Odell et al. 1981; Oster et al. 1983), or specialized cell adhesion molecules (Edelman 1986). Must this not influence the direction of evolution? These two phenomena–**integrity of structure** and **topological restriction of development**–suggest that an embryo can only be transformed in a limited number of directions during the process of development and evolution. That is the fundamental message about form that Richard Goldschmidt's book *Physiological Genetics* (1938), derived from Spemann (1938), underscored so well.

Some examples of developmental mutants show the discontinuous and often spectacular nature of possible structural change. Consider the *cyclops* mutant (Bowen et al. 1966) in males of the brine shrimp, *Artemia salina* (Fig. 5.1). After the fourth instar, the lateral eyes move forward and fuse together as a single large compound eye by the ninth instar. During this fusion the ganglia and nerves of the two optic stalks fuse; the resultant eye is rather similar to the normal medial eye of the cladoceran *Leptodora*. A quirk of development has caused a

Figure 5.1: Dorsal view of normal living male *Artemia* (top) and cy-clops male (bottom) (drawn after Bowen et al. 1966).

structure to change from that characteristic of one taxonomic order to another! The development of the vertebrate limb shows similar quantal steps. Alternative symmetries and structures, demonstrated by manip-ulations of development and repair, can be spectacularly different. An excision of the posterior part of the chick (*Gallus gallus*) wing bud re-sults in the elaboration of only a single skeletal element: a humerus, or a humerus fused with a reduced radius. If, however, the anterior part of the bud is excised the posterior half develops nearly normally, elab-orating part of the humerus, the ulna, and digits 2, 3, and 4 (Hinchliffe and Gumpel-Pinot 1981). Clearly, developmental mechanisms are well organized, depend upon tissue interactions, and often involve discrete steps (Maderson 1975: Alberch 1980).

A developmental notion of macromutation springs from the nature of development described above. If a simple transplant places toes on wings or replaces scales with feathers, why couldn't evolution occur in major steps? Thus some have either directly interpreted developmental change, via such phenomena as tissue interactions, as the vehicle for major evolutionary jumps (Schindewolf 1936, 1950; Goldschmidt 1940; Lovtrup 1974; Maderson et al. 1982), or at least see them as the possible stuff of major saltations (Frazzetta 1970; Alberch et al. 1979; Gould 1980*a*).

Development, Genes, and Selection: Components of the Evolutionary Ratchet

The developmentalist point of view presented above leads to a model, which I call the **evolutionary ratchet**, generating some of the constraints in evolution. The model characterizes an organism as being a product of a long evolutionary history, where the evolution of timing, rates, and localization leads to a complex developmental process that can be disrupted less and less easily as time progresses. Any spatiotemporal interactions in the developing phenotype which, when accumulated, cannot be disrupted due to integration are accumulated by the **epigenetic ratchet**, and lead to **epigenetic constraints** (see Rachootin and Thomson 1981). Shifts in tissue interactions are well known in evolution (Hall 1984) but such transitions may be difficult to achieve due to interdependencies between different germ layers. Riedl (1978) thought of changes in developmental patterns of such complex organisms as carrying a high **burden.** "By burden I mean the responsibility, carried by a feature or decision. . . . with systemization (i.e., the evolution of development) the functional burden carried by decisions increases and with this a new lack of freedom called canalization also increases" (Riedl 1978, p. 80). He argues that "patterns of decisions that have a certain degree of burden have no real prospect of being fully dismantled" (p. 93).

Two more factors contribute to the overall evolutionary ratchet. The **genetic ratchet** refers to new genes which, when incorporated, cannot easily be lost by processes such as genetic drift. Pleiotropy is one likely explanation. Incorporated genes having widespread effects on the overall phenotype may not be easily lost in evolution (Lande 1978; Kollar and Fisher 1980). Many cis-acting enhancer sequences are now being discovered, which regulate expression of genes hundreds of base pairs away. For example, a 125 base-pair DNA segment regulates expression of the yolk protein 1 gene of *Drosophila melanogaster* (Garabedian et al. 1986). This enhancer-structural gene complex may be stabilized by this positional arrangement, especially if the enhancer is required for expression of other nearby genes. Evidence from *Drosophila* suggests considerable stability of some DNA sequences involved in development. For example, a controlling sequence, the homeobox, has been found to be widespread in animals (McGinnis, Garber, Wirz, Kuroiwa, and Gehring 1984; M.M. Muller et al. 1984). Such sequences may be crucial in gene regulation and development. The domain about the homeobox apparently regulates the binding to specific DNA sequences which in turn regulate expression of genes crucial in the development of segmentation (Gehring 1987). While the DNA sequences flanking the homeobox may vary among organisms, a surprizing degree of homology has been found among, fruitfly, honeybee, and mouse. This suggests conservation of some sequences for over 500 million years.

Ohno (1973) has characterized other modes of incorporation by the genetic ratchet as "frozen accidents." Accumulation of deleterious mutants in chromosomes with low recombination, and dosage compensation in the X chromosomes of mammals are examples. Evolution after the development of dosage compensation (inactivation of one of the two X chromosomes in the female) would select against translocations of X-linked genes to other chromosomes. The X chromosome seems to always be about five percent of the total mammalian genome.

Such a model is evocative of the reverse of the adaptive landscape metaphor of topographic peaks, connoting high fitness. Instead, evolution cuts channels, forming the developmental landscape. Waddington's **epigenetic landscape** was conceived as a reflection of the limited number of stable developmental pathways possible from the combined actions of different genes involved in development. "The reactions of these genes with one another (and with the environment) interlock so as to define a developmental track which will always be followed by the antennae of flies of a certain genotype. . . . the normal developmental track is one towards which a developing system tends to return after disturbance" (Waddington 1940, p. 93). The pictorial representation of this is the steepness of the walls of the valleys, which reflect the tendency to return to the developmental track, if a deviation from the central stream occurs. Such a theory acknowledges that the evolution of internal organization is as important as the effects of the external environment. As Riedl (1978) puts it: "selection does not work from the outside only"(p. 244).

The **selection ratchet** refers to features that, once acquired, are not easily lost due to functional integration within the developing phenotype. For example, the presence of the vertebrate humerus is essential in the function of the radius and ulna. On functional grounds alone, the humerus should be the last structure to disappear as limbs

are reduced in evolution. In the construction of functional morphological models, all three types of constraints–**epigenetic, genetic, and
selectional**–will have to be incorporated into the study of morphology
and adaptation.

The evolutionary ratchets complement neo-Darwinian theory; we
only acknowledge that evolution tends to organize the phenotype and
genotype so that it begins to resist some directions of evolutionary
change for reasons of internal organization. Such presumptions are implicit in the works of the founders of the Modern Synthesis (e.g., Mayr
1963; Stebbins 1950; Dobzhansky 1970), although genetic, rather than
epigenetic, interactions have been emphasized. Maderson et al. (1982)
take such points about development and evolution to be major challenges to the Modern Synthesis. They are not. As Riedl (1978, p.
235) notes: "In terms of feedback it is a *selectional theory*. It assumes
the correctness of the Neodarwinian synthesis as a precondition, but
supplements this." The concern for constraints influences our conception of the mechanisms that regulate the nature of variation, but
neo-Darwinian theory focuses on the fate of variation, once it is generated. There is therefore no conflict at all between the two approaches;
they are complementary.

Evolution and Internalization of Developmental Programs

Parts of an organism do not evolve function, nor are they genetically determined independently of the rest of the phenotype. The entire body,
and certainly large parts of the body are tightly integrated biological
units, from the points of view of genetic determination, development,
and function. This fact is acknowledged implicitly when evolutionary
biologists view the phenotype as a unit of selection, and not individual genes, developmental tracks, or parts of the body. It is fitness
that matters, and fitness is determined by the functioning of the entire
phenotype.

The postulation of an evolutionary ratchet begs the question of
how developmental, genetic, and functional integration correspond in
a living organism. On the level of phenotypic variation, one might ask
whether phenotypic correlations among functionally related traits are
strong. The same question might be asked of developmental and genetic correlations. If functionally related traits are strongly correlated
genetically, then evolution will operate on blocks of the phenotype

that have functional significance. A correspondence between genetic correlations and functional relatedness of traits might suggest that the genotype is a coevolved unit, designed to serve the entire organism. This concept has been advocated strongly by holistic geneticists such as Ernst Mayr (1976).

It would be of great value to describe the relationships between phenotypic correlation and degrees of functional, developmental, and genetic integration (Olson and Miller 1958). Unfortunately, detailed studies are few and incomplete. A study of the cranium of rhesus macaque suggests that such a correspondence may not exist, or is at least more complex than we can uncover (Cheverud 1982). While functionally related traits are relatively tightly integrated, genetic correlations among the traits can have a completely different pattern. These correlations may arise through stochastic processes, suggesting that the integration of the phenotype is maintained by a continual process of natural selection. Cheverud notes that some of the genetic correlations do correspond to expectations of functional integration and may reflect the action of stabilizing selection along ridges or on peaks of the adaptive landscape. The nonadaptive genetic correlations provide opportunities for new directions of evolution.

The three ratchets determine the difficulty of losing a gene or phenotype at the present time, but the current situation does not indicate how the traits or genes might have evolved. The evolution of an ordered developmental program subject to the ratchets can be described as **building**. As this process continues, order becomes important and internal relationships become as important as interactions with the external environment. If the program is organized as a unit, it may then become **internalized**, that is, a genetic-epigenetic mechanism seals it off from the external environment, allowing it to interact and evolve dependencies with other such units. Once such units evolve, internal constraints will direct evolution and the units may be shuffled as organized entities with other such units. Evolution would no longer consist of infinite molding by the external milieu, as might have happened during the evolutionary building of a given unit.

A possible example of the combined action of the genetic and selective ratchets is the sequence of incorporations of evolutionary changes underlying a functioning protein. Tryptophan synthetase A is an enzyme of 267 amino acids in the bacterium *Escherichia coli* (Allen and

Yanofsky 1963). At position 210, a substitution of serine for glycine (the wild type) causes no change of activity. A mutant with glutamic at this site eliminates function. If, however, this nonfunctioning mutant encumbers a further change, of tyrosine to cystine at site 174, full function is restored. The same change, at site 174, of the wild type protein eliminates function. Thus the same mutant (tyr→cys at site 174) will destroy or save the phenotype's full function, depending upon the order of incorporation. The sequential incorporation of mutants, moreover, may result in a frozen ensemble of changes which cannot be removed except by rare double mutants. The phenotypic and genotypic basis of a functioning phenotype is therefore as much a matter of the order of genetic incorporation of change as it is a matter of function per se.

Potential complex relationships between genotype, functional considerations, and developmental order can be illustrated by the formation of the developmentally arrested dauer larva stage of the nematode *Caenorhabditis elegans* (Riddle et al. 1981). Normally the larvae pass through four molt stages, but overcrowding or starvation causes entry into the dauer larvae in the second. The duration of the dauer stage does not affect subsequent longevity. Studies of mutants suggest that the genes responsible for induction of the dauer stage are ordered in a pathway, and that the order corresponds to neural processing of environmental stimuli necessary to stimulate the developmental switch. Two of the mutant genes are correlated with chemotaxis defects, and both mutants exhibit ultrastructural alterations in specific neurones which have previously been implicated in the chemosensory response to salts. This case shows the intimate relationship between genes, developmental order, and selection. We do not know how the sequence was built up in evolution, but the disruption of a gene's function probably affects the internal organization of the order of gene action, interaction between neurosensory substances, the establishment of neuronal pathways, and the ability to monitor the external environment for the proper cue to enter the dormant state. Some dauer mutants, however, can sidestep any interaction with the external environment. In *dauer constitutive* mutants the dauer larvae stage is switched on irrespective of crowding and food conditions. This suggests that once a developmental pathway is integrated, probably by natural selection in this case, it can then be incorporated into the internal organization of the genotype and phenotype. One can imagine a harsh seasonal environment where the dauer larva is no longer an option; the developmental pathway becomes fixed by selection and incorporated into development.

One might apply these ideas to a theory of reduction of plasticity over long periods of geological time. Valentine (1969) has noted that phylum-level origins were confined to the early history of the metazoa. A combination of restricted ecological, genetical, and developmental opportunities may now preclude the origin of many new phyla. Jaanusson (1985) described the decline of variation through the Paleozoic in segment number among species of trilobite families, and the importance of commitment caused by the rise of functionally integrated structures. Living forms may be more constrained than those that were present at the dawn of the metazoa.

Critique of the Ratchet Theory

The notion of hardening of developmental programs has been seen by many as a fundamental challenge to the notion that natural selection is unrestricted in its bounds. The evolution of development is thought by many to bias evolutionary direction (Davenport 1979; Gould 1982b; Alberch 1981, 1982; Maderson et al. 1982). These arguments, furthermore, point to the creative power of combining units that are relatively independent to produce novel structures (e.g., Lovtrup 1974). While these arguments are useful and have been underemphasized in the past, it is worrisome that they will be applied uncritically to construct a new model of evolution that tends to think of change in the developmental program as a necessary or at least common progenitor to major morphological change in evolution. Minor changes in temporal arrangement of developmental timing would be the cause of large macroevolutionary jumps (Gould 1982b). Differences between major bauplans, moreover, would be ascribed to differences in embryogenesis that may have arisen by accident or for reasons of internal organization, and not selection to function in the external environment. This reasoning is premature and may lead us too far in the opposite direction.

Restrictions of evolutionary change may well be true at some levels of development. But different developmental pathways might lead to the same overall adult phenotype, which may have been selected to

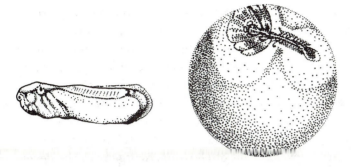

Figure 5.2: Comparison of frog embryos. On the right, *Gastrotheca ri-obambae* embryo is spread on the yolk, whereas the embryo of *Xenopus laevis* (left) is elongated showing its adaptedness for early swimming. Bar = 0.5 mm (after del Pino and Elinson 1983).

function optimally in the environment. In these systems, the burden of developmental restriction may be relatively unimportant.

Frogs are notable for their overall adult morphological homogeneity, despite great antiquity and among-species genetic distances, relative to other groups such as mammals (e.g. Cherry et al. 1978). Is this adult homogeneity determined by a singular developmental ground plan that has been hardened to the extent that no fundamental change is possible? Apparently not. Early development in the hylid frog genus *Gastrotheca* is quite different from that of other frogs, yet the adult form looks like a typical frog (del Pino and Elinson 1983). Almost all of the body forms from an embryonic disk, a group of cells that forms from fusing cells of the blastoporal lips. In two other unrelated genera, similarly located yolk-poor cells are spread around the blastopore, instead of being concentrated in a disk. The difference in *Gastrotheca* development seems due to a temporal separation of the times of closing of the blastopore and the anterior extension of the archenteron. The embryo of *Gastrotheca* looks bird- or reptile-like as it initially develops into several sheets of tissue, which secondarily form tubes after folding. This disk, however, is formed by gastrulation and is not homologous with those of other vertebrates.

The difference in the early development of *Gastrotheca* might be due to extraordinarily slow development (two weeks from fertilization to end of gastrulation, as opposed to one day in *Xenopus laevis*). This might cause the temporal separation of the closing of the blastopore

and the anterior movement of the archenteron and fundamentally rearrange tissue interactions, relative to other frog embryos. Slow development is related to an overall syndrome of egg brooding, maternal incu-
bation of embryos, multinucleate embryogenesis, and very large eggs.
Some of these have ecological significance; the environment may have
selected for the rather unusual parental traits. *Gastrotheca* females
lay eggs in water-poor habitats; this simple difference is probably the
driving force for the profound larval and developmental differences from
other frogs. The novel development of *Gastrotheca* organizes the developing embryo as an elongate disk lying above a large amount of yolk.
The alternative development of *Xenopis laevis* and other anurans produces an elongate embryo (Fig. 5.2) adapted for early swimming (del
Pino and Elinson 1983). But no macromutation changing adult development has occurred; the adult is still basically a frog. As del Pino
and Elinson (1983, p. 589) note: "the embryos of *Gastrotheca* illustrate
that it is possible to modify greatly the pattern of early development
without altering the basic adult morphology." In this case, adult stasis
cannot be said to be due to a rigidity of developmental program.

Development in the freshwater clam *Unio* provides further evidence
that natural selection can break up seemingly fundamental embryological spatial interactions, yet a typical adult is produced. Its larval
stage is highly modified relative to its closest marine relatives. Larvae
are brooded until the glochidium stage, which bears a pair of hooked
valves capable of attaching to the gills of fishes. It completes development on the gills and eventually drops off to finish development in
soft sediments. *Unio* would be expected to have spiral cleavage, a
characteristic mode of early cell division common throughout annelids,
mollusks, polyclad flatworms, and other related phyla. At the third
cell division typical groups with spiral cleavage show a differentiation
between 4 macromeres and 4 micromeres. After the next few cell divisions the spiral arrangement of cells is apparent. The embryo at this
stage is mosaic, that is, cells have specific fates which cannot be easily
reversed. In most mollusks, the first generation of micromeres divides
and forms eight cells, which eventually form the apical region of the
embryo and the prototroch, a larval feeding organ. But these structures are lacking in *Unio* larvae, due to a retardation of division in
the first tier of micromeres. The second tier of micromeres, arising (as
is normal) from cleavages from the macromeres, gives rise to the bulk

of the larva. The importance of other micromeres has been greatly exaggerated relative to the typical pattern of spiral trochophore larval development in most related groups (Lillie 1895). This alteration seems related to the development of structures of ecological relevance, such as the powerful adductor muscle used by the larva to hook the valves onto prospective fish gills. As in the amphibian case, major adjustments can be made in early development to satisfy ecological necessities, with little consequence for the subsequent adult phenotype.

The Nature of Development

All Genes are in Most Cells, but Gene Activation is Local and Timed Specifically

Development from the zygote is truly a remarkable process. A spatially complex organism develops from a geometrically simple fertilized egg. What is the role of the genome in such a process? We would like especially to understand the distribution of the genes and their effects. Are all genes in all cells? Are all genes activated in all cells; if not, are they turned off? Is this an irreversible process? We only have space here for a brief summary of major phenomena important to our general story. The reader should consult Gurdon (1974), Davidson (1976), and Stewart and Hunt (1982) for further information on the subject.

All genes seem to be in all cells, with some important exceptions such as red blood cells. At the crudest level, DNA and chromosome content seems to be constant among somatic cells. This conclusion is bolstered by other studies such as DNA-DNA hybridizations and by the partial and sometimes total developmental potential of differentiated cells. In the former case, separated DNA strands from differentiated cells are hybridized with DNA from early embryonic cells. They can be shown to hybridize at the same rate as hybridization among embryonic DNA strands (McCarthy and Hoyer 1964). Since the rate of hybridization is both a function of overall DNA content and the heterogeneity of (mainly repeated DNA) sequences, similarity can be related to sequence similarity.

In demonstrations of cell potentiality, the nuclei of differentiated cells are transplanted into earlier developmental milieus. If a complete organism develops, then we conclude that the complete genomic complement is present in the differentiated cell and the genes have not lost their potential. If a complete organism fails to develop, all genes may still be present, but some may have been irreversibly switched off. (Some embryos are **mosaic**–cells have fates determined early in development that cannot be reversed by transplantation. **Regulative** embryos, in contrast, determine cell fate by the surrounding cell environment.) Such experiments have been done successfully in the clawed toad, *Xenopus laevis*, where endodermal cell nuclear transplants into anucleate eggs result in normal development to the adult (e.g., Gurdon 1974). Similar experiments have been performed successfully in mammals and *Drosophila* (Ilmensee 1976; Ilmensee and Hoppe 1981).

Despite the genic homogeneity, a spatially complex embryo develops. Spatial inhomogeneities must be present initially, or generated subsequently to elaborate the embryo and the subsequent adult organism. The same can be said for timing. Temporal differences in gene action contribute to the development of the embryo by affecting the temporal distribution of gene products. In the cases of both spatial and temporal change, two sources of heterogeneity can be imagined: (a) properties of the egg cytoplasm, which would have initial effects; and (b) successive factors which cause varying spatiotemporal dominance of different gene complexes.

The cytoplasmic environment is an obvious source of inhomogeneity. First, egg cytoplasm contains mRNAs produced before zygote formation or gene activation in the egg nucleus. Initial dominance by mRNAs of maternal origin may eventually give way to dominance by embryonic nuclear genes, but the former may strongly influence early development. In the 16 cell stage of the sea urchin embryo, about 90 percent of the mRNA present is maternal in origin. Maternal message continues to be translated in the embryo even though gene products such as tubulin are also inherited via the egg cytoplasm (Davidson et al. 1982). The initial micromeres may be more dominated by nuclear gene activity than other cells in the embryo, as judged by their higher than average histone/nonhistone ratio. The onset of amphibian gastrulation, following a switch from synchronized to asynchronous cleavage, may be regulated by the gradual increase of nuclear dominance over cytoplasmic maternal inheritance, or by a timing system (e.g., Aimar et al. 1981).

In insects the posterior pole plasm appears early in development and is the locus of formation of germ cells in the adult. This site may be determined by localization of a messenger RNA of maternal origin (Mahowald 1968). Similar localization can be seen in the anuran amphibian egg, where a subequatorial location in the uncleaved egg is necessary for the development of a normal embryo (Spemann 1938). Subsequently other tissue layers are determined with reference to this site (Gerhart et al. 1983). Transplantation of the region to the opposite side of the egg results in duplicate normal embryos. As development proceeds, more and more such reference points provide a series of foci for developmental instructions. These organizers (e.g. Spemann 1938; Waddington 1940) are the principal part of the localization phenomena that permit the development of the embryo's spatial complexity. The process of sperm entry creates a spatial reference point which can affect the spatial pattern of subsequent embryogenesis. The rate at which such reference points develop is strongly variable. In mammals, for example, reference points seem to develop later than in anurans.

Temporal change in gene action is also an important aspect of development. Batteries of genes are successively turned on, particularly in localized groups of cells. What information is needed to turn on such batteries in higher organisms is unclear. Davidson and Britten (1971) argue that, following an initial inhomogeneity of egg cytoplasmic regulatory elements among the cells in early cleavage, alternative sets of genes are turned on by diffusible products of **integrator genes**. Activation of a given master integrator gene set could shut off the synthesis of a regulatory macromolecule, maintaining the organism in the previous state of differentiation. These regulatory events are well known in prokaryotes, and some excellent examples are now known in eukaryotes. Such on-off regulation could control a single locus or many loci. Gene enhancers may affect more than one gene in a developmental sequence.

Tissue-specific changes in chromosomal morphology are well known in the polytene chromosome puffs in the embryos of *Drosophila* and in the lampbrush chromosomes (y-chromosomes of *Drosophila* and in the chromosomes of *Xenopus*. Puffs are sites of active gene transcription and often a cloud of mRNA surrounds the chromosomal DNA. *In situ* hybridization experiments in dipterans show that polytenization of chromosomes during development is localized and genes produce increased numbers of transcripts. In *Drosophila melanogaster*, tissue-

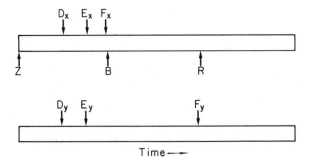

Figure 5.3: Potential complexities in developmental programs for two cell types, x and y. Z = zygote formation; B = birth; R = age of reproduction; D = time of determination; E = time of genetic expression as a differentiated cell; F = time of full cell function. See text for further explanation.

specific increases of amylase can be correlated with the degree of chromosome puffing at the locus (Doane et al. 1975). Thus batteries of genes are probably being turned on sequentially and suppressed during development. Often this results in the presence of embryonic, juvenile, and adult loci which sequentially produce different proteins serving analogous functions (as in the globins: Edgell et al. 1983). It is the localized action of specific sets of genes in spatially separated lineages of cells that partially gives development its characteristic sequential restriction of cell morphological and physiological fate.

The sequential switching on and off of successive gene batteries in insects is partially under the influence of the hormone ecdysone (Ashburner 1980). If a third instar *Drosophila* larva is tied at the midline, the posterior region cannot receive ecdysone and juvenile chromosomal puffing patterns are maintained. The anterior portion contains the ecdysone-secreting ring gland and develops normal puffing. Ecdysone probably combines with a receptor to activate early puff regions and inhibit later puff regions. At the same time, a protein produced from genes in the early puffs eventually reaches a concentration sufficient to cancel the inhibition of ecdysone, and the late puff is activated (Ashburner 1980). This simple model, verified by experiment, can be readily extrapolated to large batteries of genes, as in the Davidson-Britten model (if we ignore their postulation of an important role of repeated DNA, which seems not to hold, given more recent discoveries).

The Complexity of Timing of Gene Action

It is my purpose in the following section to show the potential complexities presented when considering the interactions of two traits, from
the points of view of the timing of gene expression and cellular function. Consider the simple developmental program of traits x and y,
diagrammed in Fig. 5.3. The traits are defined as fully differentiated
cells with complete function at times F_x and F_y, respectively. The cells
have some expression at time F (e.g., can be identified as pancreatic
cell but not fully secretory) but have been determined at time D; their
subsequent fate as a fully differentiated cell is sealed. We will ignore for
now a time before D, when the cell fate is specified but can be reversed
by the local cell environment After time $D{:}sub\ x$, other developing cells
can "know" the identity of cells of type x, either by direct contact or
by detection of a diffusable substance. Forgetting about the time of
expression, let us consider the implications of developmental evolution
with respect to timing of events D (determination) and F (function).

Even with this simple situation, the possible outcomes are complex.
Consider the cases where the developmental events D and F for the
two cell types are interrelated by **developmental constraints** and
functional constraints. Here, developmental constraints are those
where only disruption of the timing of determination causes a reduction
of fitness; cell type x must at least be determined before cell type y
can be determined. In our simple system, cell type x could produce a
substance which helps to determine cell type y. Functional constraints
necessitate a certain order of cell action. Cell type x might specify a
larval protein, which must be present and then loses function before
the adult protein is produced by cell type y. Alternatively, events D_x
and D_y might have to be coincident to function properly. Similarly,
function of both cell types might have to be initiated at the same
time for proper survival. The various possibilities of timing generate
nine possible phenotypes. Imagine a number of traits increasing to the
complexity of the simplest of organisms and you soon appreciate the
problem of developmental timing in evolution. The degree to which the
developmental and functional events D and F interact is the degree to
which the phenotype is tightly bound as an integrated developmental
system.

The importance of timing is most easily illustrated by relatively
simple developmental systems. The cellular slime mold *Dictyostelium*

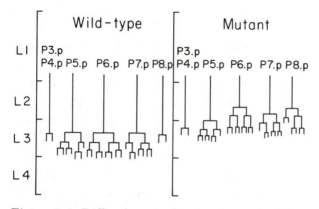

Figure 5.4: Differences in development of cell lineages in wild type and a mutant of *Caenorhabditis elegans* In the mutant strain, the extra cells generated by P8.p form an ectopic ventral protrusion posterior to the pseudovulva formed by the progeny of P5.p, P6.p, and P7.p. Vulva morphogenesis began at the L2 molt, approximately one larval stage earlier than in the wild type (after Ambros and Horvitz 1984).

discoideum responds to starvation by intitiating a ca. 24-hour program of differentiation. Within 6 hours, cells move in response to synthesis and secretion of cyclic AMP toward centers of aggregation. The cells then differentiate into spores and stalks. The developmental program is a sequence of events in which extracellular signals trigger changes in the patterns of gene expression. At each stage, new cellular surface molecules appear and probably act as mediators for extracellular signals. Removal of the signal responsible for progression from one developmental stage to the next results in the loss of gene products specifically induced at that stage. The ordered expression of genes is affected directly by the extracellular signals, which are high cell density, starvation, cyclic AMP, and the formation of cellular aggregates (Chisholm et al. 1984).

The possibility of following the fates of all cells in the nematode *Caenorhabditis elegans* makes it an ideal model system for understanding the consequences of alterations of timing of the transformation of cell fate (Ambros and Horvitz 1984). In certain mutants, cells express fates normally expressed by cells generated at other developmental stages. This can result in precocious or delayed expression of cell lineages that comprise specific body regions or tissues. Alterations in developmental timing can result in the absence or duplication of specific structures, partial sexual transformation, and changes in the number of larval stages (Fig. 5.4). Some of the alterations in cell fate transformation may be related to hormonal action on gene expression. While major transformations most likely have broadscale negative effects, minor changes may explain patterns of evolution.

Rate and Localization of Developmental Processes

To understand the developing organism at least two major components in addition to timing must be included: **rate** and **localization**. The rate of cell division and production of developmentally potent substances, such as hormones, strongly affects timing. In some cases, substances will have varying effects on different tissues; this is a mechanism commonly proposed to alter the relative times of appearance of different structures from ancestor to descendant, one of the commonly accepted definitions of **heterochrony**. The relative rates of full expression of traits can have dramatic consequences (Haldane 1932b). The relative rates of development of eye facets and deposition of pigments in the compound eyes of the amphipod *Gammarus chevreuxi* is a classic example (Ford and Huxley 1927, 1929). Various mutants influence facet color by modifying the rate of pigment deposition. All colored eyes in the adult are colorless at first. This is followed by the formation of a red pigment; facets then darken to near black by the deposition of melanin. In red *rr* eyes the onset of melanin deposition is delayed and the rate of deposition is decreased. Facets are added throughout life and later facets, even in normal black eyes, are pale. Mutants with slow somatic growth have a relatively greater amount of melanin deposited per facet, and eyes blacken earlier.

The cell's developmental fate is influenced by its location in the developing organism. This is known as **localization**, a universal phenomenon in development. Contact between cell types causes **induction**, i.e., determination of a new tissue type. Induction may require specific inducing tissue cells. In the vertebrate eye, for example, the optic vesicle will only develop when primordial tissue is placed in proximity with head epidermis. In others, a glass or biological surface yields the same developmental inductions (Wessells 1977).

Localization is intimately related to the collateral phenomena of timing and rate. For example, the great Swedish embryologist Horstadius worked out the geometrically complex early events in the development of the sea urchin embryo. This morphological work has been supplemented by a more recent understanding of the molecular biology of early development (Davidson 1976; Davidson et al. 1982). At fertilization, maternal RNAs dominate but eventually lose their influence on subsequent development, depending upon the rate of induction of nuclear genes in the new cells. After the fourth cleavage, four micromeres give rise to, among other cell types, 30 primary mesenchyme cells that invade the blastocoel cavity. Later, the gut tube bends forward across the blastocoel, preceded by strands of secondary mesenchyme cells. Filipodia extend across the archenteron, where they search for the appropriate surface before contracting to cause gastrulation (Gustafson and Wolpert 1961). The mouth is induced where the gut makes contact with the wall of the blastocoel. Like most other embryos, this event is followed by a complex series of movements and inductions, resulting eventually in the development of the adult organism. All morphogenetic movements are controlled by (a) differential cell contraction and expansion; and (b) adhesion among cells. Gastrulation is widespread in embryogenesis, even though the details are quite different. This may suggest that there is selection for developmental programs that cause new contacts between cell layers to provide positional information for further development of later structures.

Morphogenetic movements must involve not only induction, but mechanisms for cell kinesis. Presumably, substances which are transmitted from cell to cell induce contraction of microfilaments. Odell et al. (1981) suggest a model for gastrulation by assuming that a small number of connected cells in a sheet are excitable and contract on one side. The cell is modeled as a viscoelastic body whose apical end can be reduced in size by contracting microfilaments, much in the way that a draw string closes a purse. The properties of the sheet can cause an invagination to occur. Evidence from the amphibian egg suggests that some signal is necessary to cause a set of localized cells of a certain number to engage in coordinated contraction. In *Pleurodeles waltl*, the mutant *ascite caudal* shows disturbed epibolic movement during gastrulation (Bluemink and Beetschen 1981). In the early gastrula stage, ectoderm cells begin to sink in at random sites in the animal half of the embryo. In later stages of gastrulation the ectodermal pits develop into grooves. Electron microscopy shows that many cells in the bottom

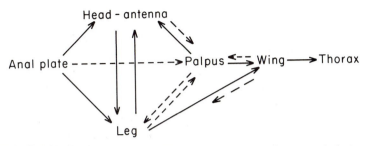

Figure 9.9: Transdetermination pathways of *Drosophila* imaginal discs. Rare changes designated by dashed lines (after Hadorn 1967).

of the pits and grooves have narrowed apices. This result suggests that some organizing signal has failed to stimulate localized contraction and proper gastrulation.

Organization, Compartmentalization, and Restriction in Development

Compartmentalization and Organization

If the evolution of developmental programs is constrained to produce combinatorial states of developmental units, then it is important to learn the degree of autonomy of different developmental units and the difficulty of evolving the breakup of the units. Such information is difficult to obtain since even the spatial relationships can only be understood with reliable cell markers permitting a sequential mapping of cell fate. With such mapping, the interaction of clones of cells of different determination can be studied effectively. We would also have to understand other properties of interaction, such as presence and effect of soluble morphogens, and the nature of cell adhesions. The establishment of developmental interactions involves not only groups of spatially contiguous cells but the nature of cell adhesion and migration and the widespread effects of single gene products over many tissues.

The simplest case of autonomy of regions would occur in embryos where individual regions are easily identified and autonomous, and relatively incapable of switching their ultimate fate as differentiated structures. The classic case of such development is found in holometabolous insects such as *Drosophila*, where the adult structures of the imago derive from **imaginal discs**, divided into nine pairs plus one. The discs are pouches of epidermal cells that differentiate at the terminal molt and each produces specific structures in the adult fly. For example, one pair gives rise to antennae and eyes, another develops into some of the abdominal structures, and yet another three pairs form the legs. The fate of the imaginal discs is determined early, and difficult to change. Serial transfers of imaginal discs can be done repeatedly between flies with low titer of ecdysone (Hadorn 1967), but the ultimate fate of the disc does not usually change. Occasionally, **transdetermination** of a given disc type to that of another can occur. Transdetermination is a quantal process; intermediate disc types are not found and the pathways of conversion from one disc type to another are predictable (Hadorn 1967; Fig. 5.5).

Autonomy of developmental regions in insect wing development further demonstrates the potential nature of independence of developmental units. **Compartments**, found in *Drosophila* and arthropods, are areas of epithelium bounded by special demarcation lines (Garcia-Bellido et al. 1973; Garcia-Bellido 1975; Crick and Lawrence 1975). They are formed from the descendants of a small number of founder cells and may be determined in fate by "selector genes" (Garcia-Bellido 1975). Compartment boundaries can be identified by inducing somatic mutants with x-rays at various times during development, and by following the subsequent fate of cell clones on wings. Cell clones become successively restricted in potential as development proceeds and are marked by the appearance of a regular sequence of boundaries across which the clones will not trespass. As development progresses a major compartment is successively divided into smaller and smaller compartments. The wing compartment boundaries are determined by some spatial signal. Differential cell division rates do not affect the ultimate shape and size of a given compartment.

Segmentation in insects may also be determined by a compartmentalization process. In the milkweed bug, *Oncopeltus*, segmentation is defined between the late blastoderm and early germ band stages. Each segment is a unit of a cell lineage and develops from a small group of founder cells set aside in the embryo (Lawrence, 1981). In *Drosophila* embryos, segmentation is first visible one hour after the onset of gastrulation as a repeated pattern of bulges in the ventral ectoderm (Nüsslein-Volhard and Wieschaus 1980). Within each segment, other developmental phenomena may determine arrangements of structures such as surface features of the cuticle (Locke 1959; Locke and Huie 1981; Lawrence 1981; Nüsslein-Volhard and Wieschaus 1980).

The nature of compartments in insects can be generalized to other organisms as **developmental fields** or regions of strong presumed developmental interaction, relatively independent of other regions. Such dependency has long been known to be important, at least as a morphological correlation. **Pearson's Rule** states that correlations among body parts are stronger with increasing proximity (e.g., Sokal 1962). In mammalian tooth development, for example, correlations between measurements in adjacent teeth are often better than in distantly located teeth (Kurtén 1954). In man, absence of the third molar is correlated closely between right and left sides, but uncorrelated between upper and lower jaws (Garn and Lewis 1963). The reappearance of the long lost metaconid-talonid dental complex in the lynx seems correlated with the reappearance of the second molar (Kurtén 1963). Some evidence also exists for a correspondence between dental developmental fields and order in development, as in the phenomenon of compartmentalization found in insects. Tooth genesis and eruption sequence is well known to be stereotyped in mammals. Polymorphisms in sequence, however, are known to occur. In cases of third molar agenesis in humans the eruption sequence P2M2 is the usual case, as opposed to the normal eruption sequence M2P2.

As will be discussed below, the spatial scale of the effects of diffusion of morphogens suggests that major effects within a developmental field can only occur over short distances. Thus an early small embryo can be divided into major realms. As the body increases in size, such morphogens can only affect progressively and proportionately smaller regions of the body. Thus there is a readily understandable mechanism in insect embryos for the progressive restriction of developmental fate of the descendants of a small group of cells.

Homoeotic Mutants–Switches of Developmental Fate

Since some developmental fields can be relatively independent of others and since the field, at the time of its origin as a few embryonic cells, might be subject to strong changes, it seems likely that major changes in developmental fate might be induced by a change of developmental instructions. Such seems to be the case in the renowned **homoeotic mutants**, which are defined as variants with alternative differentiative capacity.

The *bithorax* complex of *Drosophila melanogaster*, the best studied series of homoeotic mutants, was first discovered by Calvin Bridges in 1915. E.B. Lewis (1963) found that eight complementation groups explained the total variation. In fruit flies, as in other Diptera, the mesothorax bears the second pair of legs and the single pair of wings. The third pair of legs and halteres are found on the metathorax and the first abdominal segment lacks appendages. Bithorax mutants alter this arrangement. Apparently the form of the second thoracic segment is a sort of ground state, as mutants tend to alter the fate of more posterior segments towards this form. *Ultrabithorax*, for example, converts the third thoracic and first abdominal segments into second thoracic segments (it is lethal in the homozygote). The interaction of alleles at the eight loci is complex and not fully complementary, but there seem to be two major independent domains, whose boundary lies within the first abdominal segment (Struhl 1984). Major switches of development are common in most mutants. These must be associated with switches of determination at the time of segment formation. The effects have polarity, since mutants affect segments posterior to the second thoracic and not anterior.

The mutants seem to be the result of major alterations in the genome that apparently affect development quite severely (Bender et al. 1983). The mutant *bithorax* (first abdominal→second thoracic) is associated with the mobile repetitive element *472* while bx^{34e} is associated with the mobile element *gypsy*. In other cases large lesions of up to 73,000 base pairs (out of the 195,000 base pairs investigated) are the cause of the mutants.

Homoeotic mutations are also common in other arthropods. The cyclops mutant discussed above in the brine shrimp *Artemia salina* is a good example. To a lesser degree such variants can also be seen in vertebrates. The *ametapodia* mutant in the chicken causes the wing to develop approximately as a leg (Cole 1967), but no vertebrate system shows the simple switches characteristic of the arthropods.

Limits to the Autonomy of Developmental Fields

For an evolutionary event (mutation) to affect one unit independently of others, the effects of a gene, or set of genes, must be mainly targeted at the developmental unit. In the case of insect compartments, this is plausible, especially when the few cells giving rise to the eventual polyclone are first instructed. A mutant whose effect is specific to that time and locality might affect the fate of the entire compartment. Subsequent mutations could only affect parts of the compartment under consideration. The same could be said for more vaguely defined developmental fields, such as dental correlation fields.

The autonomy of developmental regions may be limited, which would restrict the shuffling of epigenetic units in evolution. First the pleiotropic action of mutants may cause widespread effects on the phenotype and not on just a single developmental field. Furthermore, many organisms do not consist of autonomous groups of cells whose spatial interactions are preserved throughout development, after a singular determination of differentiative fate at some crucial point. In vertebrates and other organisms, cell migrations bring groups of cells with previously determined fates into contact. Thus mutants causing transdeterminations of fate in one group of cells will result in disharmonious functional interactions with other cells with which they come into contact.

In vertebrates, cell mingling and migration in early development is far more important than in insects. This can be seen by following clones of cells in mouse chimaeras. The development of the neural crest in the chick is an excellent example of such complex rearrangement of the positions of embryonic cells (Le Douarin 1980). The neural crest is a transitory structure of the vertebrate embryo arising from the lateral ridges of the nervous primordium. It gives rise to neurones, supporting cells of the peripheral nervous system, and other tissues. During development, cells leave the crest and migrate through the developing embryo, using specific adhesion signals from other cells. The pathways of travel are crucial to the final location and fate of the cells. Even the expression of homoeotic mutant systems, such as the bithorax complex in *Drosophila*, can be spatially complex. A group of more than 20 genes have been found which controls the spatial expression of bithorax genes (Jurgens 1985).

To the degree that such intricate migration or spatially complex genetic expression occurs, the probability diminishes that independent developmental units can be transdetermined completely into functioning new units. After all, it may be possible to change the fate of a migrating cell from the neural crest, but it will find itself in a local-

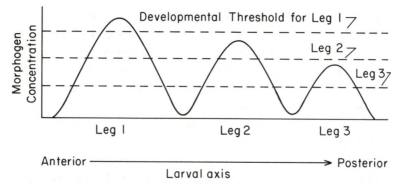

Figure 5.6: The development of a structure may be initiated above a threshold, within a concentration gradient of morphogen (model from Stern 1968).

ized region "designed" to accommodate another neurological structure. Thus the degree of cell migration increases the probability that disharmonious interactions will result in a phenotype of low overall viability.

Organizing Regions, Polarity, and Pattern

Positional Signaling and Pattern Formation

Proper development depends upon mechanisms that translate signals into spatially significant responses. Major patterns, such as segmentation in insects, are established very early in development. The question is how a group of cells "interprets" their position to produce a series of structures that are spatially correct (Wolpert 1969). It has been generally thought that some mechanism of **positional signaling** is necessary to generate spatial inhomogeneity across a group of cells. Individual cells would then be induced by "interpreting" their position at either a specific stimulus level in a spatial gradient or by their location (Fig. 5.6) at a local peak of stimulus (Stern 1968, p. 136, 159). Three principal mechanisms of inhomogeneity have been proposed: (1) concentration gradients of morphogens; (2) mechanochemical stimuli producing waves through chains of cells; and (3) electrochemical fields. Most suggestions are based upon either the presence of spatial inhomogeneities, as in electrochemical differences across cells (Nuccitelli 1983), or stem from plausible models, as in the cases of mechanochemical ap-

Figure 5.7: Generation of different tail color patterns by a diffusion reaction model (from Murray 1901).

proaches (Oster et al. 1983) or diffusion of morphogens (Meinhardt
and Gierer 1974). Direct empirical demonstrations do not permit a
choice among the mechanisms as yet. For illustration, I will show how
spatial inhomogeneities might be generated and interpreted with the
morphogen and mechanochemical models. We have discussed above a
model for gastrulation using mechanochemical interactions in a sheet
of epithelial cells (Odell et al. 1981).

Morphogens might be complex molecules or simple inorganic con-
stituents; we only require that they can be established in a spatial
gradient. These would depend upon either diffusion, or upon active
transport across groups of cells. At the time of determination of spatial arrangement, most primordial structures are about 100 cells or less
in size (Wolpert 1969). Given a simple diffusion model and given the
known developmental period of several hours, a morphogen diffusion
gradient can operate over a field of cells of a few mm (Crick 1970). A
series of **diffusion-reaction** models have been suggested where cells
at local peaks or thresholds of morphogen concentration react by developing in a specific fashion (Turing 1952; Meinhardt and Gierer 1974).
The cells are said to contain a **prepattern**, which determines the fate
of the cells depending upon the local morphogen concentration. In
particular, Murray (1981) has modeled coat color in mammals by assuming that two substances with different diffusion coefficients diffuse
and react on a two dimensional surface. The size and shape of the primordial field (set up in the embryo) is crucial in the determination of
pattern, as this determines how diffusion and local reaction can permit
given concentrations to develop at any given spot in the developmental
field. In a tapering tail, for example, an increase in size changes the
determined pattern from striped to dappled (Fig. 5.7). This variation

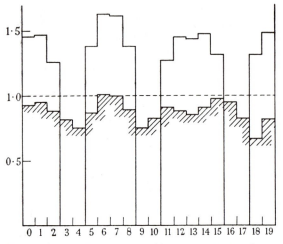

Figure 5.8: Development of spatial inhomogeneity (clear bars) following a local perturbation in concentration of a diffusable substance (hatched bars) (details of model in Turing 1952).

seems to fit the known range of patterns in felids. Murray has also successfully modeled scapellary stripe markings in zebras. The patterns are constrained by the presumed diffusion mechanism.

Peaks in concentration can be generated from simple cell-to-cell diffusion by the creation of minor local instabilities (Turing 1952). An analogy can be drawn with an electrical oscillator, whose frequency is determined by its characteristics. The oscillations will be initiated, however, by random disturbances in the circuit. The phase, but not the wave length or amplitude, will be determined by the timing of the disturbance. The latter two parameters are determined by the number of reactants, the diffusion coefficient, and the fluid through which diffusion occurs. With a small disturbance (i.e., a departure from a previous equilibrium) diffusion from one cell to another will commence, resulting in a chain reaction and a potentially long-term spatially cyclic variation in concentration. In a ring of cells, the interactions of two or three morphogens are sufficient to form regularly spaced peaks (Fig. 5.8) of concentration, given some initial local instability in the ring (Turing 1952). Simultaneous spread of an activator and an inhibitor of morphogenesis can result in spatially regular peaks with random fluctuations of activator concentrations at the source of activator production (Meinhardt and Gierer 1974). Local instability,

Figure 5.9: Himalayan rabbit raised at a temperature below 14 degrees. Temperatures listed represent critical body surface temperatures, above which no pigment is formed (after Stern 1968).

such as a contraction event of the apical end of a cell, can also generate waves of movement through adjacent cells (e.g., Oster et al. 1983). Here, too, other parameters (e.g., elastic properties of cells, rates of reaction to chemical stimuli) would determine the wave length and amplitude of response (Fig. 5.8).

Such models require that the local cells interpret correctly thresholds or relative values of a morphogen concentration or some other physical parameter. Unfortunately there is little direct evidence for such threshholds. However, on a spatial scale larger than diffusion, color pattern in the Himalayan rabbit seems determined by threshold reactions of local cells to temperature. The Himalayan rabbit occurs in a variant that has variable pigmentation, depending upon ambient temperature (Sturtevant 1913; Iljin 1927). A variable prepattern throughout the body interacts with a threshold temperature. At high ambient temperatures the rabbit is all white, while at very low temperatures the animal develops uniformly dark coloration. At intermediate temperatures (Fig. 5.9), poor blood supply to the extremities causes the skin to drop below a threshold temperature, resulting in the deposition of melanin. The remainder of the rabbit is pale.

Whatever their cause, gradients across primordia are important in the determination of development. The chick limb bud, a primordial structure that elaborates the limb, is one of the best studied systems indicating the presence of such gradients. A mesodermal structure in-

duces the formation of the **apical epidermal ridge** or **AER**. Soon
after, the rest of the ectoderm loses the ability to form **AER**. A maintenance factor produced by mesoblasts just below the **AER** is required
to maintain its activity. Excision of the posterior part of the **AER**
causes many abnormalities to develop in the limb, but excision of the
anterior part of the **AER** causes few abnormalities, save the elimination of anterior structures (Hinchliffe and Gumpel-Pinot 1981). If a
posterior part from another embryo is implanted in the anterior part
of another chick, the second chick will develop the mirror-image digit
pattern (anterior to posterior: 432234). Such experiments lead to the
inference that a **zone of polarizing activity** exists in the bud posterior, and that a morphogen diffuses anteriorly. Implants at different
distances suggest that the field of reaction is about 300 *microns* long
(Summerbell 1981). Dorso-ventral and proximo-distal concentration
gradients are also required to provide sufficient information to form the
limb (Meinhardt 1983). It is not clear whether formation of digits is
induced at concentration peaks, or whether thresholds of concentration
are involved. Since it is known that very different mechanisms occur,
for example, in the formation of interdigital spaces in reptiles (differential cell proliferation) versus amphibians (interdigital cell death), we
should expect spatial developmental mechanisms to vary among taxa.

   Gradients also seem important in the determination of segment
patterns in insects. In *Drosophila melanogaster* the early establishment of segmentation is followed by patterns affecting alternate segments, which in turn is followed by the establishment of gradients
within segments (Nüsslein-Volhard and Wieschaus 1980). Segment polarity mutants cause deletions in posterior denticles. In the milkweed
bug *Oncopeltus* the chain of segments sets up something analogous
to a wave pattern; structures within segments are determined by the
phase in relation to other elements. The segmental gradient in two
successive segments can be likened to a gradient of values (10,9,...
2,1,0/10,9,... 2,1,0). Experimental apposition of values 2 and 8 results in the intercalation of positions 1,0,10,9. If positions 5 and 5 are
apposed no intercalation occurs (Lawrence 1981). In insects, some of
the anterior-posterior difference in cuticular structure may be due to
antero-posteriorly oriented basal cytoskeletal extensions, or feet, which
contract and cause transverse ripples (Locke and Huie 1981). The initiation, extension, and contraction of the feet may be hormonally con-

trolled, perhaps by ecdysteroid in the hemolymph. As discussed above, the exact pattern of structure might be determined by physicochemical laws more than by any shaping by natural selection. This mechanism for controlling form is called fabricational noise and will be discussed in Chapter 6.

Evolution of Discontinuous Traits

The quantal nature of pattern determination raises two important points for development and evolution. First, variation in morphogen concentration, or number of concentration peaks in a field, may have an important effect on the phenotype. Second, the evolution of increasing and decreasing numbers of units may involve crossing and readjustment of thresholds to prevent the formation of partial structures. As mentioned above, developmental programs are evolved to generate complete, rather than partial, units. This suggests the action of strong stabilizing selection on morphogen concentration gradients during pattern formation.

To understand the problems caused by natural variation in developmental gradients, let us consider a simple field of length S, with a series of sinusoidal waves of morphogen concentration, with wavelength l. Assume that no partial structures are formed and that the number of repeated structures corresponds to the integer nearest to S/l. Variation in S/l should be the outcome of variation of flux of morphogens, size of the field, and variation in the reacting cells. How much variation in a developmental signal can occur before it is no longer possible to permit the vast majority of individuals to develop the appropriate number of segments or digits?

If variation of other traits is any clue, the coefficient of variation of either S or l is likely to be no less than 2 or 3 percent. This sets a limit on the organism's probability of producing the correct number of repeated structures (Maynard Smith 1960). Above a repeat number of 5-7, natural variation precludes the possibility of getting it right, at least more than 95 percent of the time. For example, in order to produce the exact number 30 in over 95 percent of the population, the coefficient of variation of S/l must be less than 0.85.

This consideration provides a mechanism for a common observation that quantal structures vary from the mode more often in large repeats than in small repeats. We often relate this to the functional

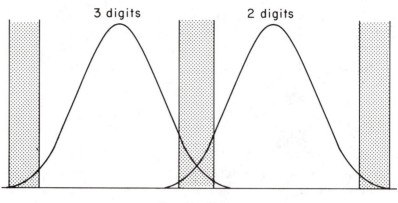

Genetic Scale

▦ Intermediate Zones

Figure 5.10: Model of polygenic digit determination in lizards with threshold effects determining phenotypic expression.

importance of producing an exact small number of structures, heads or digits for example, as opposed to an inexact larger number of units, such as hairs. The difference in determination may, however, be due to a developmental constraint combined with natural variation.

There are cases where numbers of units vastly greater than 5-7 are determined exactly. Segment number in leeches is fixed at 33. Regeneration experiments in some polychaetous annelids suggest that, when a few segments are removed, the regenerative process reestablishes the correct number. This suggests that more than a simple overall diffusion-reaction mechanism operates to ensure an exact number of repeats. Segmentation in annelids and insects seems often to include patterns involving interactions within segment pairs (e.g. Nüsslein-Volhard and Wieschaus 1980); this can give more localized control on segment number.

An appropriate genetic model regulating the determination of complete units would incorporate stabilizing selection on developmental mechanisms. Concentrations typically falling near a threshold between the determination of different structures would often cause serious errors in development. One would expect, in a population with modal digit number n, a distribution of effects, with the mode centered farthest away from any thresholds (Lande 1978; see Chapter 3). Genotypes would be expressed as different phenotypes depending upon their respective relationships to the physiological threshold. This would ensure that any small genetic variation from the mode would still result in production of a finite number of structures. Such a normal distribution of effect is illustrated in Fig. 5.10.

Consider selection from n to n-1 digits. This must have occurred in lizards such as the genus *Bachia*, where digit reduction is associated with the evolution of the fossorial habit. As selection proceeds for reduced digit number, the mode of genetic variants will be shifted to a predominance of genotypes near a threshold of determination between n and n-1 digits. Near this threshold, we would expect that incomplete units will be more common in the population, along with phenotypes with n and n-1 units. Incomplete units, such as segments and vertebrate skeletal articulations, can be common in natural populations (e.g., Wells 1957; E.E. Williams 1950).

During the evolution of digit loss one should expect a difference in phenotypic evolutionary rate. At first, the center of the genotypic frequency distribution will be centered on those genotypes determining n digits, and far from the threshold for n-1 digits. Selection at this stage cannot be a very powerful process owing to the low diversity of phenotypes. But when the distribution shifts such that the mode is near the threshold, deviant phenotypes are more frequent and selection will work more rapidly. If we assume a normal distribution of effects along a phenotypic scale, with the mode centered between the thresholds (Fig. 5.10), the rate of change should be proportional to xh^2s where x is the position on the genetic scale, h^2 is the heritability and s is the selection coefficient. When the distribution is centered on the threshold, the rate is proportional to $0.395\ h^2s$. The rate of change is approximately six times faster when the center of the distribution is farthest from the threshold (Lande 1978). This provides a mechanism for rapid transition from one digit number to another. Using a conservative value of 0.1 for h^2 and 0.01 for s, change could be effected in 29,000 y. A normal selection model, combined with a threshold model of determination, can therefore effect rapid transitional evolution between dominance by different discrete phenotypes.

The evolution of unit number must be intimately involved with the determination of thresholds. How does the presence of thresholds influence the rate and pattern of change of a system regulated by pattern formation? This has long been a subject of great interest.

Goldschmidt (1938) discussed the phenomenon of phenocopies or environmentally induced variant phenotypes which resemble major mutant phenotypes. These variants are usually induced by either heat or cold shocks. Although the majority of the variant phenotypes are not genetically transmissible, a small proportion of the variation has a genetic basis.

Waddington (1956) demonstrated that artificial selection for phenocopies can produce strongly discontinuous changes in laboratory populations. Using temperature shock, he induced a phenocopy of *crossveinless*, a mutant with a strongly altered wing. Artificial selection for the phenocopies resulted eventually in flies that developed crossveinless wings in the absence of the temperature shock. A similar experimental design produced phenotypes with metathorax partially converted into mesothorax (Waddington 1956). This change, termed **genetic assimilation**, can be explained by the cumulative action of many genes moving developmental determination past a threshold. The developmental "defect" is now no defect at all, as selection has produced a harmonious combination of genes producing a deviant phenotype.

Evolution in most situations will not involve induction of such phenocopies. Furthermore, natural selection should not be expected to necessarily favor the phenocopies, even if they are being induced by a sudden environmental change. The phenomenon of genetic assimilation, however, indicates that previously available and genetically controlled developmental programs can be mobilized and perhaps favored by natural selection.

At the level of embryogenesis, these discrete jumps suggest the presence of prepatterns which can be induced to determine new and discrete phenotypes in the population. Sohndi (1962) performed a selection experiment on a laboratory *Drosophila subobscura* population with predominance of the *ocelli-less* mutant, which lacks ocelli and bristles on top of the head. After 40 generations there was a noticeable increase of response. At this time females had essentially the wild type configuration of ocelli and macrochaetes. With further selection, neomorph setae arose between orbital and ocellar setae, a pattern never found in the Drosophilidae but typical of the Aulacogastridae. Selection may have involved decreasing amounts of a precursor that first determined the *ocelli-less* phenotype, then the wildtype, and finally the wildtype plus neomorph arrangement. The wild type cells in normal population may therefore have the prepattern, but may not be able to produce the new phenotype, due to a lack of sufficient morphogen.

These threshold effects should be distinguished from major developmental reorganization. The thresholds have already been built into the determination of the phenotype to ensure proper development, though some may have been suppressed during evolution. Natural selection works within the available developmental constraints. All the variation has been evolved and exists currently. The pattern of character appearance is, nevertheless, saltatory. Such saltatory changes can occur as a matter of normal selection processes and need not require major genetic reorganizations or speciation.

Developmental Organization and Macromutations

Atavistic Features: The Whisperings Within

Atavisms are coordinated, often incomplete, structures that appear as developmental anomalies and resemble ancestral character states of the taxon to which the individual belongs. We presume that such developmental anomalies often induce already evolved features that are normally suppressed.

The degree of organization and apparent antiquity of atavistic features is often surprising. Consider the atavistic partial limbs of sirenians and cetaceans. The fossil record of the Cetacea is spotty, but early representatives (e.g., *Protocetus*, Middle Eocene) had a welldeveloped pelvis and sacrum, indicating the likely presence of almost complete hind limbs. Modern whales lack hind limbs. Early Eocene forms have a dentition resembling both carnivorous mesonychid terrestrial mammals (the presumed ancestors of whales) and middle Eocene forms (Gingerich et al. 1983). Andrews (1921) described a humpback whale, *Megaptera nodosa*, with hind limbs over a meter long. The femur was nearly complete, even though it is normally an internal and diminutive cartilaginous element. Specimens of the sperm whale *Physeter catodon* have been discovered with femur and partial phalanges, even though the femur is normally a rudiment (Lande 1978). If the atavisms of whales and sirenians are any indication, the capacity to reexpress limb elements long lost in evolution suggests that some measure of developmental organization has been retained. Although the final development of a complete limb has been suppressed, the genes or epigenetic subprograms specifying the structures themselves appear to remain to some degree of completeness.

Other atavisms relate birds to their reptilelike ancestors. Kollar and Fisher (1980) placed chick epithelium in contact with mouse (*Mus musculus*) molar mesenchyme. The result was a great surprise. A variety of dental structures appeared, including perfectly formed crowns with differentiated ameloblasts, depositing enamel matrix. For some of the structures to have been produced, the bird epithelium could not have lost, after at least sixty million years, the appropriate mechanism to induce the production of dentin in mouse mesenchyme. This result is grander in scale, but similar to Kurtén's (1963) discovery of the reappearance of an associated group of dental structures, sometimes including the second molar, in the lynx (*Lynx lynx*), otherwise unknown in the Felidae since the Miocene. The absence of complete teeth could be explained either by loss of genes to completely specify the structures, or by some difference in the inducing mesenchyme. But the genes needed to specify the organized structures have not been completely lost.

Perhaps the most celebrated inference of an atavism was reported in the chick (Hampé 1959). Avian evolution involved a reduction of the fibula, relative to presumed reptilelike ancestors. During the development of the avian leg, the fibula is shortened relative to the tibia, while the proximal tarsals fuse with the fibula. These adult features are typical characteristics of the avian limb. Hampé altered the course of development with a series of experiments, such as grafts of fibular primordial tissue, or by insertion of a barrier to equalize blastema size between the tibia and fibula. Two remarkable things resulted from these manipulations. First of all, the chick formed a structure resembling a complete fibula. But when the fibula was completed, it apparently induced contact distally to the joint, reminiscent of *Archaeopteryx*. Muscular development is also altered in an ancestral direction (Müller 1985). In turn, *Archaeopteryx* seems intermediate to the reptilian condition, where a series of ankle bones exist instead of two fused elements at the ends of the tibia and fibula, respectively. Again, the evidence supports the notion that the genes to specify structures have not been lost; what has been lost is a switch mechanism which controls their expression.

The common reexpression of such traits suggests the retention of a large degree of organization in either the genome or the epigenetic programs, despite the loss of the phenotypic structure. If the same genes are retained, this would be superficially surprising to anyone expecting that a relaxation of selection should produce a corresponding relaxation and eventual elimination of the genes for the structure (Kollar and Fisher 1980). The result indicates that there really are no genes specifying only one structure. The genes, rather, must have pleiotropic effects on other characters of selective advantage for the organism (Lande 1978). It suggests that some of the organization among genes may be used in yet other structures. The epigenetic reaction to a developmental manipulation, therefore, is somewhat automatic.

As Waddington (1940, 1942) notes, the elaboration of the trait, including the tissue interactions involved, might eventually become so important that the genes involved might shift with no disruption of the epigenetic outcome. Genes that influence and help to effect the same epigenetic program might come and go. He refers to this process as canalization of development; the developmental program assumes an importance in itself. In cases where a structure is suppressed in evolution, but reexpressible, it is still possible that the reexpressed epigenetic programs might still be used in the determination of other structures. The preservation of potential for reexpression of the phenotype therefore may not necessarily indicate that the genotype has been exactly preserved. There must therefore be an analogous situation to pleiotropy which has only genetic implications. To distinguish between preserved genes and developmental programs which have widespread influence, we should establish the terms *genetic pleiotropy* and *epigenetic pleiotropy*.

Developmental and Evolutionary Pathways–
Jumps Across the Breach?

The preceding sections indicate that the developmental phenotype is a strongly interactive unit, and that evolutionary direction is channeled by the previous evolution and internalization of developmental programs. The question arises: can major new phenotypes spring into being by simple rearrangements of developmental interactions, and can evolution incorporate such major macromutants?

Atavisms have recently been a center of attention among evolutionary biologists (e.g., Lande 1978; Riedl 1978; Gould 1982*b*; Alberch et al. 1979; Raff and Kaufman 1983). This interest reflects the feeling that something important must spring from the fact that major traits, apparently long lost in evolution, can be reexpressed by simple manipulations such as rearranging tissue interactions. As Gould (1982*b*, p. 343) notes: "But the genome embodies an extensive set of latent capacities, some the echoes of distant ancestors . . . Small changes in rates often activate these potentials, and the result is not only a surprise, but often a major one." The proximate phenomena of tissue interactions are therefore believed to represent the ultimate limitations that development imposes on evolutionary directions.

The induction of major atavisms seems consistent with the argument that changes in early development can rearrange morphology and may produce major discontinuous macroevolutionary jumps. If an increase in calcium can induce a chick to make ankle bones, the long lost structures of reptilelike ancestors, then why can't minor changes cause major saltational jumps in evolution? This impression is only reinforced by major mutants, such as the magnificent *cyclops* mutant of *Artemia salina* discussed above.

The expectation of major jumps stems from an important assumption, which is best phrased by Maderson et al. (1982, p. 307): "developmental constraints are the basis of many discontinuities and clumpings in organic morphospace. The sparse and clumped distribution of morphology does not represent a set of optima constructed by natural selection from a set of unbounded possibilities." If this is true, then the major problem of morphological discontinuity among taxa is solved readily. Maderson et al. go on to say that such a point of view is "not congenial" with the traditional notions of the Modern Synthesis, due to the inferred rapidity of origin of novel structures and the nonadaptive nature of evolutionary change under developmental constraints.

The problem with this argument is the fact that the structures themselves, while determined as units, must be integrated functioning units to permit the organisms to survive in the external environment. The evolution of the unit is a different process from the determination of number, symmetry, or position of the units. The evolution of the unitized switch of the dauer larva of *C. elegans*, as discussed above, may have involved a gradual accumulation of the actions of many genes. A developmental switch gene would arise only after the gradual evolution of the developmental unit is complete.

The same can be said for any structure. Gradual evolution will form the structure, but only then can other mechanisms switch on or off the developmental program invoking the structure's presence (Mayr 1963, p. 220). The quantal nature of development is thus somewhat deceiving. A long process of gradual evolution may be behind the current presence of a quantally determined developmental program. Once the program is evolved, then its burden, in the sense of Riedl (1978), will determine whether it can be shuffled out of the organism's total development or rearranged to cause the combinatorial evolution of already extant developmental units. The same type of misinterpretation was behind the belief that butterfly mimicry consisted of single-step mutations (Punnett 1915; Goldschmidt 1945). Because the entire mimetic morph seemed to segregate as one unit, Punnett believed incorrectly that one gene, accumulated in one step, determined the mimicry. In fact, gradual evolution had resulted in eventual linkage of many genes whose aggregate effect, in combination with unlinked modifiers, produced the mimetic morph. Current quantal status does not indicate quantal evolutionary steps.

The significance of atavisms stems from this same consideration. It is often easy to induce novel structures by simple experiments. But this act only induces what evolution has already created. It does not, and could not, propel the organism forward into a new morphological realm. This is not to say that a possible reshuffling might not lead to occasional novel combinations and a rather distinctive organism— a minor hopeful monster, so to speak. But the reinduced structures have already evolved, and this evolution may have been through the cumulative action of many genetic modifications. The major work, in other words, has already been done. That is why the new structure appears to be so organized.

The importance of atavisms is overstated because of the fallacy of looking at a highly evolved system, backward toward ancestral states that have been suppressed in the evolution of new taxa, and then mistakenly inferring a creative force. The complete suppression of these states may have taken millions of years and many intermediates. But a bird is not just a reptile with suppressed features. The evolution of the derived states characterizing the class Aves involves the origin of many new structures, though certainly preexisting materials and developmental programs must have been modified. If lizard epidermis is placed in contact with chick dermis, it does not produce feathers, only scales! Although it is of great interest that the genes specifying ancestral structures are often not lost, mere suppression does not create organized new structures. Looking backward is not the same as moving forward.

The prospect for the single-step origin of wholly new developmental descendants is also falsely enhanced by examining the results of embryological manipulations. Because the formation of so many structures is initiated by tissue contacts, it is possible to create novel structural rearrangements. But the presence of claws on wings comes only from an isolated manipulation; the wider effects of a mutation with such an effect on the entire phenotype cannot be understood properly with such experiments. Broader investigations must be done to understand whether such changes can be isolated, or whether they have other deleterious effects.

As a general rule, major developmental mutants give a picture of hopeless monsters, rather than hopeful change. Epigenetic and genetic pleiotropy both impart great burden to any major developmental perturbation. Thus it is unlikely that mutants affecting any fundamental prepattern in development are likely to produce a functional organism. Genes that activate switches in prepatterns are not sufficiently isolated in effect on other parts of the phenotype to expect major saltations. The cyclops mutant of *Artemia* is lethal. The homoeotic mutants of *Drosophila melanogaster* suffer similar fates. The bithorax complex must function perfectly since it controls, at the least, the fate of segments posterior to the second thoracic (E.B. Lewis 1978). The *ultrabithorax* mutant converts the first abdominal and third thoracic segments to second thoracic-type segments. But the gene is always lethal in the homozygote. Disruptions, i.e., mutants, have drastic effects on other parts of the phenotype. The *apterous* mutation, for example, causes reduction of wings and halteres. But it also is the basis for nonvitellogenic oocyte development, failure of development of the larval fat body and precocious death, though these effects are alleviated somewhat at lower temperatures (T.G. Wilson 1981 *a, b*). The *engrailed* mutant strongly changes the prepattern for the normal formation of the *Drosophila* sex comb. A secondary sex comb is formed. But the *engrailed* gene also affects the formation of wing veins and development of the scutellum. Pleiotropy reduces the likelihood of major switches that escape tremendous drops in viability due to correlated changes.

The problems with the evolutionary potential of atavisms is illustrated by a relatively minor atavism in the guinea pig (Wright 1934, 1935*a*, *b*). A polydactyly mutant, *Pollex*, increases the number of toes in the hind limb of the heterozygote from four to five, the presumed ancestral pentadactyl vertebrate limb condition. The homozygote for this allele dies, usually before birth, and usually has even more toes. Other abnormalities are also apparent in a large number of embryonic organs and incoherent development is especially prominent in the limb buds, brain, and visceral arches (Scott 1937). The introduction and fixation of such an "atavistic gene" into a population would be highly unlikely. Highly complex syndromes of this sort are common in the mutations of major effect studied in other vertebrates. As in *Drosophila* they arise from the complex pleiotropic effects of the mutant genes (Hadorn, 1961, pp. 140-148, 182-202).

The organization in development is a two-edged sword. On the one hand the presence of organization indicates that evolution must move along the valleys of Waddington's epigenetic landscape. But organization also is so intricate that major jumps are precluded because of the high probability of producing a major disruption of proper development. Even Waddington's theory of canalization of development (1942) requires gradual reorganization of the developmental process. Hadorn's (1961) monograph of lethal developmental mutants puts it well (p. 304):

> It must also always be kept in mind that normal development requires the co-ordination in space and time of a large number of individual processes. A mutation which causes the slowing down or the speeding up of a single process may easily induce a standstill of development. In particular, if the development of a primordium depends on the organising function of another rudiment and the inducing stimulus is delayed, the competence of a reacting system which follows its own rhythm may have disappeared.

> The process of development must make enormous demands on the harmonious co-operation of the numerous individual

processes which are originated in the genetic substances of
the chromosomes.

Thus, the accumulated evidence suggests that major developmental mutants are of minor portent in evolution. The side effects are drastic. Moreover, the impression created by atavisms and other homoeotic mutants is misleading. The sudden appearance of change only indicates current developmental organization. not a necessary path to macroevolutionary jumps.

It might be argued that hopeful developmental monsters are significant in evolution, but nevertheless too rare to be seen in the laboratory. This may be true and is a frustrating problem since special environmental circumstances could indeed have made such monsters frequent and subject to rapid selection during unusual times in the past. But any geneticist interested in major developmental mutants would be delighted to find viable hopeful monsters in the laboratory, given the various tricks usually necessary to keep developmental mutants in laboratory cultures. But, alas, major developmental mutants are invariably sickly and show pervasive deformities. From both theoretical and empirical points of view, hopeful monsters have led only to hopeless mooting.

Although major jumps are probably excluded from the arguments presented above, evolution by quantal, if rather small, jumps, is far more probable. The phylogenetic and developmental analyses of Alberch (1980, 1981, 1983) working with salamanders, suggested that the previous evolution of developmental programs does result in constraints on the pattern and arrangement of already evolved units. Thus if reduction of the head or the limb is to occur in evolution, certain directions are developmentally more probable (Alberch and Gale, 1983). But the between-population and among-species changes found (Fig. 5.11), for example in *Bolitoglossa*, are rather easy extrapolations of within-species variants (Alberch, 1980, 1983). The stuff of phenotypic evolution is small developmental units of relatively low burden on fitness. Some of these, such as doublings of structures early in development, might eventually have strong evolutionary significance (Raff and Kaufman 1983).

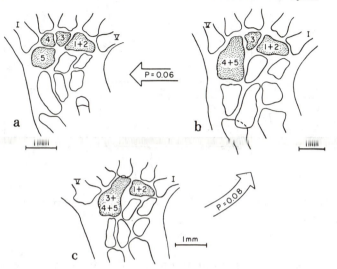

Primitive Condition in Plethodontidae   Primitive Condition in *Bolitoglossa*

Derived Condition in *Bolitoglossa*

Figure 5.11: Patterns of variation in distal tarsal arrangements (stippled) in *Bolitoglossa*, a tropical genus of salamanders. Intraspecific variation can be extrapolated to interspecific variation (from Alberch 1980).

Interaction of Developmental Units with Natural Selection in the External Environment

Adaptive evolution involves a conflict between function in the external environment and the maintenance of an organism with orderly development. Developmental mechanisms must operate to form quantal units. These two considerations lead to the expectation that variation in form within and among species should reflect developmental constraints, interacting with the external environment. In order to elucidate the interaction, we at least need information on developmental patterns and identifiable selective constraints.

Articulations in skeletons should be excellent material for such a study. Joints have definable shapes and must function properly for the organism to survive (or at least jump or twist its neck!). Are regularities of distribution the result of developmental constraints or selection? E.E. Williams (1950) studied the variation in the cervical articulations

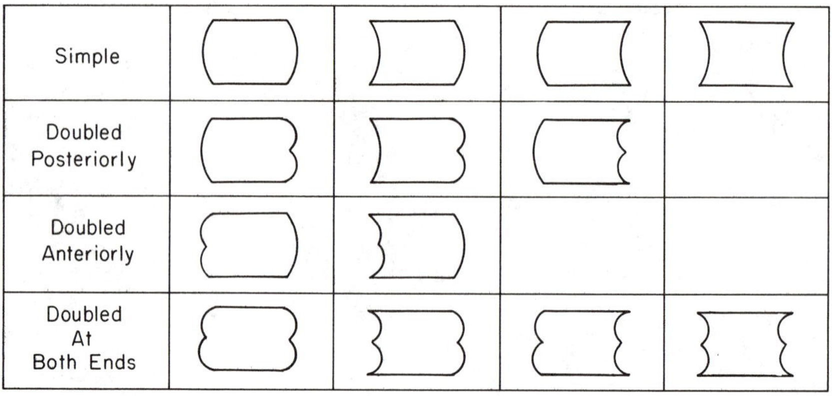

Simple				
Doubled Posteriorly				
Doubled Anteriorly				
Doubled At Both Ends				

Figure 5.12: Some of the types of centra known to occur in turtles (After Williams 1950).

among many species of turtles and found many regularities which could be ascribed to the interaction of natural selection with the constraints set by the particular joint types. In turtles the neck is flexible but the rest of the vertebral column is rigid. The neck is therefore the appropriate area for concentrated study.

Fig. 5.12 illustrates part of the wide array of possible centra. Some notable regularities arise when comparing species from many families. If a joint involves centra with a single convexity or concavity there is no variation in the nature of the joint; it is either fully convex or concave. Any structural intermediates would limit the neck's mobility. By contrast, double joints often show intermediates and are often hard to classify. Doubles function by restriction of lateral motion. The same function can be served by changes in angulation of zygopophyses or by mere broadening of the appropriate cervical joints. Changes in doubleness are thus less important in the first place and more easily compensated.

The occurrence of types of joints in the neck is highly nonrandom. Given the 8 joints studied, there should be 2^8 possible combinations of simple convexities and concavities. However, only 16 are observed. Certain centra are never amphicoelous (double concave), joint VI is procoelous in all but two families, and the cervicodorsal joint is either absent or always procoelous. Williams argues that these regularities relate to function. The presence of the procoelous condition, and the absence of opisthocoely, must be correlated with the fact that this joint is the connection between the mobile neck and the immobile cervical column, the most important center of motion in any testudinate. The constancy of joint VI within species and even superfamilies also suggests some functional limitation.

This study raises some important questions. First, the relative action of joint types in articulation seems related to their constancy of form. During ontogeny, cartilage is replaced by bone. Joint cavities are usually determined by in situ lysis of cartilage. Joint areas are first determined, then the nature of the joint. The degree of precision of determination seems related to whether joint function will be impaired. This, in turn, suggests selection as a contributing mechanism to the developmental mechanism of joint determination.

The patterns of cervical articulations present a possibly more complicated picture. On the one hand, certain joints are conserved. A biconvex centrum, always present in the neck, may be a necessity, but its location can be variable. These phenomona would seem to be due to natural selection, since no plausible developmental mechanism would freeze an articulation in one fixed mode. The arrangement of joints, however, is strongly nonrandom and may be interpreted differently. Williams found that certain cervical formulas fixed in one species were to be found as variants in other species. The sixteen patterns mentioned above may be related to developmental constraints arising from the presence of developmental fields, inferred by Williams. If we consider the most frequent condition to be the ancestral character state for the group, then deviations can be shown to be of discrete combinations of joint types. Although the general observed increase in procoely from the ancestral condition may be due to natural selection, the path may be controlled by developmental constraints.

The influence of developmental constraint on the course of evolution is evident in the development of the amphibian limb. Alberch and Gale (1983) studied the development of digits in the clawed toad *Xenopus laevis* and in the salamander *Ambystoma mexicanum*. A different developmental order of digit appearance was observed. In the salamander, digit formation was in the order $(1-2) \rightarrow 3 \rightarrow 4 \rightarrow 5$. In the toad, however, the order of formation was found to be $(3-4) \rightarrow (2-5) \rightarrow 1$. The last digit to differentiate in either case can be suppressed by the application of colchicine, a mitotic inhibitor, during development. This implies a simple developmental mechanism, perhaps reduction of the

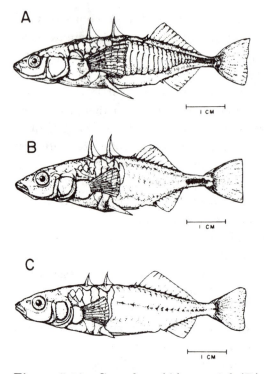

Figure 5.13: Complete (A), partial (B), and low (C) morphs of the stickleback *Gasterosteus aculeatus* (from Bell 1981).

size of the developmental field. The reduction in frogs, furthermore, seems consistent with the pattern of digit loss found in cases involving the evolution of derived forms with strongly reduced size (Alberch and Gale 1983). This suggests an overall relationship between the pattern of morphological change and the developmental constraint.

Alberch and Gale's results also suggest that developmental rules must be established for each major group, before effective hypotheses can be formulated and tested. In this case, the problem also extends to digit formation, which is enabled by interdigital cell death in reptiles (Fallon and Cameron 1977), but by differential cell growth in amphibians (Cameron and Fallon 1967). This creates strong difficulties in understanding the role of development in extinct groups with no close living relatives.

The study of the ontogeny of the threespine stickleback *Gasterosteus aculeatus* provides an opportunity to combine a developmental and functional analysis (Bell 1981). Lateral plates (Fig. 5.13) are modified scales that occur as single bilateral rows. During development the complete morph passes through stages that resemble, in sequence, the low morph and the partial morph. Bell concludes that the complete morph is the ancestral state, and that the other states arose through neoteny. This variation is of strong selective significance, as the anterior lateral plates provide the major structural support for dorsal and pelvic spines. In turn, the spines aid in resistance against avian and salmonid predators (Reimchen 1983). More recently Bell (1987) has examined the occurrence of pelvic vestiges and demonstrates that, while natural selection is behind reduction, developmental constraints channel change through a limited set of directional simplifications that often involve paedomorphosis.

The preceding examples illustrate two main points. First, a knowledge of developmental variation is essential if we are to understand evolutionary directions. Natural selection, however, can be intimately involved in the building and control of the units that are themselves determined by developmental mechanisms. In some cases, the order of development of frog digits, for example, it seems clear that there is no immediate adaptive reason why a particular digit is lost first in evolution. In other cases, however, reduction of given parts will have functional significance relative to the external environment.

The developmental constraints cited above all can all operate within the context of neo-Darwinian evolutionary mechanisms. If there is a difference, it is the emphasis on the mechanisms designed to form discontinuous structures and developmental mechanisms generating variation. The recognition and understanding of discontinuity is of great importance for the understanding of evolution.

Change of Developmental Programs: Heterochrony and Joint Responses

Order and Form in Ancestors and Descendants

Any violation of the biogenetic law is due to the overall process of **heterochrony**, which de Beer defined as a difference in order of appearance of structures between ancestor and descendant. This must entail a phylogenetic change in the onset or rate of development of a structure, relative to another. In other words, descendant structures

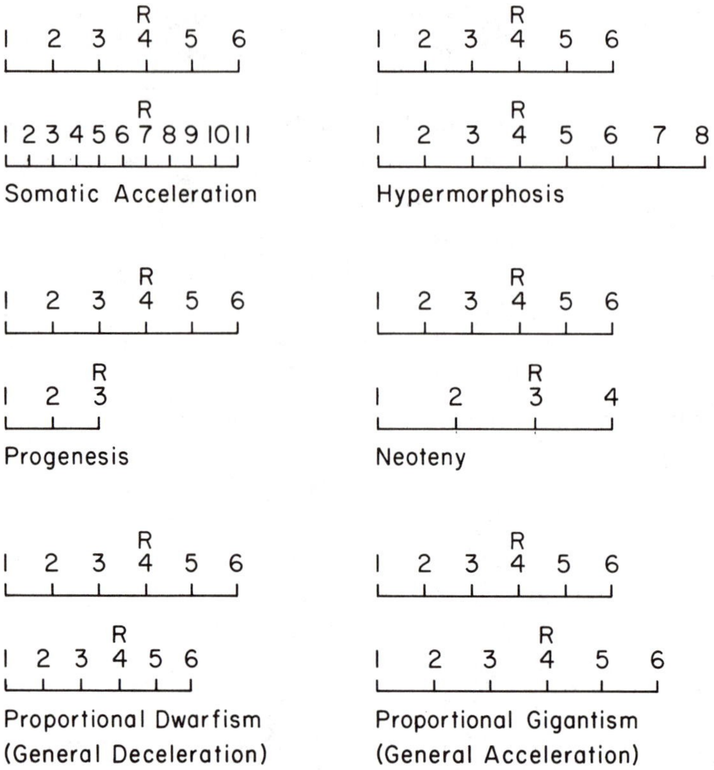

Figure 5.14: Classification of heterochrony, showing outcomes of delay, no change, and acceleration of somatic growth and reproductive maturity. Outcomes of heterochrony are shown on sliding scales.

are accelerated or retarded in appearance in development, relative to those of the ancestor. In making such comparisons, some milestone such as the onset of reproduction is employed. It is unlikely that any developmental milestone is very useful as an absolute marker for comparison. The time of reproduction, for example, is itself subject to selection, and varies substantially within and between populations (e.g., Stearns 1976; Charnov 1982). There is no reason to believe that when selection adjusts the age of reproduction, the relative appearance of other structures will be maintained. Indeed, the time of reproduction, overall growth rate, and the relative or absolute time of appearance of structures might all vary in evolution.

To describe changes in order between ancestor and descendant, de

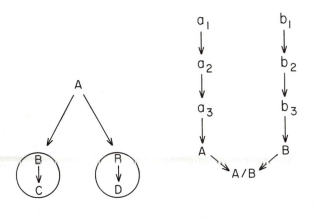

Joint Response   Independent Response

Figure 5.15: Two possible developmental interrelationships. Left: Joint response, where a single stimulus initiates two separate developmental programs. Right: Joint effect of two chains, where the final phenotype is the sum (A/B) of effects of two independent developmental pathways.

Beer devised a cumbersome system which Gould (1977) simplified to the two major processes of **acceleration** and **retardation** of either the time or appearance of a somatic structure or reproductive maturation. The effects of the two processes yielded respectively the patterns of **paedomorphosis** and **recapitulation**. In paedomorphosis, descendants have more juvenile character states at a standard marker of development such as reproductive maturity. In recapitulation, descendants have more "adult" (i.e., terminally added) states at the standard marker. Cases of recapitulation can be consistent with the biogenetic law. Fig. 5.14 summarizes some of the various heterochronic states verbally and with a sliding scale showing the descendant's change in either size, time of reproduction, or appearance of a series of numbered developmental stages (see Alberch et al. 1979 for a quantitative approach).

Joint and Independent Responses

Two different situations of dependence can be imagined. Consider the presence of two independent chains of reactions, whose end products, in combination, produce the visible phenotype. As the two chains can be modified independently, let us term this the **independent response**. Consider a simple model of correlated action, which we shall term a **joint response**. A substance (e.g., a hormone) passes through a cell membrane, and stimulates the cell to produce a substance which, in turn, switches on a battery of genes. This substance actuates genes in dispersed target cells of varying type (Fig. 5.15). We also assume that the genes are switched on after a given amount of hormone reaches the two cell types. A higher rate of production of the hormone results therefore in the earlier induction of gene action in the two cell types.

If a polymorphism for rate of hormone production occurs, genotypes that cause higher hormone production would be expected to switch on the target genes earlier in both cell types. This is the expected correlated response in evolution. This may be an important mechanism for **widespread parallel evolution** (Alberch 1980).

But what if it is disadvantageous for one of the cell types to be switched on earlier? We might expect selection for a change in the rate of the response of the cell. The stimulation-inhibition mechanism described above for *Drosophila* suggest possibilities for the evolution of changes in local response. Protein hormones, for example, can be ingested by endocytotic vesicles at the cell surface, to eventually be destroyed. Different cells will have varying reactions to the same hormone binding to the cell surface. Such differential adjustments might delay response in one cell type, to the degree that the gene battery would not be switched on any more rapidly than before and would become temporally decoupled from other responding groups of cells. Correlated responses are therefore not guaranteed in the evolution of life cycles.

Metamorphosis in salamanders illustrates a chain of necessary reactions, each of which is subject to modification by natural selection. Three systems are required (Etkin 1970). The hypothalamus produces a hormone which stimulates the pituitary gland. The pituitary produces a hormone, TSH, which stimulates action of the thyroid gland. The thyroid, in turn, produces thyroxin, which can induce a response in target organs such as the skin. The target organ must be competent to respond to thyroxin. Any disruption or alteration of this chain will delay or eliminate metamorphosis. Evolution, however, could adjust the responses of different organ systems so that the modification of metamorphosis could be organ specific. This is a specific mechanism of evolutionary alteration of a joint response.

The mechanisms of neoteny in amphibians are known to employ different portions of this chain of reactions. In the axolotl, the hypothalamus fails to function (Blount 1950). Injection of thyroxin into the hypothalamus induces metamorphosis. In the genus *Typhlomolge*, a blind, permanently gilled cave salamander, the thyroid is nearly absent, making a shortage of thyroxin the likely mechanism of neoteny. Other salamanders, e.g., the blind salamander *Proteus anguineus*, apparently have normal hormone systems, but the target organs are incapable of response. An addition of thyroxin fails to induce metamorphosis in these forms.

In the axolotl, many characters appear to be juvenile, relative to the presumed metamorphosing ancestor. But the axolotl develops to resemble the metamorphosed ancestor in many other characters such as hemoglobin. Therefore all characters are not affected as if an entire jointly responding developmental program were arrested by an all-pervasive mechanism. Even though thyroxin can induce metamorphosis, and even though the failure to metamorphose is vested in a single gene with two alleles (Tompkins 1978), the different target organs have evolved an altered response.

At the phenotypic level, the evidence for developmentally correlated change has been best developed by David Wake and his colleagues and students, in their studies of plethodontid salamanders. The neotropical genus *Bolitoglossa* comprises a range of species with strongly varying morphological and ecological features. In highland environments, species tend to be large in body size and have the least amount of interdigital webbing in the genus, while lowland species tend to be small and arboreal and have the greatest amount of webbing. Extensive webbing stems from a failure of the digits to grow out of the embryonic pad. Independent digit action is reduced in favor of the total hand or foot action associated with movement of these organisms across wet or smooth surfaces in arboreal subhabitats (Wake and Brame 1969; Alberch 1981). The smaller size and webbing both contribute to adhesion and suction to flat surfaces such as leaves (Alberch 1981).

In some arboreal species, the total phenotype seems to represent a correlated morphological response based upon an arrest of overall development (Wake and Brame 1969). Species in the rufescens group have greatly reduced dentition, poorly developed skull bones, reduced phalangeal elements, mesopodial fusions, and extensively webbed and flattened hands and feet with sharply pointed digital tips and short tails. In other species the adaptations of the hands and feet are more independent of the rest of the body. Paedomorphic characters in populations of the Californian black salamander, *Aneides flavipunctatus* are similar to the latter case, in that a series of pigmentation shifts result from arrested development, but the change goes on independently of others such as those in the vertebral column (Lynch 1981; A. Larson 1980).

Alberch (1980) summarized the osteological features of the presumed ancestral terrestrial and derived arboreal forms in *Bolitoglossa*. Species bearing the ancestral state have a defined arrangement of tarsal elements (Fig. 5.11), with a small minority (six percent) of another arrangement. A study of a derived arboreal species shows the predominance of yet another arrangement, with increased fusion. The only variant found in populations of this species is identical to the most common arrangement in the species with the ancestral condition. This suggests that the change can arise from intraspecific variation, and that the polymorphism is of small but discrete differences and predictable transformational direction. This is consistent with most known developmentally relevant polymorphisms which tend to have predictable transformations from one mutant to others (as in the *bithorax* series of *Drosophila*).

Restricted variation is also found in the skull elements of *Bolitoglossa* (Alberch 1983). In some species the prefrontal bone is lost, as is the third phalanx of the fourth toe. This is most likely due to a truncation of development. In the derived *B. occidentalis* the only variation observed is in the presence of the prefrontal, making it an atavism. Alberch suggests a general model of phenotype determination quite similar to that of Lande (1978).

While changes in the rufescens group derive from an overall arrest of development, in other species change is focused on the hands and feet. The latter changes suggest that the entire phenotype is not an overall developmental unit so tightly bound that it cannot be broken up by natural selection. Perhaps strong selection tends to favor larger-scale developmentally correlated changes. If selection is relatively weak, genetic correlations may have time to be broken up.

Some insight on the evolution of a developmental program can be gained by examining the plethodontid salamander genus *Aneides*, which derived from the (American) western branch of *Plethodon* (Larson et al. 1981). Here a gradual evolutionary series can be seen in ten morphoclines for which the ancestral states can be determined. Nine of these represent dental or jaw characters. Relative to *Aneides* the maxillary and mandibular teeth in *Plethodon* are greater in number, short and conical in shape, and fill a greater proportion of the maxillary and mandibular rami. The morphoclines in *Aneides* show a gradual transformation toward enlarged dentition and increased skull strength, associated with the arboreal habit. Unlike *Bolitoglossa*, arboreal forms occupy a perch and so adhesion is not important. In ground dwelling salamanders of this group, the animal subdues its prey by holding it against the ground. This would be a relatively inefficient method for arboreal forms, as maintenance of perch takes precedence over other activities. This would suggest selection for strengthened jaws and teeth, which would increase the efficiency of seizing and holding prey.

The success of this group involves two innovations: (1) rearrangement of carpals and tarsals, providing for a redistribution of forces to facilitate climbing; and (2) the fusion of premaxillary bones, providing the basis of a strengthened jaw. Both innovations seem to be discrete steps. The overall morphoclines are consistent with the interpretation that the jaw and skull modifications evolved gradually in a series of separate transformations of many small morph features. Larson et al. argue that the intergeneric transformation, while showing a series of intermediates, was rapid relative to the history of either genus.

This transformation series provides evidence that a developmental program can be built gradually, and that the intermediate steps are likely to consist of a harmonious arrangement of the independent characters. Geographic/ecological range extension seems to be important in such evolution. If the most derived forms were subject to evolution by arrest of development, it is possible that the resultant forms would be, in a sense, preadapted for more terrestrial existence.

In conclusion, evidence from the study of salamanders gives a complex picture, ranging from clear nonadaptive joint responses due to evolutionary arrests of development to gradual assembly of functionally harmonious traits into ontogenies molded by natural selection. Even in the case of joint responses, however, intraspecific polymorphism of discontinuous traits of small magnitude is the stuff of evolutionary change.

Ontogeny, Phylogeny, and Some Evolutionary Trends

Haeckel's biogenetic law was meant to be a grand scheme, embracing all previous ontogenetic theories, such as von Baer's law (Gould 1977). It asserted that all stages in development were indications of the once-living adult forms of ancestors. Thus the zygote was to be interpreted as the original single-celled animal ancestor. Swimming larval forms were to be understood as the free-living ancestors of now sessile descendants whose terminally added stages were adaptations for a benthic existence. This scheme ignored completely the adaptive significance of larval dispersal or the functional importance of earlier stages in ontogeny, not as a primordially adult structure, but as an important adaptation for juvenile existence.

The universality of the biogenetic law was refuted by the demonstration of rearrangements of the order of appearance of structures between ancestor and descendant (Garstang 1922; de Beer 1958; Gould 1977). This is best encapsulated in de Beer's famous example: teeth evolved before tongues, yet tongues appear before teeth in mammalian development. The law still strongly influences our thinking, however. A cynical de Beer (1958, p. 7) noted: "It is characteristic of a slogan that it tends to be accepted uncritically and die hard." True though this may be, discussions above tend to suggest that ontogeny and phylogeny might very well be intimately related, perhaps sometimes to the degree that the biogenetic law may hold.

One special set of cases may predict evolution by terminal addition and the subsequent preservation of order. If, as an organism grows larger, a defined sequence of morphological changes best suits it to function within the environment, then those stages might be added in an evolutionary sequence and preserved as adaptive solutions to an ontogenetically ordered series of changing environments. The microenvironment might change for two reasons. First, the organism might be exposed to a qualitatively different microenvironment as it grows larger. A sea fan may live in the turbulent boundary when small and particle feeding would be omnidirectional. As the colony grows up above the bottom, it might be exposed to a unidirectional mainstream current and require a different orientation for its polyps. Alternatively, the organism might actively change habitat during ontogeny. The Caribbean suspension feeding snail, *Vermicularia spirata*, starts out life as a free-living, high-spired and typically coiled snail. Eventually, though, it attaches to hard substrata, the shell uncoils, and the animal lives as a fixed passive suspension feeder (Gould 1969*b*). If terminal addition were the mechanism for the assembly of such a series of changes then phylogeny would assemble the ontogeny by natural selection.

Mollusks and other creatures secreting accretionary skeletons (e.g., brachiopods, corals, sponges) are natural candidates for such evolutionary trends, since the interactions of the shell with the environment are intimate and change continually as the animal increases in size. Phylogenetic trends often indicate a process of terminal addition. This can be seen, for example, in the evolution of the *Athleta petrosa* lineage (Gastropoda, Eocene, Texas). During its history, body size increases steadily (W.L. Fisher et al. 1964). The following trends are notable (Fig. 5.16):

1. Increase in amount of parietal callus.

2. Anterior accentuation and posterior loss of spiral lirae on body whorl.

3. Ornamentation has the sequence: cancellate→nodose→spinose.

4. Peripheral elements of ornamentation reduced progressively in number.

The appearance of successive stages in ontogeny is accelerated relative to whorl number during the phylogeny. The trends include changes in continuous characters such as linear dimensions, and numbers of spines, and qualitative changes in ornamentation patterns. Thus very different changes can occur in different characters, suggesting very different patterns of developmental, genetic, and adaptive mechanisms within the same evolving lineage.

The Miocene-Pliocene scallop *Chesapecten* has a similar pattern of addition, although intermediate stages in some cases have been dropped (Miyazaki and Mickevich 1982). As scallops increase in size, they often pass from dependency upon byssal attachment to hard surfaces (e.g., sea grasses) to dependency upon free movement and swimming. This ontogenetic change in habit is reflected in corresponding

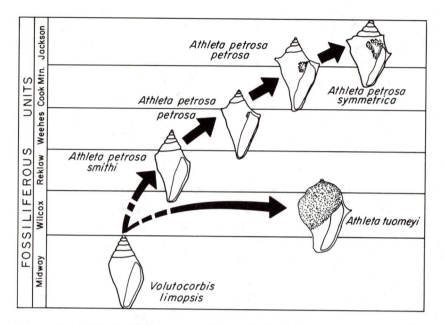

Figure 5.16: Probable instance of terminal addition: expansion of parietal callus deposits in *Athleta petrosa* stock (Eocene, Texas).

changes in, among other characters: closure of the byssal notch, equalization of anterior and posterior auricle length, movement of the posterior adductor toward the center of the disk, decrease of rib strength and number, and increase of the umbonal angle. Through the history of the *Chesapecten* lineage, the evolutionary sequence of character state change of adults follows the direction expectable from ontogenetic change. This is especially clear in the byssal notch and in the equidimensionality of the auricles. Some characters, such as position of the adductors, do not change as expected. Some other trends, such as decrease in rib number between successional species, are relatively constant through ontogeny. The pattern of rib loss during the ontogeny of the most derived species does not mimic a progressive reduction of ribs. Rather, rib number is reduced in the smallest individuals where ribbing is detectable. Therefore, a strict mechanism of terminal addition has not been maintained, though the general model of terminal addition is a useful concept in this lineage.

The second major law of ontogeny-phylogeny is that of von Baer, discussed above and in Chapter two. This law asserts that character states early in ontogeny are more general (i.e., ancestral), while states later in ontogeny are most likely to be more special (or derived). The embryos of related forms resemble each other only due to the lack of change of earlier embryos in phylogeny, and not because they represent stages achieved by adults of ancestors. Though the law derives from a conception of archetype (represented by the embryonic state), it still makes sense from developmental and phylogenetic perspectives. This law is likely to have more generality than the biogenetic law, since the stringent requirement for retention of all intermediate stages is relaxed. The biogenetic law, however, can be considered a special case of von Baer's law, as a phylogenetic-ontogenetic order could still be built up through terminal addition (Gould 1977).

The likely modification of later stages in development, relative to earlier stages, is supported by the argument above for a developmental ratchet. Stages incorporated earlier into a developmental sequence may be more difficult to eliminate. Early stages may serve as epigenetic inducers of later stages. Genes affecting early stages of the development of a given structure, such as the vertebrate limb, might have pleiotropic effects on other developmental systems. Structures appearing early in development might be functionally important, to the degree that they literally support later structures (Lande 1978). Thus the loss of a femur might occur only after more distal and functionally dependent parts were lost in evolution. Finally, simple developmental mechanisms might cause distal, and later formed, parts to be lost before proximal, and earlier formed, parts. The order of digit loss in vertebrates, for example, might depend mainly upon the order of appearance in development, which, in turn, may not have an adaptive explanation (Alberch and Gale 1983). Whatever the specific explanation, the evolution of limb loss in vertebrates is consistent with the loss of successively more proximal-early developmental structures (Lande 1978).

Although a fair correspondence can be drawn between developmental phenomena and von Baer's law, the law applies most specifically to the overall pattern of ontogeny, rather than to individual developmental systems, such as the vertebrate limb, or determination of pattern in mammalian coat colors. Therefore, it seems reasonable to formulate a modified ontogenetic-phylogenetic law, which I term the **Law of Distal-Terminal Transformation**. This law makes the following assertions:

1. Many major structures have some form of integration through development.

2. The development of the structure involves an order of appearance of substructures (as in the proximo-distal elaboration of the vertebrate limb).

3. When an order exists, evolutionary modifications tend to favor developmentally later stages first.

4. Genetic, functional, and epigenetic considerations suggest that distal parts should also be developmentally last, and therefore most subject to modification. Developmentally later stages, in general, probably are of lower burden and are most subject to change.

Paedomorphosis or Recapitulation: Which Has More Potential?

G. de Beer (1958) suggested that paedomorphosis had more evolutionary potential than what he called gerontomorphosis, or the extension of a life cycle by addition of developmental stages (hypermorphosis). The addition of developmental stages is associated with ecological specialization and, therefore, evolutionary restriction. Gould (1977) used the Irish elk, *Megaloceras*, as an example of the largest representative of a clade with less potential than a smaller form with more diminutive antlers. If increasing size can be equated with increasing specialization, then Cope's rule–that size increases during the evolution of a clade–can also be interpreted as a trend toward increasing specialization (Stanley 1973c). Decreases in size would move the clade towards more generalized forms with greater potential for evolutionary change.

It is the purpose of this section to suggest that, on developmental grounds, there is reason to believe that the alteraction of early development will have more global effects on the organism. This will occur in those cases where development is sequentially compartmental, as in insects (Garcia-Bellido 1975). However, mutants of early developmental stages may also result in low viability, since such mutants will affect the entire body. On the ecological level, however, there is no reason to believe that mutants affecting early stages of development will necessarily have greater ecological significance in evolutionary change. This depends on the nature of the morphological change.

Accepting a relationship between the addition of stages and restrictive specialization implies that a neotenous descendant is inevitably more unspecialized and therefore of greater evolutionary potential. G. de Beer (1958) argued that the insects might have arisen from an arrest of development of a Myriapodlike ancestor that otherwise would have added segments with limbs during development. Since, as discussed above, the number of segments is determined *en masse* by a prepattern early in development, the question of paedomorphosis may be irrelevant. Any change in segment number, be it an increase or decrease, involves simultaneous, not sequential, determination at an early stage of development.

While the hypothesis of enhanced ecological potential of paedomorphic forms is occasionally plausible, it is not clear that it is universal. Common examples tend to be in vertebrate lineages where size increases over long periods of time. Since the earlier representatives of the clade lie phylogenetically closer to the more general ancestors that gave rise to a great variety of vertebrate groups, we tend to think of earlier, more "juvenile," and smaller representatives as being more ancestral and with greater evolutionary potential. This must be true from the taxonomic perspective, since taxonomic rank, a tautologically integral component of one's perception of evolutionary potential, increases inevitably as more plesiomorphic ancestors are considered. Because we know, from our retrospective view, that many descendants arose from given stem groups, it seems reasonable that paedomorphic events should increase evolutionary potential, in the sense of taxonomic potential. But this may not be true of ecological potential, nor may it be true of even overall morphological potential at smaller taxonomic levels.

Does an addition of ontogenetic stages automatically decrease ecological and morphological potential? Consider the *Hinnites* group of pteryoid bivalve mollusks. During ontogeny, animals pass through a shelled stage resembling a scallop, with many ribs and typical auricles. Later, however, the scallop cements to the bottom and continues to grow in an irregular oysterlike form. We can assume that this represents a case of hypermorphosis. As an exercise in fortune-telling, consider the potential of this species if progenesis occurred. A scallop would develop, much like other scallops. It is hard to imagine any unleashing of evolutionary potential. But what if an apparent trend toward hypermorphosis continued. An oysterlike form might give rise to a wide variety of irregular forms adapted to varying hard and soft substrata, embracing the current broad distribution of oysterlike bivalves. Indeed, one possible descendant might evolve a symbiosis with algal symbionts, such as zooxanthellae. This would yield an analogue to the rudistid bivalves, a Cretaceous group that dominated tropical reefs, and often reached sizes of a meter or more. The hypermorphic or accelerative addition of ontogenetic stages might lead to a large radiation of sessile forms. Since this group has been around since the Jurassic, we can conclude that such potential does not guarantee a radiation like that of the rudistids.

Some of the celebrated neotenous amphibians give no clear indication of enhanced evolutionary potential. They are restricted to habitats of equal or even greater ecological-morphological specialization. The fixed aquatic habitat of neotenous salamanders such as the axolotl gives them no obvious latitude for evolution over terrestrial forms. Neoteny in salamanders such as the genus *Typhlomolge* places them in an evolutionary cul de sac as specialized, depigmented and blind, cave forms. Neoteny can therefore be a dead end in evolution. It should also be remembered that such neotenous forms are "adult" in a wide variety of other characters. True cases of progenesis are probably rare.

A similar argument can be made for the crustacea. The Cephalocarida are the most primitive crustaceans known currently. The degree of specialization of the first three pairs of limbs is far less than in the other extant crustacean groups. Throughout the history of the free living arthropoda, the degree of limb specialization has increased sigmoidally (Cisne 1975). Taxon survival, however, has not decreased with increasing degree of specialization (Flessa et al. 1975). This may stem from a similar lack of correspondence between limb specialization and ecological specialization. The cephalocaridans are specialized ecologically and are confined to suspension feeding on soft sediments (Sanders 1955). Morphologically specialized decapods, however, range over a large variety of habitats from soft muds (e.g., many species of fiddler crabs) to the interstices of coral heads (e.g., some grapsids). Crabs range from deposit feeding (fiddlers and hermits) to carnivory (many other species), but single species can embrace the gamut. Ecological potential is therefore not necessarily correlated with degree of ontogenetic specialization, since a despecialization does not guarantee more ecological directions.

The cherished belief in the potential of paedomorphs probably stems from the still powerful influence of the biogenetic law and von Baer's law. We still envision development as the unfolding of a phylogenetic series. It is only natural to believe that paedomorphosis peels off the layers of more specialized descendants and therefore begets ancestors with great potential for giving rise to a variety of descendants. Such a prejudice ignores the commonly pervasive evolutionary change of all stages of development.

Although paedomorphosis may not beget increased evolutionary potential, we are left with the apparent fact that many extinctions in the fossil record seem to eliminate relatively specialized forms, and subsequent postextinction reradiations stem from relatively small-bodied, unspecialized ancestors (Stanley 1973c). A possible explanation is the **extinction of advanced forms** hypothesis (Fig. 5.17). Here we argue that highly derived forms have evolved strongly specialized life cycles, and are unlikely to increase potential by heterochrony. Primitive groups, however, may survive and coexist with the advanced forms. During an extinction, we would expect that more advanced forms, being more restricted both ecologically and geographically, are liable to disappear preferentially. Primitive forms, having survived the extinction event, will now invade a variety of new habitats and have an opportunity to radiate via a combination of cladogenesis and adaptive evolution.

The preferential extinction of advanced forms is probably a common feature of cycles of extinction and reradiation. A useful example comes from G.R. Smith's (1981) summary of the history of Late Cenozoic freshwater fishes of North America. Though the Mississippian fauna was relatively little affected, the Pleistocene caused extensive extinction in the western mountain regions. The most severe extinction is associated with the disappearance of the Glenns Ferry stage of fossil Lake Idaho, where 28 species were reduced to 10-14. This extinction focused upon more specialized forms such as the large whitefish *Prosopium prolixus*, the most derived form in terms of size and jaw morphology. Forms that survived were more generalized and primitive

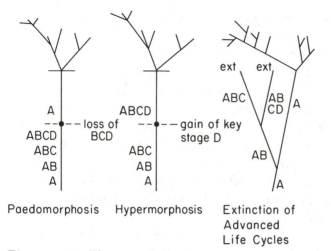

Figure 5.17: Three models of increased evolutionary potential. A, B, C, D represent character states acquired by terminal addition.

salmon, trout, smaller whitefish, suckers and sculpins.

This model is also supported by the iterative evolution of the plank-tonic foraminifera in the Cenozoic (Cifelli 1969). Mass extinction passed ancestral globigerinid forms and focused upon more specialized morphologies. The progenitors of the succeeding radiation of special-ized forms were derived from the surviving primitive generalists, not from specialists who were highly modified by a process such as pae-domorphosis. In sum, the real issue of evolutionary potential is the question of selective extinction and potential for radiation of forms who had not changed until an opportunity arose, for leaving a new diversified array of descendants. Evolutionary potential stems from opportunities presented to surviving ancestral forms, and not usually from paedomorphosic shifts by advanced forms.

The Main Points

1. Development places constraints on the direction of evolution and imposes an internal organization that may be as important in evolution as interactions with the external environment.

2. Structures arise in development through a complex sequence of changing and localized gene expression and spatial interactions. Developmental mechanisms have evolved to determine complete

structures; various threshold mechanisms suppress incompleteness of form. Models of evolution, therefore, must incorporate thresholds and deal with discontinuity in evolutionary change of traits.

3. Evolutionary mechanisms must incorporate the historical constraints of previously evolved spatial interactions and account for completeness of structure. Although development imposes constraints on evolutionary directions, developmental interactions can be broken up by natural selection.

4. It is unlikely that major saltational jumps in evolution are likely to arise even though many structures are seemingly invoked by simple switch mechanisms. Such switches have both genetic and epigenetic pleiotropic effects of sufficient magnitude to suggest strongly deleterious side effects.

5. Most likely, currently discontinuous developmental events were built up gradually and in continuous interaction with the external environment. Upon completion of the evolution of the developmental program, it may have been internalized, that is, insulated from direct interaction with the external environment.

6. In cases where such a process is involved with ontogenetically changing morphology, a rapid shift toward juvenile forms may simply select for a phenotype that is adapted permanently to an environment suited to the juvenile stage of the ancestor. The biogenetic law–that ontogeny recapitulates phylogeny–probably never holds completely, but is most likely to apply when a series of successive phenotypes during ontogeny is each adapted to a different environment. This seems likely in organisms with accretionary external skeletons, such as the mollusks.

7. Paedomorphosis is believed to be a major mechanism for increasing evolutionary potential. Known paedomorphic forms seem no more ecologically versatile than their ancestors. It is more likely that highly derived (= specialized) forms become extinct, and their ancestral, more ecologically unspecialized ancestors survive to give rise to subsequent radiations of specialized descendants. Increased specialization does not necessarily involve increased ecological specialization.

Chapter 6

The Constructional and Functional Aspects of Form

You cannot fly like an eagle with the wings of a wren.
William Henry Hudson

The Performance of Organisms

The Engineer's Ideal

The relationship of form, function, and evolutionary history raises many of the most fundamental problems of evolutionary biology. The subject is complicated by the role of form in both scientific and cultural-religious thinking. Design in organisms has long symbolized both the guidance of a divine hand and an eternal invariance, since perfection is the essence of the divinely created form. Both perfection and nontransmutability played an important role in the resistance to materialistic concepts of evolution. I lost the illusion that animals were perfect upon seeing my sister's dog trip over a doorstep.

Darwinian evolution does not invite a simple alternative to essentialist perfection. If the world were simple and organisms had boundless evolutionary potential, perhaps form might evolve to be the realization of an engineer's design, aimed at fulfilling a series of functions, such as vision, locomotion, etc. We are attracted to this image of perfection, despite our knowledge of vestigial organs and polymorphism (the latter often implies more than one solution to the same overall environmental problems). This attraction derives from three factors: (1) A precise match of form in *some* cases to the expectations of pure design, constrained by materials and various mechanical or energetic laws; (2) the remarkable fit of organisms, or at least particular structures of organisms, to the environment; (3) an approach to the study of form that predicts an optimal form, given an environment and a series of boundary conditions.

An example of the first class, the **functional paradigm** approach, is the form of the lens in the eyes of some trilobites. Most trilobites have typical arthropod eyes with facets and ommatidia (Clarkson 1979). The eye consists of a series of units, arrayed in a honeycomb pattern of varying shapes and orientations. The orientation of the array can define the visual field of the extinct organism (Clarkson 1966, 1979). One presumes that the overlapping fields of the individual units were integrated to give a mosaic image.

Two genera have schizochroal eyes, where the lens is composed of two units of calcite. The symmetry axes of the lens crystals are parallel to the direction of light transmission. Clarkson and Levi-Setti (1975) discovered a remarkable resemblance (Fig. 6.1) of the upper units to the glass lens designs of Descartes and Huygens, published three hundred years ago, but certainly several hundred million years after the trilobites had invented them! The human designs were attempts to reduce spherical aberration. The lower units in each lens, missing from the human designs, compensate for light transmission through seawater.

The prediction of form from engineering principles was formalized in paleontology by Rudwick (1961, 1964). In paleontology, functional morphology is usually as efficacious as the existence of living analogues to the extinct organisms being investigated. For example, Stanley's (1972) excellent study of the functional significance of changes in bivalve faunas through time depends upon a knowledge of living bivalves. Rudwick's prescription was meant to solve the problem of inferring the function of structures of extinct organisms with no close living analogues. He believed that, for every function, there was a **paradigm structure** that could be predicted from various first principles (Rudwick 1964). "This is the structure that can fulfill the function with maximal efficiency under the limitations imposed by the nature of the materials" (Rudwick 1961, p. 150). If the paradigm fits the structure in the fossil organism, then we conclude that the structure's function has been found. If Rudwick's approach is correct we can (1) deduce function in unknown structures; and (2) provide a convincing demonstration that the structure performs the function in an optimal way.

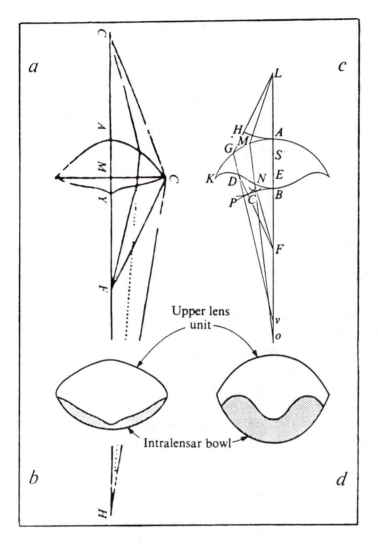

Figure 6.1: Original construction of aplanatic lenses by Descartes and Huygens (A and B), as compared with lenses of the trilobites *Dalmanitina socialis* (C) and *Crozonaspis struvei* (D) (from Clarkson and Levi-Setti 1975).

Such a fit was evidence to Rudwick of adaptation, or an historical process that led to perfection of form according to function.

One of Rudwick's primary examples was a detailed study of the form of the prorichthofenids, a group of bizarre Permian brachiopods, which typically have a deep conical ventral valve and a much smaller and flat dorsal valve that apparently could seal off the opening of the other conical valve. In a brilliant demonstration of functional inference, Rudwick (1961) proposed that the functions of filter feeding and large particle exclusion necessitated rapid movements of the dorsal valve to create powerful currents from which food particles could be collected by mantle surfaces. Spines would be required to exclude large particles. Rudwick built a scale model that demonstrated the plausibility of his functional argument.

Rudwick's work has not withstood further analysis. The shell flapping mechanism required a musculature unknown in brachiopods (R. Grant 1972). The unusual morphology has been interpreted more successfully as reflecting a symbiosis with algae, perhaps zooxanthellae (Cowen 1983).

Rudwick's proposal is a pure form of the argument of **optimality in adaptation**. If it is taken to the limit, a whole organism could conceivably be constructed from the same principles. Thus, a mammal would be a perfect machine, deducible from a sum of individual functions. I should like to focus upon three important assumptions of the argument, and show their limitations. These assumptions are: (1) a perfect fit to the environment demonstrates adaptation; (2) the evolutionary change of form is unrestricted by various constraints (Rudwick did not hold this extreme a view); and (3) unique optimal solutions can always be deduced. While the limitations reject the pure argument, they often lead to a **restricted optimality argument**, applicable only to a narrower range of specific structures and functions that often serves well in evolutionary biology and has been unduly maligned in recent years. Rudwick's (1970) descriptions of the geometrical problems of logarithmic spiral growth in the articulation of brachiopod valves show his use of such restricted optimality approaches.

The Fit to the Environment

The match of form to function is used by Rudwick as *prima facie* evidence for adaptation. This implies that a current static situation

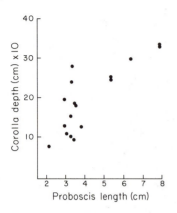

Figure 6.2: Relationship between proboscis length of a number of syrphid fly species and flower corolla depth (after Gilbert 1981).

reveals an historical process. Is this valid? It might be if (a) the paradigm approach were appropriate and (b) evolution via natural selection were such that perfection is always achieved. The universality aspect of the argument is troubling. After all, Darwin (1859, p.199) wrote that "Natural selection will not produce absolute perfection . . . . [Natural selection] must act chiefly through the competition of the inhabitants with one another, and consequently will produce perfection . . . only according to the standard of that country." But what are the standards of the "country" in which a given structure arose? If we see a current match of form to function, can we be sure that it reflects past adaptation?

Consider the data in Fig. 6.2, showing a positive correlation between tongue (labellum) length of species of syrphid flies and the corolla depth of flowers. Tongue length is associated with matters of importance to the fly. The proportion of pollen in the diet increases with increasing tongue length. Moreover, longer-tongued flies tend to visit flowers with longer corollae, because the latter tend to contain more nectar than flowers with shorter corollae. The match of many species in the figure can therefore be interpreted as an evolutionary fitting of tongues to solve the food gathering function. Why don't all species of flies have long enough tongues to invade the largest and most nutritionally useful flowers? We might invoke the pressure of interspecific competition to exploit the broader spectrum of resources. But something is wrong with any such argument. The flies surveyed have come from an urban garden filled with species usually not visited by the flies. The match is not entirely fortuitous, but it certainly does not involve any long-term adaptation of tongue to corolla. Having been thrust into a new situation, the organism has chosen its best environment. With no evolution at all, a fit of (at least one structure of an) organism to environment is achieved admirably.

This example is meant to separate the argument of optimality from adaptation. As a static concept, optimality is a testable hypothesis. One specifies a series of boundary conditions and a series of equations governing the optimal performance of a function. The boundary conditions may often be difficult to establish, but one might safely conclude that structure x_2 is optimally suited to perform behavior y. But this does not prove adaptation, in the sense of an historical process that perfected the structure from a less perfect state x_1. One must have historical evidence of the process of change. This is what we usually lack.

It might be argued that the overall perfection of an organism is proof positive that the fit of an organism to an environment is no accident. If a running, eating, reproducing species lives indefinitely in harmony with its environment, can we not conclude that the totality of the organism's form could not have been produced by a chance choice of this organism in its environment? After all, a horse does well in a farmer's field because it evolved in a similar habitat, not because the farmer chose the one organism in the world that happened to survive in his field. Such a form-habitat matching argument may be true, but as more structures are considered, any optimality argument will tend to break down. Conflicting constraints, uncertainty about interaction effects among various structures, and genetic and developmental correlations, will all become more and more intractable. As fewer and fewer structures are considered, the possibility of a functional approach becomes more and more real, and the dangers of the matching argument increase as well. I can't convince you that man, overall, has adapted to live underwater, but a few aspects of man's phenotype–hairlessness and the pelvic skeleton–have been used in a serious attempt, at least to the author, to take an environmental fit argument (e.g., hairlessness implies swimming) and to draw evolutionary conclusions (Morgan 1982).

It is unlikely that the simple alternative of choosing one's optimal environment, based upon randomly acquired structures, is a viable

explanation for the fit of form to environment. But an interaction process between the organism and the environment must substitute for the
usual dualism of an environment and a variable species, with natural
selection culling out the poorly adapted forms. Lewontin (1983*b*) sees
evolution as an interactive historical process based upon a combination
of environmental culling and the organism's set of abilities to select its
own environment. He states four principles:

1. Organisms determine what is relevant.

2. Organisms alter the external world as it becomes part of their
   environment.

3. Organisms transduce the physical signals of the external world.

4. Organisms create a statistical pattern of the environment that is
   different from the pattern of the external world.

Exaptation, Adaptation, and Preadaptation

If optimality is not an argument for adaptation, as commonly supposed,
how does one study the process of adaptation in the first place? We
must clarify some terminology before discussing this difficult question.
As defined in Chapter 3, adaptation refers to an historical process in
which improvement of function is an aspect in the context of a model
assessing the organism's performance. We do not deny the absence of
random change, and constraints on the evolution of form. Adaptation
only implies a continuity of a function–structure relationship. We invoke no implication of functional perfection but must assume a trend
toward that perfection to infer adaptation. It is also conceivable that
as such a trend occurs, a set of complex interactions among different
structures within the organism continually changes the functionally
optimal form. There may be no goal, only a treadmill.

The definition of the historical borderline in time between two distinct periods of structure–function relationships has long been the subject of terminological debate and confusion. Consider the following
example (Bock 1959). The skimmer, *Rynchops nigre*, has a basitem-
poral articulation in the skull that is a true diarthrosis. The medial
brace formed by the articulation absorbs shock as the mandible hits
the water. We can imagine the following historical sequence to result
in the evolution of the structure:

1. A change in feeding habit, brought on by an abundance of fish living just below the air-water interface who cannot detect predators skimming along the surface above.

2. Selection for increased musculature to deal with the stress on the lower jaw during prey capture.

3. A correlated response involving an increase of the medial process of the mandible for muscular insertion. The medial process thus "approaches" the base of the skull.

4. The proximity of the mandible permits a diarthrosis to develop, perhaps in different places in different bird lineages.

Bock argues that the initial morphological state before these steps constitutes a **preadaptation**, which is that state where a structure's present form enables it to discharge its original function but also enables it to assume the new function whenever the "need" arises. The term preadaptation is, therefore, a retrodictive statement. It is a loaded adaptive gun, ready to fire when the necessary situation appears. The exact evolutionary trajectory is unknown, but a general direction can be predicted once the new selective need arises. Some functional continuity exists between the preadapted and adapted structure.

Gould and Vrba (1982) proposed the term **exaptation** to characterize the evolutionary process where structures are coopted and modified for a completely new function. In a sense, all adaptation must follow an episode of exaptation. After all, no structure is believed to arise de novo, designed to serve a given function. The process of adaptation follows exaptation, and involves the evolutionary change of the new structure–function relationship. A molecular example of exaptation is the evolutionary relationship of lysozyme and lactalbumin (see Jacob 1983). Lysozyme is used by vertebrates to hydrolyze the mucopolysaccharides of bacterial cell walls. Lactalbumin has nearly an identical amino acid sequence but is a principal constituent of mammalian milk. In humans it constitutes the β chain of the mammary gland enzyme lactose synthetase. The abrupt change of function represents an exaptation.

While exaptation is a useful formalism, it is quite difficult to apply unambiguously when the function and structure both have changed

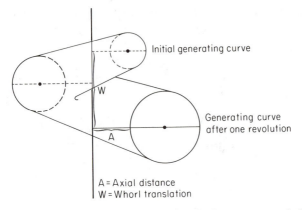

Initial generating curve

W

Generating curve
after one revolution

A

A = Axial distance
W = Whorl translation

Figure 6.3: Raup's geometrical abstraction of the coiled shell.

radically, but perhaps not enough to be considered completely new functionally. For example, the evolution of the Cetacea involves the strong modification of the limbs into flippers. If the functional change from walking to swimming is regarded as completely new, then the rearrangement of skeletal elements in the limb is an exaptation. But if **locomotion** is considered to be the function, then one might regard the change as simply part of the larger process of preadaptation. Indeed the predictive concept of preadaptation is quite useful, since it delineates the potential pattern of adaptation, should the function change slightly. Preadaptation does not refer to an exact evolutionary path, but rather to a range of possible paths, given the selection pressures (Bock 1959).

Theoretical Morphology

Morphological variation can be often be simulated with a fairly simple set of geometric rules. **Theoretical morphology** is the field devoted to specifying such rules and predicting the range of possible morphologies. To be useful and conformable to phylogenetic limitations, the generating equations must correspond to some form of biological growth process, or the geometric rules must be not be so omnipotent that they can create forms patently beyond the means of the organism. The algorithms are also accretionary, that is, they describe what the organism must do next to add a new growth increment.

The most important empirical advance in this field was Raup and Michaelson's (1965, Raup 1966) analysis of the coiled shell, based upon the knowledge that many organisms grow according to a logarithmic spiral. Growth according to a logarithmic spiral causes no change in form as size increases. Thus the rules of growth are independent of the location of the current accretionary growth increment. This simple situation does not always apply–spatial information must be provided (McGhee 1980). The growth of some planispiral mollusks can be simulated with a computer when the following parameters are known (Fig. 6.3):

1. Shape of Generating Curve–cross section of the spiraling form.

2. Whorl Expansion Rate–the factor by which the size of the generating curve (ratio of successive diameters, if a circle) increases after one complete rotation.

3. Axial Distance–the distance between the generating curve and the axis of coiling.

4. Generating Curve Angle–the angle between the the geometric and biological generating curves.

To simulate a conispiral form, one more rule is required:

5. Whorl Translation Rate–the rate of movement of the generating curve parallel to the coiling axis.

An animal cannot be found for all combinations of these parameters; actual coiled shells of groups such as gastropods and ectocochliate cephalopods occupy only a small part of the conceivable morphospace. Common correlations also exist between the parameters. In gastropods, for example, increasing whorl translation rate is usually correlated with rather low whorl expansion rate. Thus, the common abalone-shaped shells (e.g., *Haliotis*), with high expansion and low translation rates, turbinate shapes with intermediate values, and high-spired forms (e.g., *Turritella*), with high translation and low expansion rates, are common morphotypes (see Linsley 1977). The important question is: Why is the morphospace not occupied fully?

Bayer (1978) suggests that the analysis of geometric rules may reveal "weak points" where ultimate shape change will change radically with minor deviations from the program. For example, a small deviation along the coiling axis can change a planispiral into a helicospiral growth pattern. In contrast, a small deviation in an already helicospiral pattern does not change the shape qualitatively. Such an "error" in a growth program might have important consequences for a cephalopod whose hydrodynamic stability and streamlining depend upon the maintenance of the planispiral condition. Such errors would also create great difficulties for bivalves and brachiopods, due to the necessity of articulating two planispiral valves. Bayer claims that such errors are more commonly seen in typically helicospiral gastropods than in planispiral shelled cephalopods, and this reflects the presence of a strong regulating system in cephalopods to avoid such errors. On the other hand, helicospiral cephalopods exist, and are even rather common during certain times in geologic history. Pressure for strong regulation may here be reduced.

Some studies of gastropod form illustrate the potential power of the theoretical morphological approach in analyzing functional problems. Cain (1977) looked at the overall spectrum of snail shapes, as indicated by spire index (height/maximum shell diameter) ratios. A bimodality in spire index (0.5 and 0.3) is found in many snail faunas. He concludes that this must be due to natural selection, whose action results in two modes of overall shape, each conferring stability to an animal that must carry a house on its back, while often crawling in a water current. Intermediate-spired forms may be at a disadvantage since they cannot easily deploy the shell without having the center of gravity rather high above the point of attachment to the substratum. This creates a torque that slows crawling speed (see Linsley 1977). Lower-spired forms, such as abalones, have the center of gravity near the substratum. Moderately high-spired forms can deploy the shell nearer the substratum, as it is possible for the plane of the aperture to converge on the axis of coiling, thus reducing the amount of inclination necessary to maintain a tangential aperture and increasing the amount of detorsion during development (Linsley 1977, and references). Both these forms are faster than intermediate-spired forms. Thus where crawling speed is at a premium, one might expect bimodality of form. Very high-spired forms, such as *Terebra*, drag the shell along the substratum and are therefore quite slow.

Peter Williamson (unpublished) has illuminated this problem with a consideration of the theoretical morphology of the coiled shell and the constraints involved in the fabrication of the coil. From the point of view of shell strength, natural selection will enforce the fabricational restriction of logarithmic spiral coiling to that subset of morphologies where the whorls touch. This constraint greatly limits the range of overall morphologies. If one sampled this range randomly, one would be likely to get a bimodal set of spire index values. This suggests that, though function may be an integral determinant of whorl touching, spire index variation can obtain a range of equally acceptable values, from this one functional point of view. Other factors may cause an animal to have a given length or height, thus setting its form within the overall field. This overall conclusion is tentative since entire faunas adhere to one mode of spire index value that can differ from a random sampling of the range predicted by Williamson. Moreover, as mentioned above, crawling speed is maximized at two modes of shape. Clearly the overall story is more complicated. Nevertheless, Williamson's fascinating reinterpretation calls into question any hasty conclusions of completely functional significance for even rather complex multimodal patterns of forms in a group of related taxa.

The Constructional Limitation of Form

While the trilobite lenses studied by Clarkson are clear examples of functional perfection, optimality criteria cannot usually be deduced in so simple a manner. At least four major problems hamper a pure optimality approach:

1. Lack of a unique solution owing to phylogenetic eccentricity–different functional–phylogenetic constraints: We know that organisms perform the same functions in radically different ways. Consider the function *feeding on macroalgae on a wave exposed shore*. There may be an optimal form for such a function, but we know that many divergent organisms employ entirely different sorts of apparatus to gather, crush, or tear apart food. How does one compare the Aristotle's lantern of sea urchins, for example, with the radular and buccal apparatus of snails? Either we admit that both serve the function well, or we are left with the choice of absurd conclusions that: (a) one is perfect, and therefore the other will eventually be modified in this direction; (b) neither is perfect and both will converge to yet another structure, with further natural selection. The solution to this is obvious. Both serve the function well, but have evolved within a unique set of

design constraints deriving from different evolutionary histories. The problem would have to be restated as follows: Given a foot, gill, and radula, what morphology should a snail have to consume macroalgae on a wave exposed rocky shore?

2. A Series of Equally Suitable Solutions. Even within the same design constraints, the optimal solution may be a series of morphologies, all equally suited to perform the function. A variety of toxic substances might serve equally well in harming a predator, particularly if there is a threshold of toxicity above which the predator will be completely discouraged from trying another attack. Thus marine organisms employ metals such as vanadium (Stoecker 1978), hormones like prostaglandin (Gerhart 1984), sulfuric acid (Stoecker 1978), and a variety of other compounds. Similarly, a variety of warning colorations and patterns will be equally effective in associating visual stimuli with the bad experience of the poison. Thus mimicry rings of butterflies converge on the same bright colors, but rings with more than one color or pattern exist in a region and probably are equally effective (J.R.G. Turner 1981).

3. Differential Response to the Same Microenvironment. The problem of scaling raises difficulties in relating organisms of differing sizes to the same environmental conditions (Koehl 1976). A good example of this problem arises in the attempt to understand the flow regime around an organism. Body size and velocity of flow are both important in determining the nature of flow. The Reynolds number, Re, summarizes flow characteristics about an object and estimates the relative importance of viscous and inertial forces. If U is the water velocity, l the characteristic length (it may be the diameter of the organism, colony branch, feeding structure, etc.), and u is the kinematic viscosity, then

$$Re = Ul/u$$

As examples, an invertebrate larva, 0.3 mm long, moving at 1 mm sec^{-1}, would have an $Re = 0.3$, while a tuna swimming at the same speed would have $Re = 3 \times 10^6$. Viscous forces dominate as Re decreases. As Re increases, however, inertial forces dominate,

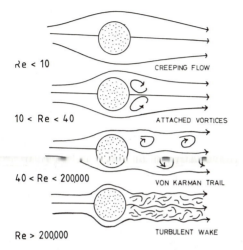

Figure 6.4: Flow patterns perpendicular to the long axis of a cylinder (after Vogel 1981).

increasing the potential importance of turbulence. When Re exceeds 2000, flow about the object (organism) usually becomes turbulent (see Vogel 1981 for an excellent introduction).

A change in flow regime can be seen easily with change in Re as water flows past a cylinder (Fig. 6.4). At low Reynolds numbers vortices are absent in the wake of the cylinder. As Reynolds number increases vortices and then full turbulence develop in the wake. Note that Re can increase linearly with current velocity *and* cylinder diameter. A variety of organisms of widely varying size, therefore, will have considerably different downstream water flow patterns. The nature of flow downstream from the organism may determine the nature of food capture by passive suspension feeders. In the gorgonian, *Leptogorgia*, Leversee (1976) found that brine shrimp could remain in downstream eddies for as long as 15 seconds. Polyps of these colonies could feed on the downstream side of the branch. Re was estimated to be about 50. By contrast, eddies at higher Re might not be so predictable and food supply might be less reliable for downstream capture. This problem would be of special importance in colonial organisms which made a permanent commitment to feeding structures in the up- or downstream direction of the colony.

The differential microenvironment created about organisms with varying forms cannot always be related to a simple parameter such as Re. Organisms in flowing regimes often are constructed of widely varying materials and, therefore, are constrained to deal with a given flow regime in disparate ways (Koehl 1976; Koehl and Wainwright 1977).

4. Conflicting Selection Pressures Within the Same Microenvironment. Even when the microenvironment can be specified properly, it is not always obvious what selective pressures are presented to the organism. A set of responses that solve one problem for the organism may have disadvantages in other autecological dimensions. The paleoautecologist must be able to scale the environment such that these conflicting selective pressures are characterized properly.

An excellent example of such conflicts can be extracted from our discussion above of sessile colonial organisms in flowing water. In order to maximize capture of plankton, a stiff fan-shaped structure normal to the current, with feeding individuals on the upstream side, will usually be best. This structure, however, maximizes tensile stress and subjects the colony to the danger of breakage in strong currents. The degree to which an organism solves this tradeoff must be a function of the material of which the organism is constructed and the current velocity.

Constructional morphology, an approach to functional morphology developed by Adolf Seilacher (1970), is an attempt to integrate the various determinants of morphology into a coherent system, useful for evolutionary and functional considerations. He suggests three principal controls on morphology: (1) **Architectural Constraints**; (2) **Ecological, Adaptive, and Functional** aspects; (3) **Historical-Phylogenetic** aspects. Raup (1972a) has added (4) chance; and (5) ecophenotypic effects as causes of morphological variation.

Architectural constraints refer to those aspects of an organism, built into its overall biology, that either (1) determine form, irrespective of functional considerations, or (2) modify the evolution of form, interacting with function. Seilacher refers to the first as **fabricational noise**. Certain aspects of form come not from any solution to an environmental problem. They arise rather from the physical principle behind the formation of the structure or certain biologically controlled processes of formation. The second form of constraint refers to those instances where such fabricational noise cannot be factored out simply, but interacts with functional controls in a complex manner.

To test the hypothesis of fabricational noise, a **constructional paradigm** (Seilacher 1979) is first established, which is used to predict a structure. Theoretical morphology as described above is used to predict one structure or a range of forms. The closeness of match of the structure to the constructional paradigm is thought to be an indication of the efficacy of the fabricational noise hypothesis. Similar patterns found in widely divergent organisms, involved in strongly differing functions, are likely candidates for such fabricational hypotheses. Deviations from the constructional paradigm call for alternative constructional hypotheses or other explanations.

Seilacher (1973, 1979) gives a variety of possible examples of fabricational noise. He likens overall form of urchins and sand dollars to an encrusted liquid-filled balloon, under some gravitational stress. The overall form, therefore, may have various geometric peculiarities that have nothing to do with functional superiority over other possible forms. The surface sculpture of the burrowing mole crab, *Emerita talpoida*, and the detailed suture patterns of some ammonites are claimed to be "pull-off" structures, that is, ridges whose form is determined as the skeletal secretory tissue is finally pulled off from the newly formed and still malleable skeletal material. This is consistent with the discovery of epidermal "feet"–cytoskeletal extensions that participate in deposition of arthropod cuticle; their contractions are known to cause the formation of transverse ripples in the soft cuticle, which later harden (Locke and Huie 1981). Divaricate color patterns and sculpture in bivalves may also be examples of a fabricational contribution to morphology. Both may be controlled by a similar underlying mechanism of material transport and deposition (either pigment or calcium carbonate), but the structures can be employed to serve a function. Divaricate ridges aid in entry into the sediment (Stanley 1970) while divaricate color patterns are probably of use in camouflage.

Phylogenetic controls and fabricational constraints are often confounded since phylogenetic history may include the use of a particular developmental mechanism that works on a given physical formation principle. In Chapter 5 we discuss a variety of developmental processes that are likely to generate such fabricational noise. Color patterns generated by diffusion-reaction processes, for example, are liable to take on a finite set of forms, governed not by function, but by the geometric features of the reaction field. Developmentally regulated patterns need not be uniquely phylogenetic, however. For example, mammalian coat color patterns, hypothesized to originate in the early embryo (Murray 1981), reflect overall principles that might apply to color patterns in other groups.

Historical differences, to a degree, can be considered separately from phylogenetic controls. A group may have originated or evolved under a unique set of circumstances. This may have led to a set of adaptations unique to that local area. Subsequent movement of the taxon into a very different environment may have resulted in modifications of form, but the stamp of previous circumstances may still be present. This may be because: (a) there is present a set of encumbered problems in losing the structure; (b) not enough time has elapsed to lose the structure or no strong loss in performance is involved if the structure is retained; and (c) the structure has been modified to perform other functions, but its historical context still is recognizable.

It would be hard to estimate the difficulty of losing a morphological structure in a lineage. In Chapter 5 we discuss a variety of developmental justifications for the presence of a **burden** of loss of a structure, based upon genetic and developmental pleiotropy. But these can rarely be measured. The issue of time is equally difficult and can be invoked ad hoc to explain the presence of everything from vestigial structures to organs that have been lost partially, such as eyes in fishes living in cave pools. Consider the question of hypotonicity of marine teleost fish, who maintain an osmotic strength one third that of seawater. They must continuously osmoregulate in a medium to which nearly all other marine creatures are approximately isotonic. Some have argued that this reflects the origin of the group in brackish water of a salinity corresponding to the tonicity of fish body fluids. Subsequent osmoregulatory adaptations have been, therefore, superposed upon this accident of history. While this hypothesis may very well be true, it is probably impossible to test.

Optimality, Adaptation, and Constructional Morphology

Optimality presumes a fit to the environment in which the organism is found. It is a static concept, independent of history. In contrast, the essence of the concept of adaptation is its historical context. Thus optimality does not necessarily imply adaptation. The example given above for syrphid flies of a match of form (labellum) to environment (flower corrola) is an example of the difficulty of translating a static situation directly into an historical inference of adaptation. Does optimality therefore, imply anything, and is it a useful concept in evolutionary biology?

Optimal foraging theory makes a series of predictions, which, when satisfied, carry the connotation of excellent fit of function to environment. But one cannot exclude the possibility that this fit involves a behavioral matching to the appropriate environment, precluding the necessary inference of gradual evolutionary modification to fit the environment. Certain features, however, have built in them an inescapable necessity of adaptation. If, for example, a consumer forages optimally, under a variety of experimental conditions, we must conclude that some evolutionary process has built in a capability to change the rules as new environments are encountered. Some excellent examples of such fits can be found (e.g., see Pyke 1984).

The pure notion of optimality in form or behavior presumes that no genetic, developmental, historical, or other constraints prevent evolutionary progress toward the optimal state. But all that we have learned about evolution flies in the face of this pure notion, which is not seriously believed by most evolutionary biologists. The quote of Muller at the beginning of Chapter 3 probably represents a general feeling, even if we believe strongly that much of the beauty of an organism rests in the functional significance of its anatomy.

Lewontin (1978), and Gould and Lewontin (1979) have mounted the most concerted attacks against optimality. The argument is quite old, since interpretations of perfection, molded by natural selection, were commonly ridiculed around the turn of the century. To find a function in every structure was the objective of any "good" Darwinian. Ironically, this faith has usually proven well invested. Anatomical structures, usually thought to be functionless curiosities at first, have often proven to have essential functions. Most biologists have not thought that structures are perfect, only that they do something!

But Gould and Lewontin (1979) point out that history can cast a strong imprint on form, to the degree that we may spend an embarrassing amount of effort trying to prove a functional **superiority** for a structure that has arisen in a nonfunctional context. Lewontin (1978) gives the example of determining the selective significance of the one-horned Indian rhinoceros, versus the two-horned African rhinoceros. He claims that a biologist devoted to optimality would try to infer a reason for the difference, when it is likely that it is unrelated to natural selection. As the presence and size of horns may have selective significance (Clutton-Brock and Harvey 1979) and the particular case is uninvestigated, perhaps Lewontin could have picked a better example. Gould and Lewontin (1979) are on somewhat[1] safer ground in using a nonorganismal example, the spandrels of San Marco Basilicum in Venice, a set of triangular spaces created simply because of the mounting of a dome on a series of rounded arches. The spaces are decorated because they are there. The spandrels were not created for the decorations! The examples of fabricational noise given above are somewhat better, and lead to a useful approach to some of the obvious problems of the action of natural selection in a world of organismal eccentricities.

While Gould and Lewontin's point is well taken, an examination of Maynard Smith's (1979) criteria for satisfying optimality considerations rather dismantles their apparent belief that all functional evolutionary biology is blown apart by problems of historicity and constructional limitations. An optimality theory requires something to optimize and series of boundary conditions describing the system. Gould's criticism (e.g., 1980*a*) of the selectionist's approach to evolution is an inaccurate caricature in that it assumes the unconstrained possible shaping of form to function. While there may indeed be cases where such an outlandish assumption cannot be dismissed, most realize that organisms can be made of skin and bone, are constrained to be denser than air, etc. These constraints must be included in the boundary conditions for an optimality argument as Rudwick (1964) did explicitly. This approach is similar to the comparative method, often used qualitatively by systematists (e.g., Mayr 1983). It is true that some

---

[1]The mainly angelic decorations in the spandrels are actually not very good examples, as several have an overall rectangular aspect which fits rather clumsily into the triangular space.

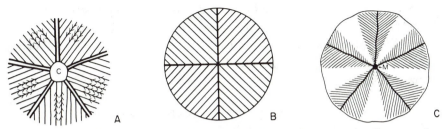

Figure 6.5: Comparison of three gathering schemes. A. Crown of a melocrinitid crinoid. B. Ideal road layout for a banana plantation. C. Food grooves on the sand dollar *Arachnoides* (after Cowen 1981).

cases seem like unfettered perfection, especially in foraging theory. It is also true that some unfortunately rationalize poor fits to optimality theories and then invoke other factors. But there is no reason why historical and fabricational constraints cannot be included as conditions in any optimality theory.

Gould and Lewontin also complain mightily about the rampant use of atomism to explain form in terms of natural selection. "An organism is atomized into traits and these traits are explained as structures optimally designed by natural selection for their functions" (1979, p. 585). To completely, and inaccurately, demolish the adaptationist program, the small minority of workers [2] who believe that gene action can be mapped simply to emergent morphological and behavioral phenotypic traits, are often lumped with the more level-headed majority, who believe that genetic and developmental flexibility is very often sufficient to permit the evolution of structures plausible in function and understandable by models of performance.

The invocation of fabricational noise can be abused much in the same way as optimality arguments. Cowen's (1981) functional study of the camerate crinoid family Melocrinitidae is based upon a pattern of arm branching that Seilacher (1979) likened to fabricational noise when found in a sand dollar. But Cowen's analysis demonstrates otherwise. From the Ordovician to the Late Devonian, arm branching becomes more complex and increasingly resembles a theoretical layout of roads for optimal harvesting on a banana plantation. The pattern demands a rigidity of arrangement of transport found in the rigid cam-

---

[2]For example, Richard Dawkins, who, by the way, shows himself (1983) not as quite the reductionist Gould (1983b) claims.

erate crinoid crown, but not in more flexible crinoids where arms can be rearranged, depending upon current conditions. The optimal solution for the plantation consists of the pattern in Fig. 6.5. The following functional analogy between crinoid particle handling and banana gathering can be established:

1. Tube feet trapping food–cut banana with machete

2. Pinnule (ciliary) transport–hand carry baskets

3. Ramule (ciliary) transport–trucks on dirt roads

4. Arm (ciliary) transport–trucks on arterial roads

5. Digestion in calyx–processing plant

Given the historical documentation and conformance to an optimality model, this example satisfies our requirements to infer adaptation.

Constructional Optimality and Theoretical Morphology

Constructional morphology provides boundary rules for the development of **restricted optimality arguments,** or **constructional optimality.** Consider the primitive state within a clade. We take the biological limitations, both architectural–fabricational and phylogenetic, and use them as restrictions within which optimality is calculated. Under this restriction, organisms are not so much perfect as they are "the best possible under the circumstances." I will freely admit that this type of optimality argument invites the danger of boundless rationalization, should an initial pure optimality argument fail to work. Thus if an organism fails to be an optimal performer, one can invoke a series of phylogenetic limitations. This problem is akin to statistical hypothesis testing, where one has no a priori way of knowing what level of confidence is best, and may adjust the threshold level, if an initial attempt to prove significance fails. One must be prepared to make the hypothesis first and be willing to abandon the enterprise if the hypothesis fails. We are not all so courageous.

Theoretical morphology provides an excellent starting point for the study of restricted optimality. A biologically meaningful theoretical generating function is used to produce a range of shapes. Then a set of optimizing principles is used to ask: Within the fabricational limitations of the generating function, what range of shapes would serve a function or functions. This has been a common approach in studies of the coiled shell where the range of shapes of actual organisms is far more restricted than could be generated theoretically (e.g., Raup 1966, 1967; McGhee 1980; Chamberlain 1980).

Ontogenetic change in the form of brachiopods, a dominant Paleozoic (yet still extant) group of marine epibenthic bivalved animals illustrates the argument. The animal secretes a pair of external valves that are typically bilaterally symmetrical and articulated along a hinge region. Using a lophophore to gather and sort particles, brachiopods depend upon water currents to deliver food to the aperture created by the gaping valves. Each valve grows according to a logarithmic planispiral modified by a successively decreasing whorl expansion rate (McGhee 1980). If there is a correspondence between a developmental process and growth according to the modified logarithmic spiral, then the constructional limitation involves articulating the valves for tight closure. In addition to this constructional constraint, the functional considerations of efficient feeding and stability of the shell in a current should combine to explain the main aspects of overall form in most forms.

As has been found for ammonoids (Raup 1967) the biconvex brachiopods occupy a small subset of the conceivable range of geometrical space available to them (McGhee 1980). Very few brachiopod valves have whorl expansion rates of less than 100. This follows the requirement that for distal articulation there can be no whorl overlap between valves. Strong modifications of growth and form at the hinge area satisfy the requirements of hinge articulatory considerations. These limitations of form are related to the limitations of the mode of growth. But form is further restricted according to functional considerations relating to feeding and respiration. One would expect that the animal would maximize the access of outside waters to the internal soft parts, particularly to the lophophore, which is involved in both feeding and oxygen uptake. This is to be expected since the body volume increases with the third power of length, while filament area increases only with the second power. Large brachiopods partially offset this geometric limitation by having increasingly convoluted lophophore systems (Rudwick 1970), but one might expect a tendency toward a spherical form, which would give the maximum surface area per unit volume. Because

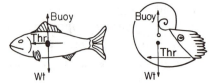

Figure 6.6: Center of buoyancy, balance, and thrust in a typical fish and planispiral shelled cephalopod. Buoy = buoyant force, Thr = thrust, and Wt = weight. Closed dot = center of mass, open dot = center of buoyancy (drawn after Chamberlain 1980).

of the limitations on articulation imposed by the growth system, this apex of efficiency cannot be obtained (McGhee 1980, p. 67), but the majority of biconvex brachiopods cluster in the geometric region of minimum internal shell volume to shell surface area, given the overall constructional restriction. Such a constructional–functional analysis also tends to highlight outliers, such as the strongly flattened shells of strophomenids and productoids.

Ectocochliate (externally shelled) cephalopods illustrate how constructional morphology can be combined with optimality to produce the restricted optimality models necessary to investigate function. Using the parameters mentioned above, Raup (1967) found that coiled fossil cephalopods ranged in shape over a highly restricted range relative to the domain of conceivable shapes. Can this restriction be related to function and to constructional morphology? We presume that buoyancy, balance, strength against implosion, and locomotion, particularly in the special context of jet propulsion characteristic of cephalopods, are functions to be examined most closely.

First we must discuss the limitations of the ectocochliate shell. Cephalopod shells behave like blunt rounded objects moving through a fluid medium. Most of the drag generated relates to a few simple characteristics of the shell: (1) the size of the umbilicus; (2) the width of the shell relative to the diameter; and (3) the aperture size (Chamberlain 1976, 1980). The coefficient of drag C_D, a linear index of the relative energy required to propel the animal through the water, increases with whorl expansion rate (increasing aperture), increases with distance from the coiling axis (increasing size of umbilicus), and increases as the apertural shape changes from compressed to circular.

The swimming of ectocochliate cephalopods such as *Nautilus* is complicated by the presence of two spatially separated body parts of quite different buoyancy: a mainly gas-filled shell and a soft body-filled terminal chamber. As a result, the center of mass differs from the center of buoyancy. Because the center of buoyancy is usually above the center of mass, the animal has considerable static stability. When fluid is expelled from the hypnome to propel the animal, however, a rocking motion is induced (Fig. 6.6). As it moves, *Nautilus* rocks back and forth like a pendulum. The drag exerted by the shell is at best three times greater than swimming fishes, mammals, etc., and typically ten times greater. The combination of relatively inefficient jet propulsion, shape, packaging of muscle power, and limitation of rocking during bursts of movement prevents ectocochliates from ever achieving the swimming velocities of fishes, whose center of mass and buoyancy are approximately coincident and whose packaging of muscle power is considerably greater (see literature summarized in Chamberlain 1980).

Ectocochliates were probably nectobenthic in habit, much like the living *Nautilus*. The angelfish are piscine analogues of shape and locomotory habit. This would be in contrast to laminar drop shapes (e.g., tuna) that perform sustained swimming, or fusiforms (e.g., barracuda or squids) that swim in short rapid bursts. From this, one concludes overall that ectocochliates are shaped and live according to a compromise struck by the inherent limitations of their design and the possibilities offered by the environment.

Are these animals optimally designed to minimize drag, given their overall limitations? Fig. 6.7 shows the effect of whorl expansion rate (W) and whorl position (D) on the shell drag coefficient of shells with circular cross sections. Drag increases dramatically with increasing D, especially as early whorls fail to be obscured by successive whorls. Within the region of relatively low drag, two peaks of minimum drag are apparent, but Fig. 6.7 shows that ammonoids of roughly circular crosssection mainly fall near one of the peaks. Why this peak and not the other? Apparently because shapes near this peak confer superior static stability. As we have already built in an inherent restriction of swimming efficiency by using a coiled form, the morphology we see is a compromise between, as Chamberlain (1980, p. 325) puts it: "the largely antithetical morphological requirements for optimizing static stability and hydrodynamic efficiency." This analysis demonstrates that swimming efficiency and static stability are at least elements in

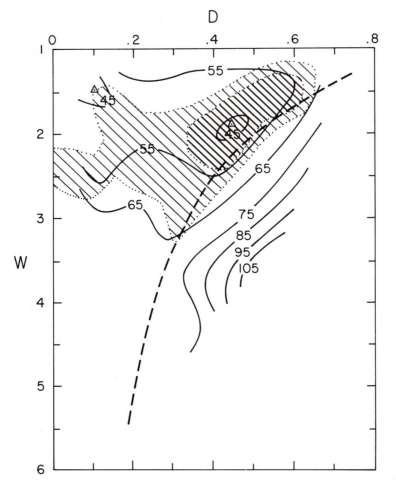

Figure 6.7: Plot of whorl expansion rate (W) versus whorl position (D) of depressed ammonoids, with contours of the drag coefficient of the shell plotted (lower values indicate lower drag). Dark-shaded area is the range of morphologies of most actual species, while light-shaded area indicates total ammonoid domain. (after Chamberlain 1980).

the form variation of coiled cephalopods. One could not perform this analysis without the contributions of theoretical morphology and restricted optimality models, based upon biomechanical principles and phylogenetic restrictions.

The overall pattern of cephalopod evolution lacks a trend in relative taxon abundance that can be interpreted as adaptation. Indeed, the dominant Paleozoic nautiloids already fall on the functionally optimal peak. With time, as the nautiloids lose out to the ammonoids, the latter replace the former in modal shape. Whether this is due to a simple replacement as the nautiloids went extinct, an active competitive replacement, or occurred in response to some other factors, is unknown. It is by no means clear that overall changes in the cephalopod fauna over time had anything at all to do with the swimming efficiency of coexisting taxa. There may be an exception in late Cretaceous times, when a clear increase in swimming efficiency may be related to an increase in predator abundance (P.D. Ward 1981). Those lineages without streamlined shells developed sculptures characteristic of predator avoidance (P.D. Ward 1981). The question of adaptation in any genus is effectively hidden in smaller scale historical events that may be beyond the reach of any empirical effort at documentation.

I have discussed a few examples in some detail to show that the Gould and Lewontin argument against optimality relates much more closely to a straw man not considered today by working functional morphologists. Most acknowledge that their evolutionary materials have limitations and seek to formulate optimality models within these limitations. While Jacob's (1977) metaphor of evolutionary tinkering is a useful way to conceive of evolutionary progress, it is equally clear that the tinkering can be quantified and studied systematically.

While constructional optimality studies are useful in the context of function, they do not illuminate completely the overall question of adaptation. One can formally entertain the hypothesis that the restricted geometric range of form seen in brachiopods may conform with function, but that the morphologies appeared de novo, with no period of modification toward better performance. Alternatively, a large number of more poorly performing taxa may have appeared and become extinct, leaving the residual, which we see in the functionally optimal part of the geometric space. While I cannot entertain this alternative seriously, it does indicate that the nature of the adaptive process is not solved by constructional optimality arguments. The element of history is absent, except on a grand, and not illuminating, scale, as in the cephalopods. Therefore, to ask in fossil studies "does the function suit the structure?" is not necessarily to ask "did the structure derive from adaptation?" It is, therefore, worthwhile to note what *isn't* demonstrated by restricted optimality arguments:

1. **Environmental trigger**: Did the environment change and induce a change in form, or did form progressively improve in a constant environment?

2. **Taxonomic level**: Was improvement steadily occurring within evolving lines, or did improvement require selective extinction (Stanley 1975), speciation (Eldredge and Gould 1972; Stanley 1979), or unique types of evolutionary novelties (Goldschmidt 1940, Gould 1980a).

3. **Historical accident**: Did change stem from a unique historical event?

These questions require a record of history and a viable mode of historical inference.

It is rare to be able to find a case in the fossil record where all of these questions can be answered effectively. Most fossil lineages are documented too sparsely and functional morphology is too poorly known to understand the significance of change where relevant. Moreover, the mode of evolution is usually poorly understood (e.g., speciation versus phyletic evolution) as will be discussed in Chapter 7. Finally, the precision of environmental reconstruction necessary to understand functionally significant changes is usually absent.

An exceptional instance is the evolution of bivalved mollusks, which have an excellent fossil record and can be related functionally to their respective substrata (Stanley 1970). In the Miocene-Pliocene sediments of the mid-Atlantic Coastal Plain, extensive collections have produced a series of well studied lineages of scallops (e.g., Ward and Blackwelder 1975). The shell of scallops have been studied in detail and the functional significance of variations in the byssal notch, valve symmetry, and the location of adductor muscle scars have been investigated quite successfully (e.g., Stanley 1970). Most interesting is the genus *Chesapecten*, whose evolution and cladistic relationships (see

Chapter 2), functional morphology, and intraspecies versus interspecies morphological variance have been well studied (Miyazaki and Micke-vich 1982).

From the mid-Miocene to the Pliocene (Yorktown formation) the genus shows a continual trend consonant with morphologies departing from byssal attachment and toward increasing commitment to a swim-ming, free-living mode of life. The variation seen among individuals within a named species is of the same qualitative nature seen between species, so the question of taxonomic level of origin of novelties can be referred to a context easily understood in terms of evolution within populations. Overall the trend of morphological change seems corre-lated with a deepening of the nearshore marine basin sediments from which the group is collected. Thus the evolutionary change can be said to involve adaptation to an environmental change.

Adaptation: Is Improvement Inevitable and Continuous?

The use of restricted optimality implies improvement, if by a standard somewhat more restrictive than the optimality criteria that might be concocted in a "from the ground up" design. Is improvement con-tinuous? Are advanced mammals really better than therapsids? Are placentals superior to marsupials? This thinking comes from the pre-Darwinian concept of a *scala naturae*, but it survives in our think-ing about adaptation. Although the measure of adaptation must be against an optimal form, the restrictive optimality argument also im-plies that it must be against a starting set of boundary conditions. If one looks millions of years ahead, it is not obvious that our descen-dants' performance should be compared with our own. Their measure of improvement must be against their own most immediate descen-dants.

The potential for lack of overall improvement in performance was demonstrated nicely in a series of experiments for tolerance of yeast to toxic substances (Paquin and Adams 1983). Periodic selection in nonsexual forms resulted in successive sweeps of adaptively superior genes, associated with newly arising mutant clones. Any given clone was found only to be superior in resistance to the previous clone, but not necessarily superior to much earlier clones. This suggests that adaptation has a context: The genotypes that are present in the pop-ulation at the time the "superior" genotype is favored by selection. In the worst case, overall performance can actually decrease over time.

The performance of complex structures will rarely drift or decrease, so long as the basic organismal structure is about the same. Most improvements in function are liable to be transitive; that is, if improvement a is superior to improvement b, and b is superior to c, then a will be superior to c. An eye with a primitive lens will likely be supplanted by one with a more advanced lens. Such an eye will likely be complemented by complex musculature. These advances are interactive and would likely result in cumulative improvement. Cases of intransitivity are liable to arise when many structures work together to improve overall performance, e.g., to function in a dry milieu, but, over millions of years of evolution, a variety of different and functionally and developmentally unrelated structures are coopted to improve performance. Under these circumstances, one might imagine a taxon achieving dominance by virtue of its impervious skin, while a distant descendant might beat its contemporaries by improvements in its metabolic efficiency. Such a complexity of adaptations would not immediately suggest that the first adaptive change resulted in an organism much better in performance than a distant descendant. That is why it is difficult to guess whether mammals are superior over dinosaurs. There is no uniform standard to compare adaptations, even if our prejudices tend to favor fur, mammalian skin, and mammalian metabolism.

The Open Door to Change: Key Innovations

We have considered above the analysis of fabricational and functional aspects of form, based upon limitations imposed by developmental and architectural constraints, and shapes predicted from functional considerations. But certain evolutionary novelties bring a form into a wholly new realm of evolutionary potential, permitting a great expansion of form diversity in a group. Such a change may involve (1) a continuous change of a given type of form, past a threshold that has qualitatively different implications for function; (2) acquisition of a new element, which can be readily generated by the available developmental apparatus, but whose presence has far-reaching functional significance; or (3) combination of available elements with significance corresponding to (2). One can also imagine (4) a wholly new structure, appearing as a highly improbable novelty.

Consider a cladogram, complete with character changes registered at the nodes. We might identify a **key innovation** (Liem 1973) when the following two criteria are both satisfied.

(1) The character change is followed by a large ensemble of other character changes in a set of derived taxa, whose numbers are strikingly larger than the sets of derived taxa following different character changes on other parts of the cladogram.

(2) The character change can be shown to have key functional significance in permitting the entire ensemble of derived taxa with their concomitant derived and functionally potent character states, which otherwise are not likely to exist (see Lauder 1981).

In a sense, it is irrelevant how the key innovation was acquired, but one presumes that local circumstances of natural selection, combined with various other constraints and stochastic processes, have yielded a taxon with the change. These local circumstances provide an opportunity but don't guarantee the future of the taxon in question. In other words, a key innovation is necessary, but not sufficient for a subsequent radiation.

Thus the two major criteria for identifying key innovations require an understanding of adaptedness and history. The functional criterion is based upon the constructional optimality argument presented above. But history must play a role as well. Conceivably, there is a large ensemble of innovations that functionally might open many new doors to morphological diversity, but one or more historical factors may have prevented their translation into an adaptive radiation. A novelty with great potential, for example, may appear and be fixed in a very small population, but will likely disappear. In contrast, the same novelty's appearance in a widespread and populous taxon would assure its survival and potential to give rise to other changes. The immediate factors permitting or causing an adaptive radiation can be placed in two overall classes of explanations: (1) After the key innovation is acquired, the taxon can invade a wide variety of previously extant, but unoccupied modes of life (e.g., Lull 1929, p. 246); or (2) after the key innovation is acquired, inferiority of competitors, a change in environment, or extinction of competing forms allows the radiation to occur (e.g., Simpson 1944, p. 193). In both cases we assume that a tendency for ecologically significant cladogenesis exists.

The degree of change required to achieve the acquisition of a key novelty in a taxon will vary depending upon the overall requirements for functionally significant change, complexity of interactions among parts of the organism, and genetic/developmental mechanisms required for the rise of the novelty. Organismic complexity may allow a wide variety of functionally and developmentally permissible states, thus permitting a smooth transition through many intermediates, of eventually large magnitude (Lauder 1981). In other situations, only slight reconstructions of existing structures permit the change necessary for drastic shifts in habit (Liem 1973).

Jaanusson (1973, 1981) points out that many important innovations appear as discontinuous morphological changes. In fossils, the alternative morphs may coexist in a population but are often different enough to be assigned to different taxa. This is true for example in graptolites, where coexisting forms with different numbers of stipes are probably temporally transitional polymorphisms from the same species, different to the degree that they have been assigned to different genera (Skevington 1967). As discussed in Chapters 3 and 5, known genetic and developmental mechanisms easily account for such discontinuities. Nevertheless, some of them may bridge important functional gaps and have great significance for future evolution. Jaanusson (1981) calls these **dithyrial** (two door) **populations**.

A rather interesting case that lacks the historical requirements for designation as a key innovation (the group bearing the novelty is near extinction and has not given rise to a large variety of derived taxa) is the movable joint present in the maxillary bone of bolyerine snakes of Round Island, near Mauritius (Frazzetta 1970). The acquisition of the joint probably involved a transitional stage where individuals both bearing and lacking the joint were present in the population. Rather simple genetic–developmental mechanisms would permit the appearance of such a novelty, but its fixation might have led the group towards the evolution of a wide variety of dietary adaptations.

The degree of jaw diversity in several endemic radiations of cichlid fishes in African rift valley lakes may have depended upon changes in the pharyngeal jaws, and would thus constitute a case of a key innovation (Liem 1973, 1980). As mentioned above, a key innovation is a necessary but not sufficient cause for a radiation in morphological diversity. The particular breeding behavior of the cichlids and the stability of the lake systems may have been the important elements that fostered the radiation. The anatomical changes involve: (1) a synarthrosis between lower pharyngeal jaws; (2) a shift of insertion of the two fourth levator externi muscles; and (3) the establishment of synovial joints between the upper pharyngeal jaws and the basicranium. These changes freed the premaxillary and mandibular jaws to evolve numerous specializations in the collection of diverse foods (Liem 1973). It is of interest that a primitive neotropical form, *Cichlia ocellaris*, has a basipharyngeal joint, and lacks a complete synarthrosis in the lower branchial jaw. The fourth levator externus is weakly developed and is usually inserted on the lower jaw. In a few specimens a part of the muscle is attached to the dorsal aspect of the fourth epibranchial. Both the joint changes and the muscle insertion change require genetically trivial changes.

A kinematic analysis of *Petrotilapia tridentiges*, a rock scraper with versatile jaw movements, was shown by Liem (1980) to have eight distinct types of jaw movments. Based upon these movement types, the other African cichlids can be grouped into four overall classes, each with a characteristic subgroup of these types. These four classes, furthermore, are different overall in feeding behavior (pursuit hunters, ambush hunters, epilithic algal feeders, variable feeders).

While the key innovation described by Liem may explain the degree of jaw diversity, it may be insufficient to explain the magnitude of speciation of the cichlids in certain lakes. Behavior linked with color recognition may be far more an important explanation of the degree of speciation. Therefore it would be incorrect to state emphatically that the key innovation was the cause of the total number of species. Indeed, the innovation may not be the cause of rapid speciation at all. Rather, the innovation permits a certain amount of morphological diversification to occur. Therefore we conclude that the conditions of the lakes and the behavior of the fish are necessary to the degree of speciation. The key innovation may have coincidentally permitted some morphological radiation with trophic significance.

The particular historical context of the evolution of a key innovation cannot necessarily be related to its subsequent potential success in permitting evolution to lead a taxon into a variety of new forms and lifestyles. Thus the evolution of siphon fusion in bivalved mollusks occurred in a specific environmental context that may not bear any relationship to the ultimate advantage leading to the rise of the descendant clade, which was probably invasion of a wide variety of shifting substrata and protection against predators (Stanley 1968; Vermeij 1983). Natural selection leading to the key innovation may be compatible with the overall success of the clade. To the degree that the latter is true, the success of entire clades is merely an expression and an extrapolation of the success of the acquisition of small scale adaptations. But it is also possible that selection of the clade bearing the key innovation occurs at a higher hierarchical level, since the factors in survival and expansion are different from the circumstances that promoted the adaptation. In the case of mantle fusion and cichlid jaws, this simple separation is hard to imagine, but it may be true in other cases.

Problems arise, however, when considering evolutionary radiations of groups where a large complex of morphological, physiological, and biochemical features are apparently involved in the group's success. This is true of groups such as the mammals where many innovations combined to spell success and whose interactions, for many mammalian features, do not place any particular innovation before another in overall importance in explaining the group's evolutionary success (Kemp 1982). The relationship of single innovations to large morphological complexes reaches a pinnacle of confusion in the use of the term *bauplan*. In McGhee's excellent study of brachiopod form, bauplan is used to describe the general growth pattern of a pair of articulated valves, each growing according to a modified logarithmic spiral. This feature alone creates the many constraints used in McGhee's functional model. But Gould (1983*b*) uses the term to apply, for example, to the complex of adaptations that make up the Felidae. This includes a set of mammalian features, plus additional felid synapomorphies. A pussycat sees an ecological challenge in a wholly different way than a clam! But the reason for this may reside in the possession of a large number of characters. As I will show in Chapter 8, such bauplane arose gradually and the pace and order of aquisition of features may have been partially due to historical accident. I discuss in Chapter 5 the similar evidence for the gradual construction of some developmental programs. This gradual construction of a character complex should not be equated with single features that constrain evolution, such as logarithmic spiral growth. I assume that the latter is the result of a fairly simple cellular growth mechanism. The term bauplan seems to be used by Gould as an emblem for all notions of character complexes that constrain future evolution. But this is usually confusing without any historical information or functional, developmental, or genetic details.

The Study of Form

Allometry

Allometric Relationships **Allometry**, the measure and analysis of proportional changes in traits with size change, is probably the most commonly employed means of studying form (see Huxley 1932; Cock 1966; Gould 1966). The relationship between two measurements is characterized and comparisons are then made, often with functional hypotheses in mind. We assume that two different homologous points can be located on a set of organisms, as termini for each of the two measurements. Using an appropriate technique of statistical association, the following general equation is established to relate the first measurement X_1 to the second X_2

$$X_1 = bX_2^k$$

where b is the **allometric coefficient** and k is the **allometric exponent**. If k equals one, then shape, as estimated by the two measurements X_1 and X_2, does not change with increasing size, as estimated by either X_1 or X_2. If the exponent differs from 1, then some change of shape with increasing size must occur. If k departs from 1, then it is more convenient to plot $log\ X_1$ versus $log\ X_2$. The slope of the linear plot equals k, while log b is the intercept. Linear regression can be used but it must be remembered that an error in X_2 causes a downward bias in the estimate of k.

Five main types of allometric relationships can be established, and these are often confused with one another (Cock 1966).

1. **Static** relationships are established among individuals of a population that are all of the same age.

2. **Cross-sectional** studies come from a set of individuals each of known, but different, age.

3. **Longitudinal** studies take measurements for the same individual at different ages. The exponent k is estimated from measurements taken from a group of such individuals.

4. **Mixed cross-sectional** studies take single measurements from a group of individuals of different ages, but no exact criterion exists to measure age.

5. Finally, **interspecific** comparisons employ representative values from each of several species. These values should be taken from individuals of similar age, sex, and developmental state.

These different comparisons are not equivalent, and usually yield different estimates of k and b. This is especially true in interspecific versus intraspecific estimates. For example, the relationship between facial and cranial length of horses in ontogeny is not equivalent to interspecific differences in fossil horses. The exponent is higher in the phylogenetic series and the value of b differs as well (Reeve and Murray 1942). In the interspecific case a change in b may be explained as an evolutionary increase in the size of the facial primordium, which accommodates the coordinated evolution of hypsodont teeth. Usually estimates of k from static comparisons are smaller than for longitudinal studies. In some interspecific cases, the value of k differs when species are compared at different taxonomic levels. Although brain weight increases with the 0.75 power of body weight for the species of mammals as a whole, the exponent for species within genera drops to 0.5 (Harvey and Bennett 1983). Such comparisons often involve great differences in overall size range, so that different mechanisms may be at work when looking over three orders of magnitude in size, as opposed to lower level taxonomic comparisons that may be done within one order.

Comparisons of the exponent k may be used occasionally to draw conclusions on growth patterns and mechanisms of gene action in intraspecific and interspecific studies. In some breeds of dogs, horses, and pigeons, Darwin (1859) noticed that the young of various breeds more closely resembled each other than did the adults. The young of the tumbler pigeon, however, differed from the young of the wild rock pigeon in almost the exact proportions as did the adults of the two breeds. Similar differences have been recognized subsequently by many geneticists. Sinnott and Dunn's (1935) classic paper on genetics and shape differences in plants revealed shape differences that were established in the embryo, as opposed to others that became more exaggerated with increasing age. In the latter case, it is possible that the allometry lines will, therefore, converge with decreasing overall size and age. In the former case, the shape difference is established in a developmental stage previous to the one used in the measurements that are plotted. Here, the allometric exponent may be the same, but displaced with a different intercept value of b. It is possible that the shape difference was established in a period of allometric growth in this earlier developmental stage, or was achieved in one quantum developmental step.

While studies often focus upon differences in the exponent, k, of the allometric equation, variation may also occur commonly in b, with k being identical for all measured individuals (J.F. White and Gould 1965). For example, Kurtén (1954) found that the allometric intercept for trigonid length versus crown length in teeth of badgers, *Meles meles*, changed during the Holocene. Trigonid length decreased with no ontogenetic change of shape towards the present. The *creeper* gene in fowl probably acts at an early stage of development, probably during morphogenesis. From the seventh day onward, the position of the allometry lines, taken in a longitudinal study, for various skeletal dimensions is altered but k is unchanged. This is common for other chondodystrophies in vertebrates (see Cock 1966, 1969).

Gould (1972) quantified **White's criterion**, the relative difference in size at which shape on two given allometric regressions is the same. If b_i is the allometric intercept for different regressions, i, then

$$S = b_1 / b_2^{(1/1-k)}$$

This estimator was usefully applied to the evolution of the Jurassic oyster *Gryphaea*, whose shape evolution has been the subject of much controversy. In the basal Lias, descendant and larger *Gryphaeas* can be shown to be geometrically scaled up models of their smaller ancestors. For descendants (younger): $b = 0.237$; while for ancestors: $b = 0.210$. The exponent is 1.766 in both cases. From this we calculate S to be 1.17. This matches closely the ratio in maximum size between the descendant and ancestral oysters, which is 1.21. In contrast, the evolution of coiling in the later Lias consisted of a series of shifts in allometry lines, which retained the same shape at larger sizes in descendants. Gould argues that an evolutionary increase in size provides the primary impetus for changes in shape, as shells of the same shape become more unstable as they grow larger (Hallam 1968). When the actual increase in size lags White's criterion, evolution proceeds via paedomorphosis. This happens in the later Lias, where the ratio of maximum sizes for latest to earliest Gryphaeas considered is 1.76. But $S = 7.42$. The spinatum shells would have attained the same shape as angulata shells had they grown to be 7.42 times as large, but they are not even twice as large in reality.

The value of the allometric exponent can be predicted in some cases from biomechanical and metabolic principles. The simplest cases of models of scaling arise from conflicts posed by differential expansion of body parts, stresses, or organismal needs. Consider the differential expansion of bone cross-sectional area S and body weight W, as a function of overall length L. The following equations are to be expected:

$$S = k_1 L^2, W = k_2 L^3$$

Therefore:

$$S/W = (k_1/k_2)L^{-1}$$

Thus, cross-sectional area must increase at a slower rate than body weight. This would pose a significant problem for a terrestrial vertebrate of considerable weight. As might be expected, limb cross-sectional area is commonly found to increase relative to body length, with an allometric exponent greater than 1. In other cases, such as trees, the height/diameter ratio will continue to increase to a point where elastic buckling under wind stress becomes a problem. McMahon (1973) has found that trees do not exceed more than a fourth of a critical buckling height.

Allometry and Developmental Constraint

> Your nose grows as the rest of you grows because you're all
> one piece! (from song by Mr. Fred Rogers)

Huxley (1931, 1932) discussed a possible link between developmental constraint and form, and concluded that it may have important implications for the interpretation of form in terms of function and evolutionary trends. If we consider a longitudinal study of shape, the value of the allometric exponent and intercept should be explained partially by mechanisms of gene action and development. If this is so, then trends among taxa may be due to the same genetic and developmental mechanisms. Are size changes in phylogeny simply an extrapolation of ontogeny (Gould, 1977)? To the degree that this is true, any evolutionary trend under the influence of functional differences among phenotypes of different shape must be constrained by developmental–genetic effects. Allometry has, therefore, been a primary battleground where proponents of nonadaptive constraints have argued with supporters of entirely adaptive explanations for shape differences in phylogeny.

An excellent example comes from the work of Huxley (1931) on the relationship between antler size and skull size (or body height) in different species of deer (Cervidae). All of the measurements yield an approximate straight line on a log-log plot. This is true to the degree that the spectacular extinct Irish elk (actually a deer), *Megalocerus giganteus*, appears on the plot as a "typical" species (Gould 1974). The question is thus: Is the Irish elk a captive of its unusually large size via a genetic–developmental constraint, or is the overall relationship between antler size and skull size among the species simply regulated by natural selection on overall form, based upon some as yet poorly understood biomechanical or other selective constraint? Huxley (1932) argued for the former, though he later seemed to abandon this position (Huxley 1960).

A classic study of evolutionary relative growth in titanotheres (Osborn, 1929; Hersh, 1934) shows a typical interpretation of interspecific allometric relationships based upon developmental constraints. Using Osborn's data, Hersh found the following, when comparing horn length to skull length: (1) Species within a genus, consisting of an ancestor-descendant series, have the same values for b (allometric coefficient) and k (allometric exponent); (2) genera are distinguished from one another by differences in b and k; (3) b is a decreasing function of k as phylogeny proceeds. Hersh (1934, p. 548) notes that "If we were to suppose that the value of k for the relation between horn-length and skull-length, instead of being so unusually high, were more nearly a common value in the neighborhood of 1, then a titanothere with a skull length of 800 mm would possess a horn about 1.5 mm long. Even if k should have the quite high value of 2.75, such a specimen would have a horn not longer than about 5 mm, which, considering the broad oval base of the titanothere horn, would most likely fail to be detected as more than a change of contour in the fronto-nasal region of the skull."

He used this to reach conclusions of far-reaching importance: (1)

The mammalian organization may have a potential mechanism for horn growth. In many groups, however, the horn does not develop because of the allometric relationship, which reflects a developmental process. Therefore, (2) "the titanotheres of early Eocene times did not have horns because they were not large enough. As mutations for larger size occurred and were selected..the ontogenetic mechanisms for the production of horns was able to produce first incipient horns, and with the passage of generations horns of progressively increasing size were produced" (Hersh 1934, p. 548). Finally (3): "the horns in their incipiency were not directly adaptive. But once the horns had reached a sufficient size to be used as organs of offense and defense, we might reasonably conclude that the animals with larger horns, as a consequence of the presence of horns, were selectively favored at the expense of smaller-horned animals" (p.550). At this point selection for larger horns might drag along a correlated selection for larger body size. Hersh concluded that the process would end when the newborn titanothere has a horn so large that it would damage its mother during birth!

I reproduce this argument in some detail as it illustrates beautifully the concept of developmental constraint in evolution, so popular again, a half a century later. Hersh and Huxley conceived of evolutionary trends as regulated by developmental programs, to the degree that they might be thrust by a correlated evolutionary change into a new realm that is incidentally adaptively significant.

Can geometrical comparisons be so easily translated into developmental and evolutionary change? If development does constrain proportions within developmentally united blocks, then one should be able to dissect an allometric relationship within a clade and see an overall similarity in trends over great and small taxonomic distances. After all, the constraint hypothesis presumes that development locks in a certain overall pattern of growth. Unfortunately, variation is expectable enough that it would be difficult to quantify the magnitude of a deviation from a given allometric exponent that would be sufficient to falsify the limitation of developmental constraint.

Let us return to the antlers of deer. Is the allometric relationship between body size and antler size a product of a genetic-developmental constraint, or is it shaped by natural selection? Arguments by Gould (1977) and Lewontin (1978) support the former. Small deer such as the muntjac have small antler size relative to overall body size. But larger species such as the reindeer, or the wellendowed Irish elk, display proportionally far larger antlers. Clutton-Brock et al. (1980) demonstrated that large deer species tend to be more polygynous (more females mated per male) than smaller species. Larger species, with proportionally larger antlers, tend to form larger breeding groups, than do smaller species. This suggests a role for intermale combat. But there is still a significant allometric relationship between shoulder height and antler length within groups identified for similar overall breeding size. The slope of the within-group relationship is less than the one established when species from all groups are considered together. Among other possible explanations, this may be due to a general overall advantage of larger deer and their antlers in mating success over smaller deer. The developmental constraint hypothesis may also be invoked, but it is surely not an hypothesis that seems any more likely than a number of others.

In some instances, trends among closely related species seem to cross the overall allometric trend. Such cases, when explained in ecological terms, would appear to falsify the developmental constraint hypothesis. In the primates, testis weight follows an overall allometric relationship with body weight. Significant deviations occur, however, from the overall trend. These can be related to breeding system (Harcourt et al. 1981). Males of monogamous and harem genera have a smaller testis than those where several males compete for the same female. In the latter case, the contribution of greater volumes of sperm is selectively advantageous, as it dilutes the chances of contributions from matings by other males. Species with larger testes also have a proportionally larger volume of seminiferous tubules as opposed to merely increased supportive tissue. The relationship cuts across taxonomic borders. These results suggest that there is an adaptive reason for an overall correlation between body size and testis size, but that species-specific differences have selected for significant deviations from the overall trend.

A similar case can be made for deviations from the overall brain-size body-size allometric relationship found in small mammals and primates. Folivores tend to have smaller brains relative to body weight than do frugivores (Harvey and Bennett 1983). This may be related to the more elaborate ensemble of behaviors required to locate dispersed and clumped sources of fruit, as opposed to more homogeneously distributed leaves. The trends seem to cross freely the overall allometric relationship, suggesting that developmental constraints cannot be that strong. Indeed, the breakdown of such overall allometric curves into distinct and virtually nonoverlapping subgroups suggests that the impressive linearity of larger-scale groupings is deceiving. The primates hav notably larger brains for their body weights, which further suggests that the allometric relationship is not frozen by an overall developmental program (Lande 1979b). The relatively small brain relative to body weight of the gorilla (*Gorilla gorilla*) results probably from its folivorous habits, as opposed to the better endowed and frugivorous chimpanzee, *Pan troglodytes* (e.g., Harvey and Bennett 1983). In other cases, however, such a prediction does not hold well. On the basis of such reasoning, carnivores would be expected to have larger brains for their body sizes than ungulates, but this is not the case (Radinsky 1978). The overall slope for all species may have other selective explanations, but may have some developmental restrictions as well.

The problem of the overall relationship of brain weight to body weight in the mammals has been a subject of controversy, even over the value of the exponent. Based upon recent evidence, the value of the exponent when comparing all species, is 0.75 (Martin 1981). This suggests the possibility of a relationship with metabolic rate, which also scales on body weight with the same exponent. It is possible that the brain acts as a coordinator of bodily activity correlated with overall metabolic rate.

Allometric exponents relating brain to body size among species at different taxonomic levels often differ. It is the usual case that the value of the exponent, comparing a given measurement with overall body size, decreases as the species compared come from lower taxonomic levels (Gould 1975; Clutton-Brock and Harvey 1979). This has been investigated by Harvey and Bennett (1983). The relationship cannot be easily explained with either a selectionist or developmental constraint hypothesis. If some singular selective or developmental factor tightly regulated the relationship, then the exponent should not differ with taxonomic level. Lande (1979b) and Harvey and Bennett (1983) relate this to a difference in rate of response of brain and body size in the same evolutionary series. If body size responds more rapidly than brain size in an overall trend of size increase, then we would expect a lag time until brain size adjusts to the overall scaling value.

The best possibility of learning the role of genetic and developmental constraints on form would involve a detailed genetic analysis of correlated characters during development and under natural selection. Correlations among characters have both nongenetic and genetic components and strong intercharacter correlations do not necessarily imply that the two characters are controlled jointly by the same genes (e.g., Atchley and Rutledge 1980; Cock 1969; Sinnott and Dunn 1935). The overall phenotypic correlations among traits can be small because both traits have low heritability, and because phenotypic and genotypic correlations are of opposite sign (Atchley and Rutledge 1980).

If we accept a simple model of transmission, with effects on traits explained only by allelic differences, then the response of a trait to selection is:

$$D_1 = h^2 s_1$$

where D_1 is the change in mean of the trait 1 in a population over one generation, h_1^2 is its heritability, and s_1 the selection differential. The correlated response for trait 2, D_2, during one generation of selection on the first trait is: $\gamma_{12}(\sigma_2\sigma_1)$, where γ_{12} is the correlation of the additive genetic values of the two traits (Falconer 1981; Lande 1979b for assumptions). Cheverud et al. (1983) studied the quantitative genetics of age-specific trait values in the mouse and found that ontogenetic gene effects most commonly caused an individual to be larger or smaller throughout ontogeny. It was much less common to find genetic effects that were opposite at different ontogenetic stages, or to find effects unique to a given stage.

Lande (1979b) used the literature on genetic correlations between brain and body weight to conclude that short-term selection experiments would involve mostly change in overall body size with changes in brain size largely a genetically correlated response. The allometrical relationships of closely related species and populations within a species suggest that the genetic regression of brain on body size permits a successful prediction of the magnitude of the brain response when size is increased either by selection or genetic drift. If the laboratory data are extrapolated to longer term evolution, Lande concludes that during the long-term allometric diversification within most mammalian orders there has been more net directional selection of brain size than body size.

The genetic correlation between brain and body size may vary, depending upon the ontogenetic stage upon which selection acts (see Atchley 1984*a, b*). The genetic correlation between brain and body size is far higher in early postnatal growth in mice than in later growth (Atchley et al. 1984). The genetic correlation between brain size and body size decreases in mice as brain size growth in mice slows and eventually stops. Riska and Atchley (1985) suggest that this ontogenetic change in correlation may explain the increased value of the allometric exponent with increasing level of taxonomic comparison mentioned above. Body size evolution among higher taxa may include rapid shifts involving selection on earlier stages of growth that share more genetic growth determinants with brain size. Evolution among closely related species may involve adjustments of body size, which can be accomplished by selection on genetic components influencing later growth. Although this is an attractive hypothesis, the predicted allometric slope from selection on the earlier stages of growth can range from 0.43 to 0.76, using an estimate of one standard error. Unfortunately, this range embraces the slopes seen among all taxa (0.75) and those observed among more closely related taxa (0.4 and up, Harvey and Martin, unpublished). The analysis, therefore, may suggest that pleiotropy is indeed important as an evolutionary constraint. It does not prove, however, that the overall brain–body-size relationship is frozen by a long-standing genetic correlation that cannot be overcome. Riska and Atchley interpret the gorilla's large body size relative to brain size (R.D. Martin 1981) as a continuation of later postnatal growth relative to the chimp, whose relative brain size is considerably greater (see also Shea 1983). If the chimp were the ancestor of the gorilla, this reasoning might be justified, but the phylogenetic relationships of these two forms is in hot dispute, and one is certainly not the ancestor of the other. The fact that in such closely related forms body size can change so much relative to brain size suggests a degree of evolutionary lability that would permit fine adjustments of brain to body size to satisfy functional considerations. Moreover, it is not always possible to predict genetic correlations based upon developmental expectations. In the fowl, no genetic correlation exists between length of the tarsometacarpus and carpometacarpus, despite the serial homology (Cock 1969). Intercorrelations among skeletal elements are often difficult to understand and impossible to predict (Cock 1969; Cheverud et al. 1983).

The extension of laboratory measurements of genetic correlations among traits to interpretations of field distributions and even inter-specific studies is tenuous at best. In *Drosophila*, extrapolations from laboratory measurements are possible to a degree (Robertson 1962). Lines of *D. melanogaster* selected for change in wing, leg, and thorax length showed genetic correlations between wing and leg, but not between either and thorax length. In *D. subobscura* and *D. robusta* geographic races, wing length and leg length tend to covary, whereas head width and thorax length do not. The similarities in experiments and geographic variation may be due to genetic mechanisms whose effects are translated through developmental apparatus of the imaginal disks.

Cheverud et al. (1985) have suggested an interesting approach to partition phylogenetic constraints from other factors. Total variation in trait values among a group of related species can be divided in a phylogenetic component of the variation, due to inheritance from an ancestral species, and a residual, which may be explained by other factors. This approach showed that 50 percent of the variation in sexual dimorphism in weight is due to "phylogeny." Size and diet may account for the residual. One cannot imply from this, however, that the size difference that permeates the whole lineage is due to a developmental constraint. After all, the sex difference in size might be strongly constrained by sexual selection or natural selection, which might have been completed before the stem species of the group gave rise to the descendants used in the analysis. If an ancestral species had two eyes and all descendants had the same number, it would be reasonable to hypothesize that there was no genetic variance available for adaptive change, but it would be invalid to claim that the presence of two eyes arose for nonadaptive reasons. Cheverud et al. seem to be falling in this trap. It is also possible, however, that a high genetic correlation between the sexes might constrain the response to selection on dimorphism; this would create a sort of inertia (Cheverud et al. 1985). Therefore, Cheverud et al.'s analysis certainly calls for further inspection of the dimorphism.

In conclusion, the available evidence is firm enough to cast considerable doubt on the developmental constraint hypothesis as an explanation for allometric relationships in form. While no one would doubt that developmental constraints must have a strong influence on short-term selective processes, it seems likely that most of these constraints are breachable over the long term. The intriguing evidence presented by Riska and Atchley may, however, lend credence to a hypothesis of constraint through pleiotropy, but much more information is required before a quantitative assessment can be made.

It is worthwhile to come back to the famous study by Hersh on titanotheres. McKinney and Schoch (1985) have reanalyzed his data and find that extrapolative growth along a constant allometric trend does *not* account for the trend of body size and horn size increase (part of this was recognized by Hersh). One requires shifts in the allometric intercept and possible changes in the slope. These may be adaptive responses to more massive body shape, which might increase the stress applied to the horns. This suggests a reversal of the prejudice cited by Gould and Lewontin (1979); Hersh believed in developmental constraints and perhaps saw them when they weren't present! Hersh (1934) apparently exaggerated the allometric slope by over 80 percent, probably to ensure that a modest change in size would produce a disproportionately large increase in horn size. McKinney and Schoch suggest that this was done to counter the vitalist streak that was common in evolutionary biology at the time. Oddly enough, a belief in a developmental constraint was used to bolster neo-Darwinism! Let us hope the approaches of Riska and Atchley (1985) and Cheverud et al. (1985) will lead us toward the appropriate course where unbiased methods will cut through our prejudices.

Other Approaches to the Comparisons of Forms

Form Deformations and Transformations   Despite the difficulties of establishing ancestors in complex clades, some investigators have been interested in formal analyses of the geometric deformation required to transform one related form into another. D'Arcy Thompson (1915) suggested the method of grid deformation, which placed a rectangular coordinate grid on one species and showed the transformation to a related form by a regular deformation of the grid. This technique is pictorially illuminating and is useful in convincing one that many organisms are related by fairly simple transformations of dimensional proportions. Unlike comparisons, such as bivariate allometry, it gives a comprehensive picture of the entire deformation of form. Its disadvantage, however, lies in its inability to report the deformation quanti-

tatively and its limited applicability to most form deformations, which
are far more heterogeneous than can be represented by simple grid
deformation (Bookstein 1982; Benson and Chapman 1982). The deformation of a grid is also misleading in that it presumes a set of
equally plausible morphologies along the deformation trend, whereas
intermediates may often be functionally, genetically, or developmentally implausible.

Recently, several suggestions have been made for the quantitative
analyses of transformations between forms via the analysis of defor
mation. An important aspect of these methods is the establishment
of **homologous points**, or **h-points** (Siegel and Benson 1982) on
both forms. The deformation to be depicted, therefore, has at least
a phylogenetic, if not an ancestor-descendant aspect. The deformation analysis estimates transformation by the degree of displacement
of the points. As is well known the establishment of such homologous
points can be highly speculative. For example, it would be commonly
supposed that the pseudangular process of the cynodonts (advanced
mammal-like reptiles and probable ancestral group of the mammals)
is a point homologous to the angle of the mammalian dentary. Fossil evidence, however, suggests that the mammalian structure is newly
derived, independent of the pseudangular process (Jenkins et al. 1983).

Resistant Fit Theta Rho Analysis (RFTRA) attempts to
highlight local form changes by calculating an overall scale for transformation, based upon those areas that have not changed very much.
First a series of homologous points is digitized for each form. Then
the forms are adjusted to be the same overall size by calculating a
scale factor, based upon all proportional changes calculated from relative distances between the homologous points. To do this, a matrix of
values is calculated corresponding to proportionate distances between
all homologous points on the two figures. For a distance from one ho-
mogous point to any other, the median proportional change between
forms is selected. Eventually, as many medians are produced as there
are homologous points. Then the median of these medians is used as
the scale factor to convert on form to another. The angle needed to
rotate the figure to fit the other is estimated with the same repeated
medians procedure.

The scale and rotation factors thus obtained are used to multiply
the first figure and this transformation is superposed upon the second.

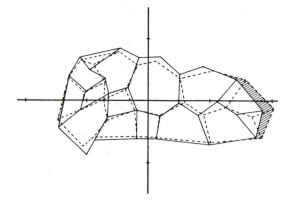

Figure 6.8: The application of Resistant Fit Theta Rho Analysis to a comparison of a male and a female of the ostracode *Costa edwardsii*. The shaded area indicates the inflation coincident with the larger genitalia of the male (after Benson and Chapman 1982).

Lines are drawn between the proportionally scaled and changed homology points from the transformed first figure and those of the second figure. The angle and distance of these lines is an estimate of the degree of shape change. The RFTRA technique emphasizes those parts where major shape changes have occurred.

Figure 6.8 shows the technique as applied to shape change between males and females of the ostracode *Costa edwardsii*. Note the posterior inflation of the male which is emphasized relative to the remainder of the form, which does not change in shape very much. This comparison shows an essential aspect of the technique: while it attempts to estimate deformation, it uses two descriptors (θ and ρ) more as a convenience to arrive at a deformation estimate that emphasizes locations on the form where contrasts in shape are the greatest. No attempt is made to think of the deformation in terms of a continuously deformed grid, as in the D'Arcy Thompson approach.

Bookstein's **Biorthogonal Analysis** produces an estimate of deformation that is much closer to the D'Arcy Thompson approach. The major components of strain that produce the change from one form to another are described in terms of two principal orthogonal deformation axes, describing maximal and minimal strain, and defined throughout the form at homologous points (Bookstein 1978, 1980). This technique can also be used to calculate summary descriptors of change of popu-

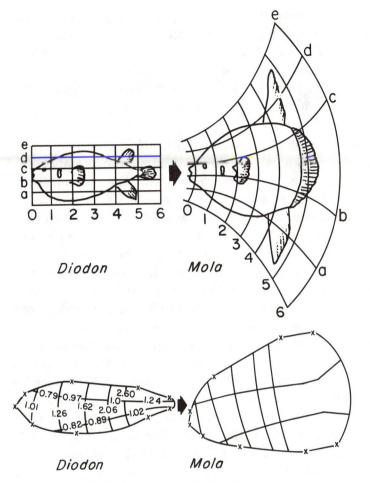

Figure 6.9: Deformation analysis of the transformation of shape between two tetraodontiform fish. Top pair shows change from *Diodon* to *Mola* using the qualitative grid of D'Arcy Thompson. Lower pair shows use of biorthogonal analysis, where numbers on grid for *Diodon* indicate relative strains to produce *Mola*'s form (after Bookstein et al. 1985).

lations of homologous points where shape change is approximately the same (Bookstein 1982). This technique takes better advantage than the RFTRA technique of the available information from the form and shows deformation in far greater local detail (e.g., Fig. 6.9). The benefit of more detailed knowledge is balanced by the disadvantage of a large amount of information needed to visualize change.

Simplifying Form Change Into Synthetic Variables Direct measures of deformation seem to be by far the most intuitively revealing estimators of form differences between related taxa. The techniques described above, however, are recent developments and such approaches have been given relatively little attention aside from the common citation of D'Arcy Thompson's famous grids. Form differences have been much more commonly estimated by calculating synthetic variables that either attempt to get the essence of form variation, or attempt to recalculate axes along the greatest multivariate directions of form variance. Multivariate morphometric techniques seem to dominate morphometry today (see Bookstein 1982; Blackith and Reyment 1971).

A simple technique that illustrates this approach is **fourier analysis** of closed figures, such as the outline trace of a bivalve mollusk. Every location on the periphery of a closed figure can be assigned a vectorial angle and distance, relative to a center point. The overall form can then be described by the sum of a number of cosine functions of varying amplitudes and periodicities (see Kaesler and Waters 1972; and Brande 1979, for paleontological applications). The meaning of the variables is usually obscure and typically has no direct correspondence to any growth mechanism. Lohman (1983) has developed a technique called **eigenshape analysis**, which resolves shape variation into a series of eigenshapes. The first eigenshape accounts for most shape variation and can be used to describe form variation among taxa (e.g., Malmgren et al. 1983).

More commonly, **multivariate morphometrics** is used to take a large number of measurements and simplify the overall variation into a variation along a smaller number of dimensions that best describe the variation. **Principal components analysis**, or PCA, will serve as an example (the reader should consult Neff and Marcus, 1980, for an excellent and relatively nontechnical discussion of many techniques). If n measurements are taken on x organisms, the x organisms can be as

a cloud of points (each being an organism) plotted in the multidimensional space . With two dimensions only, and a normal distribution
for data, the cloud would take the form of an ellipse. PCA attempts
to fit a set of orthogonal axes that run parallel to the major axes of
the cloud. In the case of two dimensions, the new axes would be the
two principal axes of the ellipse. A given point can then be replot-
ted in terms of the new axes, which maximize the ability to visualize
the variance of the overall data set. A principal component of a data
matrix is a new variable produced from the linear combinations of the
original variables, using the elements of eigenvectors determined from
the variance–covariance matrix (Neff and Marcus 1980).

A typical result is as follows. Variation along the first principal
component (new axis, if you will) corresponds to variation of body
size among the organisms. Variation along the second corresponds to
variation in shape (Blackith and Reyment 1971). One can immediately
see that the subject of allometry–the change of shape with overall size–
can be obscured by the recalculation of these synthetic variables. Also
different means of coding, e.g., mean-centering, normalizing standard
deviations to unity, etc., can result in very different results (Neff and
Marcus 1980, p. 53). The technique, if used judiciously, can identify
covarying characters, and thereby reduce the overall need for many
measurements. It might also identify a series of measurements whose
covariation may have developmental or functional significance. For
example, in a study of an hemipteran, Blackith et al. (1963) were
able to define fields of form covariation, in terms of molt and body
region (Fig. 6.10). An allometric enhancement of the mesothorax was
definable, and is easily related to its specialized role in adult flight.
Similarly, the enhancement of the posterior abdomen in the last molt
is associated with sexual maturation.

Like most multivariate approaches, the technique is not designed
to test a model of form change, nor is it explicitly designed to provide
a simple mapping between form change and the final PCA plot. For
example, differences in form that are functionally significant may find
expression in cases where among-organism variation is rather low in
magnitude, relative to other more variable and less important traits.
Thus, a functionally important difference may be found expressed most
heavily along the fourth principal component and, therefore, will tend
to be ignored. Furthermore, a complex form change may be broken

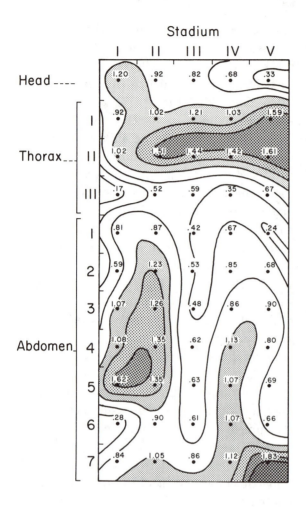

Figure 6.10: Contours of k of the allometric equation, for the hemipteran, *Dysdercus fasciatus*, relating length of a given part to the total body length, as development proceeds. Darker shading indicates higher values of k. Segments and developmental stages are numbered (after Blackith et al. 1963.

up into components that are not geometrically interpretable. If a part of the body is deformed from one organism to the next, it is entirely conceivable that variable elements will plot on one PC axis, whereas less variable elements will plot on others, even though they are part of the same region of the organism.

In summary, synthetic variables, particularly the ones obtainable in multivariate morphometric approaches, can often be profitably used to interpret among forms. As most techniques have varying outcomes depending upon sometimes overlooked differences (e.g., using a correlation matrix versus a variance–covariance matrix). But, commonly, the relationship between form variation and the new synthetic variables is difficult to establish and may even obscure the discovery of variation in form.

The Main Points

1. As a departure for studies of adaptation, a model must be devised that relates form, function, and evolution. The simplest case would involve a function that could predict, from a set of first principles, a unique form. If this form is found in a matching habitat, we might conclude that the form's appearance involved adaptation.

2. This supposition is limited by several problems and features of organisms. First, behavior and habitat selection bring an organism into a milieu most harmonious with its form. One cannot be sure that the form as currently observed evolved in the current habitat.

3. The prediction of form from "first principles" is hampered in several respects. First, historical limitations of organisms will preclude the appropriate engineer's solution. Second, there may be more than one optimal solution, or even a series of such solutions. Third, the evolution of form might involve a continuing shift in the optimum. As size increases, aquatic organisms may encounter new hydrodynamic regimes, and the optimum form may therefore change.

4. Constructional morphology is an integrated approach to the evolution and functional significance of form. It explains form as

the net result of phylogenetic historical factors, functional influences, and the constraints imposed by the synthesis of biological materials.

5. Theoretical morphology is the field which devises (usually) algebraic rules for growth and development. If the mathematical algorithm used to simulate growth matches biological development, then the range of forms generated can be compared with the range found in nature. With independent criteria for performance, theoretical morphology is a useful tool to assess the functional significance of those biological forms that have appeared in biological history.

6. Using phylogenetic history, a restricted optimality model can predict improvements in performance based upon adaptive evolutionary change from an ancestor with a specific set of traits. Such predicted restricted optimal forms may be quite different, relative to forms evolving from other ancestral ground plans.

7. Some innovations, key innovations, may have been crucial in the origin and radiation of major groups.

8. Allometry is the study of changes of shape with increasing size. An exponential equation is typically used to describe the relationship between two linear dimensions.

9. Allometric relationships have been cited as evidence for developmental constraints. Although a few cellular and quantitative genetic models have been devised to account for a rigidity of allometric relationships, the presence of trends across the overall allometric trend suggests that a strong developmental constraint may be lacking and that the dictates of function may contribute to most well known large scale allometric trends.

10. The analysis of covariation of traits can be combined with cladistic analysis to map the stability of character combinations during the history of a taxon. Certain relationships may survive long periods of speciation and morphological change in the history of the taxon. Such analyses can be the basis for hypotheses explaining the long-lived relationships as the result of genetic correlations, epigenetic constraints, or functional interactions.

Chapter 7

Patterns of Morphological Change in Fossil Lineages

> But we see now through a glass darkly, and the truth, before it is revealed to all, face to face, we see in fragments (alas, how illegible) . . . so we must spell out its faithful signals even when they seem obscure to us and as if amalgamated with a will wholly bent on evil.
>
> *Umberto Eco, The Name of the Rose, following Corinthians, 13*

Introduction

Darwin (1859) predicted at first writing of *The Origin* that the rate of evolutionary change would be irregular. In *Tempo and Mode in Evolution* (1944), Simpson asked (p. 3): "How fast, as a matter of fact, do animals evolve in nature?" He confirmed that the rate of evolution is highly uneven, and that bursts of morphological change are highly correlated with periods of cladogenesis. In this chapter I shall demonstrate the extent to which the variability in rate of morphological evolution can be documented with fossils, and I will set limits on the applicability of paleontological data when testing various models of the tempo and mode of evolution. In particular, I shall show that some apparent paradoxes presented by the fossil record are best explained by limitations of the data. The most interesting of these I term **Haldane's paradox**, or why does the recorded rate of change in fossil lineages seem so slow, while laboratory experiments can produce rates of change that are orders of magnitude higher? The punctuated equilibria model of Eldredge and Gould (1972), which has been supposedly

329

tested by many, will be shown to be usually untestable with data on fossil lineages.

Neontologists might think that the rate of evolution would be routinely estimated as the rate of change of morphological features such as size or number of spines. But surprisingly these kinds of data have been collected rather sparsely and were not reviewed in great depth in Simpson's seminal monograph. Rather, **taxonomic longevity** has been used more frequently by paleontologists as an estimate of evolutionary rate. Charles Lyell set the pattern for this approach in his *Principles of Geology* (1830-1833), by documenting the rate of disappearance of extant taxa as one sampled strata of increasing age back into the Cenozoic Era. Using mollusk species, he assumed a regular loss and, when discovering an abrupt irregularity, concluded correctly that some of the early Cenozoic record must be missing in Britain (Rudwick 1972, Chap. 7). Lyell dismissed fossil mammals as useless in resolving Cenozoic stratigraphic questions, because their disappearances back into time were so rapid that extant taxa could be identified as fossils only in the younger Cenozoic strata. This very comparison among mammals and mollusks has been revived twice in this century (Simpson 1944; Stanley 1973a). Simpson assumed that the rate of morphological evolution was inversely related to taxonomic longevity. If taxa were short lived, then morphological evolution was rapid and rapid changes of species-level rank were occurring within a phyletic lineage. If morphological evolution was slow, then such change was too sluggish to produce many changes of significant taxonomic rank, hence taxonomic longevity would be considerable. In this Chapter we will focus on more direct studies of morphological change, but the taxon-based orientation of most paleontological research should be kept in mind.

The Stratigraphic Record: How Much Do We Have?

What Can We, and Can't We, See?

Just how good is the fossil record? Fossils were the prime instruments used to date rocks until the advent of radiometric dating. Even today, the faunal zonations of groups such as ammonites and foraminifera can give a more accurate relative date for local correlations than radiometry. But we also know that preservation is uneven, periods of nondeposition and erosion exist, and embarrassing gaps preclude complete records of change, even in the more continuous deposits of the deep sea. Is the record of sufficient quality to estimate rates of evolution and to locate sites of speciation?

To document evolutionary change in the fossil record, we would require the following: (1) a complete and continuous sedimentary record of a high and constant rate of deposition so that fossils could be collected throughout long local geological sections; (2) a good sedimentary record throughout the paleogeographic range of the taxon; (3) abundant and wellpreserved fossils distributed continuously throughout each of the local stratigraphic sections; (4) time markers; and (5) indicators of the paleoenvironment. If any one of these requirements has not been met, we cannot precisely estimate evolutionary rates and patterns. The virtually insoluble problem of discriminating evolutionary change from localized ecophenotypic responses is a continual impediment to determining whether morphological change truly represents evolution. For example, Bretsky and Kauffman (1977) found a progressive size decrease in a Paleocene lucinid bivalve mollusk, *Myrtea uhleri*, preserved within a regressive sedimentary sequence. A shape change was correlated with the size change, but juveniles also had the same shape throughout the fossil sequence. Was the trend merely one of increased mortality thus leading to elimination of adults, or an evolutionary trend toward more "juvenile" morphologies?

The Completeness of the Sedimentary Record

No place on earth has a complete biological record of Phanerozoic time. Even if a habitat appropriate for a certain group of species was present for a long time at one geographic location, periods of non-deposition or erosion would reduce the completeness of the stratigraphic record. Even when rocks are present, deformation or metamorphosis may have destroyed all but the faintest remnants of fossils. In southern New England, in the United States, for example, early Paleozoic sedimentary rocks are common but strongly metamorphosed, thus fossils are rare. As a result of localized metamorphosis and nondeposition, periods of the order of 10 to 100 million years are typically unfossiliferous in a given region. If evolution has any geographic component, then major losses of transitional forms are inevitable. This loss is compounded by uneven exploration and problems of fossilization. An indication of the

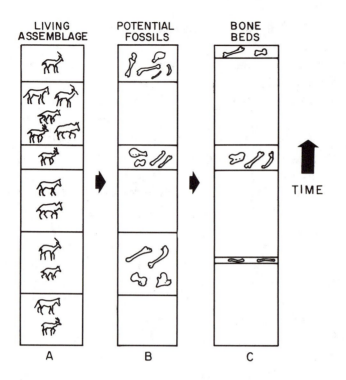

Figure 7.1: Schematic representation of loss of the representation of time intervals, as living organisms die (A) and are transported to a site of deposition (B), and as these deposited fossils are either buried or eroded (C).

potential gravity of the problem is the positive correlation between total rock volume and fossil taxon richness (Raup 1976b; but see Sepkoski et al. 1981). This positive correlation may mean that taxon richness is simply a matter of sampling bias.

Even when the stratigraphic record seems to be continuous over a given time span in a given region, parts of the fossil record are missing. Some gaps are readily identified, since strata may lay upon tilted and older beds that have been uplifted and eroded. In marine deposits, such obvious unconformities, however, usually indicate that the entire environment was not present in the area during the time period, and usually represent retreats of the sea. So nothing of the habitat and evolutionary record has been lost. More worrisome are those stratigraphic gaps that represent nondeposition or even erosion, while there has been no obvious change in the environment. This is especially likely in those terrestrial environments where fossils are preserved in stream gravels and sands that were not the typical living habitat of the organisms (Figure 7.1). Preservation in terrestrial environments of this sort is bound to be incomplete, and different local stratigraphic sections are likely to have sediments representing different spans of time. It would be instructive to discuss the problem of correlation among local geologic sections in greater depth from a detailed geological perspective, but we only have space for some generalities. The reader is referred to Shaw (1964) who discusses the problem of correlating a series of geological sections, each of which is stratigraphically incomplete but in different portions of the section.

Gaps can be identified by missing time-diagnostic horizons, such as volcanic ash layers, or by fossil biotas. Schindel (1982a) defines the three main types of information that would be useful to know from a given section:

1. **Temporal scope** is the total span of absolute geological time represented by the section. This information is essential for time correlations, and for calculations of evolutionary rates.

2. **Microstratigraphic acuity** refers to the time represented by each fossiliferous stratigraphic sample (Schindel 1980). The degree of acuity is calculated from short-term rates of sedimentation, using modern environmental analogues. Large errors are likely.

3. **Temporal stratigraphic completeness** refers to the fraction of the temporal scope represented by strata, as opposed to gaps. To make this measure meaningful, an absolute time period must be specified. Absolute time can be measured by radiometric dating, biostratigraphic correlations with radiometric markers, or magnetostratigraphic correlations. Thus we can ask, for periods of x years, how complete is the record? If a resolution of one year is required, for example, the fossil record is usually woefully incomplete. But completeness for greater time spans can be useful for some kinds of evolutionary studies. Thus, for example, a stratigraphic section may be only 50 percent complete using a resolution of one year while it is 100 percent complete at a 100-year resolution because there is at least one year present from

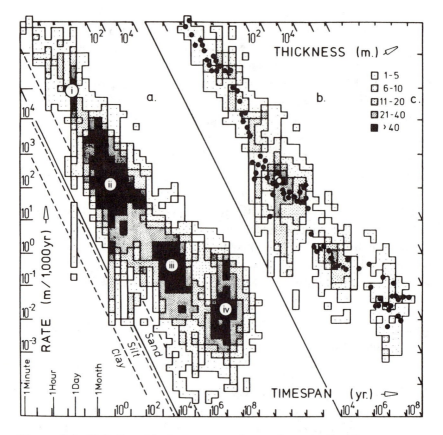

Figure 7.2: Relationship between estimated rates of sediment accumulation versus time scale over which rate was measured (from Sadler 1981. Reprinted from the *Journal of Geology*, volume 89, pages 569-584, with permission from The University of Chicago Press).

any 100 year time interval. In addition to these measures, we require a measure of a fourth item not mentioned by Schindel.

4. **Geographic stratigraphic completeness** is the proportion of the biogeographic range for any one time that is represented by sediments. Most evidence suggests the presence of strong geographic variation within species. There is also the possibility that a variant arising in a small area may then spread throughout the range of a species. Therefore, this measure can be crucial, but has not been studied adequately.

How complete is a sedimentary record? Sadler (1981) compiled over 25,000 published records of sediment accumulation and demonstrated that the time span (temporal scope) over which the sedimentation rate is measured explains much of the variance in rate. The estimated sedimentary rate is inversely related (Figure 7.2) to the temporal scope over which it is measured! This can only mean that long-term calculated rates have incorporated significant periods of nondeposition or erosion. Calculated rates from thinner sedimentary columns are more likely to include particular portions representing continuous sedimentation. As the stratigraphic section to be measured decreases in thickness, the estimated sedimentation rate should converge on the true sedimentation rate for the appropriate environment (Sadler 1981). As longer time spans are usually represented by thicker sedimentary sections, sediment compaction would also cause an underestimate of the rate of sedimentation. This factor is a relatively minor component in explaining the overall variance of measured sedimentary rates (Schindel 1982a).

The inverse relationship between estimated sedimentary rate and temporal scope makes our chances of a complete sampling of evolutionary time increasingly remote when a finer time scale is required. For example, if we wanted a year-by-year record, the stratigraphic record used by Simpson (1944) to study the evolution of fossil horses is only seven percent complete.

Temporal stratigraphic completeness can be estimated as the ratio of the accumulation rate, S, to the average rate of accumulation for a given smaller time scale, S_*. Alternatively, completeness can be estimated from the ratio of S to the average rates from a modern sedimentary environment similar to that represented by the rocks in question. Because of the overall negative relationship between estimated sedimentary rate and temporal scope, as the desired time resolution increases, the proportion of intervals at that time resolution represented by sediments will decrease. Temporal stratigraphic completeness, C, is:

$$C = S/S_*$$

The gradient of rate to time span, m, is related to completeness. We estimate m from the empirical relationship between temporal scope and estimated sedimentation rate. Thus, $S = m(lnt)$, where t is the temporal scope. If t_* is the time span of a given arbitrarily smaller time unit,

$$\ln S/S_* = m \ln t - \ln t_*$$

Thus, completeness is estimated by

$$C = S/S_* = (t_*/t)^{-m}$$

Sadler (1981) gives a useful example, based upon the depositional rates of platform limestones, the source of much of the invertebrate Paleozoic record. Consider a 900-m limestone section deposited over 15 million years. What is the completeness of the section for intervals of 5 my? Based upon available data, $m = -0.35$ and $C = (5/15)^{.35} = 0.68$. In other words, one out of every three 5-my intervals will not be represented by sediment. Is it any wonder that large gaps exist? Darwin was quite right in believing that at least some of the major gaps in the fossil record might be explained by lack of preservation. Most would admit that 5 my is enough time for much to happen; for such a loss to occur a third of the time is to guarantee significant losses.

Many studies of the rate of morphological evolution involve relatively short time intervals of well-preserved sediments. Of eight such studies tabulated in Schindel (1982a) only two have temporal scopes over 12 my and four are below 2 my. In a study by Williamson (1981) the overall sedimentation rate estimated from the total section is much higher than expected from the average relationship of lacustrine sedimentation rate to temporal scope. This is probably due to the proximity of the sedimentation site to a tectonically active area. Williamson was thus able to document morphological changes over a period of only 0.4 my.

Figure 7.3 gives stratigraphic completeness for the eight studies. At the level of 10^5 years, most of the samplings are virtually complete, though two show large gaps. The famous bed-by-bed *Kosmoceras* study by Brinkmann (1929; Raup and Crick 1981, 1982) is less than 30 percent complete even at this coarse time scale. At the level of 10^4 y, most of the studies are less than 50 percent complete. That is to say, less than half of the expected intervals of 10^4 y could be sampled. This is alarming as much evolutionary change can happen in 10,000 years. It takes only modest selective pressures to shift many morphological features to a significant degree over such time scales, given the relatively short generation times of many invertebrate groups. These estimates are filled with significant errors, since (1) they assume that the total

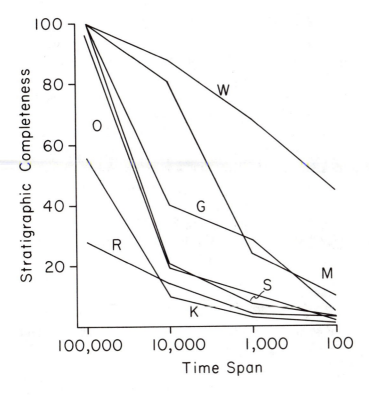

Figure 7.3: Relationship between stratigraphic completeness and time span represented by the deposit. (data from Schindel 1982*a*. W = Williamson 1981; M = Malmgren and Kennett 1981; G = Gingerich 1976; S = Schindel 1982*b*; O = Ozawa 1975; K = Kellogg 1975; R = Raup and Crick 1981).

column has been compacted by a factor of two (Schindel 1982a) and (2) the short-term sedimentation rate used from more general data to calculate completeness may not be representative of the particular study.

These few analyses can only lead to provisional conclusions, but they also represent cases where preservation was believed good enough to investigate rates of evolution. The cited examples thus are probably better than the typical situation for the fossil record as a whole. The generally poor representation of 10,000-year intervals is discouraging, as speciation probably acts on this time scale in many cases. The situation is, in reality, far worse as we have not taken geographic stratigraphic completeness into account. The relative completeness seen between the 10,000- and 100,000-y levels of resolution suggest that only general trends may be observable with fossil data.

In some cases major increases in evolutionary rates may be correlated with changes in sedimentation regime. Barrell (1917) first discussed the general pattern of shifting wave energy and sea level as related to potential cycles of sedimentation. He argued that a general **base level** depth existed, above which erosion was common. During a lowering of sea level, this baseline would shift seaward, and records of evolutionary responses to shallowing would tend to be lost.

Speciation would be expected to occur for the most part in the shallow parts of marine basins, which are ecologically and geographically marginal (e.g., Jeletzky 1955; Eldredge 1971; Emiliani 1982). These are the very sites where geographic and temporal stratigraphic completeness are likely to be minimal. The more complete central basins may be sites of more continuous sedimentation, but they may not be indicative of the general tempo of evolution. It thus seems likely that speciation would not be commonly observable in the fossil record. This is amply confirmed by the many careful studies showing the apparent effects of cladogenesis, but failing to show exact patterns of ancestry (e.g., Lang 1919; Kaufmann 1933; Fisher et al. 1964).

Variability of the Rate of Evolution

A Measure of Evolutionary Rate and Haldane's Paradox

Rates of morphological evolution were compiled, principally for vertebrates, long before the bias of stratigraphic completeness was discovered. We owe most of our knowledge of these rates to George Gaylord Simpson who compiled data principally for horses in *Tempo and Mode in Evolution* (1944). These rates were usually based upon differences in sizes of various structures between taxa living at different times, with relatively poorly understood cladogenies below the level of the genus. The differences were usually accompanied by some unknown degree of cladogenesis as Simpson indicated. By the time of the writing of Simpson's classic, it had long been known that even modest selection pressures could shift morphological traits in populations by large degrees in only a few thousand generations. As early as 1915, Punnett had reported calculations showing that a favored genotype could rapidly increase in frequency in a population. It was thus of great interest to see how such changes occurred over the temporal scope of millions of years available to paleontologists.

Simpson's compilation and other data on fossil reptiles were used by Haldane (1949) to calculate proportional rates of change for linear measurements. Over a time, δt, a structure may increase in size from x_1 to x_2. The proportional rate of change is:

$$\frac{\delta \ln x}{\delta t} = \frac{\ln x_2 - \ln x_1}{\delta t}$$

For example, between the fossil horse taxa *Hyracotherium* and *Mesohippus*, paracone height increases from 4.67 to 8.36 mm. The proportional rate of evolution is, therefore, 0.5823 divided by the time, 16 x 10^6 y, or 3.6 percent per my. Haldane proposed a convenient measure, the *darwin*, which equals a change by a factor of e per million years. On a semilogarithmic scale (time versus *log* length), a rate of one darwin would correspond to a difference of one logarithmic unit per million years. For the above example, the rate is 1.036/2.718, or 0.38 darwins. The overall estimate of vertebrate evolutionary rates on the order of a few percent per million years has been generally reported since Haldane's estimates (Van Valen 1974, Kurtén 1959a). Some much larger rates have been reported for invertebrates. Swinnerton's (1940) study of the Liassic *Ostrea-Gryphaea* lineage yielded a size change of 1 percent in about 3000 y, while Teichert's (1949) study of a Permian crinoid yielded a rate of change in bulk of the basal plates of about one per cent per 600 y!

Haldane also calculated the time, T_{SD}, for a trait to change by one standard deviation in size as:

$$T_{SD} = Vt/(\ln x_2 - \ln x_1),$$

where V is the coefficient of variation. For the evolution of the para-cone from *Hyracotherium* to *Mesohippus*, it would take half a million years, assuming a generation time of three years. More recent esti-mates for other mammals and foraminifera give results of similarly low magnitude (Charlesworth 1984a). This is remarkably slow, considering laboratory selection experiments that can accomplish the same change in a few tens of generations (Mather and Harrison 1949; Robertson and Reeve 1952). Thus Haldane's paradox was raised: Why is evo-lution so slow? Haldane argued that the data from the fossil record did not strongly support the action of natural selection in affecting phenotypic evolution. He suggested mutation pressure as a possible alternative hypothesis. This would be in strong contrast to Simpson's belief that natural selection had been behind the evolution of hyp-sodonty in horses. The calculation does not take into account the joint evolution of different characters, but it is not entirely clear that this would make a great deal of difference (Charlesworth 1984b).

Using a quantitative genetic model, Lande (1976) calculated the minimum selective mortality necessary to generate the commonly ob-served rates of evolution in the fossil record. He assumed that the character under selection follows a normal distribution, and that fit-ness is a normal function of phenotypic value, which can be estimated along a linear scale (e.g., length). Truncation selection is used as the mechanism of directional change. Heritability cannot be calculated for traits belonging to fossil taxa, but a range of values known for analogous living forms might be useful in setting bounds within which drift might be excluded as a force in evolution. Lande assumed that a genotype–environment interaction is absent, and that heritability, h^2, and phenotypic variance both remain constant during the period of change. If z is the amount of phenotypic change, and σ the standard deviation, then a calculation of $z\sigma$ (see Lande 1976 for details) permits a further calculation of the proportion of the population culled each generation to cause a given rate of evolution from natural selection alone. The effective population size, N^*, at which there is a five per-cent chance that genetic drift can cause a change of z over t generations is

$$N^* = (3.84h^2t)/(z/\sigma)^2$$

If the effective population size is greater than N*, then drift can be excluded as a likely hypothesis. There is a likely bias in the estimate of σ. If we are combining a time-averaged sample of fossils, where the mean value of the trait changed significantly over time, σ will be inflated.

Lande (1976) calculated, for Tertiary mammals, that the mortality necessary was only about 1 death per million per generation. Observed rates, moreover, are slow enough to have been caused by genetic drift occurring in effective population sizes of 10^1 to 10^0. If stabilizing selection was weak, horse evolution could have occurred by genetic drift between the browsing and grazing adaptive zones described by Simpson. With moderate stabilizing selection, however, directional selection would almost surely have been required to produce the trend.

Reyment (1982a) made a similar calculation for phenotypic evolution in a fossil ostracod lineage. In the upper Cretaceous of Morocco, a transspecific evolutionary event, from *Oertliella tarfayensis* to *O. chouberti*, can be identified by a change in surface ornamentation and other characters, which exhibit a transitory polymorphism over $1\text{-}2 \times 10^5$ y. Length of carapace also changes, but Reyment's calculation shows that, assuming a heritability of 0.2, selection could have been accomplished with one selective death per 10^{11} individuals per generation. Drift could have occurred in populations as large as 2.7×10^6, which is far larger than thought to be likely for ostracod populations. Reyment questions whether such a univariate expression of change is meaningful, and suggests that stronger selection may have been required to change the entire morph from the ancestral to the descendant state. Reyment (1982a) inferred higher selective mortality, on the order of tens per hundred thousand per generation, in another study of megalospheric proloculus size in the Cretaceous foraminiferan, *Afrobolivina afra*. Depending upon the heritability, N* falls within the range of 10^4 to 10^5.

Charlesworth (1984b) estimated the genetic load associated with directional selection needed to account for observed rates of change, when the simultaneous load of stabilizing selection is taken into account. Based upon published cases of vertebrate and foraminiferal evolution, the cost of directional selection is usually quite small—on the order of 10^{-10}—and is still on the order of 10^{-6} if variation in direction of evolution is taken into account. He argues that these estimates are

342   Patterns of Morphological Change in Fossil Lineages

on the conservative side and that directional load is probably overestimated. If we consider multiple character evolution, genetic correlations among characters will reduce the load, relative to that calculated as the sum of independent directional selection on single characters. Charlesworth's calculated loads for single characters are so low as to make the evolution of several independent characters possible with little load.

The Incomplete Record and Measures of Evolutionary Rate

Haldane's paradox can be viewed in the light of our discussion of the temporal completeness of the sedimentary record. An apparent inverse relationship exists between temporal scope and the measured rate of sedimentation. This can be explained by the presence of gaps in deposition, whose cumulative effect is an apparent slowing of sedimentary rate as more sedimentary column is accumulated and more gaps in sedimentation are added. By analogy, is it possible that Haldane's paradox can be explained by the addition of periods of rapid evolution, and periods when virtually no evolutionary change or reversals occur? The slow rates of change, after all, are usually measured over periods of millions of years. Given the low stratigraphic completeness of most cases, it would be difficult to identify change over periods much less than 100,000 years. Indeed, many terrestrial bone beds in fluvial deposits can be shown to be agglomerations of about 1,000-10,000 y (Behrensmeyer 1982). The theoretical studies of Kirkpatrick (1982) and Petry (1982) show that conservative values of selection intensities and genetic variation could easily result in large scale changes within the cracks of the geological record. Could one ever see rapid rates of evolution in the fossil record?

A compilation of evolutionary rates shows a bias analogous to the sedimentary rate relationship discovered by Sadler and Schindel. Gingerich (1983) compiled rates of morphological evolution (Figure 7.4) for laboratory selection experiments (averaging 50,000 darwins), historical colonization events (ca. 400 darwins), faunal changes following Pleistocene glaciation (ca. 4 darwins), and of small rates from the vertebrate fossil record, such as those used by Haldane and Lande cited above. Overall, measured evolutionary rate is related inversely to the time scale over which the change was documented! This result suggests that much of the basis of our belief in the low rate of fossil evolutionary

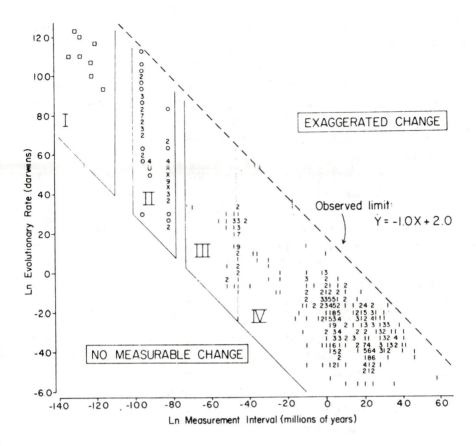

Figure 7.4: Inverse relationship between temporal scope and estimated rate of morphological evolution. Domains I to IV correspond to rates from laboratory selection experiments, historical colonization events, recovery from Pleistocene glaciation, and fossil invertebrates and vertebrates, respectively (from Gingerich 1983).

change is incorrect, especially because the average ratio of the initial
to the final states used to calculate the evolutionary rates is about
1.2. As Gingerich argues: "Organisms differing by a factor of much
more (or less) than 1.2 are so different (or so similar) that they are
rarely compared in calculating rates, regardless of the time available
for one to have changed into the other. The net effect of such a stable
difference between initial and final morphological states over all time
intervals studied is to make interval length the principal determinant of
rates. The greater the time separating similar initial and final states,
the slower the inferred rate of change."

The negative relationship between evolutionary rate and temporal
scope might be an artifact, since there is an apparent homogeneity of
proportional morphological difference, about 1.2, used in the calculations of evolutionary rate (Gould 1984a). If a difference of 1.2 is used
over 1 year or 1 million years of time, then a negative relationship between time and rate is inevitable. One is effectively plotting k/t as a
function of t, where k is a constant, and equal to 1.2.

This result represents much more than a mere artifact of plotting.
As Gingerich (1984) notes, change is not constrained to 1.2 and varies
substantially. Moreover, it is doubtful that workers looked only for
those proportional differences of 1.2 and then divided by the time that
coincidentally had been covered. Rather, this spectrum of rates gleaned
from the literature represents a wide range of studies. There is no
constraint that would set the average rates for laboratory populations
to be so rapid. Artificial selection *is* potent, and brings about changes
of several standard deviations within a few tens of generations. There is
a more important mechanistic argument that suggests that Gingerich's
result is more than artifactual. Short-term rates are probably biased
toward those cases where change has been observed; other cases where
change failed to occur are probably not reported. Of course the same
is true for fossil lineages. One can, therefore, assume that the data
are biased toward maximum change. If the estimate of rate is slower
in fossil finds, one can argue that this is not for lack of a search for
rapid rates. It is only the apparent mixture of rapid rates of change
with periods where no change occurs at all that produces relatively
slow apparent rates of change in the long-term fossil studies (Gingerich
1983).

Kurtén (1959a) recognized much the same bias in the rates of morphological evolution in fossil mammals. Typical rates gleaned from Pleistocene examples were 20 to 25 times as fast as those taken from times in the Tertiary. Although the rate of morphological evolution is probably somewhat greater in the Pleistocene, due to rapid environmental shift, Kurtén concluded that time scale was the major determinant. He notes that the slower rates taken from the Tertiary may be "partially or wholly spurious" because "they are without exception based upon samples millions of years apart, and the intervening histories may have contained any amount of fluctuation at higher rates" (Kurtén 1959a, p. 213).

One might argue that Gingerich's and Kurtén's results combine inappropriately a diverse assortment of evolutionary rates calculated from very disparate types of data. But a recent study by Bell et al. (1985) on a much finer scale shows a similar result for rates of morphological evolution measured over different temporal scopes in the Miocene stickleback *Gasterosteus doryssus* (T. Phillipi, unpublished). Here a presumed temporal scope of about one hundred thousand years is employed with lake varve dating. The same negative correlation between measured rate of morphological evolution and time period was found in four of six characters that were examined (Figure 7.5). Reversals in evolutionary direction are partially responsible for the overall negative relationship between temporal scope and estimated evolutionary rate (M.A. Bell, personal communication).

High rates of change in selection experiments cannot be expected to be maintained over more than a few tens of generations. Exhaustion of genetic variability (assuming population size to be quite small and selection quite strong), the pleiotropic effects of genes, or interlocus interactions will probably preclude rapid change beyond a given point. It is also unlikely that selection will be so intense and similar in direction over long periods. Variation in rates and direction of selection will appear as low rates of evolution when integrated over geologically long periods of time. The rates measured for invasions, of the order of 400 darwins, may be typical of adaptive radiations and ecologically driven changes during speciation events. These would be difficult to identify given the generally poor resolution seen in the fossil record. Differences between closely related mammal species are often of a magnitude of 0.1-$0.2e$, a difference that could arise in an interval as short as 250 to 500 y, given rates of 400 d. The data of Bell et al. show that this effect

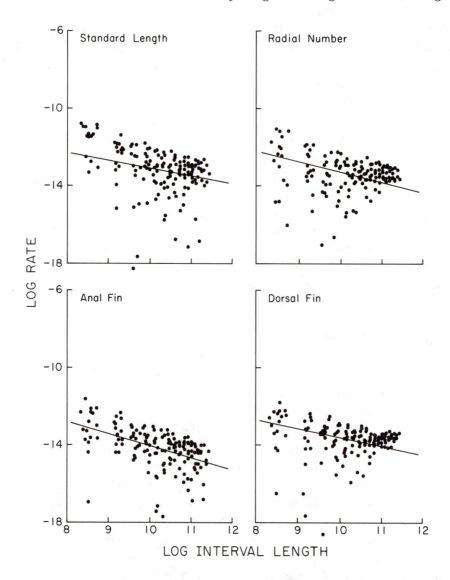

Figure 7.5: Evolutionary rate versus time interval in the Miocene stickleback *Gasterosteus doryssus*. All slopes are significantly negative, but no relationship was found for two other characters (courtesy Tom Phillipi).

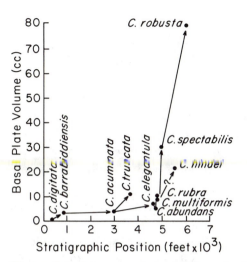

Figure 7.6: Increase of size of the principal "species" of *Calceolispongia*, as a function of stratigraphic position (after Teichert 1949).

will be registered at time scales usually at the finest level of resolution available to most paleontologists.

The fossil record *invites* measurements of slow rates of evolution; both the uneven nature of evolutionary change and the gaps of the sedimentary record tend to conceal rapid evolutionary events. Haldane introduced us to the broad swells of evolutionary change, not the more brief violent chops. It is crucial to remember that there is no qualitative difference in mechanism required to generate either rate of change. The fossil record does not falsify the notion that the evolutionary genetics of populations can be extrapolated to large scale changes, although it would be premature to say that these data confirm a homogeneity of process among changes registered over all time scales.

Variation of the Rate of Evolution

Most studies have asked the question: "What is the rate of evolution?", and most workers have appreciated that rates of phenotypic evolution are highly variable. This was one of the major conclusions first drawn by Simpson (1944). However, Simpson's conclusions were mainly based upon strong differences in taxonomic longevity, not on direct estimates of phenotypic change in geological columns. Brinkmann's (1929) classic study of the ammonite *Kosmoceras* suggested that apparent accelerations in phenotypic evolution might be correlated with diastems (small breaks in deposition), and the actual rate of phenotypic change might, therefore, be constant. Teichert's (1949) study of the phenotypic evolution of the western Australian and Timor Permian crinoid *Calceolispongia* revealed strong variation in rates of change in size and ornamentation. Figure 7.6 shows some of his evidence which assumes constancy of sedimentary rate, in order to relate stratigraphic position to time. The accelerated rate of evolution observed corresponds to Simpson's notion of **quantum evolution**, which implies occasional strong bouts of directional evolution.

In recent years, some have rightfully questioned whether patterns, usually reported only qualitatively, demonstrated either significant directional evolution, or, if no net directionality was observed, significant variation in the direction of evolution (e.g., Schopf et al. 1975; Gould 1976). Raup and Crick (1981) used a runs test to demonstrate that evolution in *Kosmoceras* was significantly directional, but suggested that much random variation in rate may have occurred as well (it should be mentioned, however, that the runs test has rather low statistical power).

Bookstein et al. (1978) used a technique called hierarchical linear modeling to examine variation in evolutionary rates, and estimated the component of phenotypic evolution corresponding to linear change with elapsed time. In the Eocene *Hyopsodus*, studied by Gingerich (e.g., 1976), some part of the change was linear, but there was also a significant component of deviation from this temporal linearity. This is to be expected in any typical evolutionary case history. Charlesworth (1984a) looked at six cases of invertebrate and vertebrate studies of fossils using modified regression tests and found significant variation in rates in five cases. In a study of tooth dimensions (lower M2) in brown bears (*Ursus etruscus* → *U. arctos*, Kurtén 1959a), tooth size first increased steadily for 0.6 my and decreased for another 0.2 my with little net change over the entire period. In the Pacific foraminiferan *Globorotalia merotumida* → *G. tumida* lineage (Malmgren et al. 1983) a sharp acceleration in the rate of change occurs in size and shape at the Miocene-Pliocene boundary (Figure 7.7). Variation between levels, however, gives no indication that the transspecific change was anything but a transformation based upon within-species variation, i.e., phyletic evolution. A recent statistical study using stepwise G-tests (Bell and

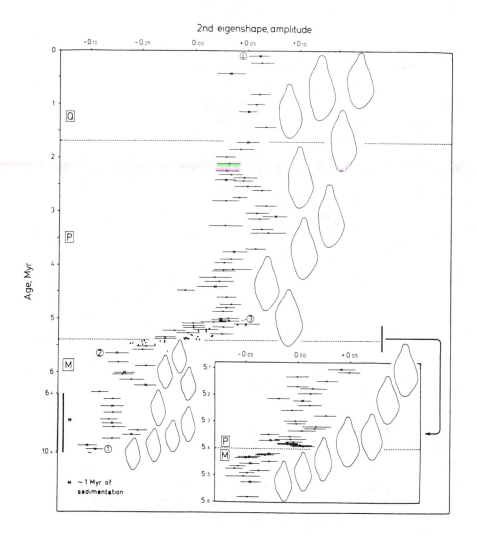

Figure 7.7: Variation in shape in the *Globorotalia plesiotumida* — *G. tumida* lineage, from the Upper Miocene to the Quaternary. Inset is a closeup of changes across the Miocene-Pliocene boundary (from Malmgren et al. 1983).

Phillipi, in preparation) of the fossil stickleback, *Gasterosteus doryssus*, shows alternating periods of evolutionary change and stasis. Overall there was no net change over the interval examined.

One might imagine a series of episodes of directional selection, with no net trend. After all, an environment might be constant over the long haul, but brief changes might cause deviations of environmental conditions away from the longer term constancy. A study of 19 bivalve "species" lineages showed little change from the Pliocene to the Recent. The magnitude of change was usually within that found for geographic variation (Stanley and Yang 1987). This result is consistent with the notion of long-term stasis, but not necessarily with the speciation requisite of the punctuated equilibria theory. Bursts of phyletic evolution could be followed by approximate stasis for millions of years. Stabilizing selection would be the explanation for stasis (Stanley 1979, p. 57; Charlesworth et al. 1982). Charlesworth (1984a) notes that even in cases where no net change occurs in the mean of a phenotypic trait, significant variation in rate with fluctuating direction is the rule.

Can we measure the rate of evolution with fossil data? To provide a convincing case at brief time intervals, we would expect that the apparent rate would converge on the true rate, as temporal scope decreases. But it doesn't, at least as inferred from data such as in Figure 7.5. This raises the disturbing question of whether fossil data can be distinguished from a random walk (Raup and Crick 1981; Bookstein 1987). In the context of random motion, stasis can be defined as a statistically lower deviation from a starting condition than expected by chance. A directional rate of evolution should not be inferred unless there is a deviation from the starting condition that could not be explained by a random walk. Unfortunately, random walks look decidedly nonrandom to the eye. Coin-tossing is often dominated by long runs of heads, or long runs of tails. It is rare to see the end point of a long random walk process be near the starting point, and individual runs often deviate in one general direction. To make things worse, estimated "rates" of change will decrease with increasing time intervals. In other words, the inverse relationship found by Gingerich (1983) could be merely a property of a random walk. An analysis of the foram data of Malmgren et al. (1983) suggests that there may be three phases of random walks at different overall rates, and not alternating phases of directional evolution and stasis (Bookstein 1987).

This argues for a healthy degree of skepticism over our current ability to quantify evolutionary rates in the fossil record. It certainly seems premature to feel confident over whether rates can be used to distinguish genetic drift from selection. We have net degrees of change (or lack thereof), and sometimes this is information enough, but rates are something else again.

Testing for Punctuated Equilibria?

The Punctuated Equilibria Hypothesis and Fossil Data

The punctuated equilibria theory (Eldredge and Gould 1972) claims that most morphological evolution is associated with cladogenesis. It is my intention in the following sections to discuss the punctuative hypothesis in the context of paleontological data on phenotypic change, typically within single geological sections. I shall demonstrate that most claims of supportive data ignore completely the most fundamental limitations of fossil data, namely, the equivalence of taxonomic rank and morphological difference. I shall also show that such change as has been documented can easily be accomodated into a scheme involving phyletic change within populations. This is not to say that cladogenesis has never occurred, or that speciation is not occasionally or even often correlated with periods of rapid morphological evolution.

Punctuation and Two Definitions of Stasis

The punctuated equilibrium hypothesis is a claim at once about paleontological data and about the nature of speciation. Two principal assertions are made about the fossil record. First, the history of a species is dominated by stasis; change throughout the lifetime of the species is usually minimal and rarely directional (Eldredge and Gould 1972). The hypothesis also claims that speciation is the major source of phenotypic evolution, because phyletic evolution is too sluggish to generate sufficient change (Stanley 1979). Change, when it occurs, is therefore concentrated at times of speciation. Thus evolution, as documented by the fossil record, consists of long-term periods of phenotypic equilibrium, punctuated by speciation events when form changes rapidly. Long-term stasis, especially when expressed as a lack of directionality (Schindel 1982b), is consistent with expectations of theoretical population genetics and could be accounted for by stabilizing and fluctuating directional selection. As mentioned in Chapter 3, certain organism controlled patterns of habitat selection would reinforce stasis by keeping the progeny of each successive generation in approximately the same habitat. Overall constancy of habitat, habitat loyalty, plus culling of extremes, would explain stasis. Frequent broad ecological tolerance of single phenotypes would also result in stasis.

Discussion of the punctuative hypothesis has been confused because of the failure to distinguish between the claim, on the one hand, that evolutionary change in characters through a geological section is usually minimal–which I call **character stasis**–and, on the other, that evolutionary change is minimal because speciation is required in order for significant departures from the typical morphology to occur–which I term **species stasis**. Character stasis means simply the condition where for a given morphological character, no temporal change occurs. Species stasis implies that morphological change does not usually occur without a speciation event. This same distinction is made between the punctuative **pattern** (= character stasis) and the punctuated equilibria **theory** (= species stasis). Some fine statistical analyses (e.g., Bookstein et al. 1978; Charlesworth 1984a) usually speak to the issue of character stasis but not to species stasis, and most of the consideration of fossil data, particularly in Gould and Eldredge (1977) also centers around the issue of character stasis and not species stasis. To deal with species stasis, one must have evidence that speciation has occurred in the lineage within which character change is measured. This is where the fossil record fails the punctuated equilibria theory.

Can Fossil Species Be Identified?

There are many co-occurring species in fossil assemblages. In both vertebrate and invertebrate communities, quite stereotyped assemblages are found to be constant ecologically and taxonomically over many millions of years (e.g., Boucot 1975, 1983). Two questions must be raised:

1. Can we investigate cladogenesis among fossils at the species level successfully?

2. How are paleontological species usually recognized?

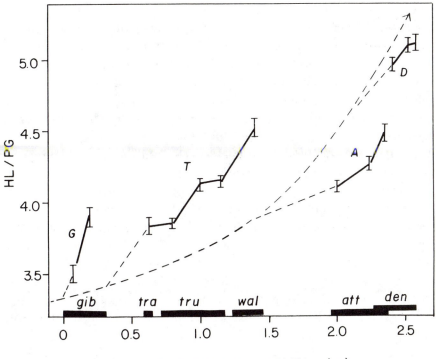

Figure 7.8: An inferred phylogeny within the Upper Cambrian trilobite *Olenus*. Four distinct iterative lineages, G, T, A, and D, were defined on the basis of several characters (illustrated here as a function of the ratio of head length, HL, to preglabellar length, PG). Within each lineage, a series of horizons (more than those depicted here) define transitions from one "species" to another (indicated by black bars at base). A conservative stock is believed to give rise to the lineages, and migration from a locale east of Sweden is believed to be the source (after Kaufmann 1933).

The first question, unfortunately, must be answered usually in the negative. Fisher and others (1964) documented the evolution of the volutid gastropod genus *Athleta* and showed that a main stem stock is accompanied by several offshoots. Despite fairly careful sampling, they were unable to provide any real evidence on the time or exact phylogenetic pattern of cladogenesis. Kaufmann's [1] (1933) classic study of the Cambrian trilobite *Olenus* in Sweden identifies four lineages or successions of taxa of species level rank. Each succession is distinct and demonstrates transitions from one "species" to the next (Figure 7.8) but an ancestral conservative stock, which supposedly gave rise to the four lineages is only hypothetical. The progenitor stock may have been geographically remote; indeed this is a common explanation given for the sudden appearances of new lineages (e.g., Kaufmann 1933; Jeletzky 1955; Palmer 1984). Few paleontologists believe that entire paleontological populations evolve gradually into descendant states throughout the range of the taxon (but see evidence in Hallam 1982). On the contrary, the geographically peripheral origin of new taxa is widely given as the explanation for apparent sudden appearances in the fossil record. The findings of the lineage studies of Fisher and others and of Kaufmann are typical of the fossil record in general. Its generally spotty preservation usually precludes complete records of splitting. The previous discussion of temporal stratigraphic completeness suggests why this might be so.

If cladogenesis cannot be documented then it is unclear how it is possible to understand the relationship between speciation and morphological change in fossil lineages. A few studies have been able to document cladogenesis. Gingerich's study of mammalian phyletic evolution (1976) shows an apparent split, but it is unclear whether this results from in situ evolution or, more likely, from immigration, because Gingerich was sampling from a single section. Grabert (1959) and Prothero and Lazarus (1980) document the splitting of lineages among foraminifera. Grabert found (Figure 7.9) increased intrapopulation variation at the time of splitting.

---

[1] Due to his part Jewish ancestry, Rudolph Kaufmann's career was ruined by the Nazis. He was forced from his job, and, after emigrating to Lithuania, was recognized by Nazi soldiers and murdered (Teichert 1946). Had he lived, he would probably have been a major paleontological contributor to evolutionary theory.

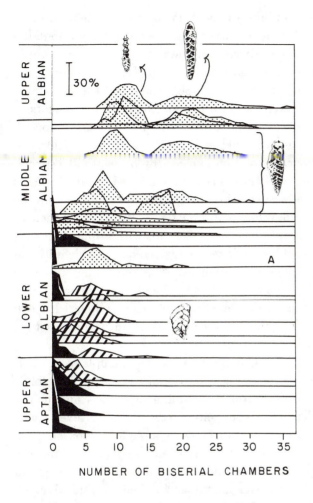

Figure 7.9: "Transgeneric" evolution between the foraminifers *Gaudy-rina* and *Spiroplectinata* (black and diagonal stripes; intergeneric transition occurs at level A). Cladogenesis is inferred to lead to the evolution of *S. annectens* (pictured at upper left) and *S. bettenstaedti* (upper right) in the upper horizons (after Grabert 1959).

Can the punctuated equilibria hypothesis be tested when no splitting of a lineage occurs? Paleontologists have long worried about the the problem of demarcation of species rank taxa when an uninterrupted record of ancestors and descendants might occur (e.g., Simpson 1944, Newell 1947). Levinton and Simon (1980) point out the inevitable tautology established between species recognition and assessments of character change. Paleontological data are used to recognize speciation by identifying changes in "species-specific characters" (Eldredge 1974). Thus, morphological evolution can readily be related to speciation since one is the source of the other's inference! McNamara (1982) has recently described how successional sequences of "species" can be related to the molding of developmental patterns. But these sequences are geographical or temporal successions of forms without any evidence for cladogenesis; the recognition of species again required a definitional association with morphological differences. It is one thing to decide that it is necessary to use phenetic differences as a convenience to recognize fossil species, as in stratophenetic approaches (Gingerich 1979; S.S. Bretsky 1979). It is quite another to relate such changes to a qualitative process such as speciation.

It is instructive to look back at the first known example of a small-scale paleontological study inferring evolutionary relationships, for it sets the stage for the problems in the naming of species and in the analysis of splitting that have confronted paleontologists ever since. The oldest paleontological publication showing a phylogeny based upon fossils was Hilgendorf's studies of the 1860s and 1870s of a planorbid species group (Gastropoda) from a Miocene fresh-water lake deposit in the Steinheim basin of South Germany (see Lindenberg and Mensink 1979; Reif 1983*b, c* and references therein). This case comes from a lake with a minor endemic evolutionary radiation. Hilgendorf described his taxa as varieties, or subspecies, and showed the phylogeny by glueing representative specimens onto cards, which were rediscovered by Dr. W.-E. Reif. Hilgendorf's inference of transmutation first derived from his observation that, in the portentously named *Valvata multiformis*, flat forms always occupied the low part, and trochiform shells the high part, of the geological section. He used overall similarity of morphs between proximate strata (i.e., the "modern" stratophenetics of Gingerich 1976!) to construct a phylogeny (Figure 7.10), which involved an "intergeneric" transition with no cladogenesis. His diagram

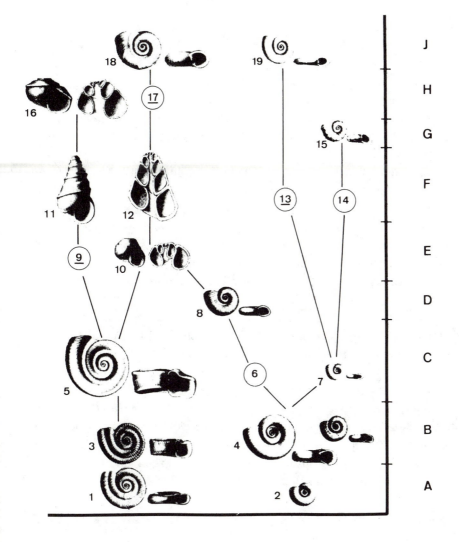

Figure 7.10: Reconstructed phylogenetic diagram for the *Planorbis multiformis* lineage, taken from Hilgendorf's 1863 thesis. (from Reif 1983*a*).

included instances of splitting as well, and he even suggested a possible case of between-lineage fusion, i.e., hybridization. Overall, the 19 taxa he named arose from a combination of phyletic transformation and splitting, though the latter was less important. He noticed that change between successional species was relatively rapid, relative to longer periods of little change (Hilgendorf 1879). Apparently, the claim by punctuativists that stasis is the rule for fossil species has a rather venerable history.

As noted by Reif (1983c), Hilgendorf adopted a compromise position on the taxonomic delineation of species. He neither believed that all morphs bridged by transitions should be assigned to the same species, nor that all differentiable morphs should be given different species names. He adopted an intermediate position, using qualitatively established thresholds of difference to assign species names. This method, of course, is precisely the same one used subsequently by nearly all paleontological systematists. It helps when the rate of evolution is uneven and breaks in morphological resemblance can be related to stratigraphic horizons. Thus Hallam (1982, p. 358) argued from his experience with *Gryphaea*: "morphological trends concerning more than two species . . . may be the expression of a simple one-to-one ancestor-descendant relationship." From the time of Hilgendorf to the present (see below), paleontologists have demonstrated again and again that differences characteristic of coexisting species can be traced by phyletic change through geological sections. This is the most elegant refutation of the punctuated equilibria argument, which claims that most species-level differences require cladogenesis.

Testing Character Stasis and Tests of the Punctuated Equilibria Theory

Cladogenesis must be recognized before one can produce data consistent with the punctuated equilibria hypothesis. To falsify the punctuative hypothesis, an extrapolation of phyletic character change must be typically large enough to diagnose populations at two horizons as being members of the same phyletic evolutionary lineage yet belonging to different phenetic species (Gingerich 1976). Rapid changes in the rate of character change in a radiolarian (Gould and Eldredge 1977 interpretation of data reported by Kellogg 1975) supposedly supports the punctuated equilibria theory. As one moves up the stratigraphic

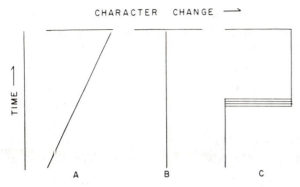

Figure 7.11: Different possible forms of character change through time in a fossil lineage. A: Gradual change in a character such as a linear dimension. B: Character stasis. C: Change from dominance by one discrete state to dominance by an alternative. The horizontal lines represent a period of polymorphism.

column, phenotypic changes occur, with short periods of character stasis. It is rather unlikely that any character would change at a constant rate for millions of years, given variation in the rate of environmental change, possible (but unlikely) exhaustion of genetic variability, and stochastic processes. Thus, in order to refute the punctuated equilibria theory, gradual character evolution should be redefined: A character's evolution is **gradual at the species level** when change up a geological section is sufficiently directional such that reversals and periods of character stasis do not preclude phyletic divergence of two morphs sufficiently different to be recognized as distinct species. (Note: for the purposes of testing the potential of gradual evolution, we accept the diagnosis of species differences with a predefined threshold of phenetic difference.) In characters with discrete states, the distinction between gradualism and stasis-punctuation is indeterminate unless one knows the genetic-ecological basis of character-state determination (see below). With this more reasonable definition, gradual evolution at the species level is a commonplace phenomenon in the fossil record.

Most studies by paleontologists in recent years have concentrated on one character in a single species (e.g., Hayami 1973; Kellogg 1975). It would be more valuable to study groups of species changing in the same geological section (e.g., Williamson 1981) and to consider a number of characters for each species (e.g., Brande 1979; Reyment 1982a, *b*; Malmgren et al. 1983; Bell et al. 1985; Stanley and Yang 1987). It would also be desirable to measure character variability. Some hypothetical results would be illuminating in assessing the punctuated equilibria hypothesis. First, consider the case of discordant patterns of change occurring among characters within the same species. Suppose (Figure 7.11), that character A changes gradually with a constant rate of change, character B shows no change, and character C exhibits periods of character stasis, interrupted by rapid periods of change from one discrete state to another. Given such data, one can conclude that no homogeneous process determines the mechanism of character evolution for all characters. Differing patterns of change might be explained by the following: (a) The genetic mechanism of character state determination differs among characters. (b) Either different characters may respond differently to the same selective agent or the characters may be influenced by alternative selection agents in the same overall environment. Such cases occur commonly in geographic variation of morphology and of allozymes (Chapters 3, 4).

In the Cretaceous foraminiferan *Afrobolovina afra*, megalospheric prolocular size changes steadily, but microspheric prolocular size is invariant (Reyment (1982*b*). In the same species, three discrete ornamental variants can be found as a polymorphism (Reyment 1982*c*). Similarly, Bambach's (1970) study of evolution in the silurian bivalve *Arisaigia postornata* shows discordant patterns of change in different types of shell surface characters.

Michael Bell (Bell and Haglund 1982; Bell. et al. 1985) has used a truly unusual case to examine the pattern of character change in a fossil stickleback. *Gasterosteus doryssus* is preserved in the Miocene saline Lake Truckee and can be collected from diatomites layered in annual varves, over an inferred and stratigraphically complete period of 110,000 years. The characters he investigated–pelvic structure, predorsal pterygiophore number, dorsal spine, dorsal fin ray, and anal fin ray number–are all known to have a strong genetic component of variation in the modern species, *Gasterosteus aculeatus*. The various characters have different mechanisms of genetic determination, which partially explains the different patterns of change (Figure 7.12). Known effects of ecophenotypic changes in the environmental factors on morphology of modern species cannot explain the degree of evolutionary change in *G. doryssus*. In all characters, gradual evolution is predominant, and

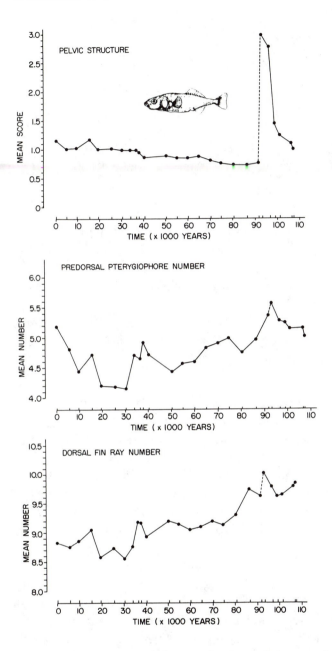

Figure 7.12: Temporal changes in three characters in the Miocene stickleback, *Gasterosteus doryssus*, in fossil Lake Truckee. Time control established by varves (from Bell et al. 1985).

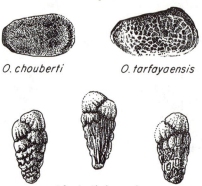

O. chouberti   O. tarfayaensis

Afrobolivina afra

Figure 7.13: Examples of discrete morphs in fossil lineages. Top: The Cretaceous ostracodes *Oertliella? chouberti* and *O. tarfayaensis* (female carapace, right side). While these are described as different species, they are members of a single phyletic lineage and passage forms have been identified. The derived *O. chouberti* does have features lacking in the ancestral species (after Reyment 1982a). Bottom: Three discrete morphs coexisting in populations of the Cretaceous foraminifer *Afrobolivina afra* (after Reyment 1982c).

nearly linear in one character (Bell et al. 1985). In some cases, the changes would represent those diagnostic of family level differences in teleosts. Rapid evolutionary bursts are seen, but the most rapid change most likely represents a local extinction, followed by immigration from another differentiated population. The patterns of change indicate that conventional paleontological samples, which usually record changes of far greater temporal magnitude than the common case here of 5-10,000 years, must usually be too widely spaced to pick up much significant evolutionary change.

This discussion highlights the inherent difficulty of understanding the mechanisms of character stasis or gradual change by using the fossil record. We do not know the mechanism of genetic determination of character states, and we cannot predict the norms of character reaction to the environment without genetic–environmental data. Genetic mechanisms behind fossil variation are difficult to infer, and Mendelian variation has almost never been found (see Best 1961 for an interesting example in trilobites). When "sudden" and qualitative changes, such as in surface shell ornamentation, are found (e.g., Hayami 1973–

intraspecific case; Reyment 1982c; Skevington 1967–transspecific examples) they can readily be related to characters that are probably determined by polygenic systems with threshold effects. Invariably, a polymorphism is observed, documenting the transition from dominance by one morph to the other. For example, consider the reticulation that appears rather suddenly as a new phenotype in the Cretaceous ostracod *Oertliella* (Reyment 1982a). Such ornamentation (Figure 7.13) can be easily "switched on," as it probably is controlled by a set of underlying epidermal cells that can be signalled simultaneously to control skeletal morphology (Okada 1981). The polymorphism for reticulation in *Oertliella* species could easily be the effect of a single gene. There are other differences between the two species, but intermediate "passage" forms for all characters have been found (Reyment 1982a). In a more recent study of a similar ostracod system, Reyment (1985) found similar arrays of morphology, but with far more cases of intermediate forms. This only points to the variety of mechanisms of genetic determination of traits, and the futility of pointing to specific character changes as examples of punctuation.

Indeed, in the original case cited (Eldredge 1971) for punctuation– a "sudden" change in file number (number of rows of eye facets) in the American Devonian trilobite *Phacops rana*–could be simply based on a transient polymorphism controlled by minor thresholds known to occur in living arthropod populations. One would be hard pressed even to demonstrate that this was a speciation event, as claimed by Eldredge. The geographically peripheral origin and subsequent spread of a variant is hardly a unique element of speciation; it is the expectable outcome of within-species evolution. No one expects variants to arise *and* be fixed simultaneously throughout the range of a species. But, given that a fixation event is likely to be missed, and that spread is likely to be rapid, most fossil cases will appear to demonstrate such an unlikely event. Could the fossil record have documented the spread of the melanic morph in *Biston betularia* through most of its range in a century?

This confusion about punctuation is exacerbated as many paleontologists use sudden changes in character state as opportunities to delineate species. For example, external ornamentation in the Cretaceous foraminiferan *Afrobolivina afra* occurs in three discrete morphs (Figure 7.13) yet the documentation of these forms is not evidence for punctuation, as polymorphism is found and the morphs are most likely determined by standard genetic mechanisms (Reyment 1982*a*). If the morphs occurred in three geographically discrete populations, or in successive horizons, it is certain that there would be three named "species" of *Afrobolivina* instead of just one. Newell's (1947) quote of another paleontologist, F.W. Millet, is perhaps an exaggeration of typical practice, but it is often close to the truth:

> In preparing your slides you should be very careful to select typical specimens only, the intermediate things are a nuisance and should be severely ignored, unless the types are absent from the deposit . . . Take *Globigerina*, the difference between *G. cretacea* and *G. lineana* is very great, but you can select specimens that will form a complete chain from one to the other.

This problem is not very new. Meyer (1878) claimed that the ability to recognize gradual evolutionary change in *Micraster* was obscured by the rampant naming of separate species by previous taxonomists.

Periods of character stasis, strong variations in the rate of character evolution, and the appearance of new morphs controlled by easily attributable and trivial genetic mechanisms are the mainstay of species-level taxonomy in the fossil record. One could not easily delineate qualitatively different taxa without such breaks. But these apparently rapid changes are usually within single lineages with no observed cladogenesis. If apparent species level differentiation can occur in such instances, then what does speciation have to do with phenotypic evolution? The punctuativists' typical test (e.g., Gould and Eldredge 1977) of the causal relationship of speciation to morphological divergence is usually dependent upon the tautologous relationship of change with species taxonomy (Levinton and Simon 1980). Species-level and genus-level differences are also exaggerated with no substantive support from the paleontological literature (see for example, Rose and Bown 1984). Stanley (1979) for example, claims that Kurtén's work indicates substantial gaps among fossil mammal genera and species. But Kurtén (letter, October 21 1981) replies: "I most certainly do not believe any such thing. I tend to regard genera as after-the-event artifacts . . . I think new genera can arise rapidly or slowly. I think the same holds for species." Intergeneric differences in mammals can be shown to be rather slight and easy to document in fossil lineages (Figure 7.14). "Splitting may sometimes increase the rate of species evolution because the wide spread of a population will cause it to encounter a wide variety of environments."

Teichert (1949, p. 49) reveals the paleontological practice of naming species tautologically by degree of morphological difference: "Tan . . . demonstrated how in Lepidocyclinidae the 'variation curve of the same species gradually shifts in a definite direction.' In *Calceolispongia* these shifts are so quick that assemblages in consecutive horizons are distinct enough to be separated species, although there is considerable overlap in morphological variants." We see here the acknowledgment of morphological variation, but the inevitable tendency to name species when a threshold of mean phenetic difference is surpassed. Werdelin's (1981) study of the evolution of the felid genus *Lynx* is a typical example of how a phyletic series of species is eventually defined by a series of taxa, based upon characters known to change rather easily in phyletic evolution and also known to be polymorphic. Figure 7.15 shows the inferred phylogeny, demonstrating a line of interspecific transitions. The genus *Lynx* itself is defined by the absence of the second premolar, a character state that reappears later within the group and is known to be polymorphic in several felid species. The multispecies transition in Europe from *Lynx issiodorensis* to *L. pardina* is gradual and species are defined principally on the basis of the relative length of the first molar, though gradual reduction in size also occurs.

The Case for Gradual Change

Gould and Eldredge's (1977) review and Stanley's monograph (1979) both suggest that there is little evidence of gradual evolution in fossils and that morphological evolution is generated by discrete speciation events. Our reconsideration of the nature of character change and a compilation of fossil investigations both permit a reassessment of fossil data. The fossil record is full of excellent examples of gradual evolution, at the intraspecific (Table 7.1) level. Continuous change in nondiscrete characters, or transitions in characters that are developmentally or genetically determined as discrete states, has been found in foraminiferans, radiolarians, ammonites, bivalves, ostracodes, trilobites, graptolites, echinoderms, mammals, and fishes. Indeed, the burden of proof is on those who claim that such evolution cannot be found

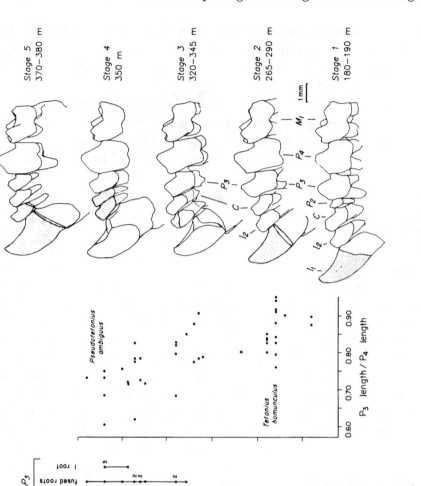

Figure 7.14: Evolution of the anterior lower dentition in the omomyid primates *Tetonius homunculus* → *Pseudotetonius ambiguus* lineage from the central and southern Bighorn Basin, Early Eocene. Shaded areas are reconstructed (from Rose and Bown 1984).

Table 7.1 Some cases of intraspecific gradual evolution.

Taxonomic Group	Age	Character	Source
Fusilinid - *Lepidolina multiseptata*	Permian	proloculus diameter	Ozawa 1975
Foraminiferan - *Afrobolivina afra*	Cretaceous	megalospheric proloculus	Reyment 1982b
Bivalve - *Nuculites planites*	Ordovician	presence of anterior fold	Bretsky and Bretsky 1977
Bivalve - *Arisaigia postornata*	Silurian	rib number, presence	Bambach 1970
Bivalve - *Cryptopecten vesiculosus*	Pliocene-Recent	ornament	Hayami 1973
Ammonite - *Kosmoceras*	Jurassic	Size, shape, ornament	Raup and Crick 1981
Ostracode - *Cytherelloidea*	Cretaceous	surface ornament	Bettenstaedt 1958
Trilobite - *Flexicalymene*	Ordovician	axial rings	Cisne et al. 1980b
Fish - *Gasterosteus doryssus*	Miocene	several	Bell et al. 1985
Mammal - *Felis issiodorensis*	Pleistocene	postcarnassial element	Kurtén 1963

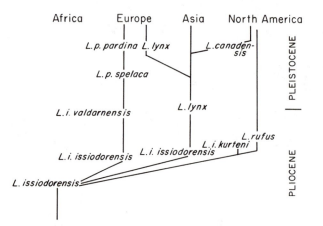

Figure 7.15: The phylogeny of the genus *Lynx*. Note the transitional subspecies and species in the evolution of *L. pardina* (after Werdelin 1981).

or is even the minority case.

Between-species boundaries in the stratigraphic column are, not too surprisingly, placed where evolutionary rate appears to have increased markedly (e.g., Malmgren et al. 1983), or where discrete changes occur in phenotypic traits which, by analogy to living populations, are likely to be determined by several genes with thresholds between states (e.g., Werdelin 1981). Transitional evolution has been documented in the absence of cladogenesis for transspecific (e.g., Reyment 1982*a*, *b*; Malmgren et al. 1983; Kaufmann 1933; Hallam 1982; Chaline and Laurin 1986) and even transgeneric (Grabert 1959; Skevington 1967; Rose and Bown 1984) evolution (Table 7.2). Most important, variation within populations accounts for the differences observed between ancestral and derived taxa (see Charlesworth 1984*a*; Reyment 1982*a*; Malmgren et al. 1983). This is best seen in the works of Kurtén and his colleagues (e.g., Kurtén 1959*a*; Werdelin 1981). Within-species dental polymorphism is the stuff of trans-specific diagnoses in mammals.

The marine protists are excellent for tests of hypotheses concerning character evolution and "species" origins. The deep-sea drilling program has greatly amplified an already large body of data. Gradual evolution, even by Eldredge and Gould's narrowly defined standards, seems common (Figure 7.16, Figure 7.17). Kennett and Srinavasan's (1983) monograph on Neogene foraminifera demonstrates that inter-

Table 7.2 Some cases of gradual evolution at the transspecific and transgeneric level.

Taxonomic Group	Transition	Age	Source
Foraminifera	*Gaudryina-Spiroplectinata*	Aptian-Albian	Grabert 1959
Foraminifera	*Globorotalia spp.*	Pliocene-Pleistocene	Malmgren et al. 1983
Foraminifera	*Bolivenoides spp.*	Cretaceous	Betenstaedt 1958
Radiolarians	*Pseudocubus spp.*	Pleistocene-Recent	Kellogg 1975
Bivalves	*Chesapecten spp.*	Miocene-Pliocene	Miyazaki and Mickevich 1982
Bivalves	*Gryphaea spp.*	Jurassic	Hallam 1982
Ostracodes	*Oertliella spp.*	Cretaceous	Reyment 1982b
Trilobites	*Olenus spp.*	Upper Cambrian	Kaufmann 1933
Trilobites	*Triarthrus spp.*	Ordovician	Cisne et al. 1980a
Crinoids	*Calceolispongia spp.*	Permian	Teichert 1949
Graptolites	*Holmograptus-Nicolsongraptus*	Ordovician	Skevington 1967
Mammals	*Ursus spp.*	Pleistocene	Kurten 1959
Mammals	*Lynx spp.*	Pleistocene	Werdelin 1981
Mammals	*Hyopsodus spp.*	Eocene	Gingerich 1976
Mammals	*Nyanzachoerus-Notochoerus*	Plio-Pleistocene	Harris and White 1979
Mammals	*Tetonius-Pseudotetonius*	Eocene	Rose and Bown 1984
Mammals	*Mimomys spp.*	Plio-Pleistocene	Chaline and Laurin 1986

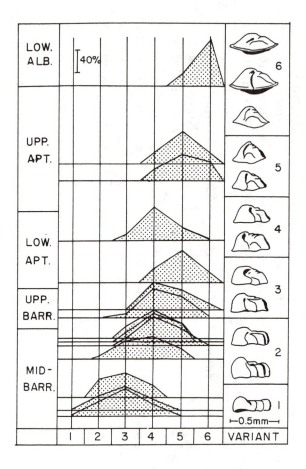

Figure 7.16: Gradual evolution of the Cretaceous foraminifer *Globoro-talites bartensteini* from the Barremian to the Lower Albian. Several variants of the species (at right) coexisted at any one horizon (after Bettenstaedt 1958).

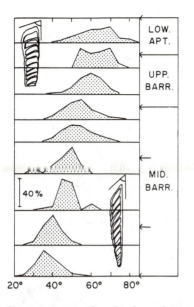

Figure 7.17: Gradual evolution in the angle between the chamber wall and the main body axis in the Cretaceous foraminifer *Vaginulina procera* from the Middle Barremian to the Lower Aptian. Arrows denote "interspecific" boundaries (after Bettenstaedt 1962).

specific phyletic evolution is commonplace. Recent studies (Prothero and Lazarus 1980; Malmgren and Kennett 1981; Malmgren et al. 1983; Reyment 1982*b*) have only duplicated the findings of the classic German works of the 1950s and 1960s (Bartenstein and Bettenstaedt 1962; Bettenstaedt 1958, 1959, 1960, 1962; Grabert 1959) and earlier work in the 1930s (summarized in Tan 1939). These older studies estimated variation at a given geological horizon and found that changes in the frequency of morphs at any one horizon accounted for the changes across horizons and between successional species (Figure 7.16, Figure 7.17). Good stratigraphic sampling can help to explain generic-level transitions (Figure 7.9). One might argue that the foraminifera are biologically unique and hence not an adequate sample of evolutionary history. But the mammalia fit this pattern as well. I would argue that foraminifera are exceptional solely from the point of view of sedimentary and geographical continuity.

An exceptional study of the radiolarian lineage *Pterocanium prismatium* shows the ambiguities that arise when complete sections are

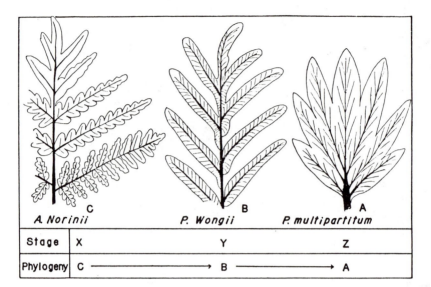

Figure 7.18: Branching sequence and phylogeny in the transitional sequence *Alethopteris norinii → Protoblechnum wongii → Psygmophyllum multipartitum*, Permian of China (from Asama 1962).

available. The lineage gave rise to the distinct species *P. charybdeum* within an interval of 50,000 y at approximately 4.3 ma. Both descendants continued to diverge over the next 0.5 million years; over ten standard deviations of difference accumulated. Over the next 2 million years, divergence slowed down considerably. This is hardly an example of punctuated equilibria, but it would have been characterized as such if one of the daughters failed to change substantially. The split in this case was not the motor of morphological evolution. Divergence continued long after. If this is the case, why should the change of only one descendant be considered consistent with punctuated equibilibria unless we in fact know that the speciation event itself is the cause of change?

Unfortunately the fossil record of plants seems not to have been studied with such stratigraphic care, and may often be too poor in occurrences for adequate sampling. Although his sampling is sparse, Asama (1959, 1962, 1981) has documented apparent phyletic evolution in a number of lineages of Carboniferous and Permian plants. For example, the Shansi flora of China can be sampled through a major change of environment recorded by transitions from Carboniferous marine carbonates to nonmarine Lower Permian coals to nonmarine red sandstones, of possible Triassic age. A sequence of 33 beds has been sampled sequentially in the Shihhotse Valley of China (Asama 1981). Through these sequences, Asama has documented the gradual evolution of simple leaves in several unrelated lineages, through enlargement, and through fusion achieved by growth retardation (Figure 7.18). Significant changes can be seen within a geological series (subsection of a period, such as the Permian). These changes are associated with overall drying trends in the environment (see also Wolfe and Hopkins 1967).

Stasis

There is no contradiction in showing that (a) interspecific differentiation derived from intrapopulation variation is an entirely plausible explanation for evolutionary change; and (b) change of a phenetic-specific level does not always, in fact, usually does not, occur. Stasis may be the dominant pattern. For example, continuous gradational evolution has been claimed to be relatively rare in shelf invertebrates (Johnson 1982; Schindel 1982c). This may be explained by the strong association of benthic fossils, such as articulate brachiopods, with particular sedimentary and hydrological regimes. Benthic species often are diagnosable by sediment type (and rock type) and become locally extinct when the particular environment disappears. An apparently constant environment thus maintains an unchanging morphology. Abrupt changes in the environment of the shelf are associated with alteration in the sedimentary regime often too severe for a species to survive.

In his extensive studies of the invertebrate faunas of nearshore mid-Paleozoic deposits, Boucot (1975, 1978) provides an environmental description which indicates why morphology of many taxa should fluctuate, but usually shows no directional change to any major degree. Community composition is remarkably constant over tens of millions of years both in terms of species composition and relative frequency of species. Even when specific taxa change, functional groups within Paleozoic shallow water environments (Walker and Laporte 1970) and within Permian vertebrate communities (Olson 1966) often remain the same over long periods of time. Boucot suggests that such physical and biotic environmental constancy leaves little room for expectations of major directional evolutionary changes. In Late Devonian nearshore marine communities dominated by brachiopods, extinctions of dominant elements usually were followed by replacements by morphologically and presumed ecologically similar forms (McGhee 1981). This constancy may lie behind the lack of common large-scale phyletic trends.

A recent study by Stanley and Yang (1987) has elegantly documented long-term stasis in 19 "species" of bivalve mollusks that persisted from Early Pliocene to Recent. When compared with geographic variation between pairs of selected Recent populations, change in 18 of the 19 lineages was insignificant. It is of interest that seven of the lineages had previously been considered to be pairs of phyletic species. This suggests some question as to species identification, and one cannot be sure just how many biological species are represented in the study. Nevertheless, it is clear that no large-scale change has occurred. This may further strengthen the argument for stabilizing selection for overall form in benthic environments.

Stanley and Yang (1987) find that intraspecific variation (differences between two contemporary populations) is of the same magnitude as differences among members of the same phyletic lineage. Both of these differences are significantly less than interspecific differences. This difference is interesting but not complete as shown by a similar study of *Mulinia lateralis* (Brande 1979). A population of this bivalve reared in a predator-free cage was as different from typical *M. lateralis* as any other species! This difference was most likely nongenetic, and demonstrates that ecologically extreme situations were overlooked in Stanley and Yang's study. In a sense, this only strengthens Stanley and Yang's case for stasis, as this makes the intraspecific contemporary variation potentially even larger than the extent of a long-term trend. It also suggests, however, that intra- versus interspecific comparison tests of variance are dubious without more geographically and ecologically diverse samples (as opposed to sample size, which is very large in the Stanley and Yang study).

Long-term stasis prevailed in a series of gastropod lineages in the eastern Turkana Basin, but a "sudden" change in several lineages occurred within a 5,000 to 50,000 y interval (Williamson 1981). As discussed above, Williamson's study is liable to be stratigraphically complete, even at the resolution level of 1000 y. The change was spectacular indeed, and radical morphological alterations were observed. Accord-

ing to Williamson, the strong and rapid changes occurred in an isolated lake within populations of large numbers. Stratigraphic evidence
suggests that the changes occurred when the habitat was shrinking.
There is some question as to the rapidity of change, since the major jump occurs at a formation boundary, which might include a gap
(Boucot 1982). Even if the time scale is correct, as J.S. Jones (1981)
notes, the period over which change occurs is relatively long for most
population–genetic processes and, therefore, is hardly a challenge to
neo-Darwinian theory. Furthermore, according to Williamson, it took
place within an isolated basin with large population sizes. Thus, even
though this basin was separated from others where no change occurred,
it does not fit Mayr's requisites for a genetic revolution, as Williamson
acknowledged. But it also does constitute phyletic evolution within
one large population. One must even be somewhat suspicious that the
change seen, in many of these Turkana gastropod groups simultaneously, including a parthenogenetic form, is perhaps an ecophenotypic
response to strong changes in lake chemistry.

Recent studies have produced some other purported cases of stasis
in mammals. One example in *Homo erectus* (Rightmire 1981) has some
statistical problems likely to be present in other work. Owing to poor
sample size and strong variability, the null hypothesis that stasis occurs
cannot be falsified easily (Levinton 1982b, but see Charlesworth 1984a).
Despite poor statistical analyses and small sample size, *Homo erectus*
has been cited repeatedly as an example of stasis (Gould and Eldredge
1977; Stanley 1979, 1981; Eldredge and Tattersall 1982). An analysis
by Wolpoff (1984) argues this to be incorrect. Nevertheless, near stasis
has been found for size in the Eocene mammal *Hyopsodus* (West 1979),
though large scale (intergeneric) gradual change has been seen in the
same geological sections for other taxa (Rose and Bown 1984).

The Persistence Criterion

Gould's ultimate (1982c) test of the punctuated equilibria hypothethe-
sis is the persistence of ancestral with derived species. Given that
coexisting discrete morphs of a single fossil species can be easily misinterpreted as members of two genera, one can see that this test can
rarely be put into practice. There is one important exception to this,
however, and that is the case where the taxa survive to the present, and
some biological confirmation can be made. Alternatively, it would be

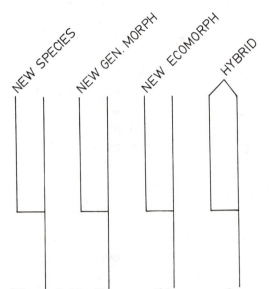

Figure 7.19: Four possible cases of taxa or morphs with overlapping ranges. Only case one conforms to the punctuated equilibria hypothesis, but all conform to the expected pattern.

useful to have a Recent standard of average species difference, against which differences among fossil taxa might be scaled. This approach, however, entertains the danger of ignoring the possibly many speciation events that may have occurred without any morphological change. It would be impossible in the fossil record to compare the number of biological speciation events with the degree of morphological change, if the former cannot be sampled adequately.

The criterion of coexistence of descendant species with parent is as strong as our ability to identify species in the first place. The following four cases would all satisfy this criterion (Fig. 7.19):

1. A rapid speciation event generates a distinct daughter species, but the parent species survives unchanged. The daughter species does not change significantly during its subsequent history.

2. Natural selection in a part of the species range rapidly generates a new and distinct morph, which comes to dominate local populations, or part of the species' range. The two morphs survive and persist unchanged.

3. Natural selection causes the evolution of a flexible but distinct response to environmental circumstances. Depending upon local conditions, the population is dominated by one of several ecomorphs.

4. Divergence occurs by means of one of the first three mechanisms, but the distinct coexistence of divergent morphologies disappears, owing to hybridization (case 1), or a relaxation of selection (cases 2 and 3).

Cases two and three would pose significant difficulties for the persistence criterion if distinct morphs could not be parsed from species differences. Distinct morphs have been commonly mistaken for different genera and families, let alone species. Our knowledge of living species and populations presents an equally pessimistic picture. Distinct ecomorphs and genetic morphs are commonplace, particularly in groups where simple mutations can produce large-scale morphological changes (e.g., Palmer 1985). The barnacle *Chthamalus anisopoma* has two distinct morphs, one of which has a bizarre recurved test. This morph is induced by the presence of the predator *Acanthina angelica*, and barnacle populations exposed to predation are wholly dominated by this morph. As fossils, populations dominated by these morphs would be regarded as mixtures of two distinct species (Figure 7.20). Even more notable is the rapid evolutionary response of two northern New England gastropod species to the arrival from Europe of the shore crab *Carcinus maenas*. Within a few decades, predation caused selection for distinct new morphs of the dog whelk *Nucella lapillus* and the periwinkle *Littorina obtusata*. In the latter case, distinct populations of thin-shelled, high-spired forms exist within a few kilometers of populations with thick-shelled, low-spired forms. This is strictly an intraspecific phenomenon (Seeley 1986), but it conforms precisely to the persistence criterion. The morphs are so distinct, that they would almost certainly be classified as different species by paleontologists. Indeed, much more subtle differences may distinguish marine gastropod species.

At this writing, I have found only one conclusive case that seems to fit the persistence criterion, and that is a study of Miocene-Pliocene bryozoa (Cheetham 1986). Although the stratigraphic sampling is not very dense, there is no change within the history of "species" (these are

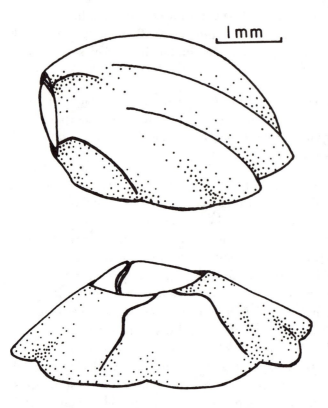

Figure 7.20: Two distinct ecomorphs of the barnacle *Chthamalus anisopoma* in the Gulf of California (courtesy of Curtis Lively; see *Evolution*, v. 40, p. 232-242).

of necessity defined strictly on the basis of morphology), which in the main give rise suddenly to distinct offshoots, the ancestors surviving and coexisting. On the one hand, the average stratigraphic resolution is 1.6×10^5 y. This certainly allows for extensive and plausible morphological change "between the cracks." But, as Cheetham notes, the apparent lack of change through the rest of the history of most of the ancestral species does not admit to any phyletic evolutionary hypothesis. The approximate time of cladogenesis is the time of morphological bifurcation in most of the cases he studied. Unfortunately, the lack of knowledge of the genetic determination of the characters leaves this case moot with regard to issues of genetic revolutions, selection, and genetic drift. In addition, one cannot rule out the possibility that speciation is rampant, but morphological evolution only occurs occasionally when a population is forced into a marginal environment and subjected to rapid directional selection. What then becomes interesting is why the character complexes evolved in the daughter species remain constant. This is, again, the issue of stasis, which I believe to be the legitimate problem spawned by the punctuated equilibrium model.

Phyletic Gradualism: The Making of a Strawman

Who Really Was a Pure Gradualist?

Despite the surge of discussion over punctuated equilibria, a devil's advocate could be led to say that its original justification or its current popularity derives from only two factors: (a) the building of a strawman, i.e., phyletic gradualism, which supposedly dominated, and crippled, paleontologists newly entering the fold of "pure Darwinian gradualism" and the Modern Synthesis; and (b) a belief in the speciation process as genetically paroxysmal and a cause of most evolutionary change that was somehow qualitatively different from other sources of evolutionary change. The proponents of punctuated equilibria highlighted a new dogma by inaccurately characterizing Darwin's original work, the attitudes of some paleontologists, and the body of work produced by evolutionary biologists in the first half of the twentieth century. It is a measure of their success that many of the architects of the Modern Synthesis and their disciples have been forced into the situation of defending a position in which they had never believed.

The strawman of phyletic gradualism was built up in two stages. First, Darwin's conception of evolution was represented as a belief in the slow and even transformation of entire species into new morphological entities. Stanley's (1981) *The New Evolutionary Timetable*, envisions Darwin as frozen into the position that even gradual change was the only sort possible:

> Even if Darwin had possessed theoretical reasons for adopting a punctuational model of evolution, for him to have advanced such a scheme would have seemed as absurd as were many of the speculative pronouncements of his predecessors [He] would have been claiming that evolution . . . was operated by a natural mechanism . . . in small, localized, transitory populations. (quote from Stanley 1981 assembled by Rhodes 1983)

> It was a measure of Darwin's desire to underscore slow, continuous modification that he violated his own philosophy of empiricism and reached back into history for what is essentially religious dogma. (Stanley 1981, p. 47).

Many have discussed the inaccuracy of this general claim (Levinton 1982*c*; Rhodes 1983; Penny 1983). Darwin certainly believed that evolution often occurred in isolates. Indeed, this was one of his two major explanations of sudden change in the fossil record. As Darwin argued:

> Varieties are often at first local . . . rendering the discovery of intermediate links less likely. Local varieties will not spread into other and distant regions until they are considerably modified and improved; and when they do spread, if discovered in a geological formation, they will appear as if suddenly created there, and will simply be classed as new species. (1859, p. 464)

Darwin would have fit well in Eldredge's (1971) paper on rectilinear evolution in the trilobite *Phacops rana*. It is also the common argument (see e.g., Jeletzky 1965) given by paleontologists for sudden change in the fossil record at low taxonomic levels.

Darwin believed similarly, at least by the sixth edition of *The Origin*, that most species-level differentiation occurred in periods very short relative to the entire history of the species (Penny 1983):

. . . it is probably that the periods, during which each underwent modification, though many and long as measured by years, have been short in comparison with the periods during which each remained in an unchanged condition.

Darwin's views are compatible with at least three of Eldredge and Gould's (1972) criteria characterizing the empirical pattern implied by punctuated equilibria: (a) New species develop rapidly and morphological change is rapid relative to the entire history of the species, (b) the change is concentrated in a geographically restricted population (although Eldredge and Gould, following Mayr's (1963) lead, would have the population be at the periphery of the species' range) and (c) the change occurs in a small subset of the entire ancestral species.

Even the first edition of *The Origin* did not accept the "purest form of Darwinian gradualism" ascribed to it (e.g., Gould 1980*b*, p. 122). In discussing his famous and hypothetical diagrammatic example of a phylogeny, Darwin rejected even rates of evolution by stating that he did *not* suppose that evolution goes on so regularly as expressed in the diagram. There is no evidence, despite the claims of Gould and Eldredge, that Darwin ever believed in absolutely constant rates of evolution. From the beginning, *The Origin* acknowledges the geographical complexities of evolution. There is an amusing irony in the fact that the neutral theory of evolution (e.g., Kimura 1983) sees constancy of evolutionary rates as proof of non-Darwinian stochastic processes in evolution, whereas Eldredge and Gould (1972) use the same constancy as a characterization of Darwinians being true to their faith!

The same degree of exaggeration is apparent in the punctuativists' characterization of the Modern Synthesis. As Levinton and Simon (1980) discuss, the Modern Synthesis is hardly a unified claim for gradual evolution at homogeneous rates. Indeed, speciation has perhaps been the subject most continually in dispute. Many believe that genetic reorganization during speciation is either predominant (e.g., Mayr 1963, 1982*b*; M.J.D. White 1982) at least occurs commonly (Templeton 1981). The work of population geneticists and other evolutionary biologists in no way conforms to the gradualist picture presented by Eldredge and Gould. Haldane (1932*a*) recognized that speciation was often sudden and noted (p. 59) that species "can arise at one leap."

Stanley (1979, p. 72) impugns the Modern Synthesis for relying upon phyletic evolution, a supposedly slow process, to generate the be-

wildering amount of morphological diversity in evolution. On the broad
scale, many paleontologists and neontologists have been impressed with
the slow rates of change in fossil lineages. As noted above, Haldane
(1949) was deeply impressed with the apparent slowness of morphological evolution. But this questioning of evolutionary rates is hardly
unique either to punctuativists or to Haldane. Indeed, Simpson (1953)
mentioned this very troubling issue:

> The morphological difference between modern opossums
> and some Cretaceous opossums is slight, but some 60 mil-
> lion years of evolution occurred between them. If the miss-
> ing pre-Cretaceous sequence changed at a comparable rate,
> the transition from a reptile to an opossum can hardly have
> taken less than 600 million years; it probably took several
> times that long–in short it must have occurred in the pre-
> Cambrian, which is certainly absurd. (p. 351)

Simpson concluded that rates of phyletic evolution varied tremen-
dously (from quantum evolution to slower rates). Stanley differed in
believing that some special feature of speciation was necessary to accelerate evolution; but certainly the problem of variable evolutionary
rate has long been appreciated.

In concocting their strawman of phyletic gradualism, Eldredge and
Gould (1972) characterize the tone of textbook representations of paleontological thought by selective quotation. After misquoting Darwin
as seeing speciation as (p. 89) "a long and insensibly graded chain
of intermediate forms" they argue that "Our present texts have not
abandoned this view." But what do the texts actually say? Moore,
Lalicker, and Fischer (1952), one of the two preeminent texts of the
1950s and 1960s, present the two alternative models of transformation
within lineages and splitting, i.e., speciation. But what do these authors say about the tempo of change in fossil lineages? On page 35
they suggest the following:

> The fossil record indicates that evolutionary rates are by no
> means uniform either in different stocks or in the same stock
> in different times. Some organisms exhibit such extremely
> sluggish evolutionary modification that they seem to stand
> still, whereas others undergo amazing sorts of morphologi-
> cal change with geological abruptness. Also, paleontologi-

cal study of various lineages indicates that periods of slow, gradual adaptive change alternate with "explosive" spurts of evolution. These accelerated bursts develop descendants which differ markedly from their ancestors in form and function. The lack of known fossil connecting links beween various orders and classes of the animal kingdom suggests that their origin may lie in this type of accelerated evolution or abrupt divergence.

I reproduce this paragraph completely as it shows a complexity of interpretation never acknowledged by Eldredge and Gould, who reproduce Moore et al.'s figures, but curiously miss this clear statement. The same can be said of the other important text of the period, W.H. Easton's (1960) *Invertebrate Paleontology*. It is true that Easton expounds on the utility of evolutionary series and claims that there are transitional forms. But he illustrates only one example of species/genus level evolution in his entire text, the evolution of a Cretaceous cheilostome bryozoan group. This study shows both cladogenesis and transformational series of "species." The phyletic chains of species show parallel evolution. Again, the real world of fossils, as seen by paleontologists, doesn't quite fit the gradualist demonology erected by Eldredge and Gould.

Paleontological Practice: Species Stasis Enmeshed in Broader Trends

Is there a reason why phyletic gradualism would be thought by punctuativists to be such an important ideological target, beyond the apparent error of attributing its credo to either Darwin or the Modern Synthesis? Indeed, have paleontologists been deceived by the phyletic gradualist "dogma" and thus been affected in their respective researches? This question is approached by Penny (1983), who notes that "phyletic gradualism as defined by Eldredge and Gould . . . appears more consistent with orthogenesis than with Darwinism" (p. 73). Phyletic gradualism, as described by Eldredge and Gould, comes from the old paleontological and morphological traditions begun in the early and mid-nineteenth century. These traditions searched for gradual change at levels of resolution of tens of millions of years.

The punctuation–gradualism dichotomy seems peculiar when it is

juxtaposed with paleontological practice at the species level. Paleontological species are erected in the hope of dating rocks with biostratigraphic correlations. Most studies of invertebrate microfossils
and macrofossils yield sequences of fossil "species" whose recognition
is based upon a threshold of perceived phenetic distance (Kaufmann
1933; Newell 1947). Such practice tends to select for the definition
of paleontological species as entities with implied geographic homogeneity and temporal constancy. Cases where the gradual evolution of
characters was used to accomplish biostratigraphic correlations (e.g.,
C.T. Smith 1945) are rare relative to species zone correlations. Strati-
graphers may have believed that there were intermediate forms, but
this rarely entered into their actual practice, which focused on any
characters defining some or another discontinuity.

Punctuation, therefore, is the typical paleontologist's intuitive bias
in biostratigraphic practice. The underlying objective–to give species
zone names to time horizons–directs paleontologists to search for static
taxa, or at least to pretend that they are static. Cisne, Molenock,
and Rabe (1980) documented how this sort of bias led a paleontologist to identify two distinct "species" of the trilobite genus *Triarthrus*
with supposed stratigraphic significance, but the two forms were end-
members of a continuum of variation. Major sedimentary breaks may
be used as convenient breaks for species delineation (e.g., Stenzel 1949).

The prejudice in favor of punctuative-type species sequences in constructing biostratigraphic zones is well stated by the invertebrate biostratigrapher Jeletzky (1955, p. 485):

> Equating all kinds of evolutionary changes observed, such
> authors tend to express in terms of conventional taxonomic
> units and to name as species individuals possessing even
> the finest morphological distinctions without due regard to
> their intergradation or the stability of their morphological
> distinctions in time and space. This attitude may also well
> be caused by the widespread but unfortunate feeling that
> a species is an essentially **morphological, static and at
> the same time smallest recognizable category** [my
> emphasis].

As Jeletzky so aptly recognized, paleontologists were punctuative in
approach. Neither Darwin nor the Modern Synthesis did much to
change the paleontological practice of "variety hunting."

On the one hand paleontologists see the problems of species recognition between major paleogeographical realms and over time (e.g., Shaw 1969; Imbrie 1957; Newell 1947). They must name species, however, according to the rules of the International Commission on Zoological Nomenclature. The resulting frustration is illustrated by Shaw's 1968 Paleontological Society presidential address: "I have come to the conclusion that the concept of species is misleading, inappropriate, and useless in professional paleontology" (Shaw 1969, p. 1085).

Paleontological research rarely is done at the level or with the stated intent discussed by Eldredge and Gould. This is why the historical argument of the belief in phyletic gradualism at the species level is so vacuous. Gould and Eldredge (1977), for example, ridicule one of paleontology's favorite examples of gradualism, the evolution of *Micraster*, because it consists of just three successional species from few horizons, sometimes having reversals in character trends from species $1 \rightarrow 2 \rightarrow 3$. But Rowe's classic study (1899), as much as it is cited in some paleontology textbooks as a case of evolutionary transition (e.g., Raup and Stanley 1971), was *not* intended to demonstrate gradual evolution. It was designed instead to demonstrate that unique morphologies could be used to delineate stratigraphic horizons (see Rudwick 1972). This study was an attempt to distill singular entities, corresponding to species names, from a wealth of usually gradational variation. Consider Rowe's remarks:

> We must either make a species out of every trivial variation, or mass certainly obviously allied forms into groups . . . It is only by this means that passage-forms [transitional forms between defined species] and mutations can be intelligibly arranged. (Rowe 1899, p. 541)

> In reviewing this Protean genus . . . it would be easy to place a series on a long table, and to show an almost imperceptible transition from one form to another, and yet from that same series to pick out specimens which would serve as museum-types of the several well-known species. (p. 497)

This is not to say that Rowe did not claim modes among the types. But, what is important, he did acknowledge a great deal of morphological variation, yet chose still to name successional species in a summary table, probably characterized by arbitrary differences. Rowe's judgment of gradational differences between "species" was confirmed in later studies (Kermack 1954; Nichols 1959). Rowe also recognized transitional change within one species, an observation omitted by Gould and Eldredge (1977).

The typical skeptical attitude of paleontologists toward the species level usually shifted emphasis away from small-scale studies of morphological change in a stratigraphical column. Paleontologists believed that the species in paleontology was essentially an unworkable level of paleontological research, hence Simpson's (1944) proper claim of ignorance of paleontological evidence at this level of resolution. It is a telling point that the famous *Genetics, Paleontology, and Evolution* (Jepsen et al. 1949) was conspicuously lacking in studies attempting to show small-scale change in the fossil record. Indeed, few such careful sampling studies had ever been performed. Brinkmann's (1929) classic study of the ammonite *Kosmoceras* is a glaring exception. Cloud's (1948) seminal paper on evolutionary trends, discernible from paleontological data, deals only with trends on the scale of millions to tens of millions of years. Rudwick (1972) has argued that Darwin's emphasis upon the supposed incompleteness of the fossil record may have discouraged paleontologists to collect fossils in closely spaced stratigraphic intervals to document transitional evolution. On the other hand, D.E. Allen (1978) argued that mid-nineteenth-century paleontologists and geologists, in developing their new profession, used an inductive approach not conducive to the interdisciplinary approach required to link with Darwinism. In any event, paleontologists were not thinking at the small-scale level of species. As Weller (1960, p. 556) remarked: "species have been little employed in attempts at evolutionary reconstruction . . . phylogenies generally are based on the presumed relations of genera."

Although Eldredge and Gould are incorrect in claiming that paleontologists have been assuming and searching for phyletic gradualism at and below the species level, their argument does reveal the confusion between the time scales considered by paleontologists and those examined by neontologists. It was no accident that the 1980 meeting on Macroevolution, at the Field Museum in Chicago, was held at a time when population geneticists were surprised at what was to be considered by paleontologists as a "short" time interval for evolutionary change at the species level; tens of thousands to hundreds of thousands of years was a typical answer. For workers accustomed to a literature registering speciation events within hundreds to a few thousands of years, and considerable morphological change in field populations within decades to years, this was a bit of a shock. Paleontologists took a neontologist's eternity to be a "geological microsecond."

Consider Henry Fairfield Osborn's (1934) approach to the evolutionary paleontology of the Proboscidea. Taking grand sweeps of lineages from the Eocene to the Pleistocene, a period of forty million years, he nevertheless felt able to criticize Bateson's mutation theory and suggested his own theory of the nature of variation and differences among evolutionary novelties. Even though his time resolution was usually several million years, he was able to assert confidently in the picturesque language no longer found in the current pages of the *American Naturalist*:

> . . . the first grand result is the replacement of all... hy
> potheses of "discontinuity" and of breaks between species.
> These time honored difficulties melt away like a block of
> ice in the glowing sun of observation of the actual modes
> of phylogenetic origin of adaptations. (1934, p. 213)

Osborn's scale of time resolution could never have resolved the short-term dynamics of the evolutionary process. His study, nevertheless, sketched a wonderful picture (Figure 7.21) of gradual change.

I conclude that Eldredge and Gould's (1972) phyletic gradualism is not a belief held by neontologists or Darwin or even a formula for a research program practiced extensively by paleontologists working at or below the species level. Eldredge and Gould, both educated as paleontologists, were trained to believe that large-scale trends were the stuff of evolutionary study (Gould 1980a acknowledges Simpson 1953 as his textbook source when he learned evolution). It was only natural to misinterpret these trends as being microevolutionary, not because paleontologists ever investigated evolutionary change at small scales, but because they maintained a fuzzy connection between grand evolutionary trends and short-term concepts used by neontologists such as natural selection. Although I disagree with Eldredge and Cracraft's (1980) characterization of Simpson's confusion with the connection, I do think that they expose most paleontologists and neontologists in a common confusion among time scales and evolutionary events.

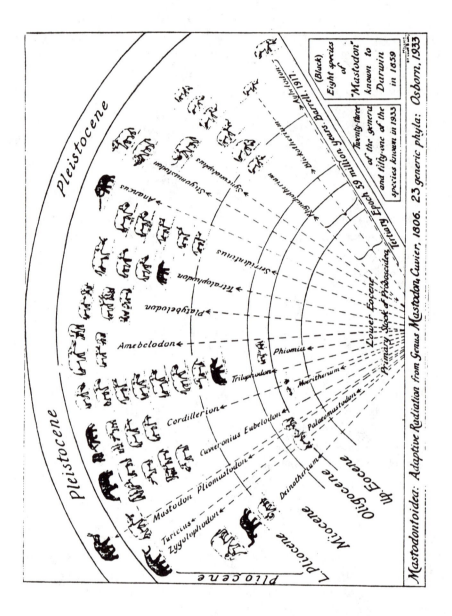

Figure 7.21: H.F. Osborn's conception of evolution in the mastodons (from Osborn 1934).

Taxonomic Longevity and Evolutionary Rate

Cherry and colleagues (1978) discovered a remarkable relationship between genetic relationship and taxonomic level in frogs. It was possible to obtain a linear relationship between genetic distance, based upon microcomplement fixation, and ascending taxonomic rank, suggesting that there is a correspondence between genic and morphological differences. But these workers showed that such a comparison cannot be extended from one group to another. The degree of morphological difference between man and chimp is much greater than among frogs, when scaled against the degree of biochemical differentiation. Intergroup comparisons of morphological differentiation are, therefore, poorly indicative of genic difference; relationships found within members of a clade may be the result of group-specific genetic characteristics, perhaps in gene regulation.

Paleontologists commonly assume that equivalent taxonomic levels in distantly related groups can be employed profitably to contrast rates of morphological or genomic evolution. As mentioned in the chapter introduction, taxon longevity has been used in this way (we discuss the measurement of taxonomic longevity in Chapter 8). Simpson (1944) constructed frequency distributions for molluscan and mammalian families and found that mammal families were of far shorter duration. From this he concluded that mollusks evolved more slowly. Stanley (1973a) did a more detailed analysis using bivalve mollusks and came to the same conclusion. He asserted that it was likely that mammalian evolution was accelerated by interspecific interactions, which were presumably less intense among marine bivalves. Stanley (1973a) found that the rate of first appearances of Cenozoic mammal families decreased steadily toward the present, while bivalve families instead showed an increase in rate of appearance from the Ordovician to the Neogene. To Stanley (1973a, p. 492) this was evidence that there was an "absence of restraints on diversification."

Figure 7.22 shows an analysis of bivalve appearance rate at the family and superfamily levels. Until the Jurassic, there is a close correlation between appearance rate at the two levels. The post-Triassic shows a significant difference; the rate of appearance of families continues to increase in the Cretaceous and Tertiary, while the rate of superfamily appearance declines, much like the pattern for mammalian families. Overall, there appear to be two stages of decline, a Paleozoic stage

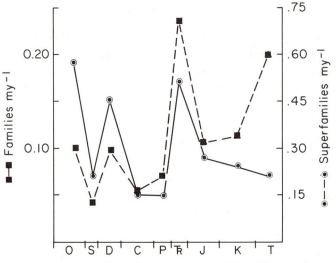

Figure 7.22: Rates of appearance of families and superfamilies of Bivalvia (data from the *Treatise on Invertebrate Paleontology*).

followed by a post-Triassic decline. These results call into question the assumption that two taxa as different as mammals and bivalves can be blithely compared as if there was some absolute meaning to the taxonomic level of family. In this case, which taxonomic level of bivalves is equivalent to mammalian families: family or superfamily?

Can we relate morphological evolution to taxon longevity? Within a taxon, this seems to be a reasonable assertion, as in the case mentioned above for frogs. But it is not clear that one can relate groups as different as mammals and mollusks on the same taxonomic scale. Speciation events may not be as detectable in morphologically simple groups, such as protobranch bivalve mollusks, whose shells are relatively simple and often devoid of sculptural detail. A study by Schopf et al. (1975) further suggests the futility of intergroup comparisons of taxon longevity. A positive correlation was found between extinction rate and the number of morphological terms used to describe the morphology of a taxon. The data fell into three groups; most notable are the mammals and ammonoids, with the highest number of terms and extinction rates. At the least, this result suggests a degree of difficulty in making intergroup comparisons as it calls into question the possibility of using uniform criteria to establish common taxonomic levels across a group of taxa of strongly varying morphological complexity.

The correlation of genetic differentiation with morphological diversity shown above for frogs might suggest that it is possible to extrapolate rates of evolution to the degree of taxonomic longevity. If we have a phyletic series of "species" and there is some uniform notion of the degree of morphological differentiation associated with species differences, then species longevity should be associated with the rate of morphological evolution. Unfortunately there are no careful studies to determine whether criteria for phyletic species recognition are at all uniform during the history of a clade. A good example of the possible problems that might arise comes from a summary of mammalian species longevity through the Cenozoic (Kurtén 1959b). Paleocene and Pleistocene species longevities were both significantly less (by a factor of approximately 5) than in Neogene mammals. Of special interest is the Pleistocene mean value of only 0.62 my, the minimum longevity recorded by Kurtén. If longevity, and by extension, speciation rate, is an accurate measure of evolutionary rate, then one must ask why it was so low in the midst of the major Paleocene radiation of mammals, and equally low in the Pleistocene when no new major groups appeared. In the former case, brief longevity may indeed be correlated with rapid morphological evolutionary rate. But in the latter case, sudden extinction might contribute strongly to reduced longevity

In summary, taxon longevity or rates of appearances at the same taxonomic level should not be used for comparisons among taxa of great biological difference. Unfortunately, we have no data base to know how different taxa can be before such comparisons are invalid. The morphological and genetic differences appear to be scaled totally differently among phyla. Moreover, it may also be invalid to use taxonomic longevity even within a single major group at different times during its history. Criteria for taxon recognition may change substantially. Comparisons should be restricted to closely related groups in which the morphological criteria for fossil recognition at a given taxonomic level are known to be homogeneous.

The Assembly of A Complex Bauplan

The major phyla, and sometimes different taxonomic classes or orders, are often thought of as a series of *baupläne*, with no intermediates. The concept of *bauplan* springs from idealistic morphology and, therefore, has a typological connotation. We all recognize, nevertheless, a series of distinct body plans at the phylum and class levels, and the term can still be used to designate these groups. The presence of *baupläne* raises two important issues: (1) Is the very organization of a *bauplan* a source of evolutionary inertia? In Chapter 5 we have defined the possible sources of this inertia, in developmental, genetic, and adaptational terms. (2) Are the gaps among *baupläne* achieved in single crucial (and nonadaptive) leaps or is the process gradual and cumulative?

It may be sufficient to document change through a geological section to justify the phyletic nature of single-character evolution, but an attribute such as "mammalness" cannot be restricted to one character, even if the definition of the group can be made in this way (Simpson, 1960; Crompton and Jenkins, 1973). The success and ecological position of mammals owes as much to homeothermy, reproduction, and a complex central nervous syste, as it does to the acquisition of a new jaw joint and more complex teeth. This is after all the crux of any macroevolutionary question.

1. What is the pattern of acquisition of the many features?

2. What controls the pattern and rate of acquisition?

3. Once acquired, what holds the plexus together, if anything?

These macroevolutionary questions can be treated independently of the issue of speciation. The pattern of acquisition of various mammalian traits can be entirely coincident with speciation events or may consist of entirely within-species phyletic evolution. But as long as the mammals, with all of their unique characters fully formed, did not arise in a single speciation event, the questions still have relevance.

Three end member patterns can be imagined for the acquisition of the characters making up a complex group.

1. **Saltational Hypothesis**. First, all characters could have been acquired at once. I don't believe that anyone takes Schindewolf's (1936) claim that the first bird hatched from a reptile's egg very seriously, without accepting some clever definitions of bird and reptile. Gould (1980a) has revived the notion of hopeful monster, but certainly not at the level of wholly new organized body plans. He (1984c, pp. 185-186) argues, nevertheless, for the "enormous

reservoir of potential for rapid evolutionary change" in developmentally significant mutations.

2. **Independent Blocks Hypothesis**. The features that characterize groups such as the mammals might be ascribed to distinct blocks, and these blocks might have evolved independently. As time progressed, interactions between the blocks occurred but only because of coincidental response to the same new environment, although later functional interdependencies among the blocks may have arisen. This has also been termed **mosaic evolution** (see De Beer 1958).

3. **Correlated Progression Hypothesis**. Different morphological complexes can be distinguished, but they have a sufficient interrelationship such that each system appears to "evolve in a loose correlation with all the other systems" (Kemp 1982, p. 313), maintained by function.

I support Kemp's (1982) hypothesis that correlated progression is a major component of mammalian evolution. Over time, the system has congealed from both a functional and developmental point of view, to the point that evolution to maintain the whole is probably as important as evolution designed to adapt independently any particular block to a change in the environment. I emphatically do not imply that this will be the case for all groups. Indeed, since theory does not predict unique solutions in most cases, our approach to such questions must be largely inductive. Mosaic evolution is a major component of the rise of *baupläne*, and, in some cases, single key innovations may have set the stage for the evolution of a large suite of associated features. For example, the molluscan *bauplan* may have derived from the acquisition of a hard dorsal integument, by a flatwormlike ancestor.

The evolution leading to mammals is unique in that a remarkable gap in morphological organization seen in living forms between mammals and reptiles is bridged in the fossil record with an array of intermediates (Crompton and Jenkins, 1968, 1973, 1979; Kemp, 1982; Parrington, 1971). This is particularly interesting, given the considerable evidence that additional discoveries are likely to be made with increased sampling (e.g., Jenkins et al. 1983). As it now stands, the fossil record of the mammallike reptiles (synapsids) documents approximately 130 million years of change from their earliest appearance in the early Upper Carboniferous to the first appearance of the "true" mammals in the Upper Triassic. Because the continents were not separated during this period, faunas were geographically relatively homogeneous and incomplete sections in one location can, therefore, be filled in, at least at a coarse scale, with collections from other localities. The earliest mammals, found in South Wales and northeastern Arizona, already show a diversity of jaw structure (Jenkins et al. 1983), so we are far from knowing the complete story.

The synapsids made their first appearance in the Late Carboniferous. They dominated the reptilian faunas until the end of the Triassic, but then declined, to be followed by dinosaur dominance in a second age of reptiles in the Jurassic-Cretaceous. During this period their apparent descendants, the mammals, played a minor role in terrestrial faunas and were typically diminutive in size. The shift from dominance by obvious reptilians such as the pelycosaurs in the Permian to groups much more mammalian in character in the Triassic is believed to be associated with a general adaptation to terrestrial life (Kemp 1982). This would imply that the groups extant in the Triassic were superior to their predecessors at surviving the exigencies of terrestrial life. We must remember, however, that this superiority did not guarantee a place of dominance in the terrestrial world. Indeed, the dinosaurs seem to have forestalled the rise of the mammals for a considerable period of time. Trends seen within a group do not imply superiority to anything other than their own antecedents. It is unfortunate to think of the mammals as superior to the dinosaurs, or, for that matter, to think of the dinosaurs as inherently superior to the mammallike reptiles.

The distinction between advancement within a clade and interclade superiority is crucial. The former can be argued on functional morphological grounds, while the latter can only be argued from historical evidence of interclade competition, which is often weak or nonexistent (Benton 1983a, b). Even within the synapsid clade, "advancement" can only be considered within the context of the characters that are associated with terrestriality per se. Many unique features acquired by members of the clade may have momentarily been more important in survival than the features associated with the overall trend toward "mammalness."

Many characters that we associate with "mammalness"–heart and double circulation, sweat glands, hair, lactation and parental care, an

impermeable skin, production of a hypertonic urine, and a larger number of mitochondria per cell– cannot be documented meaningfully at
this time from the fossil record. We are thus left with those characteristics that can be inferred directly and indirectly from the cranial
and postcranial skeleton. We paint a picture with a palette of few colors. Skeletal characters can reveal much about physiology and activity,
but not everything. We have, for example, a very incomplete picture
of temperature control. Even among the extant homeothermic mammals, large differences exist in the absolute value of temperature, even
if the temperature is maintained relatively constant. In fossils, there is
a strong case for homeothermy in advanced mammallike reptiles and
dinosaurs, based upon the lack of seasonal growth rings in bones, other
features of bone structure, and predator-prey ratios characteristic of
endotherms (e.g., Bakker, 1974, 1977). Homeothermy could have been
accomplished at first by increased body size and by the evolution of
mechanisms to retard loss of the heat gained in exercise (McNab 1978).
An increase of metabolic rate could have occurred separately. In the
cynodonts, a group of advanced mammallike reptiles, the presence of
a secondary palate and the complex mastication apparatus both point
to a high metabolic rate, requiring concomitant features enhancing the
rate of food acquisition.

Most research (e.g., Crompton 1963; Crompton and Jenkins 1968,
1973) has focused upon the changes in the jaw and dentition. This
was partly due to their relatively good preservation and to the importance of dentition and mastication to the efficient processing of food
required to service the mammalian life style which includes a high
metabolic rate. The record of transitions– from the Pelycosauria, to
the Therapsida, to the Cynodonta, to the Mammalia–is impressive, as
it includes a complete graded series ranging from the primitive reptilian skull with (1) alternate tooth replacement and relatively simple
homodont teeth, (2) articular-quadrate jaw joint, and (3) relatively
simple jaw musculature, to the mammalian condition of (1) single replacement of complex heterodont teeth, (2) dentary-temporal jaw joint,
and (3) complex jaw musculature, which reduces the mechanical load
on the articular-quadrate joint (Crompton 1963). A host of important features evolve in concert with these three major systems, most
notably the various muscular support structures such as the temporal
fenestra (Frazzetta 1969).

The most interesting transition can be seen from the perspectives of jaw articulation and mastication. The evolution of efficient food processing requires complex occlusion and the ability to move the lower jaws in several directions (Crompton 1963). In the advanced cynodont, *Thrinaxodon*, the postcanine teeth did not occlude in the sense of extensive intermolar contact. This was precluded by the reptilian tooth pattern of alternate tooth replacement (Crompton and Jenkins 1968). In more advanced cynodonts, occlusal facets essentially characteristic of the later mammals were established by gradual wear. Later, the faceting was genetically determined. Such changes in occlusion were accompanied by a change in the jaw musculature that permitted greater latitude in movement and increased mastication strength. Such strength would have exerted significant forces on the reptilian jaw joint, which functioned as a third class lever. The evolution of the jaw musculature increasingly reduced the load on the reptilian jaw joint, and the opposing action of the temporalis and masseter muscles permitted the lateral control necessary for complex chewing. In the cynodonts, a trend for reduction of the post dentary bones eventually led to the possibility for the mammalian style jaw joint, which, in several transitional forms, existed in line with the older reptilian joint. This is the most unlikely transition of all: the coincidence of two fulcra (Jaanusson 1981). In modern reptiles and mammals, it would seem like an impossible transition for the location of a jaw articulation to switch from one fulcrum to another without a saltation.

In conjunction with this already remarkable series of changes, the evolution of the mammalian jaw joint is intimately involved with the evolution of the middle ear bones. The articular and the quadrate jaw bones in reptiles are the homologues of two of the mammalian middle ear bones, the malleus and the incus (Hopson 1966). As the new mammalian dentary-temporal jaw articulation arose, the post dentary bones decreased in size and were freed for cooption as part of the middle ear apparatus. Allin (1975) argues that, along with the trends in synapsid evolution toward masticatory efficiency, some features just anterior to the reptilian jaw joint indicate an auditory function. If this is true, then the evolution of hearing and chewing is an excellent example of Kemp's (1982) concept of correlated progression. From the point of view of functional morphology, this series of changes demonstrates that functioning structures cannot be predicted "from the ground up."

In order to assemble a model of functional change, one must start with the reptilian joint, and the constraint of a requisite for efficient hearing.

Our account of the changes occurring in the evolution of mammals demonstrates the complex interrelationships among the various parts of the skull, most notably muscular supports, dentition, and jaw articulations. The evolution of the mammalian middle ear (at least as we see it now) could not progress without significant change in the various features associated with the evolution of efficient chewing. It seems unlikely that any part of the skull could have evolved very much without an effect on the functioning of another. The best example is the need to reduce the mechanical load on the reptilian jaw joint, as masticatory forces become more intense and complex (Crompton and Jenkins 1968). Changes in the postcranial skeleton, the evolution of dorso-ventral flexibility for example, also helped contribute to the general need for an increase in food gathering efficiency. The latter was encumbered by still another evolutionary change incorporating higher metabolic rates that generate the need for a higher rate of food intake. As Kemp (1982) noted, it is difficult to take a modern mammal and dissect the crucial mammalian feature; they are all so interdependent. It is, therefore, not surprising and rather gratifying to see the fossil record show a correlated progression of change.

This general account of transitional [2] change can be made more precise by defining the phylogenetic relationships of the various synapsid groups involved in the evolution of mammals with a cladistic analysis. We can then ask the question: Were the various mammalian characters acquired in independent groups or can we speak of a correlated progression of features. Kemp (1982) has associated the features acquired by the various groups with a cladogram depicting genealogical relationships. Current data can only articulate the big picture; we only have a general cladistic skeleton, and cannot say with the evidence whether transitions consist of minute quantitative changes or major qualitative changes. Intermediate morphological stages are usually absent below the family level, but the continual discovery of new forms suggests that

---

[2] In discussing the rise of the mammalian condition, we trace a line of descent through a very complicated cladal structure. As a matter of shorthand it becomes convenient to say that "the synapsids gave rise to the mammals," but surely this must actually mean that "a species with the characters defining synapsids gave rise to a descendant bearing an additional character or set of characters defining the mammals."

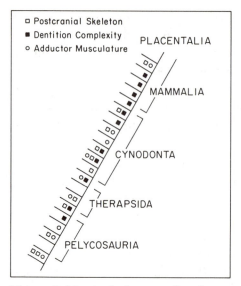

Figure 7.23: A cladogram for the genera in the lineage leading to the therian mammals. Major new acquisitions of skeletal features leading to the mammalian condition are indicated (after Kemp 1982).

missing data, rather than saltations, are the likely explanations. As an example, evidence until recently would have suggested a possible major quantum change, perhaps based upon a developmental mutant, to explain the origin of the angular region of the mammalian jaw, from the reptilian pseudangular process. Recent early mammalian fossil discoveries reveal an incipient process that may have been the progenitor of the angular region (Jenkins et al. 1983).

The pattern of acquisition of mammalian characters illustrated in the cladogram (Figure 7.23) shows the piecemeal nature of mammalian evolution. Significant incremental changes toward mammalness in the size of the jaw adductor musculature, complexity of the dentition, and the postcranial skeleton, all occur in the majority of internodal steps in the cladogram. In the case of the interaction of jaw musculature, dentition, nature of the jaw joint, and the middle ear, these changes could not help but be interactive. Intermediate stages in the process are therefore functionally harmonious, the result of adaptation. Some other features, however, may have been acquired according to the model of mosaic evolution. The evolution of the kidney, for example, may have been in response to similar selective pressures posed by the terrestrial environment on feeding, but the acquisition need not be intimately correlated with changes in the skull.

As the record of the evolution of mammals is a story filled with intermediates, two important and unresolved questions come to mind.

1. Why did the process take so long–about 130 million years?

2. Is there any functional problem posed by intermediate forms?

The case of the mammals demonstrates that both questions, even if they have been major concerns in the history of evolutionary biology, may have little meaning when considering organizational changes of this magnitude.

The question of tempo is nearly impossible to solve. We often forget that the complex of taxa involved in mammalian evolution had important and unique features other than those upon which we focus in understanding the transitional stages leading to the mammals. In some cases these unique features may have interacted significantly with intermediate character complexes and may have either led members of the taxon in question down a dead end path, or at least decreased the chances for evolution toward the mammalian condition.

For example, the dinocephalians constituted the majority of the early Late Permian therapsid faunas. They were large animals with a number of features that functionally may have precluded an easy advancement to the mammalian condition. Unfortunately, we have no idea as to how such conflicts might influence whether a group can be a progenitor of further advanced forms. In the context of terrestriality, there seems no doubt that the overall path toward the mammals involved adaptation, but other conflicting adaptations may have been present in many of the groups. Unless some clear model of selective value of these diversions and retardants can be established, it becomes fruitless to argue why the rate was so slow (or fast, for that matter). In many ways, the diversions are more interesting than the main mammalian line itself. Too often, we ignore the fact that much of the diversity of life simply is precluded from being ancestors by being burdened with one or many features that restrict evolutionary direction and potential. It is, therefore, equally fruitless to view the evolutionary trend toward mammalness as "progress"; mammallike reptiles didn't know what they were going to be, and true mammals do not remember what their ancestors were! Once mammalness was completed, however, it does make sense to ask why they have done so well with the array of features they have acquired over the past 130 million years.

The question of intermediate forms continues to be an intriguing issue, since they are often absent in modern populations. Too often, an organism is chosen as an intermediate and an attempt is made to see what function its intermediate structures might have had. *Archaeopteryx* is probably the best example of such a transitional form. Since it is neither bird nor reptile, a believer in natural selection is often forced to make the case that the intermediate is functionally harmonious and could survive, but was then inferior to either its ancestors or descendants and, therefore, lost out in evolution (e.g., Simpson 1944, p. 92; Bock 1979). While this is plausible, it ignores the historical context within which the intermediate arose. *Archaeopteryx* is not half-bird and half-reptile. It is *Archaeopteryx*! It can be viewed as a mosaic of reptilian and avian features, which is to say that it was an organism in its own right with a complex evolutionary and ecological history. Like the case of mammal evolution, the intermediate stages were integrated organisms with specific traits that were not passed on to distant descendants, but may have enhanced survival in a particular milieu. Thus although it is a fascinating and viable question as to exactly how *Archaeopteryx* used its feathers and forelimbs (e.g., Ostrom 1974), it is not clear to me that the fossil record is complete enough to expect that we will ever learn, for example, whether flight in birds started from "the ground up" or "from the trees down" without some extraordinary new discoveries or insight.

We might ask if this sort of gradual accumulation of characters, found in the broad picture of mammalian evolution, can be seen at lower levels, such as orders or families. As noted above, interfamily transitional forms are often absent, probably as the result of inadequate finds rather than saltation. But sufficient evidence exists to examine the question of whether evolution within a mammalian order is of the same type as on the level of Mammalia.

The mammalian order Carnivora first appeared in the middle Paleocene and was represented by members of the family Miacidae until the end of the Eocene, some 20 million years later. After the extinction of some coexisting carnivorous mammals, the miacids radiated into what produced most of the modern families of the Carnivora (canids, felids, viverrids, mustelids, ursids) as well as two extinct families. Using factor analysis, Radinsky (1982) demonstrated that the differences that now distinguish modern viverrids, canids, felids, and mustelids were less pronounced in the Oligocene, when these families made their first appearance in the fossil record. Cranial differences in representatives of the living families can be related to differences in prey-killing, but not necessarily to differences in prey type or size. But only some of these distinctive features can be found in Oligocene ancestors of the modern families. Radinsky suggested that body size differences established early in the history of the group may have been an important factor in the radiation. As body size is correlated to prey size in carnivores (Rosenzweig 1966), adaptations to prey handling may have been triggered by or coincided with body size differentiation. These adaptations may have been entrenched with the accumulation of further traits, when the miacids survived the late Eocene extinction of many early carnivores, perhaps even as a matter of chance. From a quite different specific context, size variation in spores may have been the initial variation that was later evolved into morphologically distinct gametes, and ultimately, seeds in plants (Tiffney 1981). Heterospory is represented initially in the fossil record simply by spore size variation within a single sporangium in the Devonian *Chaleuria*. Thus later complex morphological diversification may have been initiated by simple differences in size.

The evolution of the angiosperms provides another example of the piecemeal evolution common in the evolution of major new groups. As Tiffney (1981) noted, many of the features thought to be characteristic of the angiosperms were already present in their remote ancestors. Thus, if there is a feature such as "angiospermness," it did not arise suddenly, but rather by a combination of correlated progression and mosaic evolution. Tiffney outlined a scheme of potential correlated progression remarkably similar in approach to that of Kemp's analysis for mammals. Many of the individual features of angiosperms appeared in various fossil groups, but only one line seemed to accumulate them all. Stebbins (1983b) briefly summarized the literature and reached the same conclusion for the origin of birds, amphibians, and several other groups. Indeed, Schindewolf (1936) is exactly correct. The first bird did hatch from a reptile's egg. But this is only true when we use some particular feature to define the birds. The class Aves was nearly already fully formed the moment it was "born." In some cases, the birth of a new group can be marked with a key innovation that must have permitted an extensive radiation. But an examination of the Carnivora reveals no keystone innovations that can readily define "carnivoreness" in one step (Radinsky 1982). The same can be said in the evolution of the mammals. A.B. Smith (1984) notes that the distinct nature of the extant echinoderm classes is not nearly so true in the Paleozoic. This has led to a spurious conclusion that ophiuroids and asteroids are the most closely related classes.

This discussion demonstrates, in some well-documented cases, that what we now consider a fixed *bauplan* in given groups of organisms arose gradually by accumulation of features that may or may not have arisen by correlated progression. In some groups, the doors to success were opened by key innovations, but this seems not to be the case for the mammals. This perspective, obtainable only from the fossil record, undermines a nearly essentialist viewpoint, which treats different *baupläne* as integrated entities that resist change and tends to ignore their gradual origins. Gould states (1983*b*, p. 80): "the cluster of cats exists primarily as a result of homology, and historical constraint. All felines are alike because they arose from a common ancestor shared with no other clade. . . All feline species have inherited the unique cat *Bauplan*, and cannot deviate far from it as they adapt." Genealogically, of course, all felids have descended from a common ancestral species. But this omits the essential nature of the evolutionary process. The "*cat bauplan*" was assembled gradually, first as "mammalness," and later as "catness." After this set of adaptations was congealed by both functional interactions and incorporation of traits into developmentally linked complexes, it is no surprise that current conditions do not now permit much of a deviation from the basic organization that we use to define cats. The entire process of adaptation and crystallization of the cat *bauplan* involved a presence in the same ecological milieu, that is, hunting, seizing, and eating usually mobile prey. In writing of present constraints, Gould leads the reader away from the likely adaptive origin of cat traits. It is thus meaningless (loc. cit.) to dismiss correlations of cat morphology with the present environment and to argue that "genealogy, not current adaptation, is the primary source of clumped distribution in morphological space." How does one distinguish between genealogy and adaptation? The dichotomy is false.

The fossil record reveals a very clear pattern. From the lowest level of resolution of morphological change in phyletic lineages at the species level, to the highest level, we see the gradual accumulation of novelties that contribute to form the distinctive and morphologically separate clades that exist today, or have existed at any time in the past. The unique genetic properties of various traits strongly influence the pattern of change seen in a geological column. On higher levels, we see traits assembled over time and trait groups gradually congealing into total organisms with the evolution of various developmental programs. Some of these programs make any change very costly to the organism, but it is a matter for empirical investigation to determine this cost. But it is clear that the evolution of *baupläne* are often gradual and that the present distribution of groups tightly clustered in morphospace is in large part a reflection of this gradual congealing. Developmental constraints may regulate the congealing, but adaptation may often be a major element in the evolution of developmental constraints. Since the organism evolves mechanisms to stay in its favored environment, the gradual congealing is a reciprocal process of (1) culling of variants arising as mutants and (2) the increasing commitment of the organism to a given lifestyle that keeps it locked into the same selective regime for long periods of time. It is this very reciprocal process that weakens the claims for the isolated importance of historical or developmental constraints on evolution. Organisms adapt partially by sealing their own fates through a continually increasing commitment to a given lifestyle. As in the mammals, there is patent evidence that this increasing commitment is associated with improved performance. Adaptation is not divorced from history; it is an integral part of history.

Despite the host of mammallike reptiles, the mammals prevailed. This raises the question of intraclade superiority. In any such situation there are two alternative explanations for success of the derived condition:

1. The sinuous path taken toward the mammalian condition produced a taxon with more survival value than its antecedents, which had relatively more primitive character states for "mammalian" characters, but its own synapomorphies, which involved a unique evolutionary path on one side branch of the clade.

2. Stochastic Extinction. It may be that all members of the clade

happened to become extinct and the surviving one is falsely taken
to be superior. In the case of mammalian evolution, the latter
explanation is much weaker, for the advanced mammal- like reptiles had means of dealing with the terrestrial environment that
were functionally superior to earlier forms. The net path to the
mammals, moreover, contributed to the development of a functionally integrated phenotype that served the needs of efficiently
finding and processing food. It is, of course, true that no given
group in the main line of progression evolves for the sake of the
ultimate and most progressive form, but the latter could not have
arisen without the appearance first of the former.

Cladogenesis may hasten the advancement toward a derived condition, as in the case of mammalness. Consider a single phyletic lineage,
acquiring gradually the specializations necessary to achieve the mammalian condition. At any one time, the taxon will have a given set of
ecological restrictions and will, therefore, have limited opportunities for
novelties to arise from the extant morphological material and the ecological exigencies that challenge the populations. But if cladogenesis
occurs, it is likely that more types of environments will be encountered
by the group, and that more particular adaptations may bring the total
group into exposure with a larger range of ecological challenges. This
may hasten the chance that one of these taxa, assuming all have the
same degree of advancement, will encounter the situtation that will
lead to further progression. Cladogenesis, therefore, does probably
hasten the rate of progressive evolution. But this process fits neither
the notion of directed speciation nor species selection, since we are still
considering a core path of progress that leads to the derived and advanced state. It is not speciation per se but phyletic evolution in the
isolated lines that results in the fixation of given novelties. Cumulative
change may thus be hastened by cladogenesis.

The arguments presented above view trends in evolution as the result of complex adaptations acquired mainly by phyletic evolution. In
the case of the mammals, the functional integrity of the many observed
evolutionary changes strongly argues against the model of punctuated
equilibria and species selection. There is no compelling evidence that
phyletic evolution is insufficient to generate and congeal the series of
successively acquired novelties that eventually comprised the mammalian condition. Whereas, as mentioned above, cladogenesis increases

the chance of encountering more opportunities for new character combinations and new ecological situations, the line leading through to
the mammals can still be viewed as a sinuous phyletic path. Selective extinction may easily, of course, be the reason why the mammals
are around today, and the pelycosaurs and therapsids are only history. But the cumulative evolution of traits leading to the mammals is
an excellent case of complex adaptation that requires no decoupling of
within-species versus among-species processes, as suggested by Stanley
(1975).

The species selection model (Eldredge and Gould 1972; Stanley
1975, 1979) argues that speciation generates a series of morphologies
that are random with respect to a trend, and that selective extinction
determines the trend. This can hardly be argued very seriously in the
case of the mammals, or in other groups where integrated functional
morphological complexes can be readily identified (e.g., Miyazaki and
Mickevich 1982). In such cases random generation of morphologies
about a trend seems functionally absurd (Levinton and Simon 1980).
In the mammals, change can only be visualized in highly restricted
directions, given the constraints of facial muscles, pressure on the jaw
joint, and the overall selective pressure of terrestriality. Stanley (1979)
suggests the possibility of directed speciation, where species arise only
in a morphologically biased direction. But this only associates tautologically a phyletic pattern of evolution with a series of species delineated strictly and arbitrarily by morphological changes. In principle,
nothing excludes species selection from being an important aspect of
evolutionary trends. At present, however, there is no strong evidence
for the general importance of the process, and it certainly is not necessary, given the possible rates of phyletic evolution.

Of course, it is possible that many independent lines of evolution
toward the mammalian condition developed, but that random extinction spared all but one or a few. While possible, this alternative still
begs the question of the likely directional progression of all the independent lines in the first place. While some minor features might have
been different, the major trends towards homeothermy, processing of
food, efficient locomotion would probably have been solved in similar
ways. Therefore, a somewhat different type of mammal might have
arisen if history, in the form of extinction, might have been somewhat
different, but similar constraints on phyletic evolution would have led

in predictable directions. A possible example of a different outcome might be imagined for the evolution of the bones of the mammalian middle ear. Somewhat different directions of evolution might certainly have resulted if other bones had been exapted for hearing.

The German paleontologist Rudolf Kaufmann (1933), in his studies of evolution within the Upper Cambrian trilobite genus *Olenus* in Sweden, articulated the concept of *Artabwandlung*, whose meaning might be conveyed by the term "intraspecific directional modification" (see Teichert 1949, p. 49). The four stocks all show similar phyletic trends in size, shape and ornamentation, which can be seen partially in Figure 7.8. Here, where some definitive evidence exists, it is clear that phyletic trends within elements of the clade can be seen as the net evolutionary trend of the entire clade. This situation is by no means unique. In the evolution of the horses, trends towards hypsodonty and monodactyly can be seen in many lines, particularly in *Cormohipparion*, *Hipparion*, *Neohipparion*, and *Nannippus* (Simpson 1951; Woodburne and MacFadden 1982; MacFadden 1985). In this latter case, the process of *Artabwandlung* can be placed in a context of functional morphology. These examples suggest that there is nothing compelling about the species selection model, since net trends seen for the whole clade are duplicated within phyletic lines. History might be slightly different in the different lines, but functional constraints provide an important predictable guidance to the directions of morphological evolution. In some cases, the continuing interplay between the rise of new structures and functional constraints may make for a host of unique solutions; in these cases interclade extinction might play a very important role in the group that finally "wins out." This would be an intermediate, and acceptable, position between those who believe in either the primacy of phyletic evolution or species selection in adaptive trends.

Kaufmann's concept of *Artabwandlung* corresponds to Vavilov's (1922) law of directed series. Vavilov argued that when representatives of related taxa are placed in similar environments, similarity of biology would result in similar reactions. Thus adaptive evolution in parallel lines may result in similarity of form owing to a similar "predisposition," which might include a variety of genetic and developmental constraints. Unfortunately, we are only at the beginning of an understanding of the relationship of functional, genetic, and developmental integration. This must be a major area of research in the coming years.

The Main Points

1. In large measure, the incompleteness of the sedimentary and fossil record strongly biases our perception of evolutionary processes.

2. The completeness of the sedimentary record depends upon the desired temporal resolution. For short periods of time, such as a thousand years, most intervals will not be preserved in a geological column. This effect is so pronounced that sedimentation rate appears to be related inversely to the overall period of time represented by the geological column. Periods of interest to students of speciation and population genetics can be shown to be poorly sampled by most geological columns.

3. Most estimates of evolutionary rates seem paradoxically slow, and could be explained by genetic drift. This paradox seems related to the under-representation of short periods of time, when rapid rates of evolutionary change could be delineated. The longer periods of time typically sampled by paleontologists represent periods of rapid evolution, reversals, and no change. As a result, evolutionary rate appears to be inversely related to the time period represented by the geological section, much as is found for estimates of sedimentary rates.

4. The punctuated equilibrium theory asserts that the major part of a species history is characterized by stasis, and that evolutionary change is usually concentrated in speciation events. Most discussions of stasis, however, focus on change in individual characters. The coarse time resolution of most fossil studies biases us towards a perception of stasis.

5. Because species are identified only by morphological difference the association of speciation with morphological change is usually a tautological exercise. This is especially true in studies where there is no evidence for cladogenesis, yet speciation events are documented by sudden morphological change. Given the limitations of species identification and the time scale resolution of the fossil record, the punctuated equilibrium theory cannot be

tested in the fossil record. By now, however, at least a couple of cases exist that seem to fit the requisites of the punctuative model.

6. It is possible to test for the efficacy of character change. By now, a large number of studies demonstrate the common occurrence of phyletic evolution of a sufficient magnitude to produce change on the order of specific and generic differences. The variety of population genetic mechanisms affecting evolutionary traits and the range of genetic mechanisms controlling traits make it fruitless to look for the pattern of stasis and ascribe it to any particular process. There are a few well documented cases of long term stasis, particularly for shelf invertebrates. Mammals and marine protists seem to have common directional phyletic trends.

7. There is little fossil evidence giving us insight into the origin of major new body plans. The accumulated evidence for the evolution of the mammals suggests that a series of adaptations were acquired gradually and piecemeal. The origin of the mammals seems to have involved the interplay of a variety of ancestral constraints (e.g., the osteological configuration of the reptilian skull) with adaptive changes that improved performance in a terrestrial environment.

Chapter 8

Patterns of Diversity, Origination, and Extinction

The sea lost its serenity.
Yukio Mishima, The Decay of the Angel

Introduction

A paradox arises from the large-scale perspective afforded by the fossil record. Phyletic evolution and adaptation appear to be ubiquitous; yet adaptation cannot always be equated with success, relative to supposed coexisting but inferior forms. The rise of vagile shell-crunching predators in the Mesozoic was accompanied by an expected overall mechanical resistance to predation in the marine benthos (e.g., Vermeij 1977, 1983). But this is not always the case. Are the dinosaurs to be considered inferior to their successors, the mammals? If so, one has to explain why "mammalness," when once achieved in the Triassic, took over 100 my to manifest itself in worldwide dominance. Why did the dinosaurs fail to succumb immediately, if the mammals were so superior? Is this the way to frame the question at all?

The great time period involved in the gradual evolution of "traits" such as mammalness also invites difficult questions. How can such an overall development transcend the Carboniferous–Triassic time span, which included the greatest biological catastrophe–the Permian–of the Phanerozoic? It isn't clear that we can ever answer these questions, but a biologically meaningful history of the earth is essential as the historical backdrop for major evolutionary events. We have seen from our discussion of form and function that history plays a key role in the nature of the adaptive process. We must know how the organism's past history shapes its reaction to its present environment, but we must also know how the immediate environment imposes itself upon the organism.

Paleontologists have mainly used the very presence of the fossil organisms themselves as the principal evidence of the relevant events in earth's history. The sudden disappearance in geological sections of groups of species was the basis for the catastrophism of Cuvier in the 18th century, and the nature of mass extinctions still figures prominently in current debates about the biological history of the earth. Major turnovers in fossil biotas are often far clearer than associated physical evidence, which may involve indications of retreats of the sea, arrangement of the continents, or changes in climate. The fact of the turnover is much easier to establish than the causal factor, whose exact identification can usually only be vaguely inferred by means of imperfect correlations. Even in the cases where obvious correlations occur, it is not usually clear whether the change is "important" enough to induce the change in the biota, without the post facto establishment of the correlation with an extinction. Unfortunately, the time scale for such changes is probably unapproachable by the neontologist, who can only observe "normal" extinction (Diamond 1983). To examine larger scale changes, we are thus left with the tools of the geologist at our disposal.

What do we know about such large-scale patterns? Because paleontologists are best at recognizing given taxa by diagnostic characters and can also establish an approximate date of occurrence, facies (rocktype indicating the overall environmental regime), and paleogeographic location, we can collate a series of snapshots of the biotic character of the earth at various times and places. These pictures taken together reveal the following general patterns that must be important in our explanations of macroevolution.

1. Correlated origins of different clades.

2. Correlated diversifications of different clades.

3. Near simultaneous extinctions of groups seemingly unrelated in ecology or genealogy.

4. Replacement by ecologically similar forms after extinction.

5. Occasional long-term constancy in the morphological types and quantity of coexisting taxa, despite some degree of turnover.

These patterns appear at many levels of time and taxonomy. It is the purpose of this chapter to discuss the quality of the overall evidence for these generalizations and their explanations.

The Quality of the Data

Broad-scale trends in taxon richness usually derive from a complex data base produced by many workers using different systematic methods and collecting techniques, and encountering variable modes of preservation. Such inherent problems raise substantive issues as to whether it is possible to infer any clear patterns through the haze of potential collecting and preservational biases. As we will see below, some disagreements over major historical events turn around interpretations of potential sampling bias.

The Taxic Approach

We mentioned in Chapter 2 some of the problems that an unevenly applied systematic philosophy creates in the interpretation of the fossil record. A taxon can be assigned a given rank by virtue of its species richness, phenetic difference from other taxa, or phylogenetic relationships with other taxa. No standard can relate a given taxonomic level of one group with another, although criteria that are internally consistent within a group might be imagined.

In studies of changes of taxon richness over time the problem of bias becomes more complex, as the number of species is often the objective, but usually taxon richness can only be easily assessed at some higher level (e.g., families, orders). This is particularly a problem, as it is well known that certain taxonomic levels are "natural" for one group, but rather arbitrary for another. In principle, it ought to be possible to calibrate a given taxonomic level against the species level by using taxon–species ratios appropriate for the time period (see Van Valen 1973a). This approach is weakened by the difficulty in identifying species in fossil groups, but would be preferable to bias-uncorrected tabulations, such as ordinal level richness over time. For example, the ratio of orders to families decreases significantly in the Mesozoic (Sepkoski 1984). Thus a previous report of a long-term steady state level of ordinal diversity (Sepkoski 1978) becomes ambiguous in interpretation. Does it represent species richness (species diversity aspect of taxonomic rank), or temporal constancy in the degree of morphological diversity (phenetic diversity aspect of taxonomic rank)? Similarly, the ratio of families to species decreases by a factor of two from the Mesozoic to the Cenozoic (Valentine 1969; Sepkoski 1984). An additional problem is the asymptotic climb of numbers of families as a function of increasing species numbers. During a modest extinction, that is random with respect to families, the number of species going extinct will be underestimated if a linear relationship between families and species is mistakenly assumed. The acuteness of the problem becomes still more obvious when we consider more detailed data.

Rowell and Brady (1976) followed the diversification in a Cambrian biomere, which represents one of several iterative radiations of marine benthic trilobites from cooler-water Olenid progenitors (Palmer 1984). In the case of the polymeroid trilobites, an increase of generic richness correlates well with species richness. In the inarticulate brachiopods of the American Cordillera, however, species increase in number while the number of genera remain constant. If the tabulation were reported on the generic level alone, we would have a far different picture than that expected for species. Unfortunately, few of us usually appreciate the magnitude of the problem presented by the use of various taxonomic levels. The information we get at different levels must be interpreted with care.

Preservational Biases

Paleontology students quickly learn that fossil collecting is as much art as science. Geological sections rich in fossils are common, but usually not available in a form ideal for random statistical sampling. Of course, not all gaps in fossil occurrence are due to poor preservation. Many paleoecological studies have demonstrated that sporadic occurrences often represent the uneven presence of suitable microenvironments, or episodic colonization by species with rapid colonization potential (Levinton 1970). The student is nevertheless confronted with obvious cases of preservational bias. In coastal plain sediments of the middle Atlantic states, obvious zones of poor preservation of molluscan faunas can be correlated with solution by groundwater; immediately adjacent areas will show good preservation in sediments with identical texture and degree of burrowing. In the nearshore clastic sediments of the Upper Cretaceous of New Jersey, one side of a hill may yield internal molds of gastropods and oysters, while the other will lack such molds but bear oyster shells.

Differences in fossil shelly faunas can be due to the stricter set of conditions often required to record the presence of animals with aragonite shells (most gastropods), as opposed to calcite (oysters). As a result, especially wellpreserved bivalve coastal plain fossil faunas have a decided peak in richness, relative to other localities. This may lead to spurious beliefs in strong endemism, where unusually fine preservation is the actual cause of the unique occurrence. The Owl Creek Formation in the Cretaceous of the Gulf coast, for example, has a malacofauna enriched in well preserved and aragonitic forms. Its endemism was halved once the nearby Providence Sand was taken into account (Koch and Sohl 1982).

Preservational bias may operate on the grand scale of the Phanerozoic. Valentine analyzed family-level diversity of "wellpreserved" marine invertebrates throughout the Phanerozoic, and noted a middle Paleozoic high, a Permian low, and a post-Permian expansion. Raup (1972b, 1976a, b) tabulated species richness from the Zoological Record and obtained similar results. But he also found a strong positive correlation between species richness and estimated world rock volume for given periods of geological time, and found a positive correlation on a smaller geographic scale using Canadian ocurrences. Tiffney (1981) reported a positive correlation between nonmarine outcrop area and species richness of fossil plants. These correlations are worrisome and raise the possibility that all patterns of species richness in the record might be artifacts of preservation. If the effect of rock volume is subtracted, a significant pattern of relationship between area and species richness remains (Sepkoski 1976). Moreover, a comparison among very different data sets based upon global and local taxon richness all have yielded a similar temporal pattern of overall taxon richness (Sepkoski et al. 1981). Most significant are the changes of within-community diversity, which parallel the broader trends (Bambach 1977; Niklas et al. 1979; Tiffney 1981). This suggests that the qualitative pattern is real, but that care is required in the interpretation of taxon richness trends.

Although the overall change of diversity in the Phanerozoic may not be due to a bias created by fluctuations in preserved sedimentary rocks, smaller-scale changes in sedimentary preservation diminish a clear understanding of crucial periods in the history of life. The end of both the Permian and Cretaceous are believed to be crisis periods for the world biota. But both time periods are marked by extensive regressions of the sea, with concomitant reductions in rocks recording events of those times. In the Permian, the problem is so severe that many groups seem to go extinct, only to reappear, like Lazarus, in the Triassic, when appropriate environments are preserved to a greater extent. The end of the Cretaceous is marked by a widespread erosional gap that precludes detailed study of biotic declines, but also creates the illusion of a disappearance of entire biotas, where none may exist (Newell 1967; Kauffman 1984). Marine regressions reduce the amount of sediment deposited which concomitantly reduces our sampling power of fossil taxa. A gradual regression, therefore, could make the number of shallow-water species appear to diminish gradually, even if an extinction event were, in fact, sudden (Signor and Lipps 1982). An estimate of temporal stratigraphic completeness suggests the unlikelihood of distinguishing episodes of extinction lasting 100 y or less from those lasting as long as 100,000 y. Sudden catastrophes would therefore be impossible to document with our stratigraphic record (Dingus 1984).

The fossil record is biased toward readily preservable forms with hard mineralized skeletons. Most studies of fossil taxon richness therefore tend to include only those taxa that are liable to be consistently well preserved (e.g., Valentine 1969; Flessa and Imbrie 1973). The vast array of soft-bodied forms are usually absent, making any attempt to completely understand the record of groups such as annelids essentially hopeless. But what is the overall quality of the record of those forms that have skeletons, and are found fairly commonly? The very fame of *Archaeopteryx* is testimony to the absence of a record of the majority of early avian evolution. Many embarrassments of preservation exist, such as the very long gap between living monoplacophora and their most recent relatives, and the living occurrence of *Latimeria* despite an immense temporal gap between it and any close relatives. The "pull of the Recent," as Raup (1978) described it, enhances the chance that poorly preserved fossil taxa will appear in the, by definition, perfectly preserved and better sampled living biota, with no close fossil relatives.

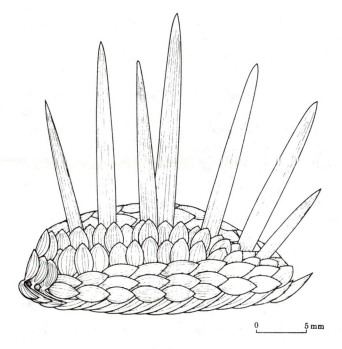

0   5 mm

Figure 8.1: Reconstruction of *Wiwaxia corrugata*, a member of a Cambrian invertebrate group with no clear membership in any extant phylum (from Conway Morris 1985a).

Combined with better preservation and comparatively easy identification of younger Cenozoic fossils, there may be a false sense of rapid increase of taxon richness as we approach the present.

Occasional sites of extraordinary preservation lend insight as to the degree to which the fossil record faithfully records even the skeletonized taxa. The Burgess Shale is a particularly extraordinary window on the Middle Cambrian. It was discovered quite accidentally in 1909 by Charles D. Walcott in the Canadian Rocky mountains near Field, British Columbia, Canada. A series of publications was followed by a later re-investigation led by Harry B. Whittington (see Conway Morris 1979). The fossils (Figure 8.1) are usually thin films on shale bedding surfaces, and mainly consist of arthropods and their allies including trilobites, the onychophoran-like *Aysheaia pedunculata*, and a host of arthropods with no apparent close living relatives. A suite of echinoderms, mollusks, brachiopods, polychaetes, and priapulids, among others, are also present. The whole assemblage was probably representative of one overall habitat type, though more recent studies suggest a series of distinct biotic reef-front assemblages (Collins et al. 1983). While the vagile trilobite epifauna is fairly similar to other less well preserved communities, the vagile infauna–dominated by burrowing priapulids, and the sessile epifauna–often dominated by probable sea pens–are both indicative of an overall preservational problem for ecologically significant groups in this time and environment. These groups are widespread in rocks of the immediate region and are not a peculiarity of one local site (Collins et al. 1983). The bias is therefore quite serious. If the unusual forms are subtracted from the overall fauna, however, the residual of typically fossilizable groups closely resembles Middle Cambrian shelf assemblages. The residual therefore at least represents a core of fossil material that is consistently represented through various states of preservation. Bits and pieces of the Burgess Shale fauna appear in the Wheeler and Spence Shales of Utah and even the Lower Cambrian Kinzers Formation of Pennsylvania. Both comfort and discomfort are therefore imparted by such unique windows into the past.

The Overall Pattern

Diversity

As mentioned above, the concordance of several independently compiled data sets at several taxonomic levels (Sepkoski et al. 1981) lends credence to a general pattern of changing taxon richness throughout the Phanerozoic. The pattern can be conveniently described from Sepkoski's (1984) recent analysis of the fossil record at the family level. The overall richness curve (Figure 8.2) is divided into three principal times of diversification: (1) Vendian-Early Cambrian interval, with a slowing of diversification in the Middle and Late Cambrian; (2) Ordovician radiation, followed by stabilization between the Late Ordovician and Early Permian; and (3) Mesozoic-Cenozoic expansion. These expansion periods are punctuated by a series of mass extinctions, which we shall discuss below. Sepkoski (1984) argues that the overall pattern is one of three sequential logistic curves; this presumes that the period of Cenozoic expansion is dampening.

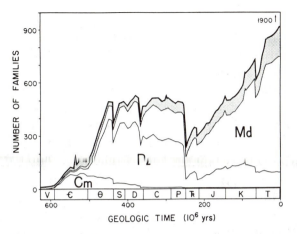

Figure 8.2: Family diversity through the Phanerozoic. Shaded area represents poorly preserved taxa. Three remaining fields comprise the relative contribution of the evolutionary faunas to the total diversity of well-skeletonized taxa (see text). The number 1900 represents the approximate number of extant families (from Sepkoski 1984).

Evolutionary Faunas

Both the expansions and extinctions are characterized by coordinated changes of unrelated taxa. This has long been recognized qualitatively, and was quantified by Cloud (1948) with an earlier marine invertebrate data set. He termed the expansions **evolutionary outbursts**. Bursts within the articulate brachiopods were correlated with periods of expansion of morphological diversity, many of whose taxa were short lived (Cooper and Williams 1952). Many newly appearing groups occur complete with their associated ecological community. As Camp (1952) notes:

> New elements usually appear in the record as migrants from an unknown geographic center. . . . When they appear they may already be accompanied by new complementary elements (i.e., herbivores plus highly developed carnivores); which makes it probable that we seldom see even the approximate time of original appearance of a group.

Darwin was right about gaps in the record.

The more recent statistical studies provide bases for more careful analysis as they correct for the different time intervals represented by geologic periods and epochs, and have the benefit of various taxonomic revisions and later discoveries. Most notable is Sepkoski's (1982) compendium of fossil marine families, which is being constantly updated.

Using a Q-mode factor analysis of occurrences of marine and terrestrial families through the Phanerozoic, the pioneering study of Flessa and Imbrie (1973) measured the coordination among family-level taxa. This analysis establishes a series of composites that behave statistically independently. Distinct associations dominated successively through geological time. Ten marine factors explained 96 percent of the fluctuations of diversity, while four terrestrial diversity associations explained 91 percent of the data. Taxonomic turnover (extinctions plus appearances) of the marine and terrestrial assemblages showed distinct peaks, as was found earlier by Newell (1952); neither the associations of taxa nor their temporal disposition could be reckoned with a random distribution. The temporal pattern of first appearances of families were also in most cases nonrandom (Flessa and Levinton 1975).

The establishment of statistically independent associations by factor analysis does not guarantee any biological significance. Further supportive evidence would be required to establish whether a given assemblage consisted of either an interacting and interdependent group of species or a group of species that responded similarly to the same environmental changes. At present we have no such evidence for interdependence.

The re-analysis by Sepkoski (1984) recognizes three distinctive major **evolutionary faunas** (EFs), which are not too different from those of Flessa and Imbrie. The **Cambrian EF** (Figure 8.2) is dominated by trilobites and inarticulate brachiopods, expands during the Cambrian, and reaches a plateau in the Middle Cambrian, followed by a long slow decline through the Paleozoic. The **Paleozoic EF**, dominated by articulate brachiopods and other classes, appears at the same time as the Cambrian EF, but radiates more slowly until the end of the Cambrian when a rapid diversification ensues in the Ordovician. After a plateau, a long decline occurs from the end of the Middle Devonian to the end of the Paleozoic, when the Permian mass extinction decimates it. Finally, the mollusk-dominated **Modern EF** appears in the Early Cambrian, then expands even more slowly than the Paleozoic EF, and increases as the Paleozoic EF is decreasing so that it is approximately 40 percent of the total diversity just prior to the Permian extinction.

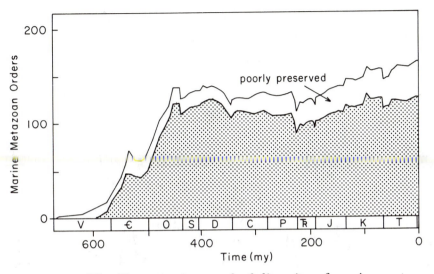

Figure 8.3: The Phanerozoic record of diversity of marine metazoan orders (after Sepkoski 1978).

Because the Modern EF suffered less in the Permian extinction, it came to dominate Triassic faunas and has been expanding ever since (Sepkoski 1984). An interesting detail is the higher expansion rate of the Paleozoic EF in the Triassic, after the Permian extinction, repeating its higher capacity for expansion than the Modern EF in the early Paleozoic. Its ultimate failure seems due more to its relative sensitivity to extinction, than to a failure to diversify.

As mentioned above, the objective of documenting changes in fossil species richness often depends upon the assumption of a linear relationship between number of higher taxa and number of component species. A comparison of temporal changes at the phylum, ordinal, and family levels (Figure 8.2, Figure 8.3, Figure 8.4) illustrates that something more than overall species richness is being documented. The number of phyla stabilized in the Ordovician and has not changed since. The number of orders stabilized by the mid-Paleozoic. In contrast, the number of families has had a far more volatile history, but overall trends are probably qualitatively concordant with the number of species.

This hierarchical response pattern, first noted by Valentine (1968), suggests that differing mechanisms may be operating at the various levels. If the phylum level represents fundamental morphological dif-

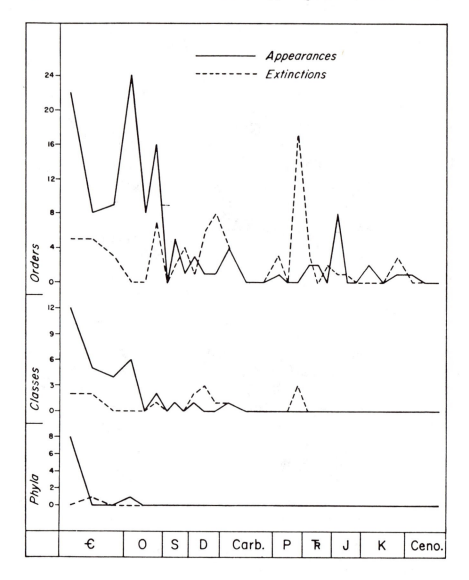

Figure 8.4: Appearances and extinctions of phyla, classes, and orders
of continental shelf benthos (from Valentine 1969).

ferences, most major divergence had ceased to occur by the beginning
of the Ordovician. The ordinal-level diversity ceases to expand soon
thereafter, but here there is extensive turnover. Stability in numbers of
order-level taxa means something different than at the phylum level.
Despite a major period of evolutionary expansion in the Mesozoic,
and the occurrence of major ecological changes in the Mesozoic marine
benthos (Thayer 1983; Vermeij 1983), no new phylum-level morphological innovations occurred, and the standing diversity of morphological
variation (ordinal level diversity) has not expanded since the Ordovician. The Silurian animal conquest of land must have been a major
event in physiological evolution, but no exclusively terrestrial animal
phyla are recognized. Ecological pressure may have induced adaptive
morphological change, which seems to have been limited after the Ordovician. This is perhaps the fossil record's most important message.
Apparently, evolution has had to work with that morphological diversity that existed by mid-Paleozoic. All subsequent evolution seems to
have been constrained by this initial pattern of dominance.

If we examine the Vendian and Early Cambrian periods of expansion, the impression of extensive morphological innovation is only increased. A widespread late Precambrian fauna, whose representatives
were first described extensively at Ediacara, Australia (see Glaessner
1984, and references therein) consist of a wide spectrum of organisms
with no obvious modern relatives. It may have been an early evolutionary expansion that failed. These were exclusively soft-bodied
forms, whose generally poor preservation precludes any certainty of
their extinction before the beginning of the Cambrian. Even in the
Cambrian, a wide variety of groups exist, with occasional obvious phylum affinities, but no clear association with living classes or orders.
Combined with the fact that all of the phyla appeared very early in
the Phanerozoic, one gets the impression that an initial period of "experimentation" with body plans was followed by extinction of many,
and the rise to dominance of the remainder.

Taxon Longevity and Lyellian Curves

As discussed in Chapter 7, taxonomic longevity is used commonly
by paleontologists to estimate the tempo of evolution within a clade.
Longevity estimates have also been used to contrast groups with differing ecology. The frequency spectrum of longevities at a given taxo-

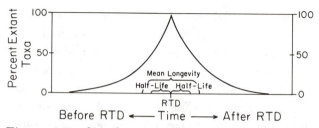

Figure 8.5: Simple model of loss of extant species, before or after a Reference Time Datum (**RTD**) in the fossil record. Assuming a constant rate of loss, half-life and longevity are indicated.

nomic level within a more inclusive taxon has also been used in various models of evolution (Van Valen 1973*b*; Levinton and Ginzburg 1984).

Longevity can be catalogued directly from the first and last occurrences of the group in question. Sporadic occurrence often makes it necessary to assume that these occurrences are not just a small part of the total duration. Estimates of longevity are biased by the minimum stratigraphic time resolution. In some cases, such as mollusks (Kelley 1983*a.b*; Miyazaki and Mickevich 1982; W.L. Fisher et al. 1964), a significant part of taxon longevity consists of ancestor-descendant sequences lacking cladogenesis; taxonomic longevity is therefore confuted with morphological change. Stanley (1979) has termed this type of longevity **pseudoextinction**.

The disappearance of given taxa with increasing time from either the present or a given stratigraphic level, the **Reference Time Datum (RTD)**, can estimate mean longevity (Figure 8.5). In the case of extant taxa, such an approach is essential as we do not have a complete record of the future longevity, and therefore have no extinction point. In the case of fossil taxa it is sometimes more reliable to examine the overall pattern of disappearance of a group of taxa from one **RTD** backwards, than to tabulate individual durations, which are subject to many sorts of collecting bias. The gradual disappearance of an entire fauna when sampled backward in time may give a smooth loss curve, as first discovered by Lyell in his investigation of mollusks. Stanley (1979) therefore coined the term Lyellian curves for plots of loss of taxa before the present. **True Lyellian curves** plot the percent of a fossil fauna that is still extant. Although there is increased error (as Stanley 1986*a* notes), it is also possible, however, to simply plot the fractional loss of the extant species as a function of time before the present, or any other **RTD**.

In true Lyellian curves, one must assume that the overall diversity has remained constant in order to estimate longevity. Lyell held to a concept of plenitude and believed that as one species became extinct, it was immediately replaced. If diversity rapidly increases towards the Recent, for example, the loss of living taxa as one samples back in time will give an underestimate of longevity, as a disproportionate number would have arisen recently. Similarly, if a recent extinction event had occurred, going back in time will yield a sudden abundance of those taxa that became extinct; this, too, will give an underestimate of true longevity. It is therefore better to tabulate the presence of species and plot a **cohort loss curve** of percent of taxa present at an **RTD**, with increasing time before or after the **RTD** (see Raup 1978). To obtain an estimate of mean longevity, some model of loss is assumed, such as a constancy analogous to radioactive decay. This analysis is to be distinguished from so-called **taxonomic survivorship curves**.

Assuming that the extinction for a given group occurs at a given rate, the rate of change in the number of taxa, N, disappearing with time before the **RTD** should be proportional to a rate constant, k (Kurtén 1959b).

$$dn/dt = -kN$$

The change in taxonomic richness with time before the **RTD** is

$$N_t = N_o e^{-kt}$$

The **half life** is therefore the time period over which N_o decreases to $N_o/N_t = 0.5$. Rearranging and taking the logarithms, where the number of taxa has decreased to $0.5N_o$, gives

$$T_{1/2} = 0.693/k$$

But k is proportional to the reciprocal of longevity, T. Therefore, longevity is proportional to $1.443T_{1/2}$. In order to get the true value of T, we must double this, as our derivation applies only to the portion of the life span of a taxon before the **RTD**, which must be symmetrical to the probability of longevity afterward. Therefore

$$T = 2.886T_{1/2}$$

This estimate differs from that of Stanley (1979) who thought T to be twice the half life. But this would only be true if the longevities and origins of taxa through time were precisely uniform, as his diagram of loss (p. 115) implies. Linear decay implies a random dispersion. Pseudoextinction will also bias the results as the durations of taxa will be artificially placed end to end. With pseudoextinction, a Lyellian analysis inflates the true value of T.

When working above the species level, complications arise as the taxonomic unit of survival consists of component species which have a given rate of appearance and loss (Raup 1978). Thus, the disappearance of a species may not result in the loss of the taxon at a higher level (e.g., family). Furthermore, a speciation event may result in either a member of the same higher taxon, or may result in a new taxon, especially given the use of phenetic criteria in defining the latter. Raup uses a random branching model to determine species longevity from generic longevity. If cohorts of genera are employed, a smoother loss can be seen in the readily preservable marine benthic groups, with increasing time from a series of **RTD**s, corresponding to the beginning of the various Phanerozoic periods. A mean species duration of 11.1 my arises from this analysis, which is somewhat greater than other estimates. The mean is far greater than some groups, most notably ammonoids (1.2 to 2 my), graptolites (1.9 my), and Cenozoic mammals (0.7 to 1.4 my).

Stanley (1975, 1979) has applied an ingenious technique to estimate speciation rate, and longevity. Assuming that speciation is constant, the number of extant species and the time of origin may be used to estimate a rate of splitting (he assumes dichotomous splitting). This approach can only work if the group is still actively undergoing cladogenesis, with a net constant rate of species increase. For bivalve mollusks, the number of species of extant families is linearly related to the logarithm of time of origin. Unfortunately, the technique has limited value, as it is questionable whether most groups have a constant splitting rate over periods long enough to get a reliable estimate. M.V. Wilson (1983) analyzed the Coleoptera, Diptera, and Hymenoptera and failed to find any such relationship. Nevertheless, the quite different longevities are similar to those reported by Stanley. It seems unlikely that any overall biases in the data will contribute to making average mammalian species longevity much greater than that of bivalve mollusks, for example.

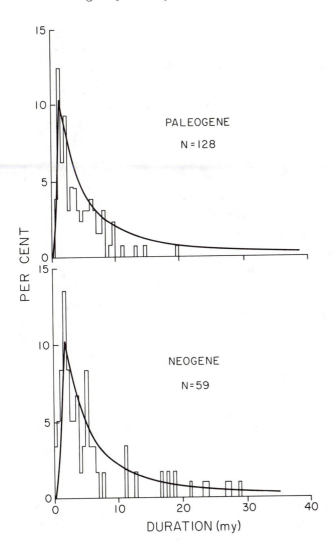

Figure 8.6: Distribution of longevities for Paleogene and Neogene species of Caribbean planktonic Foraminifera. Model of random appearance and extinction is shown for comparison (after Levinton and Ginzburg 1984).

To justify a natural order among different taxonomic groups, it would be necessary to demonstrate, within a taxon, a degree of homogeneity of species longevity over time, in comparison to overall differences with other groups. At present the available data are too scattered to give anything other than a vague impression. **Iterative evolution,** or repeated radiations following extinctions, is common enough in fossil groups to provide an opportunity to compare longevity. Cifelli (1969) described the two evolutionary radiations of planktonic Foraminifera during the Cenozoic. Globigerinid ancestors in the earliest Paleogene gave rise to a morphologically diverse clade that decreased in diversity toward the end of the Oligocene. Subsequently, another radiation in the Neogene reproduced nearly the same spectrum of morphologies. A compilation of Caribbean species longevities (Levinton and Ginzburg 1984) permits an evaluation of the quantitative aspects of the two radiations. Figure 8.6 shows the distribution of longevities for the Paleogene ($N = 128$) and Neogene ($N = 59$) radiations. They point to a remarkable similarity of modal longevity and distributional shape (Paleogene median longevity = 3.40 my; modal longevity = 1.5 my; Neogene median longevity = 3.62 my; modal longevity = 2 my for extinct taxa). Using a logarithmic correction to achieve normality, the mean of the two distributions does not differ significantly (t test, $p > .05$). A Kolmogorov-Smirnov test on the cumulative distribution, however, shows a significant difference ($p < .01$). The latter test is far more sensitive to differences of small magnitude, with large sample sizes.

Some studies of longevity suggest significant temporal variation. Stanley (1979) estimates mean bivalve species longevity to be 11 my. This is far greater than the estimate of about 1 my for Cenozoic mammals. But in some periods, mean species longevity of bivalves deviated significantly from the overall value. In the Cretaceous of the Western Interior of the United States, the mean species duration for the *Sciponoceras gracile* zone is 1.82 my, with a duration over several ecological groups of 1.82 to 2.20 my (Koch 1980). Kauffman (1978) has found similarly brief longevities in the Upper Cretaceous. Mammalian species longevity has also varied substantially (Kurtén 1959b). Paleocene (1.5 my) and Pleistocene (0.6 my) values are low, but Neogene values are 5.2 my, well within the range of bivalves.

Taxonomic Survivorship

The longevities of a group of extinct fossil taxa can be compiled into a **taxonomic survivorship curve**, which treats the group as if it were a cohort and records the cumulative loss as a function of increasing longevity. In a semilogarithmic plot the magnitude of the slope gives the mean longevity and the departure from linearity is a measure of the probability of taxon survival with increasing age. Thus a linear slope indicates that the probability of survival is independent of age, and mean age corresponds to probability of extinction for the species. Linearity would imply that some process sets the mean extinction, but variation around the mean is a random process. Van Valen (1973b) formulated a unit, the macarthur (ma), which corresponds to a taxon survival rate giving a half life of 500 years. Note that this survivorship curve and half life are different from Lyellian and cohort survivorship curves, which truly measure survivorship from a given time. As in the case of Lyellian curves, the slope yields a measure of mean longevity, so differences in slope among major taxa might indicate differences in overall biology (Levinton 1974; McCune 1982).

Some biases occur in taxonomic survivorship curves but they have not been analyzed in most studies. At higher taxonomic levels, complications arise as species are both appearing and disappearing within the same taxon. Survivorship curves constructed from such composites should be concave up, rather than linear, if the probability of species-level extinction is constant and independent of taxon age (Holman 1983; see also Raup 1978). The slope of family-level survivorship should be shallower than or equal to that for genera. The number of species per genus is a factor in survivorship at the supraspecific level, rather than the variation in longevity of component species (Arnold 1982). The survival of a genus may simply be proportional to the number of component species.

A bias also stems from the minimum resolvable time units that are employed (usually geological stages) to measure longevity. If a taxon is found in one stage, its longevity is by definition a function of the time represented by that stage. Because stage time lengths are roughly log normally distributed, a mildly concave curve can be made to look linear from the time bias. A taxon straddling two unusually long stages would appear to have a greater longevity than another straddling two short stages (Sepkoski 1975). The average length of stages differs throughout geological time (Flessa and Levinton 1975). This effect can make taxa confined to one set of stages appear to differ in longevity from another set. Also, the minimum length of time resolution must also produce a flat top near the beginning of the survivorship curve (Raup 1975).

Van Valen (1973*b*) used a large compilation of different groups, at various taxonomic levels, to show that taxonomic survivorship curves appear to be linear. A statistical test for linear decay showed that the survivorship of the ammonoids is linear, if one subtracts the initial flat top due to the bias of minimum time resolution (Raup 1975). But many of the other cases are ambiguous. Levinton (1974) examined survivorship curves for suspension-feeding and deposit-feeding bivalve molluscan genera and found a strong linearity, except for small sets of genera that were geographically widespread and tended to exist over the same long spans of time. These superposed a ledge on the otherwise linear curve. Suspension feeders had a lower rate of survival, which might be related to their relatively volatile response to changes in food in the water column. Arnold (1982) examined species-level survivorship in the Cenozoic globigerinid Foraminifera and also found linearity.

Survivorship curves sum up the longevity spectrum of a group over its entire duration, which may obfuscate a temporal pattern of changes in taxon longevity. For example, taxon longevity might be low, and then high. When summed into a survivorship curve, this might yield a cumulative distribution producing a straight line survivorship. In the development of a Cambrian biomere–a probable radiation initiated by an onshore inmigration of taxa–taxon longevity is at first quite short, then longer, then short again (Stitt 1971; Hardy 1985). The summed longevity curve would thus be misleading since it might give the false impression of some probability distribution of mortality as a function of age that is independent of time.

Stochastic Models of Appearance and Taxon Longevity

Are Appearance and Extinction a Matter of Chance?

The temporal trends and stable phases of taxon richness occur with extensive turnover at the family level. Most paleontologists have reckoned the extensive changes in the fossil record as products of forces in the environment (e.g., Newell 1952; Valentine 1971*a, b*; Flessa and

Imbrie 1973). Raup et al. (1973), however, raised the intriguing possibility that the overall temporal pattern might be explained by stochastic appearance and extinction. They performed a series of simulations assuming equal probabilities of appearance and extinction, with an upper limit on species richness. The results looked superficially like the many "spindle" diagrams of temporal changes in diversity, so common in paleontological publications.

Gould et al. (1977) used the simulations to measure shapes of clades and defined the temporal position of half of the area of a spindle diagram as **Center of Gravity**. Simulations based upon a stochastic model produced a mean center of gravity of about 0.5. This was similar to their and Sloss' (1950) analyses of many real clades, with the exception of those in the middle of a major radiation, which are bottom heavy (< 0.5). Stanley et al. (1981) argued that any stochastic branching process with equal probabilities of appearance and extinction would not permit large expansions with any acceptable level of probability. Sloss (1950) also suggested that clades probably were initiated by some process that accelerated divergence.

Though the shape of clades may reflect a random process, the timing of appearance and extinction is usually far from random. The very presence of evolutionary faunas deducible by factor analysis (Sepkoski 1984) suggests that coordinated evolution occurs among unrelated groups. Flessa and Levinton (1975) analyzed first appearance in a wide range of fossil groups and found that, unlike the simulations of Raup et al. (1973), real appearances were aggregated in time. Gilinsky and Bambach (1986) reached similar conclusions when considering diversity changes during the history of a clade. It is therefore untenable to consider the entire historical picture as reflective of a stochastic process; but this does not preclude such an explanation for parts of the overall pattern.

Can Longevity Be Explained by a Random Model?

Van Valen (1973b) suggested that challenges sufficient to drive a species to extinction might appear continuously. He used the Red Queen's dictum–you have to keep on moving to stay in the same place–as a metaphor to describe the continuous set of biological challenges presented to a species. It is also possible that species longevity is completely random, with mean longevity controlled by some characteris-

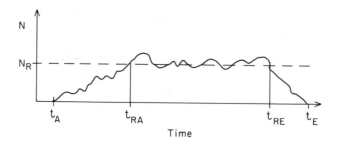

Figure 8.7: Stochastic model of the appearance and extinction of a fossil taxon, relative to a recognition threshold (see text for explanation) (after Levinton and Ginzburg 1984).

tics of the biology of the taxon independent of other species. Ginzburg (Levinton and Ginzburg 1984) devised a stochastic model that predicts the distribution of taxonomic longevities by estimating the time over which a population with a zero average growth trend fluctuates before extinction.

Consider a population growing exponentially with a stochastically varying growth rate:

$$dN/dt = [r + \sigma \epsilon t]N$$

where N is the population size, t is time, r is rate, σ is the standard deviation of the growth rate, and ϵt is the standard normal white noise process [mean $(\epsilon) = 0$, variance $(\epsilon) = 1$]. For the initial population size N_o and the critical level N_c ¡ N_o the probability density of the first passage of the process $N(t)$ through the critical level N_c shows that an increase in the variance of the population growth rate decreases the probability of surpassing the threshold for persistence, while increased r raises the probability. Therefore, taxa with large variance relative to mean growth rate will most likely go extinct rapidly, while those taxa with high r or large population size, relative to the threshold, will tend to persist.

Over the long run, it is reasonable to assume that $r = 0$, that is, no long-term positive or negative growth trend owing to deterministic forces. One then can compute the passage time from birth to extinction. The typical appearance–extinction process with $r = 0$ should look like the diagram illustrated in Figure 8.7. Here N_R is the threshold of abundance above which one recognizes the taxon in the fossil record, t_A is the time of first appearance, t_{RA} is the time of recognized appearance, t_{RE} is the time of the recognized extinction, and t_E is the actual time of extinction. The observed longevity, which is the total duration over which it is possible to collect the taxon as fossils, $(t_{RE} - t_{RA})$, is the time between the first and last passage of the stochastic trajectory through the recognition level. Period $(t_E - t_{RE})$ is much shorter than period $(t_{RE} - t_{RA})$

The first passage time model can be fitted to actual frequency distributions by taking the modal longevity, t_m, from the data, and calculating the probability density of extinction (cf. Levinton and Ginzburg 1984). Figure 8.6 shows that the overall fit is good in both Paleogene and Neogene radiations, though there is a significant excess predicted for the right tail of the distribution and a deficiency at the modal longevity relative to the actual data. Using a one-sample G-test the difference with the model is significant (first radiation: $G = 102.4$, 39 $d.f.$, $p < .001$; second radiation: $G = 56.7$, 39 $d.f.$, $p < .001$).

The reason for a significant difference between the model's prediction and the actual data may be due to at least three possible sources of empirical and theoretical error. First, some long-lived taxa may be artificially absent. If taxa are defined by pseudoextinction, then a long-lived single taxon might appear in the data as a chain of morphologically transitional shorter-lived taxa. Phyletic evolution from one morphologically defined species to another is common in planktonic Foraminifera (Kennett and Srinivasan 1983). Second, r, the intrinsic rate of population increase, is assumed to be zero. If r does not equal zero, then the variance of r would be an important component of stochastic extinction, and, hence, taxonomic longevity. If r does not equal zero, we can improve the model's fit to the data. As we have no basis to select the value of parameters, such an attempt at this stage would constitute mere curve fitting. Finally, of course, the model may be incorrect. The relatively kurtotic aspect of the real data, combined with a deficiency of long-lived forms, may suggest that some nonrandom process narrows the longevity spectrum.

Steady-State Levels of Taxon Richness

The Pattern

Sepkoski's (1984) analysis of family level taxon richness demonstrates some apparent long-term periods of relative constancy. In the mam-

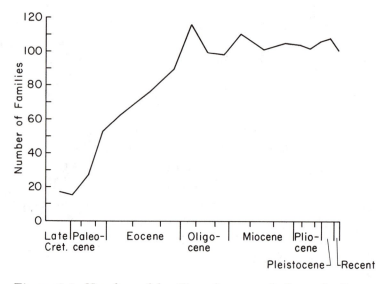

Figure 8.8: Number of families of mammals from the Late Cretaceous onwards (after Lillegraven 1972).

mals, after a Paleocene–Eocene radiation and an Oligocene decline, long-term stability in the number of families prevailed (Figure 8.8), despite extensive turnover (Lillegraven 1972). Such stability can also be detected at the species level within communities. Boucot (1975, pp. 226–237) documented a similarity in species richness over a 100 my span in Silurian-Devonian brachiopod communities despite extensive phyletic evolution and cladogenesis. Many other such examples have been documented (Boucot 1978). The fossilizable component of very nearshore marine benthic communities had essentially the same species richness throughout most of Phanerozoic time, as opposed to the changing offshore environment (Bambach 1977). Wei and Kennett (1983) recorded a period of stability of 10 my for Neogene planktonic Foraminifera, following an early diversification. Of particular interest is a detailed study of a deepening Late Ordovician nearshore assemblage, where a maximum species richness was achieved rapidly and maintained for 3.5 my, despite about 15 percent turnover per my (Bretsky and Bretsky 1976; Rosenzweig and Duek 1979). Over the period of stability, half the steady-state diversity turned over. Bretsky and Bretsky found a steep relationship between standing diversity of benthic invertebrate species and number going extinct from one stratigraphic level to the next, but only a shallow relationship of appearances to standing diversity. This suggests that extinction is the major factor in maintaining the steady state. During the colonization period, species persistence within the area was less than after the steady state was established; this may suggest something about the instability of the early habitat as it was colonized, or perhaps something about the competitive success of the later species (Rosenzweig and Taylor 1980). Persistence is not necessarily equal to longevity as there is no evidence that many of the species were endemic. Hoffman (1985) has analyzed global patterns of diversity and finds no general correlation between taxon richness and extinction rate. But the smaller-scale studies commonly show temporal constancy in taxon richness. There must have been diversity-dependent effects, as a stochastic process would lead to drift from constant taxon richness levels.

Sepkoski (1978) employed a logistic model to characterize the temporal increase in taxon richness toward a steady state. Sepkoski (1984) uses a "three phase kinetic model," employing different divergence rates and limiting values for steady-state diversity for the three evolutionary faunas. Coupled with corrections for mass extinctions, a rather good fit to the Phanerozoic temporal pattern was achieved. The Mesozoic and Cenozoic expansion only shows a hint of leveling toward a plateau. Sepkoski claims that the trend may be leveling off, however, and predicts a steady state some tens of millions of years in the future–a fairly safe prediction!

Using a difference equation approach, Kitchell and Carr (1985) questioned several of Sepkoski's conclusions. They employed a model that incorporates the negative feedback of high diversity, but uses discrete intervals, as opposed to the continuous feedback that would be present in the differential logistic equation. They find that the apparent plateaus are due more to the effects of mass extinctions, which seem to forestall the ascendancy of one EF over another. Even the large-scale Permian extinction seems only to delay the ascendancy of the Modern EF. This is an important point, as some others have suggested that the Permian extinction "reset" the biosphere, and a new evolutionary era commenced (e.g., Van Valen 1984a). Their results also suggest that an equilibrium between diversification and extinction was never reached, even if those balancing forces were continuously at work. Finally, their study demonstrates that the three EFs are interactive.

A model assuming independent diversification and extinction among the EFs cannot account for the successive rises and falls of the three EFs. Kitchell and Carr (1985) confirm Sepkoski's conclusion that the Cambrian EF has a much greater expansion rate than the other two. The lower expansion rate of the Modern EF and a less dramatic negative feedback may explain its slow ascendancy and lower family-level extinction rate.

Two alternative models may explain the steady state: (1) **Competitive Niche Subdivision Model** (Rosenzweig 1975; Rosenzweig and Taylor 1980), and (2) **Stochastic Area Effect Model** (Levinton 1979).

The Competitive Niche Subdivision Model presumes a frequency dependence between the rate of speciation and the presence of species in an ecologically saturated and environmentally complex habitat. If one species becomes extinct, then another can be expected to fill its place. The structure of the habitat therefore determines the maximum number of species that can coexist, while the number of species present is inversely proportional to the speciation rate. Rosenzweig (1978) has suggested a model of competitive speciation that is compatible with this scenario. Possible supportive evidence for this model is the convergent structure of geographically separated but ecologically similar living communities (see Cody 1974; Fuentes 1976; Crowder 1980) and the long-term persistence of similar niche structures in some fossil communities (Walker and Laporte 1970; Levinton and Bambach 1975; Boucot 1978). One problem is that most such studies report the presence of species adapted to similar regimes (e.g., a burrower, a tube dweller, etc.) in two or more communities widely separated in time. This does not prove that the community was saturated, only that convergence has occurred with adaptation of species to similar habitats.

The stochastic area effect model (Levinton 1979) presumes that the overall environment is limiting with respect to resources, and that the number of species, therefore, must be in some way related to the number of individuals or the area covered per species. Second, speciation rate is presumed to be a stochastic event that occurs at a certain constant rate, depending upon a group's particular biology (Stanley 1975, but see problems above). Finally, extinction is both a function of population size and area covered. Species with small population sizes and narrow geographic range therefore have a higher probability of extinction. Consider a species newly introduced into this region. If it succeeds in colonizing the whole region, up to carrying capacity, then its chance for extinction, owing to area-restricted catastrophes, would be quite low. Geographic speciation would occur, the number of species increases and the average geographic range occupied would therefore decrease. Eventually the average geographic range would decrease to the point of equilibrium, where extinction by regional catastrophes would balance the introduction of new species. A similar argument applies to the temporal change in population size. Although this is an equilibrium process, rapid shifts in the frequency of local catastrophes, or the speciation rate, would produce concomitant changes in the predicted equilibrium species number.

The stochastic area effect model can explain regional differences in steady-state diversity as follows. The size and nature of local habitat disruptions are larger in some locales. In such cases, the average area occupied by a species must be larger or it will soon be extinct. Differences among habitats in steady state diversity would therefore be due to differences in extinction. Järvinen (1979) found that eastern European bird species turnover increased with increasing latitude. This is presumably related to the higher frequency of severe climatic fluctuations sufficient to cause local population extinction. This is compatible with Bretsky and Bretsky's (1976) result that the maintenance of the Ordovician steady state was mainly due to extinction. It is also compatible with conclusions from analyses of fossil communities (e.g., Bretsky 1969; Bambach 1977) that nearshore habitats, where extinction rates are likely to be great due to sea level fluctuations (see Schopf 1974), have been consistently of lower species richness than offshore habitats. The stochastic area effect model also provides an explanation for the relationship between continental or marine shelf area and number of species (Schopf 1974). Larger areas permit the accumulation of larger numbers of species over time. Assuming that the areal scale of catastrophes is the same in large and small marine basins, the model would predict a greater buildup of taxon richness in large basins than in small basins. Other factors, such as temperature and salinity fluctuations, are liable to be more extreme in shallower waters and in smaller basins.

Whereas high-latitude birds show greater population overturn, it has been generally believed that offshore marine communities are more prone to extinction (e.g., Bretsky 1969), perhaps owing to the sensitivity of physiologically narrow forms. Sepkoski (1987) has recently demonstrated that the high offshore extinction rates are largely explained by the predominance of groups with high characteristic extinction rates (e.g. articulate brachiopods). In contrast, some groups (e.g., bivalve mollusks) with low characteristic extinction rates are concentrated in nearshore habitats. Within such groups, generic extinction rates are actually higher onshore, as might be predicted from the expectation that nearshore habitats are disturbed more frequently.

A notion of steady state can be extended to a consideration of morphological specialization. Specialized species would be expected to occupy smaller areas and have smaller population sizes than generalized species. This would be due to both restricted geographic range and restriction to a narrow spectrum of resources within a community. Their capacity to resist catastrophe is therefore inherently lower than more generalized forms. Specialization would therefore be expected to increase up to the point that the average restriction by specialization to a given population size and area would match the background rate of environmental catastrophes. This would provide a component of the explanation for the commonly postulated inverse association between specialization and species richness.

The two processes of diversity regulation connoted by the models are of course not necessarily independent. Both may contribute to an explanation of overall taxonomic richness. The competitive speciation model cannot be the complete answer as we know of environments, such as soft sediments, whose structural similarity and homogeneity are accompanied by extensive geographic variation in species richness (summary in Levinton 1982a). The bivalve mollusk genus *Macoma*, for example, has 13 species in Pacific Coast soft-sediment environments. Although some specialization is apparent, extensive coexistence occurs among intertidal and subtidal species (Dunnill and Ellis 1969). On the east coast of the United States only two species can be found in the same environment types, with no conspicuous increase of some other ecologically equivalent biotic component. Of course, structural habitat complexity is also a major component in explaining species richness. Within the same overall region, structurally diverse habitats, such as coral reefs and temperate rocky reefs, harbor far more species than nearby monotonous habitats. Niche subdivision resulting from interspecific competition can be seen in structurally complex habitats (e.g., the lizard genus *Anolis* in erect vegetation, E.E. Williams 1972); the carnivorous gastropod genus *Conus* in subtidal coral reefs (Kohn 1959), and even in some structurally simple habitats (e.g., mollusks in soft sediments, Fenchel 1975; Levinton 1977; Peterson 1977). On the other hand coexistence with no apparent niche subdivision has been documented in structurally complex environments (e.g., intertidal coral reef benches, Levitan and Kohn 1980). The apparent pervasive occurrence of interspecific competition in nature (e.g., Schoener 1983) does not guarantee persistent niche subdivision, but may simply involve continuous competition for the same sites with priority effects determining dominance, as required by the stochastic area-population effect model.

Biogeography, Provinciality, Diversity, and Diversification

Dispersal, Speciation, and Morphological Evolution

Dispersal and propensity to speciate are inversely related. Thus chromosomally polytypic species are those with restricted dispersal ability (Chapter 4). As might be expected, biogeographic range is also related inversely to species richness. Jackson (1974) found that very shallow water species had far larger biogeographic range than deeper water species (excluding the deep sea); the former habitat is much lower in species richness. Dispersal potential seems to be greater in nearshore habitats, an expectable adaptation to higher local population extinction rates (Jablonski and Valentine 1981). The evolution of provinciality may partially result from reduced dispersal, but the increase of regional habitat heterogeneity may also be a factor.

In the living marine invertebrates, dispersal ability can be related to speciation to a degree. Scheltema (1971) discovered the presence of long ranging (teleplanic) larvae in the surface waters of the middle and north Atlantic Ocean. Those species with commonly teleplanic larvae were found in both South America and Africa. Sanders and Hessler (1969) found a positive correlation in the deep-sea invertebrates between dispersal ability and species geographic range. A relationship between dispersal ability and regional genetic differentiation can also be seen (Snyder and Gooch 1973), but the magnitude of the latitudinal temperature gradient may play at least as great a role in differentiation

(Levinton and Suchanek 1978; Levinton and Monahan 1983). Vermeij (1982), however, showed that predation on a gastropod with limited dispersal induced significant morphological changes but failed to do so in another species with greater dispersal, presumably from habitats lacking the predator. Dispersal from adjacent habitats may increase colonization but it also reduces the chance for local adaptation.

Unfortunately, our understanding of geographic range changes in the marine environment is generally limited to invasions that are probably facilitated by marine commercial transport. Some species with planktonic larvae have moved quite rapidly along coastlines. The common periwinkle, *Littorina littorea*, probably extended its range from Nova Scotia to the mid-Atlantic within a hundred years (Brenchley and Carlton 1983). The spread occurred at a rate of about 15 to 20 km per year, which is consistent with southward larval dispersal via stepping stones. The approximate spreading rate of 10 to 20 coastal km per year suggests a limitation of spread by planktonic larval dispersal. Following latitudinal climatic shift in which a warm-water Norwegian fauna moved 1500 km northward and back, the rate of southward movement of warm-water species was also limited by the rate of dispersal, and not by the rate of climatic change (Spjeldnaes 1964).

A consistent relationship among dispersal ability, speciation, and geological duration has been found in fossil studies, principally of mollusks. The marine snail genus *Turritella* in Neogene Atlantic coastal plain sediments consists of two lineages of differing dispersal ability, speciation rate, and morphological differentiation. Dispersal ability can be judged by the size and ornament of the protoconch. The *variabilis* lineage shows evidence of strong dispersal, little cladogenesis, long species stratigraphic durations, and broader phenotypic variability within species. In contrast, the *alticostata* lineage shows evidence of reduced dispersal, extensive cladogenesis, short stratigraphic durations for species, and narrower phenotypic variability within a species (Spiller 1977). Hansen (1980, 1982) found a similar relationship among volutid gastropods. Presumed planktonic forms had greater longevities and could survive several stages of transgression and regression. In contrast nonplanktonic forms had shorter durations. Fasciolarids and mitrids had planktonic larvae, though they had short stratigraphic durations; this may indicate their relative ecological restriction. Apparently the relationship can be seen only within groups of ecologically similar taxa. Kauffman (1978), for example, found that oysters had shorter species durations than nuculid bivalves, even though the latter usually have restricted dispersal via a lecithotrophic larva (e.g., Scheltema 1971). Ecological difference may therefore swamp out the mode of dispersal in affecting longevity and speciation.

The limited evidence available from the fossil record is concordant with both theoretical expectations and work with living species. Species with short dispersal distances are more prone to become polytypic, perhaps leading to allopatric speciation. A clade with species having reduced dispersal is therefore liable to have a greater speciation rate. This creates more opportunities for divergent evolution. The reduced survivorship of species with reduced dispersal is not as easy to explain. If dispersal is reduced, gene flow from other populations will not swamp out local divergent evolution. Thus phyletic evolution is to be expected to a greater degree; some of the short taxonomic durations may reflect the successional species-morphological change bias discussed in Chapter 7. It is also possible that species produced under these circumstances are more specialized and are therefore more prone to extinction after an environmental perturbation. Finally, the reduced area occupied would enhance the probability that a given regional change would extinguish the species, relative to broader ranging species.

Because the available paleontological data follow so well a model of speciation involving polytypic divergence, followed by the evolution of incompatibility in isolation, it is incorrect to state that speciation is the cause of morphological evolution (Stanley 1979). Speciation is the result of divergent evolution, probably stemming from regionally variable natural selection, but enhanced by reduced dispersal. This raises the question of whether or not dispersal itself is under the control of natural selective forces that would help determine the degree of isolation. In the case of teleplanic (long-dispersing) larvae, successful dispersal is simply an accident since only a trivial proportion of the larvae produced in the coastal populations is liable to find its way across the open ocean. But in most cases, dispersal may be regulated to maximize success. Thus adaptive changes within populations may be easily conflated with long-term changes that might be thought to be caused by high speciation rates. The temporal increase in Tertiary snails with reduced dispersal is a case in point. As Hansen (1982, p.

372) notes: "The high percentages of low-dispersal forms among the warm-water neogastropods in apparent contrast to bivalves and most other prosobranchs is suggestive of some type of adaptive process." Thus individual adaptation could lead to reduced dispersal, which, in turn, leads to speciation. An abundance of species with low dispersal would therefore be the result of individual adaptation and not a process reducible only to speciation.

If the local population is stable and adult mortality is low, life history theory (Stearns 1976) would predict the evolution of reduced reproductive output and selection for increased investment in survival of the adult. Investment in the adult would involve ecological specializations whose rate of acquisition would be enhanced by a decrease in reproductive investment. In turn, reduced production of young would be followed by reduced dispersal, assuming the adults are fairly sessile. As the distribution of dispersing larvae attests, the nearshore environment is more dangerous than offshore sites; dispersal potential in nearshore species is far greater (Jackson 1974; Jablonski and Valentine 1981). In contrast, opportunities for the evolution of reduced dispersal often close the door to successful long-range colonization of new habitats. Species evolving both reduced reproductive investment and increased adult specialization are those whose colonists will probably not be able to find suitable habitats elsewhere or will be unable to successfully move along potential dispersal routes. Mountain passes, for example, present more formidable obstacles to (thermally) narrowly adapted tropical species, relative to the effect of passes on higher latitude forms (Janzen 1967). The Amazon River is a major biogeographic barrier, whereas a river of similar breadth in mid-latitudes such as the Mississippi River or Hudson River is no real barrier to dispersal of similar groups. The evolution of ecological specialization of the adult and production/dispersal of young are therefore interactive and must coevolve. As a result, speciation is so intimately related to local adaptation that it often becomes fruitless to argue whether or not speciation is the cause or effect of morphological change.

Do clades with higher speciation rates, or a positive speciation–extinction balance, survive by virtue of their very speciation rate? This is an argument of the **Effect Hypothesis**, proposed by Vrba (1983). For mollusks, the hypothesis doesn't fit the pattern presented above. Clades with widespread species tend to speciate at a low rate, and the geological durations of the component species are usually large. It has been generally found that fossil species with broad geographic range are quite long ranging and are far less prone to extinction than taxa with narrow geographic range (Bretsky 1973; Boucot 1975). Alternatively, clades with species of geographically narrow range tend to speciate at a high rate and species have shorter geologic durations. An inverse relationship between species longevity and speciation rate may thus result in no net difference in overall longevity between clades that speciate at very different tempos. In many cases, such speciation in fast-speciating clades results in a set of species that are morphologically similar, with no major innovations. This often occurs concomitantly with an increase of provinciality in the marine environment, such as the steady increase in provinciality during the diversification of the Silurian fauna into a highly provincial Early Devonian fauna. Such divergence has been termed **diacladogenesis** (Boucot 1978, p. 569). Increased survival of a high-speciating clade might occur if species are produced with innovations that propel it into a new adaptive zone (**metacladogenesis** of Boucot 1978), which greatly improves the chances of success. But, taken alone, this would not satisfy the Effect Hypothesis as an additional component of success from adaptation must be invoked. The increased survival of larger inclusive taxa, such as families, could be the result of increased numbers of species per family as diacladogenesis occurs (Flessa and Jablonski 1985). The data provided by Stanley (1979) on the bivalve mollusks suggests that speciation rate alone does not confer success. All of the bivalve families he considers define a linear relationship between the logarithm of time of origin and present day species richness. It follows from this that newer bivalve families do not have increased speciation–extinction differentials relative to older ones. In other words, there has been no net increase over time of clades with higher speciation–extinction differentials. Hoffman and Ghiold's (1985) analysis of family-level data shows an irregular decline of family origination rate since the Mesozoic.

Many biostratigraphers have investigated the question of speciation and biogeographic range in the context of fluctuations in sea level. Kauffman's (1973) extensive study of Cretaceous bivalve genera suggests that speciation and endemism were both maximized during periods of transgression. He related this pattern to the spreading of colonizing populations over many environments and separated areas as sea level rose. This result is generally concordant with Jablonski's (1979, 1980) study of Late Cretaceous Gulf coastal plain sediments, where maximum endemism was found during transgressive (sea-level rise) phases. On a larger scale, the entire Cambrian period of diversity expansion was a period of overall and continuous transgression (e.g., Boucot 1975). In contrast to these results, endemism is greater in Jurassic bivalves when inundation is at its lowest (Hallam 1977a). During regressive (sea-level fall) phases, maximum separation of marine basins might encourage divergent evolution and speciation. Jablonski (1980) argues that Hallam's result should not be surprising given the Jurassic situation of relative continental assembly. Under such circumstances, there would be few opportunities for isolation without regression. In the Cretaceous the continents may have been sufficiently separated that inundations brought colonists onto a series of isolated continental shelves. Regressive sequences are better preserved in the fossil record than transgressive ones. This would be particularly true of short-term events. As species with broader dispersal potential are to be found near shore, regressive sequences may have broader ranging species simply because of the relative frequency of preserved nearshore habitats. There is thus a possible bias against sampling endemic offshore species during regressions.

A Role for Population Size?

The probability of extinction should increase as population size decreases. Thus broad geographic range could reduce extinction either by virtue of large population size or owing to an extensive geographic range that is resistant to widespread habitat extinction. Population size per se, however, may depress extinction, independent of geographic range. Stochastic fluctuations in population size would more frequently cause extinction in species with small populations. Deterministic factors such as disease, changes in the productivity regime, predation, and disturbance would also fall more heavily on species with small population size. This was a major theme in Boucot's (1978) monograph on evolutionary stability. His analysis of fossil abundance suggested that common species were unusually longevous. There is often a correlation between abundance and geographic range; this makes it difficult to distinguish between the two factors in explaining extinction.

Among Pacific Pleistocene bivalve mollusks, small siphonate bivalves are more endemic than large bivalves, yet have significantly better survival (Stanley 1986a). This may be related to the presumed large population sizes of species with small body size, although Stanley presents no direct evidence to support the correlation in the groups he studied. In contrast, Pectinidae, whose geographic ranges are usually large, have high rates of extinction. Stanley attributes this to their vulnerability to predation. He also reinterprets the lower extinction rate of deposit-feeding nuculid bivalves (Levinton 1974) as the result of large population size, as opposed to trophic group. It is premature to single out predation on scallops as the only factor. In recent years, it has become apparent to marine ecologists that disease and changes in phytoplankton dominance (e.g., red tides and brown tides) are major sources of population decline. It is also unlikely that Stanley's reasonable conclusions about population size can be extended. The Pleistocene record for Pacific bivalves is marked only by relatively modest turnover. Many times in the fossil record are also marked by regional extinctions and strong changes in provinciality. This suggests that geographic range is very much an important factor in extinction.

Area, Provinces, and Diversity

During the Phanerozoic, sea-level coverage and exposure of terrestrial environments varied extensively. Paleontologists have recognized these patterns by facies maps indicating distributions of environments. Thus the marine regression in the Middle-Upper Devonian in New York State is recorded well by a sequence of sediments from fully marine to terrestrial–freshwater in origin. Changing geographic relationships of the fauna and flora have also been traditionally used to understand paleobiogeographic relationships.

The confirmation of the processes of sea-floor spreading and continental drift provide immediate explanations for many biogeographic problems. The pervasive cosmopolitanism of early and middle Mesozoic faunas can be related to a period of continental assembly. The gradual increase in provinciality from the Jurassic to the Cretaceous (Stevens 1973; Kauffman 1973) could be related to the breakup of the supercontinent Pangaea and the establishment of north-south trending continental shelves spanning large ranges of latitude (Valentine 1971a). The large present-day latitudinal variation in climate is unusual in Phanerozoic history and may date back to the major cooling event toward the end of the Eocene (Wolfe and Hopkins 1967; Shackleton and Kennett 1975; Valentine et al. 1978). Presumably, as the evidence accumulates, paleogeographic maps (e.g., Smith and Briden 1977; Scotese et al. 1979) will permit an association between continental dispersion and biogeographic subdivision. The current evidence suggests that the spatial arrangement of the continents has been continually changing, and movements have significantly affected the climatic history of individual continental blocks (discussion in Bambach et al. 1980).

Some of the Phanerozoic changes in taxon richness can be related to a large degree to changes in the degree of **provinciality**. If we take the species area effect (MacArthur and Wilson 1967) as given and constant, it is easy to calculate that, as one large province is divided in two, reductions of area are more than compensated by the increased total species richness as long as dispersal between areas is limited. The Silurian period, for example, was one of extreme cosmopolitanism, with one province of approximately 90 articulate brachiopod genera in the North Silurian Realm. In the Ludlow (Upper Silurian) two provinces can be delineated, with about 90 genera in each province. In the Devonian, the number of provinces increased to six; the total numbers of articulate brachiopod genera increased to about 350 on average (Boucot 1978, p. 577). During the Frasnian (Middle Devonian) this provinciality decreased relatively suddenly, and generic richness returned to 93.

The onset of the Permian extinction was also marked by a decrease in numbers of provinces, and the Early Triassic marks a nadir of provinciality in the Phanerozoic (Valentine et al. 1978). During the end of the Paleozoic, geographically restricted bivalve genera succumbed before more widespread genera, suggesting that the overall environmental change was filtering out those forms that define provinciality in the first place. Newell (1952) argued that the extinction was related to the major fall in sea level. Shallow marine seas were reduced from a coverage of 40 percent of their possible extent in the Early Permian to less than 15 percent in the latest Permian and then expanded to 34 percent in the Early Triassic (Schopf 1974). Reduced rates of sea-floor spreading may have been responsible for a lowering of ridge activity, depression of deep-sea bottoms and the consequential large-scale marine regression (Valentine and Moores 1971). The significant reduction in area, coupled with continental assembly of Pangea at the end of the Permian may have increased extinction rates, and would have homogenized the fauna due to the possible presence of more intershelf dispersal possibilities. In contrast, the Pleistocene reduction of area covered by the sea was far lower and on the basis of area alone the modest marine extinctions are therefore not surprising from this point of view (Wise and Schopf 1981). Area reduction itself might not be a potent agent of extinction (Jablonski 1985). Sea-level drops would hardly affect the shallow water habitat distribution of oceanic islands, where most modern families are widely distributed. Sea-level drop may just be a correlate of another change.

The changing spatial relationships generated by continental drift and sea level fluctuations must have had important influences upon climate. Valentine's (1971a) theory of climate change generated by continental assembly and fragmentation attempted to relate climate and sea level to sea-floor spreading. Periods of continental assembly were envisioned as times when interior continental climates were severe, affecting the continental shelf faunas. In contrast, times of fragmentation were times when the continents' climate was more moderate due to ameliorating marine conditions; this permitted the buildup of shallow water diversity. While the post- Permian expansion may fit this pattern, evidence from the Paleozoic does not seem to show an increase in continental fragmentation during the early-mid Paleozoic. Indeed, the continents were maximally fragmented and arrayed along the equator during the Cambrian (Scotese et al. 1979). Continental drift and arrangement nevertheless has had profound effects on climate and probably extinction. During the Ordovician and Silurian Periods Gondwana drifted southward from its Cambrian position at the equator, and came to rest on the geographic south pole. This coincides with the Late Ordovician glacial tillites that have been found in North Africa, and a large reduction in the degree of marine provinciality relative to the early Ordovician. In the Cenozoic, the spatial arrangements of the continents about the Pacific and Atlantic Ocean made for a quite different climatic history (Briggs 1970). The North Atlantic was a more enclosed basin and was far more severely affected by the late Cenozoic polar cooling. The Pleistocene initiated severe enough climates to cause a major molluscan extinction in the southeastern United States Shelf, while Pacific American faunas showed no increased extinction (Stanley 1986b; Stanley and Campbell 1981).

What is the tempo of change as provinces are homogenized? If modern mammalian faunas are any indication, it seems to be the case that readjustments and changes of diversity in isolated areas are rapid relative to the pace of change in climate and continental rearrangement. Flessa (1981) found that as much as 70 percent of the degree of similarity in mammalian genera among continents can be explained by presentday distance, longitudinal separation, and area, in order of decreasing importance. Historical effects do not obscure the present day pattern.

The effects of increasing access between biogeographic realms can be illustrated by the large-scale interchange of mammals between North and South America after the Pliocene establishment of the Isthmus of Panama, following the disappearance of the Bolivar Trough marine barrier (Marshall et al. 1982). Before the interchange there was long-term stability in numbers of mammalian families. As a probable result of North America's initial higher taxon richness, more taxa moved from north to south than in the reverse direction. In South America, where taxon richness now exceeded previous "steady state" levels by more than 50 percent, there was about a 70 percent increase in extinction rates. Descendants of the North American invaders participated in an evolutionary radiation, resulting ultimately in an overall richness higher than previous levels. Mammalian diversity is now higher in South America, in contrast to the situation previous to the exchange. This suggests that area does have an effect on regulating diversity, but evolutionary changes can impose a significant overprint on diversity.

Geographic Locus of Evolution

The origin and evolution of biotas have often been placed in a geographic context. Interpretations have usually been based upon the presence of gradients in taxon abundance, especially latitudinal increases toward the tropics (Pianka 1966). These have been used to infer centers of origin in loci of higher diversity, from which colonists have moved toward less hospitable habitats. This general argument may have some validity, but it has been called into question by evidence adduced from the fossil record. A major element of early Cenozoic southern hemisphere marine faunas appeared first in high latitudes, and then spread to low latitudes as the climate became cooler (Zins-

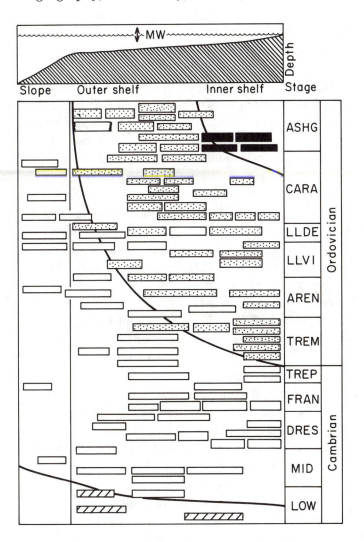

Figure 8.9: Depth distribution of successive Cambro-Ordovician fossil communities. Lines describe a qualitative fit of the time-transgressive nature of offshore spread of the communities (after Jablonski et al. 1983).

meister and Feldmann 1984). The same story has been claimed to apply in the origin and spread of early Tertiary floras (Hickey et al. 1983), but the stratigraphic correlations have been questioned (Spicer et al. 1987). A similar pattern of origins in what seem to be extreme environments has been found (Jablonski et al. 1983; Sepkoski and Sheehan 1983) in the evolution and appearance of marine benthic faunas. In Cambro-Ordovician benthic communities, new assemblages appear first in shallow water facies and then spread to offshore depths (Figure 8.9).

These results are paradoxical in the light of our explanation above of provinciality, which would predict that morphological specializations and innovation would come from more stable habitats, e.g., offshore in the marine environment. The Cenozoic high latitude heterochroneity of faunas and floras is more understandable as the overall Cenozoic world climate became progressively cooler, especially after the Late Eocene, making spread from high latitudes expectable. But the spread from inshore habitats toward offshore is more difficult to explain since offshore habitats harbor more species and clades with higher speciation rates.

We might not expect more offshore faunas to invade inshore, as they probably lack the degree of ecological flexibility required to make it in the inner shelf. The possible predominance of the reverse invasion route is not so easy to explain. In Lake Baikal, an ancient lake with extraordinary deep-water benthic endemic diversity, the deep-water forms are also believed to have arisen from shallower water, near the break in slope (Kozhov 1963). It may be that the habitat extinction rate in shallow water is sufficient to continually permit new evolutionary experiments to occur, some of which are successful enough to colonize outward. The more diverse stable habitats may have biotas too persistent to permit the survival of species much different than their ancestors. An unstable habitat may normally select against the gradual development of a specialized biota, but would not prevent the origin of new innovations.

The specialized nature of the offshore fauna may also make them more closed to innovations and more prone to mass extinction. Alternatively, the offshore biota may be dominated by groups with higher characteristic extinction rates (Sepkoski 1987). Bretsky (1969) showed an increased offshore community extinction rate through the Paleozoic. This same effect has been documented in more detail through the 5 my Frasnian (Upper Devonian). While replacements occur in both near and offshore communities, the effect was more dramatic offshore (McGhee 1981). Extinctions of the offshore fauna may provide opportunities for invasion from inshore habitats.

Mount and Signor (1985) argue that one must distinguish between the successful spread of ancient biotic assemblages and the actual first appearances of innovations. Their study of the Early Cambrian shelly and trace fossils of the White Inyo Mountains shows that first appearances occur in offshore habitats. A further examination of first appearances of Paleozoic higher shelly taxa with distinctive morphologies shows a similar lack of fidelity to nearest shore habitats (Signor and Mount 1986). This may suggest that those innovations that appear nearshore are more successful, but it may also call into question the linkage between long-term movements of whole biotas (as documented by Sheehan and others) and the first appearance of innovations in given groups.

Diversity gradients in modern biotas may be deceiving, as they can be the result of extinctions in peripheral environments, rather than a reflection of a depauperate peripheral biota, periodically colonized by a rich tropical biota. The center of diversity in Pacific coral reefs, for example, may be the result of deterioration of climate through the Cenozoic that may have mainly affected the periphery of the province (Newell 1971). Studies of coral reefs (Stehli and Wells 1971) and Foraminifera (Stehli et al. 1972) demonstrate that mean longevity of genera increases towards high latitudes. The oldest genera of Pacific corals, for example, are found throughout the province, while younger genera may be found toward the center. This may involve selective extinction of younger forms at the periphery of the province, but it may also signify that older, more widely adapted genera are more likely sources for repeated origins of descendants. This is the likely basis for Kaufmann's (1933) concept of a **conservative stock** that survives for a greater period and gives rise to clades of probably higher diversification rates. Groups with low speciation rates are ultimately the survivors in evolution and the repeated progenitors of new radiations. The unspecialized globular Foraminifera dominating high latitudes are the progenitors of repeated radiations of more specialized forms that may accompany vertical structuring of the water column (Lipps 1970).

Unspecialized forms survive well in peripheral environments, and would survive the inevitable extinction that would be brought on by a radical homogenization of water column structure.

Episodic Turnover and Extinction

Peaks in Turnover

We have discussed the nonrandom pattern of diversification of the marine and terrestrial fauna and flora. If we return to Figure 8.2 we can see five conspicuous and precipitous drops in diversity, the most dramatic occurring at the end of the Permian. Raup and Sepkoski (1982) analyzed overall extinction rate at the family level and found that four events fell outside of a one-sided 99 percent confidence interval from the overall population of measurements: Ashgillean (Upper Ordovician); Frasnian (Late Devonian); Guadalupe-Dzhulfian (Late Permian); and Maestrichtian (Late Cretaceous). The Norian (Upper Triassic) falls within the 95 percent interval. It has long been supposed that these massive extinctions must have been caused by some events related to climate, extraterrestrial influences, or changes in sea level. **Mass extinctions** are to be distinguished from a typical level of **background extinction**, characteristic of the time.

Times of biotic change can be detected if we examine turnover, a measure of the sum of appearances plus extinctions (high turnover can occur when appearances and extinctions are equal). Periodicity in turnover was observed first by Newell (1952, 1967), who found peaks of activity in the Ordovician, Carboniferous, and Jurassic. Flessa and Imbrie (1973) used factor analysis of Phanerozoic faunas and devised a measure of turnover based upon the angular change between assemblages that are each expressed as a vector among the factor axes. The peaks corresponded weakly to tectonic events. They argued that the turnover peaks may be related to biogeographic readjustments, due to rearrangement of dispersal routes and change of climate accompanying continental movements. They found no obvious correlation with area of continental inundation. Thus though some conspicuous extinctions are correlated with sea-level fall, a general explanation may have to involve several different factors in turnover and extinction. The recent analysis of turnover and extinction in Neogene planktonic foraminifera (Wei and Kennett 1983) shows a strong correlation with changes in

Cenozoic world climate.

Patterns of taxon survivorship of mollusks indicate that mass extinction may or may not be a qualitatively different phenomenon from background extinction. During normal periods, extinctions of mollusks are correlated with planktotrophic larval development and geographic range, while clade survivorship is positively correlated with species richness. During the end-Cretaceous event, however, none of these hold, and clade survival is correlated only with the geographic extent of the clade. After the event, the correlations found previously again obtain (Jablonski 1986). In the end-Cretaceous extinction of planktonic diatoms, however, a far different pattern emerges. Diatom species with benthic resting stages survived far better than those with no resting stages. Here, properties that normally would be easily related to survival of individuals can be extrapolated to taxon survival. Thus, as extinction intensity increases, some qualitative changes may emerge for some taxa, but not for all. Gould has characterized mass extinctions as a distinct "tier" to be considered as qualitatively distinct in the history of life. In contrast, Valentine and Walker (1987) suggest that increasing extinction in the Permian is, for the most part, a matter of degree, even though there are some distinctive effects for some taxa.

Both points of view are consistent with the proposition that much of the history of life is influenced by extinction. Major disruptions of habitat swept the world free of many taxa; subsequent replacements and expansions are therefore the result of happenstance and opportunity. This point of view is inconsistent with a common belief that successful survival of extinctions is always influenced by competitive superiority; that is, the adaptively "better" forms survive and proliferate. Under this hypothesis, one might postulate that the success of the placental mammals over the marsupials relates to the superior adaptations of the former, as both groups experienced the ravages of the end of the Cretaceous. We can't exclude this hypothesis, but current evidence provides no confirmation of it. In contrast, the diatom extinctions studied by Kitchell et al. (1986) suggest that such a hypothesis can apply.

Darwin worried much about extinction, particularly because there was much evidence in his time that some of the "races" of man had disappeared. It is said that when Humboldt was in South America, he heard a parrot who was the last creature capable of speaking the words of a lost tribe. The forces of civilization were causing widespread decreases of aboriginal peoples throughout the world. Darwin (1874) viewed the survival of any species during normal times as a delicate balance, which could be easily upset. Many factors, including disease and climate, could tip the balance toward population decline and extinction. He saw the influx of western man as a small but crucial factor in extinction, as "each race is constantly checked in various ways; so that if any new check, even a slight one, be superadded, the race will surely decrease in number; and decreasing numbers will surely lead to extinction" (Darwin 1874, p. 208). In the case of mass extinctions, it seems to be the case that the eradication of crucial habitat supersedes the coupe de grace that might otherwise be imposed by competitors.

Periodicity in Extinction?

The presence of clear peaks in turnover and extinction do not suggest pervasive solutions to the study of the causes of extinction. Many changes in the earth's history are plausible sources of extinction, and some major processes such as sea-floor spreading and continental drift may have had impact on climatic, dispersal, and area effects simultaneously. While area effects can be invoked in predicting the magnitude of extinctions (Permian versus Pleistocene), too often we would not be able to "predict" an extinction with geological evidence alone. Episodes of large-scale vulcanism, for example, may have caused decreases in earth surface temperature sufficient to cause extinctions, but no such event can be characterized well enough to predict the occurrence of an extinction; we are left with correlations. Many have suggested that unique extraterrestrial events have caused major extinctions (e.g., Schindewolf 1950) but for lack of evidence none has been taken very seriously. As we will discuss below, the Cretaceous mass extinction has been a focus of such concerns in recent years, due to the recent acquisition of the first substantive geological evidence for extraterrestrial effects.

Periodicity of extinction or climatic change predicted by astronomical or geophysical theories would be the most convincing way to establish a terrestrial or extraterrestrial cause of extinction. If extinctions are measurably periodic, it may be that only one credible cyclic theory would fit the available pattern. The precedent for such an approach lies with the longstanding debates over the periodicity of Pleistocene glaciations. The Yugoslav astronomer Milankovitch (Hayes et al. 1976; Imbrie and Imbrie 1979) suggested, following earlier nineteenth-century conjectures, that Pleistocene glacial advances and retreats might be regulated by changes in high latitude insolation, caused by cyclic changes in the earth's orbital eccentricity, tilt, and time of perihelion. A power spectrum analysis predicts temporal variation in climates with a series of periodicity peaks that match the Pleistocene periodicity of climate, as estimated by stable oxygen isotopic ratios (which reflect both temperature and glacial ice formation) and pelagic fossil assemblages (Imbrie and Kipp 1969).

A number of studies in recent years have taken up this theme and relate these cycles to sedimentary cycles, including some of the classic mid-continent alternations of carbonate and mudstone. Many of these cycles occurred during times when there was no significant amount of continental glaciation, and represent transgressive–regressive cycles. For example, sedimentary cycles in the lacustrine Early Mesozoic supergroup correspond to periodicities of approximately 25,000, 44,000, 100,000, 130,000, and 400,000 years. These periodicities, in turn, correspond to those expected from celestial processes, such as the precession of the equinoxes, the obliquity cycle, and the eccentricity cycle (Olsen 1986). Cyclic processes such as the precession of the equinoxes may have driven continental heating cycles that rearranged wind and climate. A cyclicity in precipitation would explain the lacustrine sedimentary cycles reported by Olsen.

Milankovitch climatic rhythms also appear in mid-Cretaceous black shale sedimentary cycles in Central Italy (Herbert and Fischer 1986). These cycles consist of alternations of carbonate and shale, with intervals of highly oxidized (red) and highly reduced (black) strata. They are particularly interesting, since they occur in marine sequences and must have reflected periods of ocean bottom anoxia, alternating with vigorous bottom mixing and high productivity in the water column. This suggests temporal swings in climate that were forced by celestial processes. Salinity and temperature variations may be the link between the external forcing and anoxic–oxic cycles.

These results may be crucial to our understanding of evolution, since these cycles occur well within the time scales of speciation for many invertebrate groups (see Stanley 1979). Since these cycles reflect both alternations of sediment type and water quality, neither the plankton nor the benthos are immune. Such cycles may set the pace of morphological evolution, and may even create periodically reversing directional selection. If the proximate effects of these cycles are long periods of stable climate, interrupted by rapid switches, then we may have a primary driving force for punctuational character evolution. On the other hand, if such cycles create continual gradual change, then directional selection may be continuous and cause a steady response in certain traits. The interested reader should consult Herbert and Fischer (1986) for a recent literature summary of Milankovitch cycles.

The earth's history has been dominated by large-scale changes in climate, arrangement of continents, vulcanism, and sea level. Fischer (1984) developed a theory connecting physical conditions with the overall pattern of Phanerozoic life. Two summaries (Vail et al. 1977; Hallam 1977b) of global sea-level change show distinct highs in both the mid-Paleozoic and Mesozoic. Periods of continental breakup, when dispersed and thinner continents resulted in smaller ocean basins, would be associated with higher sea levels. Periods of continental aggregation, when continental crust was bunched up due to collisions and ocean basins were therefore more commodious, resulted in lower stands of sea level. The temporal variation in granite emplacement matches the sea-level curve. This suggests a causal link between active continental fragmentation, vulcanism, and sea level, an environmental condition of obvious importance to the world marine biota.

Long-term fluctuations in climate are imperfectly understood, but today's conditions are not typical of all Phanerozoic time. Large polar continental ice sheets existed in the pre-Cambrian, Carboniferous-Permian, and Pleistocene, but globally equable climates prevailed in the mid-Paleozoic and Mesozoic (Figure 8.10). These periods were also dominated by great evolutionary radiations of a wide variety of unrelated phyla. The great Cambrian-Ordovician and Mesozoic expansions thus seem marked by rises in sea level, increases of vulcanism, and ameliorations of global climate.

Fischer (1984) connects vulcanism and global climate through the greenhouse effect. Increased vulcanism may have liberated carbon dioxide into the atmosphere. As these periods were of higher sea-level stand, erosion would have been minimal, and loss of CO_2 in weathering would be suppressed. During times of low sea level, low vulcanism

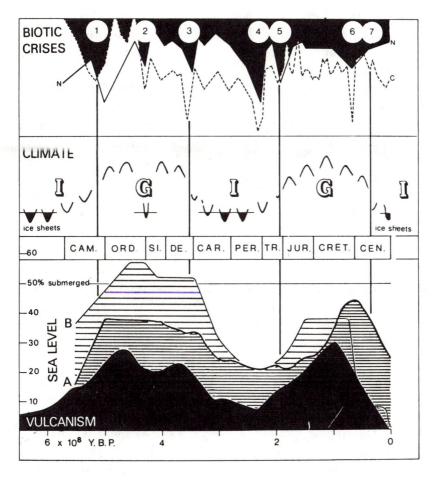

Figure 8.10: "Supercycles" of the Phanerozoic, postulated by Fischer. Sea level curves according to (A) Vail et al. 1977 and (B) Hallam 1977b are superposed on a diagram of granite emplacement, times of glaciation, times of biotic crises (numbered) as determined by Newell (curve N) and Cutbill (curve C), and a guess of climate, characterized by either icehouse (I) or greenhouse (G) conditions. (from Fischer 1984, reprinted with permission of Princeton University Press).

would reduce the liberation of CO_2, and increased weathering would consume CO_2. Thus, the mid-Paleozoic amelioration was associated with high CO_2, which, in turn caused a greenhouse effect and an increase of surface temperature. The end of the Paleozoic witnessed the termination of such conditions, and an "icehouse effect" resulted in a deterioration of climate mainly at high latitudes.

It is not clear whether these cited fluctuations are irregular temporal changes or regular oscillations. Fischer (1984) favors a 300 my cycle. The hypothesized greenhouse conditions coincide with the two major expansions of metazoan life in the Phanerozoic, but extinctions are less clear. The Permian-Triassic crisis occurred after the glaciations, and is also associated with a dramatic drop in sea level. The Cretaceous extinction is not associated with a switch from greenhouse to icehouse conditions. Cold temperatures may or may not have contributed to extinctions.

A periodicity of about 26 my in the occurrence of peaks of extinctions has been reported by Raup and Sepkoski (1984). Using a more qualitative assessment of pelagic taxa, a similar periodicity of 32 my had been previously claimed by Fischer and Arthur (1977). Using a family-level data set, a statistically significant series of regularly spaced peaks could be ascertained in the temporal changes of families among the 39 stages (250 my) since the Late Permian. As Raup and Sepkoski acknowledge, both the uncertainties of the geological time scale employed and the smaller number of time units (maximum resolution of 6 my) make such an analysis tentative, as it excludes a wide variety of factors within smaller time scales, including the Milankovitch hyothesis mentioned above. To consider an extinction important, Raup and Sepkoski used a threshold level of 2 percent. Using a threshold of 10 percent, Rampino and Stothers' (1984) reanalysis yields a periodicity of about 30 my. The time interval between peaks lies between 17 and 53 my, a variation whose importance cannot be gauged without a specific physical or biological model. Some of the variance, or regularity, may be due to biases in the time scale employed (Hallam 1984a). Mesozoic stage boundaries have an uncertainty of at least 5 my and some even larger disparities exist among different published time scales. The potential for an artificially induced periodicity can be seen in some cases, such as the Upper Triassic, where stages are each arbitrarily set at 6 my duration. Periodic behavior could also be generated by a model

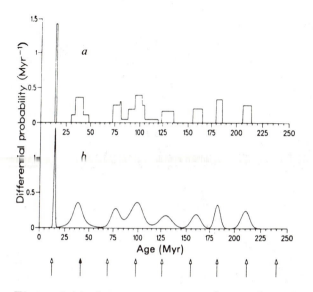

Figure 8.11: Impact craters on the earth with a diameter of > 10 km and an age of 5–250 my (from Alvarez and Muller 1984).

with a strong stochastic component (Kitchell and Pena 1984).

A recent reanalysis by Raup and Sepkoski (1986) of generic data reveals eight major episodes of extinction over the last 250 my that are more pronounced than those defined by the family-level analysis. The more recent analysis identifies two more extinction peaks, and strengthens the estimate of 26 my. Random simulations do not generate such strong periodicities. But how monophyletic are the groups? The increased strength of the generic peaks is encouraging if genera are more frequently monophyletic than are families.

Some preliminary physical evidence and astronomical models make the Raup and Sepkoski conclusion of periodicity worthy of further serious consideration, and a flurry of explanations have already appeared. Catastrophes generated by periodic extraterrestrial events seem plausible. Two independent analyses of large body impact craters (Figure 8.11) yield a periodicity of 28.4 my (Alvarez and Muller 1984) and 31 my (Rampino and Stothers 1984). The phase of this cyclicity is consistent with that of the extinctions, though the numbers are small. Two different astronomical models have been suggested as ultimate causes of the periodicity. The **Galactic Plane Oscillation** hypothesis takes into account that the solar system oscillates about the galactic plane

with a periodicity of about once in 33 my. Such movements might increase the probability of contact with clouds of gas and dust near the
galactic plane, which in turn might gravitationally perturb the solar
system's comets and thereby increase the frequency of planetary impacts (Rampino and Stothers 1984). A pass through the galactic plane
may have also influenced the earth's magnetic field and even resulted
in increased atmospheric dust (Schwartz and James 1984; Hatfield and
Camp 1970). The periodicity and the present arrangement of the solar system relative to the galactic plane, however, suggest that the
mass extinctions took place when the solar system was farthest from
the galactic plane (Schwartz and James 1984). The **Distant Solar
Companion** hypothesis argues for the presence of an unseen solar
companion whose orbit passes periodically through a cloud of comets
in the outer solar system, sending some of them on paths toward the
inner solar system and the earth (Davis et al. 1984; Whitmire and
Jackson 1984). A search is now on for this hypothetical companion.

The proximate cause of the extinctions is another matter. Presumably an increase in cometary impact might raise dust clouds sufficient
to cause a drop in insolation and a decline in climate. Much of the
evidence suggests that an increase in cold climate had something to do
with many of the mass extinctions in the Phanerozoic (Stanley 1983,
1984). If single impacts were important, then one might expect precipitous extinctions. Both models, however, allow for a series of impacts
or events, permitting extinctions to occur over prolonged periods of
time. This is a fundamental weakness for the testing of the models as
the only potentially good data we might be able to get would relate
to the suddenness of the extinctions, as will now be discussed for the
Cretaceous extinction.

The Cretaceous Extinction: Sudden Catastrophe or Gradual?

The end of the Cretaceous is not the most dramatic mass extinction
in the Phanerozoic (Figure 8.2). At the time, however, both major
terrestrial and marine elements were lost, the fauna is sufficiently modern to be understood ecologically, and some of our favorites, such as
dinosaurs and ammonites bit the dust. The discovery of a strong positive iridium concentration anomaly in rocks from localities scattered
worldwide (Alvarez et al. 1980; Silver and Schultz 1982) has raised
the possibility that the extinction may have had an extraterrestrial

cause. Positive (Playford et al. 1984) and negative (McGhee et al. 1984, 1986) evidence for an iridium anomaly has been found for the Frasnian-Fammenian (Late Devonian) mass extinction, but the extinction itself was spread over at least seven my and climatic effects are evident (McGhee 1982). There seems to be no iridium anomaly associated with the terminal Ordovician extinction (Orth et al. 1986). A similar iridium anomaly in sediments of 34 my of age occurs simultaneously with the disappearance of five dominant radiolarian species, and at the general time of a mammalian extinction (Ganapathy 1982; Alvarez et al. 1982). But the larger picture of biotic change across this boundary is gradual, with no suggestion of a catastrophe (Corliss et al. 1984).

Alvarez et al. (1980) have suggested that a massive asteroid impact caused the extinctions by blanketing the earth with dust spread along ballistic trajectories outside the atmosphere. McLaren (1970) suggested this previously for the Frasnian-Fammenian extinction. The end-Cretaceous impact hypothesis is supported by the worldwide nature of the anomaly (Silver and Schultz 1982) and by shock structures on quartz crystals (Bohor et al. 1984). The presence of impact structures on quartz suggests that an enormous crater should be present on a continent, although other origins for shocked quartz are possible (Carter et al. 1986). A wedding ring caused one of the iridium anomalies (or the extinction was caused by a rain of wedding rings), but the other Cretaceous-Tertiary cases have withstood further scrutiny (Alvarez et al. 1982, 1984). The dust cloud would exist for a time sufficient to severely disrupt climate by cutting off all light and temperature might have been expected to drop precipitously. A stable oxygen isotope anomaly at the boundary gives evidence for a sudden temperature change. All organisms dependent upon light or primary producers should have been affected. Deep-water forms not so dependent on light or warm temperature, such as nuculid bivalves, would be expected to survive.

The Alvarez theory has one strong and other weaker predictions. (1) Extinctions must follow the impact. (2) One might also expect many groups to die off instantaneously, but a less catastrophic change in temperature and light might have an effect that would be prolonged. Finally, (3) groups more prone to light stress or temperature increase would be more vulnerable (e.g., phytoplankton versus deposit-feeding benthos). The response of the sensitive groups should be geologically instantaneous.

How does the evidence stand up? As in most other mass extinctions the end of the Cretaceous was preferential as to organisms affected. Groups associated strongly (Foraminifera, coccolithophorids) or weakly (ammonites) with the water column suffered strongly, while benthic forms (e.g., bivalve mollusks) generally suffered less. Members of food webs less dependent upon plant material (marine deposit feeders, scavengers, stream inhabitants, and small insectivorous mammals) suffered less than direct herbivores (Sheehan and Hansen 1986). There is also an apparent thermal bias; mollusks and foraminifera in the tropical Tethyan sea suffered large-scale extinction, and were replaced by higher-latitude contemporaries (Stanley 1983, 1984). The question of timing is more confusing. Coccolithophores and nonglobigerinid foraminifera disappeared so precipitously (and simultaneously with the iridium anomaly) that chalks give way to clastic sediments in a knife edge contact in several sections. In the chalk of Denmark, the Maastrichtian fauna, dominated by brachiopods, disappears abruptly (Figure 8.12), with no prior warning in terms of reduced diversity, or early extinction of specialized forms (Surlyk and Johansen 1984). The sediments above the chalk are clayey and indicative of anoxic conditions. Turbidity, loss of an appropriate sediment, and anoxia may all have contributed to the abrupt extinction. Radiolitid and hippuritid rudists bit the dust during a period of flourishing radiation. The discovery of transitional vertebrate faunas in Paleocene channels cut into the Cretaceous-Tertiary boundary sediments makes the vertebrate record in Montana at least compatible with a sudden extinction. At present, the question of the North American Western Interior vertebrate fossil record is under intense debate and it is fruitless to attempt a complete discussion when the data at present are contradictory and interpretations are changing monthly (see Van Valen 1984b).

Unfortunately, the story is not nearly so simple. There is extensive evidence for significant declines in many groups well before the time of the major iridium anomaly, which would falsify the major prediction of the impact hypothesis. Hickey (1981) discussed the record of land plants and concluded that it was compatible with a gradual deterioration of climate. A dramatic change in palynmorphs does occur, however, right at the boundary (Hickey 1984). A gradual deteriora-

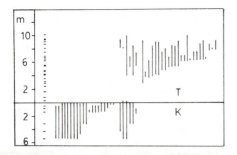

Figure 8.12: Duration chart of brachiopod species in the Cretaceous (K)–Tertiary (T) boundary sequence in Denmark. Note the Lazarus effect: Several species disappear, then reappear. On left, stratigraphic height (meters) and sample horizons are indicated (after Surlyk and Johansen 1984).

tion, well below the boundary, seems to be recorded in the record of diversity of the inoceramid bivalves and ammonites (Kauffman 1984). From the Cenomanian to the Maastrichtian stages there is a continual decrease of the rate of new ammonite generic appearances (P.D. Ward and Signor 1983). As Ward and Signor comment, the "diversity faucet" that characterized ammonite evolution was shut down in the Campanian stage, well before the Cretaceous-Tertiary boundary. The story is claimed to be the same for the dinosaurs (Schopf 1982). The dinosaur fauna of the late Maastrichtian included fewer than 20 species in 15 genera and 10 families, chiefly in the North American Western Interior. There is no good evidence that the dinosaurs were declining steadily towards this low number in the last nine million years of the Cretaceous (Sheehan and Morse 1986). In the late Maastrichtian, however, sea level decreased by 150 to 200m, making a hypothesis of increased terrestrial seasonality compatible with the ultimate disappearance. One extinction event, that of some rudistids (a group of aberrant bivalves that constructed reefs), was indeed rather brief, but also occurred well before the Cretaceous-Tertiary boundary. In one section in Spain, plankton disappear at the boundary, but two invertebrate groups do not (P.D. Ward et al. 1986). Most embarrassing for the impact theory, the fresh-water biotas seem to have emerged unscathed.

What are we to make of this? Uncomfortable as it may seem, there may be more than one process at work in causing major extinctions. The evidence from land plants (Hickey 1981), dinosaurs, and many mollusks points to a change in conditions well before the Cretaceous-Tertiary boundary. Most notable are groups such as inoceramid bivalves and ammonites that seem to be tracking an environmental deterioration before the termination of the Maastrichtian, or final Cretaceous stage. This would violate the fundamental prediction that extraterrestrial events should be followed by, not preceded by, extinctions. But the iridium layer, plus its associated faunal disappearances, cannot be reconciled with any hypothesis of climatic deterioration, unless one could establish a threshold temperature for the production of calcareous plankton. The Eocene data also suggest a repetition of an extraterrestrial event, associated with sudden disappearances of radiolarians and other groups. We are left with a compound hypothesis, at least for the proximate cause of the extinctions. Indeed, it may be premature to speak of a Cretaceous mass extinction at all, given the evidence for different and seemingly inconsistent tempos and timings of extinction. The same can be said for the Permian extinction. If we believe that sea-level drop was the sole cause, operating by an area effect, we have to reconcile the precipitous change in sea-level coverage in the last stage of the Permian (Schopf 1974), with the pattern of extinction, which was initiated earlier than the sea-level drop (P.W. Bretsky 1973). The extinction seemed to concentrate on certain ecological groups, particularly tropical forms at the end, but high-latitude groups earlier in the Permian. Most discouraging of all, what if the extraterrestrial influence occurred in a series of impacts, rather than as one big bang each 23 million years or so? Without a series of signals (e.g., a series of irridium spikes), this hypothesis is *ad hoc* at best.

In conclusion, although there is, in my view, credible evidence for a periodicity of extinction, whose likely cause would be impacts of extraterrestrial objects, several mass extinctions seem to be complicated, and were probably associated with changes in climate and sea level (that are probably interrelated). As Surlyk and Johansen (1984) and Hallam (1984b) note, anoxia may also be an important cause, as witness the extensive development of black shales in certain periods. What is lacking at present is a credible evaluation of the relative effects of these factors in both the Cretaceous and Permian extinctions. In the Eocene, low temperature may be the dominant force; it exerted visible effects on organisms such as leaf morphology (Wolf and Hopkins 1967). Even here, we cannot exclude a concomitant effect of water column homogenization (Lipps 1970) on the tropical plankton. While many extinction factors may be at work, extinctions occurring below an iridium anomaly do not falsify the impact hypothesis. At any time in the record, *some* groups must be declining.

Extinctions and Replacement by Ecological Equivalents

We have not discussed one of the major phenomena characteristic of many extinctions: replacement by ecologically similar forms. This is hard to document because of the common problem of devising criteria for ecological similarity. We are here at the interface of ecology and evolution, as groups that are sparse before a mass extinction may change significantly in ecology and morphology via evolutionary change to effectively replace another group that disappears. The mammals seem to have been mainly minor and nocturnal elements of the terrestrial fauna until the disappearance of the dinosaurs and other reptiles. (There were, however, more mammal genera than dinosaur genera at the close of the Cretaceous.) Subsequently in the Paleocene and Eocene a major radiation resulted in mammalian replacement of ecological space occupied formerly by the dinosaurs. How, then, do we draw an ecological equivalence between Mesozoic mammals and dinosaurs?

This problem appears in two alternative hypotheses explaining replacement. (1) The **Competitive Displacement model** argues that once a competitively superior group arises, it displaces its contemporaries. Alternatively, the (2) **Vacancy hypothesis** would argue that after one major ecological group disappears, another takes its place. The second hypothesis could operate in two distinct ways. One group may not be inherently superior (e.g., larger mammals over dinosaurs), but its rise is suppressed simply by the temporal priority achieved by an extant and wide-ranging group. Thus the larger mammals may not have been able to get "off the ground" without a minimum available area and habitat complexity for expansion, which appeared only after the end of the Cretaceous. Alternatively, following extinction of one group, a second that has survived a crisis may simply expand randomly to fill the place of the first.

The competitive displacement model may apply in some cases. Thayer has carefully classified much of the fossil benthic invertebrates into ecological functional groups, making resource use comparisons pos-

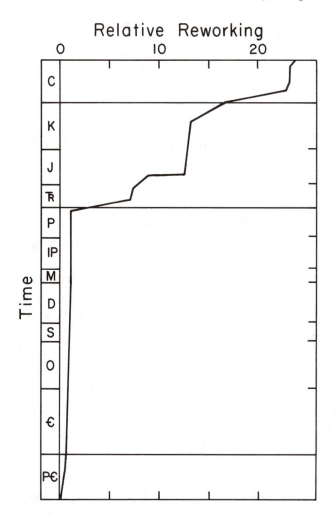

Figure 8.13: Estimate of relative amounts of biogenic reworking of shelf sediments through the Phanerozoic (after Thayer 1983).

sible. The pedunculate brachiopods, for example, decline from the Devonian toward the end of the Paleozoic; this coincides with the expansion of their ecological equivalent, the epibyssate bivalves (Thayer 1983). But many of the large-scale replacements recorded in the fossil record seem to involve disappearances, followed by radiations of either the same group (e.g., Foraminifera: Cifelli 1969; trilobites: Palmer 1984) or quite different groups of organisms (e.g., mammals replacing dinosaurs). The reclining brachiopods decrease precipitously at the end of the Permian; this is followed by a great diversification of free-burrowing bivalves in the Mesozoic. Thayer has documented (Figure 8.13) a major increase in bioturbation during the Mesozoic. It is reasonable to believe that this revolution prevented the re-expansion of the free-lying brachiopods into these habitats.

Unfortunately, replacements are difficult to document, and evidence can be interpreted differently. The brachiopod-bivalve switch in dominance from Paleozoic to Mesozoic is a case in point. Gould and Calloway (1980) examined diversity changes of both groups and found a general positive correlation between the two in degree of change of taxon richness. This was taken as evidence that the two groups could not have influenced each other negatively. Thayer (1983) has criticized this conclusion, noting that Gould and Calloway did not take functional groups (groups with ecological equivalence) into account. At the level examined by Gould and Calloway, the positive correlation may just involve overall indications of the "health" of the world biota, which might swamp the details seen in Thayer's analysis. The emphasis on diversity may also be deceiving as it does not take abundance into account. Taxon richness of pedunculate brachiopods did not decrease very much after the Permian, but individual abundance changed considerably, relative to the Pennsylvanian and Permian (Thayer 1983). The post-Permian was no longer a brachiopod world. The question of individual abundance versus taxon richness can also be considered at the community level. Bivalve-dominated faunas may often be species-poor, yet dominate by virtue of their competitive ability. In faunas of the northwest United States, two conspicuous communities demonstrate this. Subtidal mud bottoms are often completely dominated by a single mytilid (Mollusca, Bivalvia) species, *Modiolus modiolus*, while several brachiopod species exist as relatively minor components. In very shallow subtidal and low intertidal habitats, succession usually leads to dominance by *Mytilus californianus*, but several species of brachiopods can also be found, though they are usually rare and in apparent refuges.

Competition and evolutionary responses of competitors are difficult to document, except in living communities, where experimental approaches are possible (Schoener 1983). Evidence for the direct effects of one group on the evolution of another comes more easily in studies of predation. There is extensive evidence for a major Mesozoic-Cenozoic evolution of victims in response to the rise of many groups of predators, principally decapods and teleost fishes (Vermeij 1977). The best case is for the gastropods. Forms with morphological characteristics indicative of protection against shell crushing have increased since the Cretaceous. The proportion of umbilicate and trochiform species remained high and relatively constant from the Silurian to the Early Triassic. The overall aspect of the gastropods changed radically starting in the mid-Mesozoic; groups with thickened or expanded adult outer lips diversified, even though the form type had appeared earlier in the Paleozoic (Vermeij 1983). The association with the rise of predators is very suggestive. The same traits appeared independently in many taxa, making an individual adaptation hypothesis, rather than a species culling hypothesis more likely. This trend is also concordant with a temporal increase in frequency of shell repair, presumably as a response to predators, from the Paleozoic to the post-Paleozoic (Vermeij et al. 1981). The patterns seen in the fossil record can also be recorded in Recent distributions. Gastropods with features conferring resistance to shell crushing are found generally in areas where robust decapods are common (Vermeij 1977).

Some similar trends may have occurred in other taxa. Palmer (1982) documented a reduction in numbers of parietal plates in seven of the eight families of balanomorph barnacles. All four species of the drilling gastropod *Thais* on the Pacific American coast selectively attack barnacles on the margins of the parietal (lateral) and opercular plates. The trend in plate reduction occurred coincidentally with the radiation of muricacean gastropods starting in the Late Cretaceous, and the one balanomorph lineage evolving free of this pressure maintained the primitive condition of eight plates; this group lives on the integuments of marine reptiles and mammals. The minimum number of four parietals is set by the need to expand the size of the aperture with increasing growth.

In conclusion, replacement and responses to biological challenge are distinctly nonrandom processes, but the mechanisms are not always clear. In most cases of potentially competing groups, extinction seems to have preceded replacement. Here, the possible resolution power of competitive hypotheses is very weak, since it is not clear how we would prove, for example, that the group that finally wins out happened to be suppressed (e.g., Mesozoic mammals) during the period that the prior group (e.g., dinosaurs) dominated. The failure of new novelties to arise in the mammals would be an ad hoc hypothesis immune to test. In cases of gradual replacement, Thayer's use of functional groups has greatly improved our perception of trends in the bivalve replacement of the brachiopods, but there is still the nagging uncertainty that the two groups really were "ships passing in the night," as characterized by Gould and Calloway (1980). The failure of a group to reradiate is more interesting. The nonpedunculate brachiopods failed at this in the early Mesozoic; this is at least coincidental with the takeover of their preferred habitats by bioturbators.

At the end of the Permian, the Paleozoic EF suffered far more than the Modern EF. This may relate to the dominance of the Paleozoic EF by forms reliant upon warmer habitats. Once extinguished, the Paleozoic EF reradiated in both the Triassic and Jurassic more rapidly at first than the Modern EF. Yet it failed to take over despite its apparent increased rate of cladogenesis. It has been consistently more prone to extinction, despite its higher rate of cladogenesis (Sepkoski 1984). This at once suggests that sheer speciation rate (inferred from family appearance rate) is not the key to success, but it also suggests that the ascendancy of the Modern EF could not have been entirely due to their competitive superiority for limited resources. The ascendancy of the Modern EF may have been due, for example, to their coincidental origin in areas with colder climate or relative adaptability to cold; the Paleozoic fauna may have been filtered out by causes extraneous to their potential competitive interactions with other groups. But given their presence after the Permian and Upper Triassic extinctions, it is possible that the surviving groups were facing increasingly stiff competition or predation. This may be reflected in Sepkoski's observation that the Paleozoic EF finally began a slow decline in the Cretaceous after the total number of families exceeded the Paleozoic "steady state."

At this point of possible saturation, true competitive decline may have been caused by the Modern EF. This same trend can be seen in the mid-Paleozoic, when the rise of the Modern EF reaches the point that saturation is surpassed; from this time on the Paleozoic EF declines. A functional group analysis would clarify whether this reciprocal change in taxon richness may have been related to competitive interactions.

In Chapter 1, I supported the argument (Levins 1970; Lewontin 1970; Eldredge 1982; Vrba and Eldredge 1984; Valentine 1968, 1969; Gould 1982a, c) that many evolutionary processes are hierarchical. Mass extinction exhibits this major feature of evolution (Jablonski 1986). At both ecological and taxonomic levels, the effects of extinction often are reflected at hierarchical levels higher than that of the individual. The explanations vary from area covered by a taxon to ecological properties of individual organisms (e.g., dependence upon light or water column structure). By extinguishing groups at high taxonomic levels, extinction of a family would of necessity affect properties at lower levels of the hierarchy, simply because a given feature of a given group happened to "be there" when a catastrophe occurred. Thus disasters that affected Permian reclining epibenthic marine organisms because of their overwhelming presence in tropical communities (Stanley 1984) might be due to cold temperature affecting the entire tropical habitat, rather than to the individual adaptations for reclining.

This conclusion is in no sense antithetical to the modern synthesis. Adaptation confers success in a given habitat, under particular ecological circumstances–no more. It's therefore a lame argument that, for example, "The first clam that fused its mantle margins or retained its byssus to adulthood may have gained a conventional adaptive benefit to its local environment. But it surely didn't know that its invention would set the stage for future increases in diversity" (Gould and Calloway 1980, p. 395). The potential for the future is irrelevant to the adaptation (e.g., Mayr 1983). The statement "Many features must come to prominence through their fortuitous phyletic link with high speciation rates" (Gould 1982a) is more a complementary notion than a complete reformulation of evolutionary theory, which Gould admits.

The adaptations required to resist processes that operate at larger taxonomic levels often have little or nothing to do with those that operate at lower ones, even if the reason for extinction resistance can stem from the same level of the hierarchy. A clear example is mass extinction. Genetic variation in temperature resistance is a common feature of marine invertebrates, showing within-species and between-species variation. Drilling by naticid gastropods is a feature that facilitates prey capture and also could presumably evolve from within-species variation. If the extinction of many groups was due to cold temperature, as claimed by Stanley, then the former adaptation, easily constructed from standard genic variation, is the relevant consideration, not the latter feeding adaptation. At the extreme, which I would not support, one might argue that all of the species dying out during an excessive cold event were those simply lacking adaptations to cold temperature, irrespective of whether one or a hundred taxonomic families disappeared. All properties of lower levels (e.g., shell sculpture) would therefore be carried along. This is a destructive rather than a constructive process. If the evidence from predation is any indication, most progressive evolutionary trends are constructive and probably are dominated by phyletic evolution, as well as spurts of success that result from key adaptations.

Declining Extinction: An Improvement in The Quality of Life?

Sepkoski's (1984) analysis of evolutionary faunas produces the fascinating result that successive EFs are less and less prone to extinction. In the Ashgillean and Frasnian extinctions, for example, the Cambrian EF suffers more than the Paleozoic EF. In the Permian and Norian extinctions, the Paleozoic EF suffers more than the Modern EF. Excluding mass extinctions, family-level background extinctions have declined over geological time (Raup and Sepkoski 1982). This decline may have been in two stages. During the Permian, extinction may have been reset to a much higher level, followed by a second period of decline after the Paleozoic decline period (Van Valen 1984a). Alternatively, the Permian family extinction may just be a temporary perturbation.

Family-level extinction rate declines during a long Paleozoic period of fairly constant taxon richness. In order to keep a steady state, a decline in extinction must be matched by an overall decline of originations. This is born out by a reanalysis (Hoffman and Ghiold 1985) of Sepkoski's data (Figure 8.14).

Why should family-level extinction and origination rates decline

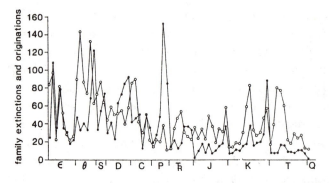

Figure 8.14: Total numbers of marine animal family originations (open circles) and extinctions (closed circles) per geologic stage (from Hoffman and Ghiold 1985).

over geological time? Van Valen argues that a decrease in probability of extinction can be explained by an increase of the ratio of positive to negative interactions among taxa. A reduction of competitive interactions might decrease extinction. Carnivores show a progressive specialization of families (Radinsky 1982), which might reflect a reduction in competition. But tiering–the presence of multiple living marine consumer layers, relative to the sediment-water interface–reached peaks of development in the late Paleozoic and in the Triassic and Jurassic (Ausich and Bottjer 1982). No long-term increase is notable. Bambach (1983), however, has shown that additional major resources have been exploited as time progresses. This suggests that the biota has been steadily invading new space; this might contribute to the decline of extinction, but it doesn't explain the decline of family-level origination.

Raup and Sepkoski (1982) suggested that the decline in extinction might reflect an increase in Darwinian fitness. This is not an argument for differences of superiority among major groups, as in mammals over reptiles. Such major shifts seem to require mass extinction. Rather, normal background extinction might have declined as fitness improved. But there is no real evidence that individual performance is related to species or family-level extinction. Unfortunately, our knowledge of extinction is limited almost exclusively to extinctions induced by man. But if the loss of the Brazilian rain forest is any indication, extinction is usually due to loss and degradation of habitat rather than to any particular properties of individuals.

Extinction may have declined as a result of the ratio of species numbers per families, which has been increasing steadily since the Mesozoic (Valentine 1969). If a family's representation in the world biota increases in numbers of species and its consequent ecological and geographic coverage, then the probability of extinction may decline (Flessa and Jabionski 1985, Jablonski 1986).

This explanation still does not provide a satisfactory answer to the decline in originations. Such a decline implies a long-term reduction in the production of novelties sufficient to define taxonomic families. In other words, the rate of origin of morphological diversity has decelerated over time. This is consistent with Valentine's (1986) characterization of the long-term decline of origin of basic ground plans. Since the Cambrian, no new ground plans have appeared, and a host of novel body plans appeared in the pre-Cambrian and Cambrian, never to reappear at the same level of diversity. Two concomitant processes may have contributed to this decline in origins of basic morphological diversity. First, a general filling in of resource space may have made it difficult for wholly new forms to take root later and spread. Second, as the shakeout of early ground plans occurred, perhaps by random extinction, the surviving ground plans may have slowly congealed to the point that genetic, epigenetic, and functional costraints precluded the rise of descendants with wholly different body plans. Heterochrony may be a mechanism for major evolutionary change, but wholly new phyla are unlikely to arise from chickens, or clams, or starfish. Our world may very well be the tangled bank conceived by Darwin, even if the exact mechanisms of constraint are more diverse than he conceived.

The world today is far different from that of the Cambrian world. We should not expect much new from our current cast of characters. They are the result of a long process of adaptive evolution, which has congealed a series of *baupläne* whose functional, genetic, and developmental interactions have all tightened to the degree that it is probably difficult to unravel the complexity of internal interactions. We may never see again the diversity of phyla that appeared so early in the beginning of the Paleozoic. It is no accident that the remnants of the many evolutionary experiments, so well-tuned and organized, have not given, and are not likely to give, rise to any new phyla. We can't speak for the future, but it may well be that the earth will see its entire biota wiped out some day. Only then is there likely to be a recreation of the great Vendian and Cambrian explosion of metazoan life.

The Main Points

1. Studies of diversity, appearance, and extinction depend upon the taxic approach, where given taxonomic levels are employed as indicators of biotic diversity. In some cases, abundance at, say, the family level is believed to be linearly related to numbers of species. In other studies, the family level is taken to be a measure of diversity of major adaptations.

2. Multivariate statistical analyses permit the resolution of marine fossil taxa into a series of evolutionary faunas that behave statistically independently of each other. The exact ecological or evolutionary significance of these associations is poorly understood.

3. The three evolutionary faunas successively achieve dominance in the fossil record.

4. Taxonomic longevity seems to vary among phyla; though average longevity may differ, the variability of longevity cannot be distinguished from a model of stochastic appearance and extinction.

5. Taxon diversity appears to be stable for long periods of geological time. The waxing and waning of the evolutionary faunas has been explained as the result of a three stage model of growth in diversity toward an equilibrium. The apparent plateaus in diversity have been differently explained as being due to a negative feedback imposed by large-scale extinction events. Some detailed studies suggest that steady-state diversity levels are maintained by a balance of appearance and extinction, but the overall issue is in need of further study.

6. Biogeographic provinciality influenced overall diversity. The Permian extinction was accompanied by a decrease in the number of marine provinces. This may have been related to a deterioration of world climate.

7. The Phanerozoic has been punctuated by periods of major extinction. Some were very severe, though it is generally not clear

whether species were just extinguished or speciation itself was depressed. It has been suggested that the major extinctions were qualitatively different from so-called background extinction; this can be seen to a degree in patterns of extinction as related to parameters such as the biogeographic range of a species. On the other hand, the major end-Cretaceous extinction showed selective eliminations that can be related to individual differences in adaptation. This is not so different from what might be expected during a fairly minor environmental shift, only on a larger scale.

8. Ultimate causes for extinction are poorly understood. Cycles related to the precession of the earth's rotational axis may cause strong changes on the scale of 10^4 years. A periodicity in extinction of about 26 my has been demonstrated for the post-Paleozoic extinctions. This periodicity has been related to several plausible astronomical periodicities, and the presence of iridium and shocked quartz at some horizons has supported the suggestion of bolide impacts as the ultimate cause of some extinctions. The actual pattern of extinction, however, is complex, and some extinctions occur before the hypothesized impacts. Unfortunately, the temporal level of resolution in the famous end-Cretaceous event is not less than 100,000 y. The proximate causes of extinction are poorly understood, but probably include changes in climate, sea-level changes, and climatic changes induced by rearrangements of continents.

9. The family-level extinction rate declined through the Phanerozoic. This is matched by a similar decline in originations. The explanation for this observation is not clear. Families may have become more speciose over time, and this may buffer their extinction, since, as time progresses, the extinction of any one species is less likely to result in the extinction of an entire family. This, however, would not adequately explain the decline in family-level origins. A gradual loss of potency to produce more families might be explained by increased competition, or the gradual increase in developmental constraints. The latter might be related to the pattern of declining rates of origin of higher taxa.

Chapter 9

Synthesis and Prospect

Introduction

In Thurber's fairy tale *Many Moons*, the king summons his advisors to fetch the moon for the princess. To the king's dismay, each advisor claims that the moon is made of a different substance and thinks it to be a different distance away. The court jester finds the solution. Obviously, each advisor sees the moon differently. Just ask the princess what she thinks. Only then will the king know what to do. As it turns out, she thinks the moon is smaller than her fingernail and is made of gold.

Evolutionary biology suffers from much the same diversity of viewpoints and expectations of solutions to complex problems. We tend to forget the importance of other perspectives and areas of study. Schindewolf saw evolutionary change through the perspective of the fossil record. Just like the princess looking at the moon alongside her fingernail, one is likely to draw incorrect conclusions. In more recent years, paleontological studies have injected a number of exciting and substantive ideas into evolutionary theory, even if the fingernail perspective is sometimes apparent. By ignoring the dimension of geological time, population biologists heretofore have largely ignored the colossal biotic changes that have swept the planet. Molecular evolutionists and students of molecular adaptation have ascended to prominence, and sometimes it seems as if they feel that a complete catalogue of nucleotide sequences will solve the deep problems of evolutionary biology. But to dismiss the new knowledge of the genome as mindless reductionism is to miss the most important window we have on organismal organization.

I wish to return to some of the general conclusions stated in Chapter

474

1, and to comment on their importance to the theory of evolution. I
will try to identify the areas of study central (and diversionary) to the
answers of some remaining important questions.

The Theory of Commitment

The Stabilization of Form

The history of life has witnessed a reduction of body plan diversity
and a stabilization of form. It is not clear how long it took for the
metazoa to arise, but a large number of phyla occur first as fossils
in a brief time span some time near the beginning of the Cambrian.
The Ediacara, early Cambrian, and Burgess Shale faunas all give an
impression of a greater diversity of body plans than we see today.
Indeed, the Ediacara fauna might be an independent experiment, only
distantly related to the rest of the Metazoa. Most of the extant animal
phyla appeared early in the Phanerozoic and the bulk of taxonomic
turnover occurred subsequently at lower taxonomic levels. I believe
that this is the single most important fact that the fossil record has
contributed to evolutionary biology. Indeed, in the Permian extinction,
the large majority of the species became extinct, but no new phyla
arose after a seeming opening of ecospace. The notion that phyla
appeared first is not just an artifact of taxonomy's correlation with
time of origin. A series of body plans took hold on the earth and
subsequent evolution has been a play on these basic themes.

Why haven't lineages bearing these basic body plans "broken out"
of the mold and given rise to new phyla? I believe that the general
model of an evolutionary ratchet (Chapter 5) is the key. Functional,
developmental, and genetic ratchets have combined to ascribe a high
burden, *sensu* Riedl (1978), to any prospective change. There has been
no drift from a set of basic body plans sufficient to preclude their identification with the phyla that became established in the early Paleozoic.
At lower taxonomic levels, however, long-term stability is also apparent. Many body plans and styles of living have remained essentially
unchanged for tens to hundreds of millions of years, despite extensive
speciation, the rise of many new groups, and significant changes in
environmental conditions.

It is therefore our task to understand the contribution of the three
ratchets to the stabilization of the number of phyla and of morphol-

ogy. If, for example, developmental or genetic constraints are most important in evolutionary conservatism, then there is some justification for taking the phyla to be *baupläne*, much as the essentialists, such as Cuvier and Owen, defined them. But we would have a mechanistic explanation for invariance. Variations within the phyla would be minor variations on a theme constrained by developmental and genetic (that is, internal) controls that prevent significant evolutionary change. But if functional considerations, particularly strong functional interactions among body parts, are the major source of stabilization, then the presence of a few body plans may simply reflect the survival of those that work. The description in Chapter 5 of the evolution of developmental programs suggests that functional, developmental, and genetic constraints evolve in concert, leading to complex phenotypic interactions. It is inappropriate to divide such patterns into adaptive and nonadaptive components.

The assembly of a *bauplan* may involve many, or just a few, steps. We might argue that in some cases singular novelties led to a fundamental reordering of evolutionary directions. If "mammalness" can be defined only with a fairly large number of strongly interactive characters, "molluskness" may have been defined by the two innovations of spiral deformation during development and a cap-shaped shell.

This subject is difficult to approach constructively, as it is easy to fall into the trap of subconsciously reifying a set of phylum-level traits into an essence that we believe to resist change simply because we know that the traits have survived. When we unite the arthropods by a flexible cuticular exoskeleton, we imply that the cuticle appeared early and remained stable for most of the taxon's history, without proving that the cuticle was retained because of its functional, genetic, or epigenetic burden. Descriptions of time course must, therefore, be accompanied by testable hypotheses on the trait's burden. Ernst Haeckel took the "tadpole" stage of development to be a fundamental indication of common ancestry of the classes of vertebrates that had never been altered in evolution. In more modern parlance one might hypothesize the tadpole to be an epigenetically constrained stage. But earlier, and later, developmental stages of vertebrates are taxonomically more diverse, as Haeckel's own work demonstrated. Thus the tadpole can be interpreted as a form retained by some functional consideration.

Despite this danger, it does seem clear that the time course of commitment seems to involve early acquisition of traits, with subsequent stability. Presumably, the stability represents the achievement of integration, which would have a high level of fitness loss if it were changed significantly. The morphology and even detailed behavior (as indicated by fossil morphology and preservation) of many groups have changed little since their appearance. Many groups appear suddenly in the record, suggesting that the evolution of many complex forms is rapid and then subsequently stable for very long periods. This early rapid evolution provides no evidence, however, for the role of species selection or nonadaptive evolution. It probably corresponds to the quantum evolution of Simpson (1944). The early rise of the phyla probably occurred in a world of very low species diversity; speciation and extinction probably had little to do with the tempo of change during the rise of the evolution of the phyla (Valentine 1986). At the beginning of the Phanerozoic, the rate of morphological evolution was a rapid response to a wide open ecospace.

The fossil record documents the longevity and constancy of many adaptations. Many of the most subtle behavioral interactions seem to have great longevity (Boucot, 1984). Coral-associated barnacles of the family Pyrgomatidae first appeared in the Miocene, and have not changed since. Their existence depends upon the ability of cyprid larvae to settle among scleractinian tentacles, and upon an intimate relationship between the coral and the adult barnacle. The initial relationship may have evolved rapidly, but once formed, this association has been stable. Even more impressive is the Coronulinae–the whale barnacles–with adaptations for life in the swift currents generated by the moving whale. Some of the basic features of this group have remained relatively unchanged since the Miocene. One could easily imagine this situation to be one of increasing commitment. The most spectacular case is the suite of invertebrates confined to deep-sea hot vents. A few mollusk species almost certainly have planktonic larvae, yet are committed to live their adult life in the rarest and most inaccessible of all ocean habitats (Lutz et al. 1984; Killingsley and Rex 1985).

The Organism's Interpretation of the Environment: Habitat Selection and Commitment

Adaptive evolution has centrifugal and centripetal components. Given sufficient genetic variation, a change in the environment can cause a rapid and extensive response via natural selection. This is the centrifugal component of evolution, which permits adaptation to new environments. There is a limit to this component since population variation is often insufficient to make an adaptive (and morphological) leap. An underemphasized centripetal component in evolution slows down the rate of evolutionary change, sometimes even in the face of environmental change. We have discussed in Chapter 5 the obvious genetical and developmental constraints that might create such a centripetal force, but there are also important forces that relate to functional interactions.

Functional constraint has several components, including (a) habitat choice, (b) functional interaction of traits, and (c) harmony of traits with the environment. Habitat choice is a major component in the stabilization of the phenotype, as it traps the organism in the same milieu. Larval selectivity of marine invertebrates assures that a dispersal stage will "find" the parental habitat. This keeps the animal in a similar substratum for many generations, but also assures that an organism well adapted to muds will not stray into sands. In cases where larvae will surely be swept to inappropriate areas, adaptations for larval retention exist, as in estuaries (Levinton 1982b). It is a telling point that Pleistocene migrations of latitudinal climatic belts have usually been accompanied by latitudinal shifts of vegetational zones, but not in a major evolutionary change in the degree of temperature tolerance sufficient to permit a species to stay put and survive the onslaught of climatic change during a glacial advance. The enslavement of organisms to their shifting environments is the essence of the fossil record, which records the shifting of sedimentary facies and their associated fossils. It is the basis of Walther's Law–one of the most fundamental precepts of stratigraphy–that sequences of environments (and associated fossils) will be identical both laterally and vertically. This can only be true if fidelity to a home environment (facies) supersedes the effects of strong directional selection to jump to a new adaptive mode.

An important component of stasis must be the relatively high cost in fitness of making the leap to a new mode. Many dispersing larvae, seeds, and adults find themselves in wholly inappropriate habitats. Thorson (1950) documented extensive wastage of marine invertebrate larvae on the wrong bottom substrata. The result is nearly always death, since the larvae and adults lack the features appropriate for the substratum. Although there is a degree of plasticity, most organisms tightly bound to a substratum are integrated in form and behavior such that they do best in one sediment type. Thus, natural selection cannot act, since all organisms are destroyed. Thus stasis by habitat choice can only be effective if mistakes are largely fatal. An impor-tant evolutionary theorem is that directional adaptive evolution in a novel habitat type will not occur if the initial drop in fitness concomi-tant with a habitat change is too great to compensate for the slow incremental increases in fitness conferred by adaptations that secure increased ability to find the organism's current preferred habitat.

Though modes of tissue interaction and the order of appearance of structures in development may well be fixed by genetic and develop-mental constraints, it seems unlikely that most of the fixed arrange-ments of external morphological traits found in long-lived fossils can be explained in this way. Most of the constancy we observe falls in mor-phological traits that are known to be heritable, and major changes can be effected in laboratory populations under favorable conditions in a few generations. This is even true of strongly canalized traits (e.g., Rendel 1959). But the movement of a founder population to a new and qualitatively different habitat constitutes the strongest directional selection we can imagine. The lack of evolutionary potential, however, is clear when one considers the very low probability of the appearance of a mutant that is suitable to the new environment. Its low frequency would diminish its probability of fixation, as its colonizing popula-tion would have an extremely low probability of lasting more than one generation. Catastrophic directional selection is therefore likely to be usually ineffectual. This would suggest that most Wrightian land-scapes effectively have only one peak: The one that the population now occupies. Stabilizing selection therefore is the predominant mode throughout biotic history.

Punctuated Equilibria and Saltations: Two Blind Alleys

Punctuated Equilibrium: The New Essentialism

The punctuated equilibrium hypothesis (Eldredge and Gould 1972) has been the basis for a large body of speculation on the structure of the biosphere and the dynamics of the evolutionary process (e.g., Gould and Eldredge 1977; Stanley 1979; Eldredge 1985; Gould 1985*a*). As discussed in Chapters 3 and 7, it ascribes a special importance to speciation as a major cause of morphological evolution. The remaining history of a species is regarded as a quiescent period, as the species has a homeostatic mechanism that resists change. Trends in evolution are therefore variations in speciation and extinction rates, and phyletic evolution is therefore an incomplete explanation of evolutionary trends. Gould (1985*a*) sees the theory of punctuated equilibrium as the cornerstone of a larger hierarchical theory of evolution, though Sober (1985) shows that punctuated equilibrium is not a necessary prerequisite for such a theory.

My discussions in Chapters 3, 4, and 7 suggest that the theory is inviable. There is no evidence that phenotypic evolution is necessarily, or even commonly, causally associated with speciation. Allozymes, chromosomes, and morphological evolution all fail to show a significant break at the species level. Speciation is caused by a variety of genetic processes which lead to intersterility. When the establishment of sterility coincides with adaptive evolution, one can indirectly associate speciation with phenotypic evolution. In these cases, however, speciation is the effect of phenotypic evolution or it is merely coincidental with phenotypic divergence.

Critics of the punctuated equilibrium theory are blamed for misunderstanding its import, which is the significance of stasis in understanding the evolutionary process. Stasis is an important characteristic of many lineages. But this does not necessarily prove the importance of speciation in breaking stasis, nor does it explain constancy of species properties because of membership in a single species. After all, many genera and families are remarkably phenotypically homogeneous (e.g., Wake et al. 1983). Rampant speciation seems not to beget phenotypic diversity in a wide variety of groups that comprise sibling species complexes. Surely one cannot credibly assert that a morphology such as that of *Lingula* has not changed much for lack of speciation. We know that shell characters can change dramatically within local populations, let alone over hundreds of millions of years. Stasis is interesting, but there is simply no body of evidence that relegates its cause to a failure of speciation.

While our knowledge of the genetics of living organisms refutes the punctuated equilibrium hypothesis, the fossil record presents a more complex picture. Because species diagnosis has always been associated with a threshold of morphological change, paleontologists have always taken speciation to be phyletic changes, beyond a given threshold of difference, or they have identified coexistent species by morphological discontinuity. Indeed, cladogenesis in fossils can only be identified by morphological discontinuity, as we obviously cannot assay for sterility. In some cases, phenotypic variation once used to distinguish generic differences has been shown to be a series of morphs within a single population. Since phenotypic change is the source of inference of speciation, the punctuated equilibrium hypothesis is untestable using fossil data. Because our scale of resolution is usually so poor, we will never know whether speciation causes sudden morphological change, or whether rapid phyletic evolution is simply closely correlated in time with the development of reproductive isolation. If the latter is true, then phyletic evolution may be sufficient to explain nearly all adaptive evolutionary trends. If punctuated equilibria merely amounts to stabilizing selection following this rapid shift, then it cannot be linked readily to a hierarchical theory involving the species level, since phyletic evolution would be the principal mechanism.

Fossil data, however, can be used to ask whether phyletic evolution is sufficient to permit extensive morphological divergence. The bulk of the evidence suggests that speciation is not required for major morphological evolution, and certainly not at the level denoting typical species differences. Unfortunately, this is not an adequate test of the punctuated equilibrium hypothesis, which asserts the predominance of speciation as a cause of morphological evolution. The evidence, however, does show that phyletic change is common and sufficient to generate large-scale evolutionary change (Chapter 7).

Longer periods of geological time tend to include more periods of nondeposition and periods of stasis or reversals in the pattern of evolution. These longer periods are those which have given us our general impression of very slow rates of evolution. Owing to this bias, we know little or nothing about absolute rates of evolution from the fossil record. Long-term stasis (i.e., a lack of a trend) may still prevail, despite a spectrum of rates.

The research emphasis started by Eldredge and Gould (1972) has had a positive effect in stimulating a number of studies of transitional evolution. But most of the debate has proceeded with examinations of character change with no genetic mechanisms in mind. As a result, traits that might be strongly or weakly canalized have all been blended together and subjected to the litmus test of stasis versus gradual evolution. This approach is naive and makes most of the arguments in the literature rather sterile. The same problem holds for the question of discontinuous versus continuous traits. We cannot hope to understand the many patterns found in paleontological data without some models of genetic determination in mind. After all, a meristic trait such as bristle number can be strongly canalized or can be quite labile and subject to rapid directional selection. Similarly, strong phenotypic shifts in shell form can be quite difficult to accomplish, or may be under the control of a simple genetic polymorphism (e.g., Palmer 1985). This ignorance leads to futile debate. Pattern does not map uniquely to process in the question of stasis or varying rates of evolution.

The punctuated equilibrium hypothesis would harken us back to the essentialism that we escaped in the neo-Darwinian period mentioned in Chapter 1. At this time, Fisher, Haldane, and Wright combined Darwin's theory with the algebra of genetics and formulated a way of looking at evolutionary change that has had an astounding degree of success. The founders of the Modern Synthesis carried this information into the practice of systematics and natural history, and found it to be an essential tool in the understanding of the evolutionary process. The empirical discoveries of the past forty years would not be very useful without this theoretical groundwork. Indeed, the force of the neutral theory of molecular evolution is its contrast with the theory and practice of preceding decades. While there still are a large number of enigmas concerning the structure of the genome and mechanisms of speciation, the population thinking founded by Darwin and carried into the twentieth century by the neo-Darwinians and the Modern Synthesis, still stands as the best framework for understanding adaptive evolution. We have tried to lay essentialism to rest; one of its descendants, the punctuated equilibrium hypothesis, deserves a similar fate.

The Nonproblem of Discontinuity

The presence and explanation of morphological gaps between organisms has been a continuing theme in biology. While preDarwinians presumed gaps, by virtue of their typological approach to biology, a succession of debates among evolutionary biologists has revived the question of discontinuity. Indeed, I began this volume with the question of why there appeared to be clumpings in morphospace. The argument was important in the debate between "biometricians" and "geneticists," where the former camp thought continuous variation to be the stuff of evolution, and the latter group saw discontinuous mutations as the driving force in evolution.

The main reasons for gaps in morphospace among members of the same phylum are divergent evolution, extinction of intermediates, and functional limitations, working within the constraints of a group of constructional plans, such as the logarithmic spiral shell. The astounding range of intermediate forms that have been discovered as fossils suggest a long history of cladogenesis, phyletic evolution, and extinction of intermediates, though rapid evolution between modes must have often been the case. Where ancestors and apparent descendants have been found in the fossil record, subsequent collections have often filled in the gap with intermediates. Integrated phenotypes in the mammals can be shown to have congealed over time, suggesting the process of adaptation in creating the gaps via divergence. In other cases, such as in the mollusks, gaps in morphospace are not filled for functional reasons. The mechanisms that generate variability, constrain direction, and create some saltatory effects in mutations are of great interest, but largely secondary to the three processes mentioned. This is not to say that all functionally harmonious forms have existed throughout geological history; historical accident has prevented any number of imaginable morphological constructs.

Recently, as has been true periodically throughout the past 100 years, there has been revived emphasis on phenotypic discontinuity, with developmental mutants taken to be the evidence for the potential for saltation. This argument has ranged from the reasonable assertion that some discontinuities stem naturally from the mechanisms of genetics and development (e.g., Alberch 1980), to the unreasonable presumption that, just because developmental anomalies are present, they might be a vehicle for rapid and saltatory evolutionary change (Lovtrup 1974; Gould 1980a). We have discounted the latter in Chapter 5, but the former follows from all we understand of the genetic determination of traits. Indeed, this knowledge has been left to us by some of the orthodox founders of the neo-Darwinian movement.

I have explained in Chapter 5 the error of taking present-day phenotypic organization and developmental anomalies and presuming that they imply saltation. This error is traditional, and mimicks Punnett's (1915) error of confusing the current genetic organization of mimetic butterflies as evidence of a saltatory origin. We now know that strong canalization among morphs, tight linkage among genes controlling the same trait, and switch genes that control batteries of other genes, are mechanisms that now control discontinuity, even if the origin of the trait's organization happened to be gradual. The vast array of studies show that even seemingly invariant traits are regulated by many genes and that the constancy has been fixed by the evolution of thresholds that enhance the canalization of the trait. This constancy must be reinforced continuously by the action of natural selection, as relaxed selection usually leads to a loss of integration of a trait (e.g., eyes in cave fishes).

While the history of many groups has not been the path that might be designed by an engineer, the role of adaptation in producing the clumpings in morphospace is quite clear. Today, after more than a decade of debate about above-species level processes, I believe that no knowledgeable scientist entertains seriously the notion that complex structures such as eyes, jaws and limbs evolved by shuffling of traits acquired by random fixation during speciation events.

The proponents of saltation in recent years have not presented a clear picture of why neo-Darwinian concepts fail to account for discontinuity, or evolution mediated by alterations in development. Examples that are commonly cited as hopeful monsters, inconsistent with neo-Darwinism (e.g., Frazzetta 1970), typically involve variation requiring fairly trivial mutations that can easily be imagined to arise within typical populations. Surely, the developmental mechanisms that have been recently popularized were discussed (with different specifics) back in the 1920s and 1930s (e.g., Ford and Huxley 1927; Haldane 1932b). Maderson et al. (1982) argue that developmental mechanisms producing variation in form are not congenial with the Modern Synthesis. But the mechanisms they discuss are not antithetical to any questions of genetic determination of variation. Indeed, P.F.A. Maderson (written communication, 1985) admits that this is so. Gould (1984b) has revived Mivart's old challenge that many structures must have been formed by saltation, because functional intermediates (half a wing, half an eye) are inconceivable, but even here, Gould has been inconsistent, even contradictory, in his arguments. He writes:

> At the higher level of transition between major organic designs . . . gradualism has always been in trouble. Although it has stonewalled with commendable tenacity. No one has ever solved Mivart's old (1871) dilemma of "incipient stages of useful structures." (Gould 1984a, p. 14)

But, elsewhere, Gould states:

> I believe that Darwinism has, and has long had, an adequate and interesting resolution to Mivart's challenge. [1] (Gould 1985b, p. 14)

These statements reveal ambivalence. Discontinuity in morphological evolution is surely an issue, but hardly one that shakes the foundations of neo-Darwinism.

There seems to be no real reason to abandon the conclusions of the neo-Darwinian era: Most mutations are of relatively small effect and larger-scale mutations, though known to occur, usually reduce fitness. Therefore, smaller-scale mutations probably are more important in evolution. Major saltations are not precluded so much as improbable. This is not to preclude discontinuity. Indeed, as I have mentioned, discontinuity has been a prime area for research in the area of transmission genetics. But no recent claim for discontinuity has been inconsistent with neo-Darwinism. The interesting notion of developmental constraint (e.g., Gould 1977; Levinton 1986; Maynard Smith et al. 1985) is an important addition to the study of adaptation, but hardly at odds with Darwinian theory, nor does it require great leaps of morphological evolution (or faith).

---

[1] Gould's second quote refers to Darwin's explanation that a structure can have two different functions during its evolution.

Organismic Integration and Character Evolution

Mosaic Evolution, Stasis, and Genetic Correlations

Although many characters evolve, it is equally true that others have remarkable stability. The latter, of course, are not used to define lower taxa such as species and genera, due to their very constancy. In the study of omomyid primates by Rose and Bown (1984), obvious transitional evolution can be documented in premolars, but molars remain relatively constant. In other lineages character change is rampant and permeates most of the organism, at least in those parts that can be examined in fossils. Change in a large number of characters has been documented in lineages of fish (Bell et al. 1985), scallops (Miyazaki and Mickevich 1982) and foraminifera (Malmgren et al. 1983). There is no reason to believe that characters are rigidly constrained, even though we do see character constancy in certain characters and character groups in particular lineages.

The evidence from the fossil record, as discussed in Chapter 7, suggests a range of rates of morphological evolution, including what I have termed character stasis. While it is indeed possible that factors such as genetic and developmental constraints shape the course of evolution, there is no particular evidence to falsify the notion that function is the major factor in constraining rates of evolution. Features that are strongly canalized are obviously constant as the result of a genetic-physiological mechanism. But laboratory studies (Waddington 1956; Rendel 1959) show that canalization itself is subject to alteration by selection. Strong directional selection can move the genotype–phenotype interactions into a new regime where constancy is now substituted by variability in a population. A constancy that appears to be shaped by developmental constraints may actually be an ephemeral mechanism to stabilize the phenotype, but alterable by directional selection.

There are functional bases for variation in rates of evolution among characters. The best example of this comes from the relationship of function to the rates and patterns of molecular evolution. Genes that encode proteins with highly constrained functional interactions are those which are most conservative in the evolution of amino acid sequences (e.g., histones). In contrast, genes that encode proteins whose change in function is more loosely related to amino acid sequence are those that evolve very rapidly (e.g., the fibrinopeptides). Silent nucleotide sites evolve far more rapidly than those involving amino acid changes (see Chapter 3). Because proteins interact with others, functional interactions probably cause a form of molecular coevolution to occur. We have an analogy with what we perceive to be functional restrictions and interactions in morphological evolution.

The nonadaptive component of discrete variation and long-term constancy in many traits has been misrepresented as evidence against the crucial importance of natural selection and adaptation. There is nevertheless a clear role of developmental interactions in imposing a centripetal force in evolution. While there is a considerable plasticity in evolution, both developmental and genetic evidence suggest that the direction of evolution may be constrained by extant interactions in epigenetic and genetic pleiotropy. Current interactions in development may preclude certain pathways of evolutionary change. As shown in Chapter 5, the dependence of one tissue upon another for induction could reduce the fitness of mutants that permit independence among these tissues. This is the basis for both the notion of developmentally based stasis, and the low probability that developmental mutants of strong phenotypic effect are the stuff of evolution. There may be units of the phenotype that are bound together, but relatively independent of other blocks. The question is what binds traits together, and whether this binding is sufficiently permanent that evolutionary direction is constrained.

Genetic pleiotropy may contribute to the construction of blocks. Some of the atavisms discussed in Chapter 5 suggest that sets of genes may have been retained that still have the capability of switching on structures that have been long absent in a given lineage (e.g., toes on horses, limbs on whales). This may be due to epigenetic pleiotropy, or to genetic pleiotropy, which precludes loss of either the genetic or epigenetic mechanism. In the latter case, we might imagine that the genes determining the epigenetic track may have changed over evolution, but the epigenetic process itself has been maintained (Figure 9.1). This, of course, does not eliminate the role of the underlying genes in evolution. The epigenetic mechanism may be the constrained phenotype but the genes are the only means of transmitting information between the generations.

While tissue interactions are likely to channel evolution in certain directions, the integral nature of the phenotype is due to the continual

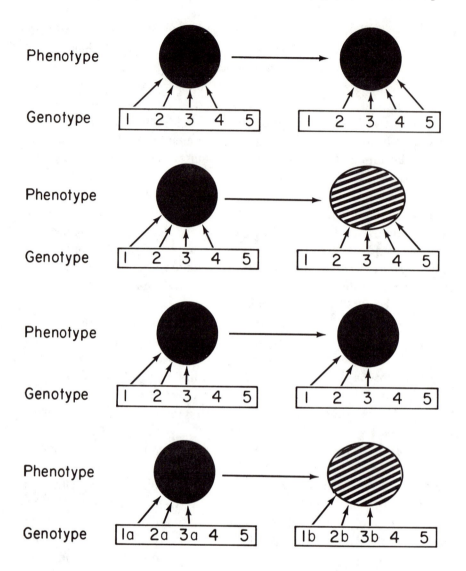

Figure 9.1: Genes, homology, and phenotypic evolution. Various alternative gene–phenotype relationships during the evolutionary change, or lack thereof, of a character. Numbers refer to different genes and subscript letters refer to alleles. A: No change in phenotype, but change in the controlling genes. B: Change in phenotype and controlling genes. C: No change in phenotype or genes. D: Change in phenotype, owing to fixation of new alleles at the same controlling loci.

action of natural selection. This is best seen when a structure loses its former function. Fishes living in caves show varying stages of eye reduction, relative to their close relatives in nearby sunlit streams. But as eyes are reduced, the intercorrelations among parts of the optic system lose their formerly strong phenotypic correlation. The regressive evolution has proceeded quite differently in different populations and has involved the effects of more than ten genes (Wilkins 1971). While the mammalian cranium is an integrated functional unit, genetic correlations among parts of the unit often do not fit the expected correlations expected from function (Cheverud 1982). This argues that stochastic processes often generate genetic correlations that are maladaptive, but natural selection maintains the functioning integrated phenotype. Therefore, the aspect of channeling suggested by the concept of developmental constraints cannot operate without the continual action of natural selection. Any genetically or developmentally constrained variants that are not functionally harmonious will be culled out by this process.

The question is how to quantify the degree of organismal integration, and to dissect its genetic, developmental, and functional components. To approach this problem, models have been developed that dissect total phenotypic variation into genetic and phenotypic correlations (e.g., Cheverud 1982). These correlations may be contrasted with those expected from functional models, and the deviation can be considered the degree to which the generation of among-organism variation creates functionally disharmonious morphologies. This approach can be extended to tissue interactions and the possible role of developmental constraints in preventing the evolution of functionally optimal morphologies.

A genetic correlation analysis might also reveal pleiotropies that are not transient and therefore constrain the directions of phenotypic evolution. Riska and Atchley (1985) have used such an argument to suggest that the evolutionary change of brain and body size are inevitably correlated in a precise way. It is not clear yet whether they are correct, but this approach is far more valuable than past studies, where allometric coefficients have been used uncritically to support the notion of constraint in evolution (see Chapter 6). One useful way of looking for such constraints is to scan for relationships among phenotypic characters that fail to change throughout a clade (e.g., Cheverud, et al. 1985). This would give information that would confirm predictions made from extant genetic and developmental correlations in experimental populations, which might, after all, be transient.

Experimental embryology may also be approached in this way. A set of developmental correlations may be used to examine the question of whether current epigenetic interactions or the potential loss of fitness when development is altered significantly might play a role in constraining the direction of morphological evolution. Although they fail to consider competing functional hypotheses, the study of Alberch and Gale (1983) on epigenetic restriction in digit loss is a model study in this area. It is hoped that future studies will focus on structures that can be examined from the genetic, epigenetic, and functional viewpoints simultaneously.

The Nature of Homology

Before Darwin, invariance among organisms was the signpost of the essence that united certain taxa and excluded others from the group. Darwinism obliterated essentialism, but substituted several possible explanations for constancy. These alternatives, convergence, parallel evolution, and homology, are the central problems of systematics today, and their understanding is necessary before classifications can have a genealogical meaning.

Homology has two very different spheres of importance. Homology has been employed mainly to establish genealogical relationships. It means little whether homologies come from genotypes or phenotypes; the more information we have, the more informative our data base, and the better the genealogy. After homologous phenotypic characters have been identified it becomes interesting, from the viewpoint of the mechanisms of evolution, to understand the genotype–phenotype relationships of the character as it changes throughout the history of a clade. Some traits, be they genes or phenotypic traits, become duplicated or reduced in number during evolution. These are *serial homologues*. Alternatively, we follow a single trait as it changes during the history of a clade. The traits in related organisms can be termed *evolutionary homologues*. It is this latter type of homology that is important in genealogical reconstruction, but serial homology must be accounted for as well, as in cases where gene duplication leads to a new phase of evolution in independent directions in two serially homologous traits.

In Chapter 2, we attempted to define homology in terms of traditional morphology. But our explorations into the nature of development and evolution, and the evolution of genes, suggests that a fundamental problem can arise when we probe these definitions. We presume that homology involves similarity of location of a trait, where "location" has both spatial, temporal, and genetic implications. But these components can be conflicting. Consider a mechanism for heterochrony proposed for the nematode *Caenorhabditis elegans* (Ambros and Horvitz 1984). The appearance of a morphological structure can be accelerated or retarded by changing the initiation of crucial cell lineages. Furthermore, changes in the initiation of cell expression can be altered by altering either rates of synthesis or transport of a number of substances. Changes in reactivity of target cells might also occur. Here, there is no predictability about which specific genes will effect the resultant phenotype. In a segmented animal, the number and form of segments might be determined by a concentration gradient in morphogen, or by a gradient of response by the target cells. A given segment may change its properties by an evolutionary change in the genes that determine the strength of the morphogen gradient, or by genes that determine the cell's interpretation of the morphogen signal. There is enough potential complexity that the form of the segment might be the only practical way to infer homology. The challenge lies at the morphological level. Are the bones, muscles, and nervous tissue of the mammalian forelimb determined by the same gene sets as those in anurans?

Cladistics and the Study of Organismal Evolution

In Chapter 2, I suggested that cladistic approaches to the inference of genealogies and the establishment of classifications are to be preferred, as they provide a great deal of information on character transformation. This claim can now be strengthened when we consider the evolution of a complex *bauplan*, discussed in Chapter 7.

Our understanding of evolution suggests a potential mapping between the taxonomic hierarchy, as established by cladistic procedures, and the evolution of organismal complexity and specialization. A phylum is characterized by a set of character states, which are found in some (perhaps modified) form in all member subtaxa. These traits must be accommodated functionally within any derived taxon. Thus the rise and evolution of the mammals must have occurred within an historical context set by the constraints imposed by the various traits of vertebrates. The derived traits characterizing the mammals are the most recent modifications of a long constructional history which first defined the chordates, then the vertebrates, then a group including the reptiles, birds, and, finally, the mammals. Although I am sure that most functional morphologists appreciate this, very few have used phylogenetic analysis to establish boundary conditions (Lauder 1981). Cladograms permit the sensible marriage of history with models involving process.

When described in this way, the ranking of groups such as the birds and mammals is no longer simply an issue of classification. The status of Aves, for example, has become a cause célèbre in the cladistics-phenetics debate, since a cladistic approach would assign this taxon to the group including crocodiles and dinosaurs, birds, and mammals. Evolutionary systematists would prefer that Aves be a class, due to its phenetic distinctness. On a superficial level of similarity this makes sense, but from the perspective of organismal evolution, the subordinate status of Aves reveals its evolutionary history and the phylogenetic factors that constrain the characteristics of birds. The subordinate status of Aves under a cladistic approach reveals quite nicely the organismal description of a bird as an "archosaur with feathers." It may be worth retaining the traditional rank, but our thinking is best clarified with a ranking system that recovers the cladistic information.

The Relation Between Phenotype and Genotype

Molecular geneticists are probing at the nature of determination of the phenotype by the genes. In nearly all cases, this amounts to an understanding of transcription and translation, and gene regulation for single genes. In a case like the *white* locus of *Drosophila melanogaster*, a visible morphological trait, eye color, is related simply to genic level changes. But in most cases structures are determined by many genes, with a complex developmental pathway. A great deal of progress has been made in simplifying the complexity, through the study of mutants (e.g., Nüsslein-Volhard and Wieschaus 1981) and surveys for persistent DNA sequences known to be important in development (e.g., McGinnis et al. 1984*a, b*; Carrasco et al. 1984). This purely descriptive effort is one of the most crucial in assuring a potential understanding of the mechanisms of morphological evolution.

On the other hand, it would be inefficient not to work at different levels of organization as well. As I have mentioned above, a brief consideration of the nature of homology leads one to believe that it is a phenotypically oriented concept. A variety of genetic mechanisms can alter the same structure in evolution. Thus, it is fair to say that we will never learn certain things about phenotypic evolution if we worry about the genes alone. The organization of the phenotype is an important determinant of evolutionary potential and limits the number of potential pathways. Thus, the investigation of constraint, be it due to function, development, or genetic mechanisms, is important. Alberch's (1980) surveys of allowable states in evolution may not give us the genetic mechanisms determining the alternative states, but they may be the most efficient way to understand some factors that limit the direction of evolution.

The new work on gene structure is important simply because it gives us a crucial window into the organization of the phenotype and gene function. There is a danger, however, that natural historians will see this endeavor as an attempt to understand the world through an extreme ontological reductionism. It would be a great pity if this were so, for there is little evidence that most advances in molecular genetics have had this atomistic objective. Schaffner (1974) demonstrates that many of the most fundamental advances in molecular genetics (e.g., elucidation of the *lac* operon) worked with hierarchical constructs and approaches that are rather typical of biological research at other levels of organization. These are the very objectives that motivated Goldschmidt's research which sought an understanding of how genes operated to determine physiology and development (G.E. Allen 1974). If anything, current molecular genetics has marched to Goldschmidt's drum, as opposed to the population and transmission genetic traditions left to us by the neo-Darwinians. Some natural historians have misunderstood this and have fought a rearguard battle which associates molecular genetics with vulgar reductionism and holds Goldschmidt as an icon of antireductionist thinking. There is an irony in this, as Goldschmidt was very much a reductionist himself, in the sense that he believed in the existence of a science, physics, that would be the explanation for biological science (G.E. Allen 1974).

Gould (1983a) has accused the founders of the Modern Synthesis of purging all points of view at variance with the concept of natural selection. There is no doubt that the exquisite intellectual simplicity and power of natural selection have been rightfully seductive to many a naturalist. But the absence of concern for factors such as developmental constraints has had little to do with any conspiracy to effect the hegemony of natural selection. Evolutionary biologists have been mainly concerned with the *fate of variability* in populations, not the *generation of variability*. Even though the neutral theory of evolution was reviled by selectionists, its concern with variability caused it to be noticed by population geneticists and to become the major issue of evolutionary biology for thirty years. In contrast, the genetic and epigenetic factors that generate variability have received relatively little attention. This could stem from the dominance of population genetic thinking, or it may be due to a general ignorance of the mechanistic connections between the genes and the phenotype. Whatever the reason, the time has come to reemphasize the study of the origin of variation. Studies of genetic correlations, epigenetic interactions, and molecular homogenization of gene families all have a place in a theory of the generation of phenotypic variability that must emerge in the coming years. I see no conflict at all between this emerging discipline and the traditional territories of neo-Darwinian theory, though it is clear that some accommodations will be made. It is a fallacy, however, to believe that these accomodations will shake the foundations of neo-Darwinian theories.

The Importance of Hierarchies in the History of Life

Biologists are accustomed to viewing biological organization in terms of a hierarchical structure. We routinely work with levels of organization such as cells, tissues, organs, and organ systems. We therefore accept implicitly that this organization must have had an influence on the course of evolution. Hierarchical levels can be defined in terms of the components at lower levels. When we say "in terms of" we mean that some complex function can characterize a hierarchical level, using elements from the lower levels. Thus the hierarchical level of cell draws its meaning from a unique determination of combined features such as mechanical properties of structural proteins, DNA replicators, RNA synthesizers, and so on. Apparently, this unique type of determination is common enough, probably due to the dictates of function, that the hierarchical level of "cell" is established.

Simon's (1962) notion of decomposability is a useful way to look at hierarchies. Here we regard hierarchical levels as constructs which respond differently from others. Having identified such levels, it is then our task to understand how the components interact to determine the level, and how upward and downward causation might effect the future of the hierarchical level.

The notion of decomposability and level characterization can be applied to the taxonomic hierarchy. Valentine's (1968, 1969) pioneering investigations showed a great difference in the rate of appearance and fluctuations of different taxonomic levels through the Phanerozoic. The phyla appear relatively early in the Phanerozoic and their numbers remained relatively constant, after an initial shakeout. In contrast families fluctuate much more, even if their extinction and appearance rates have declined regularly toward the present (Raup and Sepkoski 1982, Hoffman and Ghiold 1985). As phyla are composed of families, our particular example would take the family level to be the lowest unit of the hierarchy and our focus would be shifted towards the determination of the phylum level. In order to understand the nature of phyla, the factors leading to commitment to a given body plan must be understood, as mentioned above. Phyla may have survived as a result of a degree of success in adaptedness that would have to be defined in terms of testable hypotheses of functional success. Strathman and Slatkin (1983) argue that the survival of many phyla is improbable on the basis of a purely random model. Association with particular habitats may have assisted their survival, but there is no apparent ecological pattern to those surviving phyla with relatively few species. Alternatively, phyla may have survived simply because of the numbers of their component taxa, which have spread throughout the earth. It would be impossible today to extinguish the Chordata without a worldwide catastrophe. This view would imply the early appearance and extinction of many phyla.

Hierarchies and the Species Level

We have argued above that speciation is not a driving force in morphological evolution, in the sense that it is the principal process that generates variation or provides the necessary conditions for adaptive evolution. Speciation can occur concomitantly with progressive evolution. A separated population in a novel environment might diverge coincidentally with its reproductive isolation from the parent. During times of radiation, progressive evolution amounts to cladogenesis.

At the gene level, the comings and goings of genes may register processes occurring causally at other levels. For example, the extinction of a population might accidentally extinguish a gene unique to that deme. As a result, one might misinterpret the deme's extinction as being caused by the gene. Much in the same way, speciation events may register adaptive events, without having a causal connection. While directional selection causes phyletic divergence, reproductive isolation might arise coincidentally. But it surely would be a mistake to invoke speciation as a causal mechanism. In this case, the species and deme hierarchical levels would coincide.

But speciation does create sets of reproductively independent lines, which are then capable of independent directions of evolution. This independence is not total as two daughter species still inherit very similar genomes, developmental pathways, and functional constraints. But it does permit exploitation of a wide variety of habitats, certainly much more diverse than can be exploited by a single species. In some cases speciation is required in order for certain evolutionary advances to occur. While phyletic evolution could progressively lead to ever more dramatic secondary sexual features, or ever larger size, certain leaps could not be made in a single phyletic lineage. If two species of flower existed with corolla depths of 1 and 50 mm, it is unlikely that a butterfly adapted to the 1-mm flower will have variation sufficient to leap onto the 100-mm adaptive peak. But if a series of intermediate flower species and butterfly speciation events were available, this might occur. Interspecies variability can exceed intraspecies variability, and is therefore required to permit the living biota to exploit all possible habitats (Futuyma 1987). Note that this point has absolutely nothing to do with the efficacy of punctuated equilibria, though it does confer some importance to speciation as a necessary factor in evolution. This hardly seems controversial. After all, would we expect a single species to embrace all the variations from amoebas to man? Genetic recombination and functional considerations would preclude this.

In Chapter 4, we defined species selection and species drift, processes that are suggested by the process of speciation (Levinton et al. 1986). Species selection occurs when some some species-level property influences the relative abundance of species over time. Thus, for example, because it is inherently an aspect of "fecundity" at the species level, high speciation rate might promote the dominance of certain clades after long spans of geological time (Vrba 1980). This seems not to have been the case, perhaps because high speciation rate seems to be positively correlated with extinction rate and is also inversely related to species longevity (Stanley 1979). On the taxonomic level of phylum, the process seems to have been unimportant, given that no phylum has become extinct since the Ordovician, despite Stanley's findings of broad differences in speciation and extinction rates among the phyla. The survival of phyla may be more related to their rapidly established broad geographic ranges, and an ecological diversity sufficient to make phylum-wide extinction difficult.

Species drift may be an important process in evolution. Here, a clade increases in species richness mainly as the result of random speciation and extinction. The distribution of a given character state among a group of species is amplified, or contracts, by virtue of its happenstance association with a clade that is spreading successfully, or disappearing as the result of stochastic processes. Speciation can beget (a) a greater geographic spread and (b) the invasion of a greater variety of habitats, than would be the case if the character were associated with one species. Thus characters will be immortalized if they can spread in this way.

This effect, known as species hitchhiking, can have profound effects on the ultimate direction of evolution. Adaptation depends upon the previous groundstate of the species upon which natural selection operates. The direction of evolution that survives the vagaries of stochastic extinction depends therefore not only upon the potential direction of evolution expected within a single population or species, but also upon how traits have become distributed among species in a clade. If, for example, cladogenesis results in a given character being the most common at the time of an environmental shift, then that character state will be part of the interacting plexus that will determine the future course of evolution. Consider the evolution of the mammalian condition. While this analysis has not been performed, it may well be that the very abundance of a complex of traits among species of one group may have made it the most probable source of more advanced mammals. Cladistic analyses are therefore required to understand the backdrop of the adaptive evolutionary process.

Species hitchhiking may also be associated with evolutionary deceleration. A functionally harmonious complex of traits may spread through a clade and preclude a significant evolutionary change, simply because any further change would be functionally inharmonious with the extant group's morphology.

Evolutionary biologists routinely speak of structure x as being modified to serve a new function. This is an implicit acknowledgment that adaptive evolution is not a process wherein a population's storehouse of variability is designed expressly for every new challenge. Otherwise, one would not expect the peculiar condition of innervation of the vertebrate eye, where the retina does not have the clear optical path that one encounters in squids or even crustacea. No designer would think of making horses hooves from a set of toes, but that was the required pathway, given the history of the equids. Despite this eccentricity of origin, no processes of differential speciation and extinction have assembled complex adaptations; this is a supposed challenge of the punctuated equilibrium theory, but it is bogus.

There is, nevertheless, a place for above-species level processes in determining the fabric of the history of life. This is especially true of extinction, which has been capricious and has foreclosed many directions of evolution. The evolution of functional integration may have led many groups into evolutionary deadends. When these complexes of characters prevailed in groups that dominated the planet, then evolutionary change in ecologically similar, but rare, groups may be at a dead end. This may be behind the delay between the rise of the mammals and their eventual spectacular morphological–ecological radiation in the Cenozoic. The predominance of the reptiles may have coopted the geographic and ecological space necessary for mammalian morphological change to take root and spread. This is not to say that mammals are inherently superior to reptiles, but diversification in the mammals may have been precluded until sufficient ecospace was opened at the beginning of the Cenozoic.

Extinction, Species Drift, and Downward Causation

Extinction is a process involving bounded geographic areas. Given that species have restricted geographic ranges, the loss of a given area might drive a species or even a larger taxon to extinction and would also eliminate characters that were previously fixed and spread through the hitchhiking effect. As the geographic extent of habitat disturbance increases, this process may be more pervasive. This would be especially important if the habitat disturbance focused on a set of habitats restricted to certain environments, such as the tropics. If this were true, then features of species in these habitats that coincidentally were evolved as adaptations to this habitat might be eliminated. These features, however, may not have had anything to do with differential resistance to extinction owing to the habitat disturbance.

For example, a time of cold climate might selectively drive shelled species to extinction in the tropics. These species, however, may have had spines or other mechanical adaptations that apparently occur frequently in an area of high predation pressure. Because such features are correlated with tropical existence, the extinction would effectively eliminate ornamentation. It could not be said, however, that ornamentation had previously prevented extinction with respect to this level of habitat disturbance. This is an excellent example of downward causation, where a process affecting one level of a hierarchy has a strong effect on the elements comprising that level. Extinction has exerted this action continually in geological history.

Hierarchies at the Level of Organismic Integration

As long as a structure is tightly integrated owing to functional considerations, it is likely that the action of genes on component parts will have an upward causative effect. Owing to tight integration of components in an organ, cell, or major morphological feature, downward causation is not a major factor operating analogously to taxon extinction. As a structure loses its integration in evolution the components might be gained and lost through random processes. It is likely that traditional models of genetic variation and functional morphology are adequate to provide a framework for adaptive evolution, so long as the constraints suggested in the evolutionary ratchet model are considered.

The developmental ratchet model suggests a hierarchical framework within which to view the constraints on evolutionary change, as in the compartment model of development. In this model (see Chapter 5), determination of developmental fate is successively more localized to specific regions or groups of cells. Mutations therefore have a quite different effect, depending upon when they act in the sequence. When acting early, effects on the phenotype might be expected to be more global; when action is late in development, a mutation may be more localized. This statement redescribes in hierarchical terms the theory that mutations on early ontogeny are more significant in phenotypic change, while later changes produce less fundamental change. This framework also permits an insight into the hypothesis that major developmental mutants, acting early in development, are usually of little importance in evolutionary change; their effects are so global that loss of fitness is to be expected. Later and more local changes are less costly and therefore are more likely to be the stuff of evolution. This viewpoint at once provides a mechanism for von Baer's law and the classical neo-Darwinian view of the importance of small mutations in evolution. In this case the hierarchy to be considered is the assembly of subdivisions of the body, ranging from the effects on the entire phenotype early in development (highest level) to the local effects later on (lowest level).

This developmental model will be inappropriate in many instances, because the simple compartment model does not usually apply. Even where early developmental processes have more global effects, it seems possible to alter the fate of development without major changes in the subsequently developing phenotype, as discussed in Chapter 5. The establishment of a hierarchical system does not therefore guarantee a specific outcome.

The organism, and not the species, is the fundamental constrained hierarchical level in evolution. The organism is a tightly integrated unit, constrained by functional, developmental, and genetic underpinnings. It is this fundamental unity of the organism that determines the stasis of form, so pervasive in the history of life. Perhaps the most spectacular example is the protistan, *Tetrahymena*, whose detailed cytoskeletal form is conserved among many distantly related species, despite fundamental differences in the nature of the proteins that form the cytoskeleton (N.E. Williams 1984). The phenomenon of transspecific stasis (Chapter 4), which shows the survival of the integrity of the organism through millions of years and literally thousands of speciation events, is the best proof of the importance of the organization of the organism. Our emerging knowledge of the generally restricted nature of gene flow despite widespread homogeneity of form (Slatkin 1985) points further to the importance of processes that operate to maintain the integrity of the organism. Neither gene flow, nor its opposite, speciation, seem important in the maintenance or the breakup of this level of constancy.

The Important Hierarchical Levels: An Evaluation

The evidence presented in this volume suggests that the organism and the monophyletic group are the two crucial hierarchical levels in evolutionary biology. The integrity of the organism is the basis of the concepts of fitness (Chapter 3) and performance (Chapter 6). The theory of commitment is a theory of the evolution of the organism and its functioning.

The monophyletic group is a crucial level, as it is the basis for the historical context of evolution. A monophyletic group is characterized by a series of synapomorphies. Any descendants of a stem species must build on this set. Therefore historical circumstance determines which monophyletic groups are present at any time, and any adaptive change occurs in the restraining context of the synapomorphy sets available at the time. If flight arises in a given group, the exact pathway will be constrained by ancestral traits. Therefore the patterns of extinction of monophyletic groups shapes the future course of evolutionary trends. Because of this historical factor, which might include random extinction of given monophyletic groups, the direction of evolution cannot be predicted without a description of the constraints imposed by ancestors.

There is no theoretical way to take a monophyletic group and suggest which taxonomic level (as defined by a branch point in a cladogram) may be important in an evolutionary question. It may correspond to any of a large number of taxonomic levels. But in any case, the question usually devolves to what groups have survived the exigencies of earth history and now constitute the available materials for adaptation. In some cases, geographical spread may be the prime determinant of survival. To some degree, this may explain why phyla appear early and do not disappear. We must remember, however, that commitment may also have maintained the integrity of the organism, and, therefore, the survival of phyla. Otherwise, one might imagine a world where one could not classify organisms according to a hierarchical scheme. If characters were lost easily, then what characters would unify the higher taxa?

Great Die-Offs and Radiations

What Are Mass Extinctions?

In the mid-nineteenth century, paleontologists had two world views of
the history of life. Charles Lyell, the greatest geologist of the nineteenth
century, argued for a world that changed cyclically, and saw the number
of species as being fairly constant, with continuous turnover. By contrast, Cuvier and his followers saw the past as a series of catastrophes,
causing the disappearance of whole biotas. While Darwin's *The Origin
of Species* certainly did not see evolution as constant or without bursts
of change, it seemed to presume a world in which catastrophes were unknown. Subsequent advances in evolutionary theory have by and large
presumed a fairly static world, even if we know that many groups have
become extinct over geological history. Some of the works coming from
the Modern Synthesis have acknowledged changes in climate and geography, but none so dramatic as to cause major disappearances of entire
biotas.

A revival of interest in changes of biotas over large spans of geological time has partially vindicated Cuvier's view of geological history.
While his essentialism and disbelief in evolution can be discounted,
we now have enough evidence to document extensive turnovers, sometimes worldwide, in both terrestrial and marine biotas. Both physical
geological evidence and paleontological evidence support the notion of
great changes in the physical environment and in fossil biotas. But we
cannot safely say that we can unambiguously explain any large scale
extinction.

The modern paleontological attack on this subject began with the
pioneering studies of Preston Cloud (1948), demonstrating apparent
bursts of evolution. This was confirmed through Norman Newell's more
detailed studies (1952) and has been further documented with still
more extensive data bases. Since the nineteen fifties, paleontologists
have routinely accepted the notion of major extinctions and subsequent
bursts of evolution.

The recent studies, such as Sepkoski (1984), lend more credence
to the reality of episodes when prominent drops of biotic diversity occurred, as discussed in Chapter 8. More controversial is the conclusion
that extinction may have had a significant periodic component (Raup
and Sepkoski 1984). What is the nature of these extinctions? Are they

sudden or did they occur over long periods of time? Are they really extinctions at all, in the sense that biologists think of them–such as that of the dodo or the passenger pigeon?

Despite an intense effort to document fluctuations in diversity of fossil taxa, we still have much to learn about the nature of so-called mass extinctions. For several reasons, it is unlikely that they can be related easily to the species extinctions of the past thousand years recorded by biologists (Diamond 1983; Simberloff 1986). Most, if not all, extinctions that have been documented by biologists were the work of man (Simberloff 1986). Otherwise, we know little of what population processes and physical events contribute to extinguishing a typical species. Second, strong drops in diversity in the fossil record commonly have the properties of drops in origination rate, as well as extinction rate. Any biologically meaningful theory of mass extinctions must explain the cut-off in originations seen, for example, in Cretaceous extinctions of ammonites (P.D. Ward and Signor 1983), and in the Frasnian-Famennian drop in the origination rate of a number of invertebrate groups (McGhee 1982).

As mentioned in Chapter 8, our perception of change in the physical environment has been influenced more by observations of change in fossil biotas, than by physical geological evidence of environmental change. It is doubtful that anyone could take the physical stratigraphic record and map in all of the extinctions we can recognize with fossils. Extinctions are generally correlated with geological events such as sea-level change and orogenic periods (Newell 1967; Flessa and Imbrie 1973). But could we predict extinctions from geological evidence alone? I doubt it.

Is there anything special at all about the major drops, beyond their obvious numerical significance? Did they have a qualitatively different aspect from the others, which Raup and Sepkoski pictured as falling into the "background"? In one case, the end of the Permian, the extinction is dramatic indeed, and coincided with a major drop in sea level. It is not clear yet that this extinction, along with the others of the "big five," were much more than random prunings of the world biota, with some bias toward tropical forms. A simulation by Walker and Valentine (1987) suggests just this effect, as opposed to extinctions that have been nonrandomly targeted at given groups. In contrast, the major end-Cretaceous extinction seemed to extinguish bivalve mollusk species without regard to their geographic range, even though this factor *was* important in survival during "normal" periods of extinction (Jablonski 1986). In the end-Cretaceous event, geographic spread of a clade was all that mattered in extinction, as if presence in some multispecies spatial refuge took precedence over properties such as individual dispersal ability. Selectivity in mass extinction is further shown by relative survival through the end-Cretaceous of deeper water forms and forms relatively independent of the water column. But such selectivity would also be expected in times of modest extinction rates. At present, it is not clear whether or not mass extinctions deserve a qualitatively special hierarchical status relative to other smaller extinctions. At present, we cannot yet define mass extinctions, except by a statistical approach (e.g., Raup and Sepkoski 1982).

Our current stratigraphic precision makes it nearly impossible to be very precise about the timing of many extinctions, except in local sections. This problem plagues Raup and Sepkoski's extinction periodicity hypothesis, where the typical errors and disagreements of 5 my matter quite a bit. The timing of even single extinctions can be problematic, due to sampling error and disagreements over stratigraphic correlations. This is especially apparent in the question of the extinction of the dinosaurs, which are now popularly shown as the victims of an extraterrestrial event. In reality, the data are too poor and geographically restricted (to the Western Interior of North America) to be taken seriously. Some of the extinctions, such as those at the end of the Permian and at the end-Cretaceous, occur during substantial marine regressions, or times when deposits are rare and documentation is therefore very limited. We can only hope that further investigation will clarify many of the issues that can only be resolved by sampling. We must admit, however, that our record precludes forever an understanding of certain events. As Dingus (1984) shows, we will never know if an extinction occurs in a matter of weeks, if our sampling resolution at the end of the Cretaceous is at best on the order of a hundred thousand years! Such conclusions certainly should give some pause to even the most enthusiastic supporters of extraterrestrially induced extinctions.

Most important, we still cannot say whether there is some predictability in extinction. "The future isn't what it used to be" is an old aphorism that constitutes one credible hypothesis for the origin of extinctions. It may very well be that the several mass extinctions each had rather different origins, and that extinctions over the Phanerozoic may have had a range of causes and intensities, to the degree that all attempts to understand extinction will be bound in the worst type of historicism possible. We may be forced into inventing detailed and consistent stories for each extinction. We are thus potentially the sort described by Oscar Wilde, who saw the task of historians as "to give an accurate description of what never happened."

An intermediate stance would be to see all extinctions as having a common cause, but unpredictable in time by any physical model. We should be skeptical of such attempts, when one can find reasonable papers interpreting most extinctions as being due to cold temperature (e.g., Stanley 1984), and others using sea level or tectonism as the best correlate (Newell 1967; Flessa and Imbrie 1973). As there is no a priori reason to exclude either factor in extinction, one is still left to historical investigation of each extinction event. Given that the best evidence for extinction and its cause is nearly always the pattern and magnitude of disapppearance of the biota itself, the potential for circular reasoning is apparent.

At the other end of the spectrum are the small class of theories that have testable predictions for both the physical stratigraphic and fossil records. At present, only the periodicity theory (Raup and Sepkoski 1984) and the bolide impact theory (Alvarez et al. 1980) have these features. But even if these are the ultimate causes of extinctions, the proximate causes of extinctions may be rather like those suggested by other models. The impact of an extraterrestrial object, for example, might cause lower insolation levels, or might change turbidity in the ocean. Surface temperature might be changed as well. But all of these factors might also be changed by the processes invoked in other hypotheses, such as anoxia, sea-level change, etc. Moreover, even if we could identify the specific trace of impacts, there is no guarantee that their magnitude would be the same every time. Indeed, if we take the Raup and Sepkoski (1984) data at face value, the periodic extinction peaks suggest that the differences in risk of extinction varied by spectacular degrees. These differences in risk lead one to the same potential for invoking ad hoc hypotheses of extinction as we discussed above for more traditional investigations. This suggests that it is the nature of history inquiry, more than anything else, that weakens our ability to understand the great comings and goings through the Phanerozoic.

The Early History of the Metazoa - the Great Mystery

Paleontology has served mainly as the handmaiden of stratigraphy. Before the advent of radiometric dating, fossils were the best means of establishing a time reference system. Even today, fossils are often better time indicators than physical data such as magnetic reversals and radiometry, and many paleontologists see their primary objective as the dating of rocks, rather than the understanding of the history of life.

Because of this overall mission, paleontologists have not devoted enough attention to some of the most crucial phases of the history of life. The remarkable pre-Cambrian occurrences of soft-bodied fossils have been studied by only a handful, and the remarkable Cambrian Burgess shale fauna has been restudied far less than its significance deserves. These fossil occurrences are crucial to biologists, but they are mere curiosities to the biostratigrapher, and hence have not been central in the works of invertebrate paleontology.

This oversight must be corrected. The reexaminations of some Burgess Shale specimens (e.g., Conway Morris 1985a) suggest that many of the wellpreserved invertebrate fossils cannot be related simply to any known extant group. Indeed, the chance exists that we may discover a series of wholly new phyla. The curious fossil *Wiwaxia* mentioned in Chapter 8 is a case in point.

The Ediacaran-style fauna is an equally curious assemblage, even if the principal investigator of the fauna has attempted to classify the groups in terms of extant phyla (e.g., Glaessner 1984). Many of the fossils, however, fit no group very well, and Seilacher (1983; see also Conway Morris 1985b) suggested that they may constitute an independent experiment, unrelated to the metazoa. This is based on some morphological considerations which might be interpreted differently by others (e.g., Conway Morris 1985b).

Even if the Ediacara and Burgess Shale faunas can be related to extant groups, it is clear that this entire period needs intensive study. In the last 15 years, reexaminations of previously described fossils and new discoveries have changed our thinking about the early history of the Mollusca (see Chapter 2). We would benefit greatly from a concerted effort to restudy previously collected fossil assemblages and from some intensive new collecting. It may be time for the greater number of paleontologists to shift their efforts from the later periods in the

Phanerozoic, when the same theme may be merely repeated, not established anew.

There is a temptation to believe that there is something biologically special about the period at the base of the Cambrian. Special genetic or developmental mechanisms can be invoked (e.g., Erwin and Valentine 1984) to explain a high rate of appearance of unfamiliar forms. The need to invoke such mechanisms seems dubious. Metazoan life may have been around for as much as 500 million years by then and it seems doubtful that the proliferation of diversity of form was related to a time of unusual genetic mechanisms of evolutionary change. The forms we see in the Burgess Shale seem no less organized, no more special than those that live today. One also becomes suspicious about the potential of such hypotheses that cannot be tested in any event.

The traditional explanation of an explosion of life into an open ecospace (e.g., Valentine 1986) is consistent with our current evidence. No special genetic mechanisms are required to explain the fact that, by the time of the Permian, a major extinction does not repeat the great diversity of the Vendian and pre-Cambrian. By this latter time the surviving taxa were constrained in the degree to which they could bud off descendants of great difference. But one should not forget the possibility that the major groups of the Cambrian may have been equally constrained by functional, genetic, and developmental factors. This early period of metazoan history may have permitted the rise of a great diversity of groups through ecologically mediated divergence. By the time of the Permian, some of the richness had been lost and the survivors that dominate today are probably immune from extinction due to their geographic extent and degree of ecological diversity.

The Future - A Research Program for Evolutionary Biology

It is presumptuous to suggest an agenda for evolutionary biology. After all, we are now in a period where great empirical advances are being made toward the understanding of the nature of gene action. A series of controversies, many discussed in this volume, sit before us, waiting for further resolution. No one, moreover, does very well in predicting the future of a science. Nevertheless, I feel the necessity to take the fool's gamble and make a few recommendations of emphasis for the coming decades.

A new evolutionary synthesis will have three elements: (1) the unification of neo-Darwinian population genetics with physiological, developmental, and molecular genetics, (2) the incorporation of systematics, via phylogenetic analysis, into the workings of adaptive evolutionary theory, as promised, but never realized, in the Modern Synthesis, and (3) the injection of history into all realistic evolutionary hypotheses. The latter will involve geological and paleontological data. These three elements will contribute to a new understanding of the tangled directions of phenotypic evolution. While I do not exclude any levels of the hierarchy discussed above, the neo-Darwinian framework still stands as one of the cornerstones of our evolutionary understanding.

1. Relationship of the Phenotype and the Genotype.   For biologists, no more important task is at hand than to use the new genetic and developmental tools to produce a mechanistic understanding of the relationship between genotype and phenotype. This will lead to a revival of the field of physiological genetics and an understanding of how mutation shapes the array of phenotypic variation. Neo-Darwinians have always claimed that mutation is indifferent to the process of adaptation. But the exact process of mutational change is important, as it defines the spectrum of possible evolutionary change. Before the elucidation of the molecular basis of genetic transmission, it was inconceivable to study more than the gross surface manifestations of this determination process. As a consequence it is excusable that neo-Darwinian theory and practice tended to see such variation as a black box, spewing out the variation necessary for theoreticians to predict adaptive evolutionary change. Our few windows into the genome now tell us that this black box has a rich complexity.

Our paltry information already is becoming instrumental in our understanding of gene action. We have learned that some genes are perhaps nearly as unitary as the "one gene–one enzyme" theory, but others have a host of interactions in nearby and sometimes distant parts of the DNA. At present, we cannot say whether structural genes, flanking controlling regions, or distinct regulatory genes are more crucial in variation that is meaningful to the phenotype. It is unduly optimistic to expect short-term rewards for evolutionary biologists, but I am fully optimistic that some important systems will be worked out sufficiently well to shed light on some problems of phenotypic evolution.

We also must understand more about the structure of the genome, including the meaning of gene families and groups of closely linked genes that seem to regulate important developmental events (e.g., the bithorax complex in diptera). The work of McClintock in the nineteen forties showed that such interactive complexes might be present, and recent work has amplified this. Such information will give evolutionary biologists a clue as to the architecture which might constrain the directions of natural selection.

While molecular approaches are badly needed, the recent movement toward understanding morphological integration, through the study of genetic and phenotypic correlations, has also been rewarding. Unfortunately, the quantitative genetic approaches used to study correlated evolution among phenotypic traits cannot be related easily to mechanistic studies of genotype–phenotype relationships. Perhaps the best approach will be to use genetic correlations to postulate cellular or genic mechanisms to associate the traits, and then to explore directly those mechanisms that might produce such phenotypic correlations. Riska and Atchley (1985) have taken this approach and have produced a credible hypothesis of evolutionary change that relates correlated evolution of morphological traits to a cellular mechanism. Of course, the genetic and developmental underpinnings of this mechanism have yet to be explored. But the gap between the genetic and cellular mechanisms generating variation and the variable traits we usually measure needs badly to be closed.

2. Association of Traits and Acquisition of Derived States with Clado-grams. Although there has been a more than modest amount of aggressive selling of the approach, I believe that the attempt at producing genealogically meaningful numerical classifications, and the attendant mapping of character transition analysis on genealogically meaningful diagrams, is the healthiest movement to have arrived at the doorstep of evolutionary biologists. After 100 years of Darwinism, systematics is still largely divorced from the fields necessary to understand the process of natural selection and adaptive evolution. Until recently, systematics has failed to provide the necessary analytical framework necessary to formulate hypotheses relevant to the theory of adaptive evolution. I enjoy the irony that some of the most fervent promulgators of phylogenetic analysis have not regarded themselves as disciples of Darwin or advocates of the theory of natural selection, even if their work gives Darwinism an important potential interface with systematics.

With the cladistic approach, the acquisition of functionally important character states, and the historically constraining backdrop of other character states can be defined in a logical fashion. Hypotheses of adaptation can be formulated with the boundary conditions of phylogenetic background (Wake and Larson 1987). The distribution of character states in cladograms can reveal the tempo of acquisition of a derived character state, or of a group of derived character states. Has an interactive functioning body plan arisen by piecemeal acquisition of various traits, or has a burst of morphological diversification followed the acquisition of a single key innovation? Both of these alternatives can be investigated using cladistic techniques.

3. Study of Autochthony of Traits and Commitment. I have outlined above a theory of commitment, which invokes an evolutionary ratchet with genetic, developmental, and functional components to explain the movement of a lineage into an evolutionary rut. Some of the elements of this theory are inspired by the punctuated equilibrium hypothesis, which emphasizes the supposed dominance of stasis, unless broken by speciation. While the punctuated equilibrium hypothesis itself is invalid, we can still rescue the notion of stasis which seems to permeate the history of life. We can be grateful to Eldredge and Gould, who have focused this issue. It would also be healthy if their prose reflected the deep concern that legions of neo-Darwinians have had for this issue, including the likes of Sewall Wright, Julian Huxley, J.B.S. Haldane, and Ernst Mayr.

It is more than worthwhile to document the nature of constancy of traits, from the level of nucleotide sequences to patterns of development. We are beginning to notice a striking pattern of conservation of nucleotide sequences among distantly related taxa, including the now famous homoeobox regulatory sequence (McGinnis et al. 1984a), the chorion genes, known to be similar in moths and flies, and a neuropeptide, the "head activator," whose amino acid sequence is identical in *Hydra* and the hypothalamus of man (Bodenmuller and Schaller 1981). Why are such sequences conserved? Are they of such functional significance on their own, or are their interactions with other genes the basis of their conservatism? We have similar concerns for sequences such as the bithorax complex, which is widespread in the Diptera. Is this complex conserved because of its crucial genetic spatial interactions? At a larger scale, we wonder about the conservation of cell division patterns and tissue inductions during embryogenesis throughout distantly related taxa. Finally, we must remember that natural selection is a crucial component. Any successful approach to the study of commitment will have to understand the interaction of all three components of the evolutionary ratchet.

4. Paleontology–Pattern of Change and Search for More Organisms. Paleontologists have had a nagging feeling that they must sit on the side lines as others conceive new evolutionary theories and make significant empirical progress. While paleontological materials and methodologies obviously have their limits (don't all approaches?), I submit that progress in our knowledge of evolution will be severely limited without paleontological research. Much of the time devoted to paleontological research has been spent on proving that paleontology can show that processes occurring in the past are much the same as now. Again, one is not overwhelmed by the potential contributions of paleontology. Perhaps this attitude is the explanation for the seemingly intense desire of some paleontologists to shake the foundations of evolutionary theory with proclamations of a new dimension of theory coming from paleontology's unique perspective.

Paleontological research has two weaknesses, but it also has one strength which is crucial for evolutionary biology. The most important weakness is the degree of incompleteness of preservation. As shown in Chapters 7 and 8, we have only recently begun to appreciate just how much this might bias our perceptions of rates of evolution, the tempo of evolutionary radiations and extinctions, and the gaps in preservation of forms intermediate between taxa of great morphological distance. The degree of bias must be understood, not to increase our degree of despair, but to enable us to use paleontological data to its best advantage. A second weakness is the incomplete access of paleontologists to much of the spectrum of the biological approaches and data gathering that are routinely available to the neontologists. Paleontologists cannot study many processes (e.g., most aspects of population dynamics) and cannot survey for a wide variety of patterns (e.g., allozyme data, early cleavage of embryos, chromosome numbers and inversions) commonly obtainable from living organisms.

Paleontology's crucial importance lies in its provision of the dimension of history, to a degree that is simply not available to neontologists. It is true that evolutionary history is recorded somewhat in living forms. Cladistic analysis, for example, can give us many insights into the order of acquisition of traits in evolution. If molecular clocks have reality, and if they can be calibrated properly, then rates of morphological divergence can be obtained.

Despite these, fossils provide the crucial data needed to understand the historical context of the rise of extant (and, of course, extinct) groups. Could we ever understand the evolution of the mammalian skull without the Mesozoic fossil record? As we have discussed, models of adaptive morphological evolution will be successful only when placed in a historical context. Without such a context, how could we conceive of the necessity of, and then proceed to document, the two coincident fulcra in the origin of the mammalian jaw articulation? When tracking the homologies of characters, the closest relatives provide the best hope of giving an accurate picture of evolutionary direction. In this vein, one would prefer to know the morphologies of the inclusive group of a given taxon, in order to map out the likely pattern of character transition. Without a fossil record that is ever more complete, our knowledge of this phylogenetic context will be limited indeed.

The geological and fossil record has given us our only access to any understanding of the pattern of originations and extinctions which have led to our current biota. Our current evidence suggests that tremendous biotic turnovers have occurred, and that the physical arrangements of continents and oceans have changed to a great degree. Climate has fluctuated strongly, and some of these changes can be related directly to extinctions (Stanley 1983, 1984). In the past ten years, both paleontologists and sedimentologists have been providing the tools necessary to increase the precision of our estimates of the rates of turnover, and the degree to which the data can pinpoint certain processes. Because all evolutionary transitions work on the biota available at the time, a precise identification of the historical backdrop of evolutionary change is crucial. Moreover, we would like to know the exact processes that lay behind the apparent major changes such as mass extinctions. Some now believe that mass extinctions may be just that: massive die-offs due to a sudden environmental change induced by a catastrophic event. Such claims can only be demonstrated by careful associations of the physical and biotic evidence. Others believe that mass extinctions are more complex, and involve depressions of origination rates as well. We must look forward to a more careful analysis of the data, and more collections in the future. One can easily see that many of the macroevolutionary hypotheses operating at or above the level of species could not be explored without a fossil record (e.g., Jablonski 1986).

Paleontologists must continue to collect more detailed evolutionary rate data in closely controlled stratigraphic sections, and to pay more attention to geographic variation. While the testing of the punctuated equilibrium hypothesis has inspired a great deal of such study (e.g., Malmgren and Gingerich 1986), paleontologists have to understand within- and between-population variation, to the standard that their neontologist analogues have managed.

Most essential of all, paleontologists must maintain a faith in the unique opportunities to expand our knowledge of the earth's biotic diversity. It has become unfashionable to study rare and unusual taxa, and phylogenetic analysis has yet to become as important as it should be among paleontologists. This is reflected in the current power structure of invertebrate paleontology, which is curiously lacking in those able to examine paleontological materials and reveal new insights into the evolutionary relationships of fossil taxa, particularly the primitive forms. There is a danger that paleontology will lose its unique birthright and soon become a discipline with no unique empirical traditions of its own. The strength of the morphological tradition is more evident in vertebrate paleontology. Paleontology's unique contributions stem from its unique access. To forget this is to consign paleontology to an eternal bystander's status in the study of evolution.

Conclusion

Darwin's great contribution was to show the necessary consequences of variation in natural populations and the significance of the differential survival and reproduction of variants in the struggle for existence. The enormous diversity of living organisms, long classified and studied in terms of a variety of nonscientific approaches, could now be seen as termini of a complex genealogy. It was Darwin's genius to relate the origin of that genealogy to the natural world around him. He understood this complex subject better than anyone before him, and defined the framework for all that has followed since. It is worthwhile to remember his closing remarks in *The Origin*, which spoke of a tangled bank, with a bewildering complexity of ecological interactions. The understanding of this complexity has been the task of ecology and evolutionary biology for the past century.

Darwin also spoke of a degree of organismal complexity that we are only beginning to fathom today. In *The Origin*, Darwin speaks of the interactions within the organism, which are so complex as to make an understanding in isolation of any one component quite difficult. The organismal plexus is a genetic, developmental, and functional analogue to the tangled bank. Darwin saw the path of evolution being controlled by the "unity of type" and the "conditions of existence." It is the latter aspect that applies to the theory of natural selection, and this is much the main theme of the book. By unity of type, Darwin presumably meant those aspects of organisms that reflected their past evolutionary history. This is the very aspect of organismal biology that has been poorly explored, mainly because of our insufficient knowledge of the genetic and epigenetic determination of the phenotype. Unity of type must include those conditions that must be used as boundary conditions in any model of natural selection and optimization of form (see Chapter 6). Indeed, Darwin was implicitly aware of this and only thought of evolutionary change in the context of the background history of the group in question. He consistently thought of natural selection as working on the available materials (the phylogenetically constrained ancestors?) that might, through selection on variation, improve in performance through evolutionary change.

I don't believe that we have moved beyond Darwinism in any material way. It is true that the perspective afforded by the fossil record reveals processes involving major reorganizations of the biotic world. The pattern of extinction in the fossil record shows that major habitat disruption has eradicated many groups and has laid waste many adaptations that had arisen as a consequence of natural selection. While subsequent expansions have been adaptive, the blind pass of extinction has worked again and again to shape the ebb and flow of various taxonomic groups. If Darwinism needs to be altered in any major way it is to emphasize the apparent insensitivity of extinction to the superiority of performance that usually matters as natural selection shapes the construction of a body plan. To the degree that blind extinction cuts off certain clades from improving in performance the evolutionary process is more and more unpredictable. History does matter and all organisms record many of the eccentricities of history in the peculiar concoctions that manage to nevertheless perform well.

And now, over a hundred years since Darwin's death, we have the tools that Darwin lacked: a probe into the hereditary materials, a fuller understanding of the geological time scale and the meaning of fossils and their potential, a set of theories to predict the course of evolution in populations and among clades, an emerging knowledge of development, and a renewed interest in the construction of genealogies. Perhaps we are now equipped to truly realize the grandeur of Darwin's view of life.

The Main Points

1. Evolution and adaptation involve a stabilization of form over time. Habitat selection seems to be an important part of this stabilization, which I call commitment. While commitment is determined by function, developmental constraints, and genetic constraints, it seems clear that adaptation is the major driving force.

2. Organismic integration is a central theme in evolution, and an important focus for future research. We have little information on the relationship between genes and the phenotype, especially in traits that have been traditionally associated with major adaptive differences among taxa. We also have understudied the interaction among traits.

3. The establishment of genealogies is crucial in evolutionary hypotheses, as they set the historical framework for specific explanations. Cladistic approaches seem best suited for evolutionary hypotheses, as they organize character evolution in such a way that evolutionary trends can be assessed within monophyletic groups. Functional models cannot be properly erected without a knowledge of the historical backdrop to the newer episode of supposed adaptation. An appreciation of history is crucial to understand evolution. The establishement of a proper genealogy

is one of many ways to increase our success in dealing with the essentially historical subject of evolutionary biology.

4. An examination of the organic world in a hierarchical framework is useful as it can identify important levels of consideration. The evidence from this volume suggests that the organism is the prime level of consideration. Natural selection operates to maintain the integrity of the individual or permit its progeny to adapt to changing environments. In some cases, traits defining monophyletic groups can be traced directly to how they interact to determine the integrity of the individual. While certain processes admittedly happen on the species level, this seems to be secondary to the more basic trends in adaptation, such as the evolution of complex morphologies.

5. Because of the acquisition of a wide variety of biological and paleontological tools, many evolutionary questions can now be approached with a many-pronged attack. I strongly believe that important questions will be solved in the near future with a strong empirical approach to historical geology, paleontology, genealogical construction, development, transmission genetics, and population genetics.

References

Aimar, C., M. Delarue and C. Vilain. 1981. Cytoplasmic regulation of the duration of cleavage in amphibian eggs. *J. Embryol. Exp. Morph.* 64: 259-274.

Ajioka, J.W., and W.F. Eanes. 1986. Rapid accumulation of parasitic DNA: *de novo* versus naturally occurring insertions of P-elements in *Drosophila melanogaster Science*, submitted.

Akam, M. 1986. Mediators of cell communication? *Nature* 319:447-448.

Akesson, B. 1973. Reproduction in the genus *Ophryotrocha* (Polychaeta, Dorvilleidae). *Pubbl. Staz. Zool. Napol.* 39 (suppl.): 377-398.

Alberch, P. 1980. Ontogenesis and morphological diversification. *Amer. Zool.* 20: 653-667.

Alberch, P. 1981. Convergence and parallelism in foot morphology in the Neotropical salamander genus *Bolitoglossa*. I. Function. *Evolution* 35: 84-100.

Alberch, P. 1982. Developmental constraints in evolutionary processes. In *Evolution and Development*, ed. J.T. Bonner, pp. 313-332. Berlin: Springer-Verlag.

Alberch, P. 1983. Morphological variation in the Neotropical salamander genus *Bolitoglossa*. *Evolution* 37: 906-919.

Alberch, P. 1985. Problems with the interpretation of developmental sequences. *Syst. Zool.* 34: 46-58.

Alberch, P. and E. Gale. 1983. Size dependence during the development of the amphibian foot. Colchicine-induced digital loss and reduction. *J. Embryol. Exp. Morph.* 76: 177-197.

Alberch, P., S.J. Gould, G.F. Oster, D.B. Wake. 1979. Size and shape in ontogeny and phylogeny. *Paleobiology* 5: 296-317.

Allen, D.E. 1978. The naturalist in Britain: a social history. In *Images of the Earth: Essays in the Environment*, ed. L.J. Jordanova, and R.S. Porter, pp. 200-212. Chalfont St. Giles: Brit. Soc. Hist. Sci.

Allen, G.E. 1974. Opposition to the Mendelian-chromosome theory: the physiological and developmental genetics of Richard Goldschmidt. *J. Hist. Biol.* 7: 49-92.

Allen, M.K. and C. Yanofsky. 1963. A biochemical and genetic study of reversion with the A-gene A-protein system of *Escherichia coli* tryptophan synthetase. *Genetics* 48: 1065-1083.

Allen, T.F.E., and T.B. Starr. 1982. *Hierarchy: Perspectives for Ecological Complexity.* University of Chicago Press.

Allin, E.F. 1975. Evolution of the mammalian middle ear. *J. Morph.* 147: 403-438.

Alvarez, L.W., W. Alvarez, F. Asaro, H.V. Michel. 1980. Extraterrestrial cause for the Cretaceous-Tertiary extinction. *Science* 208: 1095-1108.

Alvarez, W. and R.A. Muller. 1984. Evidence from crater ages for periodic impacts on the Earth. *Nature* 308: 718-720.

Alvarez, W., L.W. Alvarez, F. Asaro, and H.V. Michel. 1984. The end of the Cretaceous: sharp boundary or gradual transition? *Science* 223: 1183-1186.

Alvarez, W., F. Asaro, H.V. Michel, L.W. Alvarez. 1982. Iridium anomaly approximately synchronous with terminal Eocene extinctions. *Science* 216: 886-888.

Ambros, V. and H.R. Horvitz. 1984. Heterochronic mutants of the nematode *Coenorhabditis elegans. Science* 226: 409-416.

Anderson, P.R., and J.G. Oakeshott. 1984. Parallel geographical patterns of allozyme variation in two sibling *Drosophila* species. *Nature* 308: 729-731.

Anderson, W.W. 1973. Genetic divergence in body size among experimental populations of *Drosophila pseudoobscura* kept at different temperatures. *Evolution* 27: 278-284.

Andrews, R. C. 1921. A remarkable case of external hind limbs in a humpback whale. *Am. Mus. Novitates* 9: 1-16.

Aquadro, D.F., S.F. Desse, M.M. Bland, C.H. Langley, and C.C. Laurie-Ahlberg. 1986. Molecular population genetics of the alcohol dehydrogenase gene region of *Drosophila melanogaster*. *Genetics* 114: 1165-1190.

Arnason, U. 1972. The role of chromosomal rearrangement in mammalian speciation with special reference to Cetacea and Pinnipedia. *Hereditas* 70:113-118.

Arnheim, N. 1900. Concerted evolution of multigene families. In *Evolution of Genes and Proteins*, ed. M.Nei and R.K. Koehn, pp. 38-61. Sunderland MA: Sinauer Associates.

Arnheim, N., M. Krystal, R. Schmickel, G. Wilson, O.Ryder and E. Zimmer. 1980. Molecular evidence for genetic exchanges among ribosomal genes on nonhomologous chromosomes in man and apes. *Proc. Nat. Acad. Sci. USA* 77: 7323-7327.

Arnold, A.J. 1982. Species survivorship the in the Cenozoic Globigerinida. *Proc. 3rd North Amer. Paleontol. Conv.* Vol. 2, pp. 9-12.

Asama, K. 1959. Systematic study of so-called *Gigantopteris. Sci. Rept. Tohoku Univ.* Ser. 2 (Geology) 31(1): 1-72.

Asama, K. 1962. Evolution of Shansi flora and origin of simple leaf. *Sci. Rept. Tohoku Univ.* Ser. 2 (Geology), Spec. vol. 5: 247-273.

Asama, K. 1981. Evolution and phylogeny vascular of plants based upon the principles of growth retardation. Part 1. Principles of growth retardation and climatic change through ages. *Bull. Natn. Sci. Tokyo*, Ser. C 7: 61-79.

Ashburner, M. 1980. Chromosomal effects of ecdysone. *Nature* 285: 435-436.

Atchley, W.R. 1983. Some genetic aspects of morphometric variation. NATO ASI Series, Vol. G1, *Numerical Taxonomy*, ed. by J. Felsenstein. Berlin: Springer-Verlag. pp. 346-363.

Atchley, W.R. 1984a. The effect of selection on brain and body size association in rats. *Genet. Res. Cambr.* 43: 289-298.

Atchley, W.R. 1984b. Ontogeny, timing of development, and genetic variance-covariance structure. *Am. Nat.* 123: 519-540.

Atchley, W.R., B. Riska, L.A.P. Kohn, A.A. Plummer, and J.J. Rutledge. 1984. A quantitative genetic analysis of brian and body size associations, their origin and ontogeny: data from mice. *Evolution* 38: 1165-1179.

Atchley, W.R. and J.J. Rutledge. 1980. Genetic components of size and shape. I. Dynamic components of phenotypic variability and covariability during ontogeny in the laboratory rat. *Evolution* 34: 1161-1173.

Ausich, W.I., and D.J. Bottjer. 1982. Tiering in suspension-feeding communities on soft substrata throughout the Phanerozoic. *Science* 216: 173-174.

Avise, J.C. 1976. Genetic differentiation during speciation. In *Molecular Evolution*, ed. F.J. Ayala, pp. 106-122. Sunderland, MA: Sinauer Assoc.

Avise, J.C. 1977. Genic heterozygosity and the rate of speciation. *Paleobiology* 3: 422-432.

Avise, J.C., and F.J. Ayala. 1976. Genetic differentiation in speciose versus depauperate phylads: evidence from the California minnows. *Evolution* 30: 46-58.

Avise, J.C., and R.A. Lansman. 1983. Polymorphism of mitochondrial DNA in populations of higher animals. In *Evolution of Genes and Proteins*, ed. M. Nei, and R.K. Koehn, pp. 147-164. Sunderland, MA: Sinauer.

Ayala, F.J. 1972. Darwinian versus non-Darwinian evolution in natural populations of *Drosophila*. In *Proceedings of the Sixth Berkeley Symp. Math. Stat. Prob.*, vol. 5, ed. L.M. LeCam, J. Neyman, and E.L. Scott, pp. 211-236, Berkeley: Univ. California Press.

Ayala, F.J., and T. Dobzhansky, eds. 1974. *Studies in the Philosophy of Biology*. Berkeley and Los Angeles: Univ. California Press.

Ayala, F.J., J.R. Powell, and M.L. Tracey. 1972. Enzyme variability in the *Drosophila willistoni* group. V. Genic variation in natural populations of *Drosophila equinoxialis*. *Genet. Res. Camb.* 20: 19-42.

Ayala, F.J., M.L. Tracey, D. Hedgecock, and R.C. Richmond. 1974.

Genetic differentiation during the speciation process in *Drosophila. Evolution* 28: 576-592.

Baker, R.J. 1981. Chromosomal flow between chromosomally characterized taxa of a volant mammal, *Uroderma bilobatum* (Chiroptera, Phyllostomatidae). *Evolution* 35: 296-305.

Bakker, R.T. 1974. Experimental and fossil evidence for the evolution of tetrapod bioenergetics. In *Perspectives in Biophysical Ecology*, ed. D. Gates and R. Schmerl, pp. 065 099, New York: Springer-Verlag.

Bakker, R.T. 1977. Tetrapod mass extinctions - a model of the regulation of speciation rates and immigration by cycles of topographic diversity. In *Patterns of Evolution as Illustrated by the Fossil Record*, ed. A. Hallam, pp. 439-468. Amsterdam: Elsevier Scientific.

Bambach, R.K. 1970. *Bivalvia of the Siluro-Devonian Arisaig Group, Nova Scotia.* PhD. Dissertation, Yale University.

Bambach, R.K. 1977. Species richness in marine benthic habitats throughout the Phanerozoic. *Paleobiology* 3: 152-157.

Bambach, R.K. 1983. Ecospace utilization and guilds in marine communities through the Phanerozoic. In *Biotic Interactions in Recent and Fossil Benthic Communities,* ed. M.J.S. Tevesz and P.L. McCall, pp. 719-746. New York and London: Plenum.

Bambach, R.K., C.R. Scotese, and A.M. Ziegler. 1980. Before Pangea: the geographies of the Paleozoic world. *Am. Sci.* 68: 26-38.

Band, H.T., and P.T. Ives. 1968. Genetic structure of populations. IV. Summer environmental variables and lethal and semilethal frequencies in a natural population of *Drosophila melanogaster. Evolution* 22:633-641.

Barnes, B.W. 1968. Stabilising selection in *Drosophila melanogaster. Heredity* 23: 433-442.

Barrell, J. 1917. Rhythms and the measurement of geological time. *Geol. Soc. Amer. Bull.* 28: 745-904.

Bartenstein, H., and F. Bettenstaedt. 1962. Marine unterkreide (Boreal und Tethys). In *Leitfossilen der Mikropaläontologie.* pp.

225-297. Berlin: Arbeitsgkreis deutsch. Mikropal.

Barton, N.H. 1982. The structure of the hybrid zone in *Uroderma bilobatum* (Chiroptera: Phyllostomatidae). *Evolution* 36:863-866.

Barton, N.H. 1983. Multilocus clines. *Evolution* 37: 454-471.

Barton, N.H., and B. Charlesworth. 1984. Genetic revolutions, founder effects, and speciation. *Ann. Rev. Ecol. Syst.* 15: 133-164.

Barton, N.H., and G.M. Hewitt. 1985. Analysis of hybrid zones. *Ann. Rev. Ecol. Syst.* 16:113-148.

Bateson, W. 1894. *Materials for the Study of Variation Treated with Especial Research to Discontinuity in the Origin of Species.* London and New York:Macmillan.

Bayer, U. 1978. Morphogenetic programs, instabilities, and evolution–a theoretical study. *N. Jb. Geol. Palaont.* 156: 226-261.

Beardmore, J.A., and S.A. Shami. 1979. Heterozygosity and the optimum phenotype under stabilising selection. *Aquilo. Ser. Zool.* 20: 100-110.

Beecher, C.E. 1891. Development of the brachiopoda. Pt. 1. Introduction. *Am. J. Sci.* Ser. 3 41: 342-357.

Behrensmeyer, A.K. 1982. Time resolution in fluvial vertebrate assemblages. *Paleobiology* 8: 211-227.

Bell, M.A. 1976. Evolution of phenotypic diversity in *Gasterosteus aculeatus* superspecies on the Pacific coast of North America. *Syst. Zool.* 25: 211-227.

Bell, M.A. 1981. Lateral plate polymorphism and ontogeny of the complete plate morph of threespine sticklebacks (*Gasterosteus aculeatus*). *Evolution* 35: 67-74.

Bell, M.A., 1987. Interacting evolutionary constraints in pelvic reduction of threespine sticklebacks, *Gasterosteus aculeatus* (Pisces, Gasterosteidae). *Biol. J. Linn. Soc.* 31: 347-382.

Bell, M.A., J.V. Baumgartner, and E.C. Olson. 1985. Patterns of temporal change in single morphological characters of a Miocene stickleback fish. *Paleobiology* 11: 258-271.

Bell, M.A., and T.R. Haglund. 1982. Fine-scale temporal variation of the Miocene stickleback *Gasterosteus doryssus*. *Paleobiology* 8: 282-292.

Bender, W., M. Akam, F. Karch, P.A. Beachy, M. Peifer, P. Spierer, E.B. Lewis, and D.S. Hogness. 1983. Molecular genetics of the bithorax complex in *Drosophila melanogaster. Science* 221: 23-29.

Bengtsson, D.O. 1980. Rates of karyotype evolution in placental mammals. *Hereditas* 92: 37-47.

Benson, R.H., and R.E. Chapman. 1982. On the measurement of morphology and its change. *Paleobiology* 8: 328-339.

Benton, M.J. 1983a. Large-scale replacements in the history of life. *Nature* 302: 16-17.

Benton, M.J. 1983b. Dinosaur success in the Triassic: a noncompetitive ecological model. *Q. Rev. Biol.* 58: 29-55.

Benyajati, C., N. Spoerel, H. Haymerle, and M. Ashburner. 1983. The messenger RNA for alcohol dehydrogenase in *Drosophila melanogaster* differs in its 5′ end in different developmental stages. *Cell* 33: 125-133.

Berkner, C.V., and L.C. Marshall. 1964. The history of growth of oxygen in the earth's atmosphere. In *The Origin and Evolution of Atmospheres and Oceans,* ed. C.J. Brancuzio and A.G.W. Cameron, pp. 102-126. New York: John Wiley and Sons.

Best, R.V. 1961. Intraspecific variation in *Encrinurus ornatus. J. Paleontol.* 35: 1029-1040.

Bettenstaedt, F. 1958. Phylogenetische beobachtungen in der mikropalaontologie. *Palaont. Z.* 32: 115-140.

Bettenstaedt, F. 1959. Art- und gattungsbildung. Eine untersuchung an fossilen Foraminiferan. *Natur und Volk* 89: 367- 379.

Bettenstaedt, F. 1960. Die stratigraphische bedeutung phylogenetischer reihen in der mikropalaontologie. *Geol. Rundschau* 49: 51-69.

Bettenstaedt, F. 1962. Evolutionsvorgange bei fossilen Foraminiferen. *Mitt. Geol. Staatsinst. Hamburg* 31: 385-460.

Bickham, J.W. 1981. Two-hundred-million-year-old chromosomes: deceleration of the rate of karyotypic evolution in turtles. *Science* 212: 1291-1293.

Bingham, P.M., M.G. Kidwell, and G.M.Rubin. 1982. The molecular basis of P-M hybrid dysgenesis: the role of the p element, a P-strain specific transposon family. *Cell* 29: 995-1004.

Bingham, P.M., and Z. Zachar. 1985. Evidence that two mutations, W^{DZL} and Z^1, affecting synapsis-dependent genetic behavior of *white* are transcriptional regulatory elements. *Cell* 40: 819-825.

Birch, L.C. 1955. Speciation in *Drosophila pseudoobscura* in relation to crowding. *Evolution* 9: 389-399.

Blackith, R.E., R.G. Davies, and E.A. Moy. 1963. A biometric analysis of development in *Dysdercus fasciatus* Sign (Hemiptera: Pyrrhocoridae). *Growth* 27: 217-334.

Blackith, R.E., and R.A. Reyment. 1971. *Multivariate Morphometrics.* London: Academic Press.

Blount, R.F. 1950. The effects of heteroplastic hypophysed grafts upon the axolotl, *Ambystoma mexicanum. J. Exp. Zool.* 113: 717-739.

Bluemink, J.G., and J.C. Beetschen. 1981. An ultrastructural study of the maternal-effect embryos of the *ac/ac* mutant of *Pleurodeles waltl* showing a gastrulation effect. *J. Embryol. Exp. Morph.* 63: 67-74.

Boag, P.T., and P.R. Grant. 1981. Intense natural selection in a population of Darwin's finches (Geospizinae) in the Galapagos. *Science* 214: 82-85.

Bock, W.J. 1959. Preadaptation and multiple evolutionary pathways. *Evolution* 13: 194-211.

Bock, W.J. 1970. Microevolutionary sequences as a fundamental consequence in macroevolutionary models. *Evolution* 24: 704-722.

Bock, W.J. 1979. The synthetic explanation of macroevolutionary change-a reductionist approach. *Bull. Carnegie Mus. Nat. Hist.* 13: 20-69.

Bodenmuller, H., and H.C. Schaller. 1981. Conserved amino acid sequence of a neuropeptide, the head activator, from coelenterates to humans. *Nature* 293: 579-580.

Bohor, B.F., E.E. Foord, P.J. Modreski, O.M. Triplehorn. 1984. Mineralogic evidence for an impact event at the Cretaceous-Tertiary boundary. *Science* 224: 867-869.

Bonner, J.T., ed., 1982. *Evolution and Development.* Berlin: Springer-Verlag.

Bookstein, F.L. 1978. *The Measurement of Biological Shape and Shape Change. Lecture Notes in Biomathematics*, ed. S. Levin, vol. 24. Berlin: Springer-Verlag.

Bookstein, F.L. 1980. When one form is between two others: an application of Biorthogonal Analysis. *Amer. Zool.* 20: 127-141.

Bookstein, F.L. 1982. Foundations of morphometrics. *Ann. Rev. Ecol. Syst.* 13: 451-470.

Bookstein, F.L. 1987. Random walk and the existence of evolutionary rates. *Paleobiology,* in press.

Bookstein, F., B. Chernoff, R. Elder, J. Humphries, G. Smith, and R. Strauss. 1985. *Morphometrics in Evolutionary Biology.* Spec. Publ. 15. Acad. Nat. Sci. Philadelphia, pp. 1-277.

Bookstein, F., P.D. Gingerich, and A. Kluge. 1978. Hierarchical linear modeling of the tempo and mode in evolution. *Paleobiology* 4:120-134.

Boucot, A.J. 1975. *Evolution and Extinction Rate Controls.* Amsterdam: Elsevier.

Boucot, A.J. 1978. Community evolution and rates of cladogenesis. *Evol. Biol.* 11: 545-645.

Boucot, A. J. 1982. Ecophenotypic or genotypic? *Nature* 296:609.

Boucot, A.J. 1983. Does evolution take place in an ecological vacuum? II. " '*The time has come*' the Walrus said... ". *J. Paleontol.* 57: 1-30.

Boucot, A.J. 1984. Paleobiologic evidence for rates of coevolution and behavioral evolution. Unpublished Manuscript.

Bowen, S.T., J. Hanson, P. Dowling, and M.-C. Poon. 1966. The genetics of *Artemia salina*. VI. Summary of mutations. *Biol. Bull. Woods Hole* 131: 230-250.

Bowler, P.J. 1976. *Fossils and Progress: Paleontology and the Idea of Progressive Evolution in the Nineteenth Century.* New York: Science History Publications.

Brande, S. 1979. *Biometric analysis and evolution of two species of* Mulinia *(Bivalvia: Mactridae) from the Late Cenozoic of the Atlantic Coastal Plain.* Ph.D. dissertation, State Univ. of New York at Stony Brook.

Brenchley, G., and J.T. Carlton. 1983. Competitive displacement of native mud snails by introduced periwinkles in the New England intertidal zone. Biol. Bull. 165: 543-558.

Bretsky, P.W., Jr. 1969. Evolution of Paleozoic benthic marine invertebrate communities. *Palaeogr. Paleoclimatol. Palaeoecol.* 6: 45-69.

Bretsky, P.W., Jr. 1973. Evolutionary patterns in the Paleozoic Bivalvia: documentation and some theoretical considerations. *Geol. Soc. Amer. Bull.* 84: 2079-2096.

Bretsky, P.W., Jr., and S.S. Bretsky. 1976. The maintenance of evolutionary equilibrium in Late Ordovician benthic marine invertebrate faunas. *Lethaia* 9: 223-233.

Bretsky, S. S. 1979. Recognition of ancestor-descendant relationships in invertebrate paleontology. In *Phylogenetic Analysis and Paleontology*, ed. J. Cracraft and N. Eldredge, pp. 113-163. New York: Columbia Univ. Press.

Bretsky, S.S., and P.W. Bretsky, Jr.. 1977. Morphological variability and change in the palaeotaxodont bivalve mollusk *Nuculites planulatus* (Upper Ordovician of Quebec). *J. Paleontol.* 61: 256-271.

Bretsky, S.S., and E.G. Kauffman. 1977. Morphological variability and temporal change in a Paleocene lucinid bivalve mollusk. *Bull. Geol. Soc. Denmark* 26: 161-174.

Briggs, J.C. 1970. A faunal history of the North Atlantic Ocean. *Syst. Zool.* 19: 19-34.

Brinkmann, R. 1929. Statistisch-biostratigraphische untersuchungen an mitteljurrassischen ammoniten uber artbegreff und stammesentwicklung. Abh. Ges. Wiss. Göttingen, Math-Phys. K., N.F. 13(3): 1-249.

Briscoe, D.A., A. Robertson, and J.M Malpica. 1975. Dominance at Adh locus in response of adult *Drosophila melanogaster* to environmental alcohol. *Nature* 255: 148-149.

Britten, R.J. 1982. Genomic alterations in evolution. In *Evolution and Development*, ed. J.T. Bonner, pp. 41-64. Berlin: Springer-Verlag.

Brooks, J.L. 1950. Speciation in ancient lakes. *Qu. Rev. Biol.* 25: 30-60, 131-176.

Brown, D.D., P.C. Wensink, and E. Jordan. 1972. A comparison of the ribosomal DNA's of *Xenopus laevis* and *Xenopus mulleri*: the evolution of tandem genes. *J. Mol. Biol.* 63: 57-73.

Brown, W.D., and E.O. Wilson. 1954. Character displacement. *Syst. Zool.* 5: 49-64.

Bryant, E.H., S.A. McCommas, and L.M. Combs. 1986. The effect of an experimental bottleneck upon quantitative genetic variation in the housefly. *Genetics* 114: 1191-1211.

Bulmer, M.G. 1972. The genetic variability of polygenic characters under optimizing selection, mutation and drift. *Genet. Res. Camb.* 19: 17-25.

Burgman, M.A., and R.R. Sokal. 1986. Factors affecting the character stability of classifications. Unpublished manuscript.

Burian, R.M. 1983. "Adaptation." In *Dimensions of Darwinism*, ed. M. Grene, pp. 287-314. Cambridge Univ. Press.

Burton, R.S., and M.W. Feldman. 1983. Physiological effects of an allozyme polymorphism: glutamate-pyruvate transaminase and response to hyperosmotic stress in the copepod *Tigriopus californicus Biochem. Genet.* 21:239-251.

Bush, G.L. 1969. Sympatric host race formation and speciation in frugivorous flies of the genus *Rhagoletis*. *Evolution* 23:237-251.

Bush, G.L. 1975. Modes of animal speciation. *Ann. Rev. Ecol. Syst.* 6: 339-364.

Bush, G.L., S.M. Case, A.C. Wilson, and J.L. Patton. 1977. Rapid speciation and chromosomal evolution in mammals. *Proc. Nat. Acad. Sci. USA* 74: 3942-3946.

Cain, A.J. 1977. Variation in the spire index of some coiled gastropod shells, and its evolutionary significance. *Phil. Trans. Roy. Soc. Lond.* 277B: 377-428.

Cameron, J., and J.F. Fallon. 1967. The absence of cell death during development of free digits in amphibians. *Devl. Biol.* 55: 331-338.

Camin J.H., and R.R. Sokal 1965. A method for deducing branching sequences in phylogeny. *Evolution* 19: 311-326.

Camp, C.L. 1952. Geological boundaries in relation to faunal changes and diastrophism. *J. Paleontol.* 26: 353-358.

Campbell, D.T. 1974. 'Downward causation' in hierarchically organised biological systems. In *Studies in the Philosophy of Biology*, ed. F.J. Ayala and T. Dobzhansky, pp. 179-186. Berkeley: Univ. of California Press.

Capanna, E. 1982. Robertsonian numerical variation in animal speciation: Mus musculus, an emblematic model. In *Mechanisms of Speciation*, ed. C. Barigozzi, pp. 155-177. New York: Alan R. Liss.

Carrasco, A.E., W. McGinnis, W. Gehring, and E.M. DeRobertis. 1984. Cloning of an *X. laevis* gene expressed during early embryogenesis coding for a peptide region homologous to *Drosophila* homoeotic genes. *Cell* 37:409-414.

Carson, H.L. 1975. The genetics of speciation at the diploid level. *Am. Nat.* 109: 83-92.

Carson, H.L., D.E. Hardy, H.T. Spieth, and W.S. Stone. 1970. The evolutionary biology of the Hawaiian Drosophilidae. In *Essays in Evolution and Genetics in Honor of Theodosius Dobzhansky*, ed. M.K. Hecht and W.C. Steere, pp. 437-543. New York: Appleton-Century-Crofts.

Carson, H.L., and K.Y. Kaneshiro. 1976. *Drosophila* of Hawaii: systematics and ecological genetics. *Ann. Rev. Ecol. Syst.* 7: 311-346.

Carson, H.L., and A.R. Templeton. 1984. Genetic revolutions in relation to speciation phenomena: the founding of new populations. *Ann. Rev. Ecol. Syst.* 15: 97-131.

Carter, N.L., C.B. Officer, C.A. Chesner, and W.I. Rose. 1986. Dynamic deformation of volcanic ejecta from the Toba caldera: Possible relevance to Cretaceous/Tertiary boundary phenomena. *Geology* 14: 380-383.

Carr, S.M., A.j. Brothers, and A.C. Wilson. 1087. Evolutionary inferences from restriction maps of mitrochondrial DNA from nine taxa of *Xenopus* frogs. *Evolution* 41: 176-188.

Chaline, J., and B. Laurin. 1986. Phyletic gradualism in a European Plio-Pleistocene *Mimomys* lineage (Arvicolidae, Rodentia). *Paleobiology* 12:203-216.

Chamberlain, J.A., Jr., 1976. Flow patterns and drag coefficients of cephalopod shells. *Palaeontology* 19: 539-563.

Chamberlain, J.A., Jr. 1980. Hydromechanical design of fossil cephalopods. In *The Ammonoidea*, ed. M.R. House and J.R. Senior, Syst. Assoc. Spec. vol. 18. pp. 289-336. London: Academic Press.

Chao, L., and B.R. Levin. 1981. Structured habitats and the evolution of anticompetitor toxins in bacteria. *Proc. Nat. Acad. Sci. USA* 78: 6324-28.

Charlesworth, B. 1984a. Some quantitative methods for studying evolutionary patterns in single characters. *Paleobiology* 10: 310-318.

Charlesworth, B. 1984b. The cost of phenotypic evolution. *Paleobiology* 10:319-327.

Charlesworth, B., and D. Charlesworth. 1975. An experiment on recombination load in *Drosophila melanogaster*. *Genet. Res. Camb.* 25: 267-274.

Charlesworth, B., R. Lande, and M. Slatkin. 1982. A neo-Darwinian commentary on macroevolution. *Evolution* 36: 474-498.

Charlesworth, B., and C.H. Langley. 1986. The evolution of self-regulated transposition of transposable elements. *Genetics* 112:359-383.

Charlesworth, D., and B. Charlesworth. 1976. Theoretical genetics of Batesian mimicry. II. Evolution of supergenes. *J. Theor. Biol.* 55:305-324.

Charnov, E.L. 1982. *The Theory of Sex Allocation.* Princeton Univ. Press.

Cheetham, A.H. 1986. Tempo of evolution in a Neogene bryozoan: rates of morphologic change within and across species boundaries. *Paleobiology* 12:190-202.

Cherry, L.M., S.M. Case, and A.C. Wilson. 1978. Frog perspective on the morphological difference between humans and chimpanzees. *Science* 200: 209-211.

Cherry, L.M., S.M. Case, J.G. Kunkel, J.S. Wyles, and A.C. Wilson. 1982. Body shape metrics and organismal evolution. *Evolution* 36: 914-933.

Cheverud, J.M. 1982. Phenotypic, genetic, and environmental morphological integration in the cranium. *Evolution* 36: 499-516.

Cheverud, J.M., M.M. Dow, and W. Leutenegger. 1985. The quantitative assessment of phylogenetic constraints in comparative analyses: sexual dimorphism in body weight among primates. *Evolution* 39:1335-1351.

Cheverud, J.M., J.J. Rutledge, and W.R. Atchley. 1983. Quantitative genetics of development: genetic correlations among age-specific trait values and the evolution of ontogeny. *Evolution* 37: 895-905.

Chisholm, R.L., E. Barklis, and H.F. Lodish. 1984. Mechanism of sequential induction of cell-type specific mRNAs in *Dictyostelium* differentiation. *Nature* 310:67-69.

Christiansen, F.B., and O. Frydenberg. 1974. Geographical patterns in four polymorphisms in *Zoarces viviparus* as evidence of selection. *Genetics* 77: 765-770.

Christiansen, F.B., and V. Simonsen. 1978. Geographic variation in protein polymorphisms in the eelpout, *Zoarces viviparus* (L.). In *Marine Organisms: Genetics, Ecology, and Evolution,* ed. B.L. Battaglia and J.L. Beardmore, pp. 171-194. New York: Plenum.

Cifelli, R. 1969. Radiation of Cenozoic planktonic foraminifera. *Syst. Zool.* 18: 154-168.

Cisne, J.L. 1975. Evolution of the world fauna of aquatic free-living arthropods. *Evolution* 22: 337-366.

Cisne, J.L., G.O. Chandlee, B.D. Rabe, and J.A. Cohen. 1980. Geographic variation and episodic evolution in an Ordovician trilobite. *Science* 209: 925-927.

Cisne, J.L., J. Molenock and B.D. Rabe. 1980. *Evolution* in a cline: the trilobite *Triarthrus* along an Ordovician depth gradient. *Lethaia* 13:47-59.

Clarke, B. 1975. The contribution of ecological genetics to evolutionary theory: detecting the direct effects of natural selection on particular polymorphic loci. *Genetics* 79: 101-113.

Clarke, B., and J. Murray. 1969. Ecological genetics and speciation in land snails of the genus *Partula Biol. J. Linn. Soc.* 1:31-42.

Clarke, B., and P. O'Donald. 1964. Frequency-dependent selection. *Heredity* 19: 201-206.

Clarke, C.A., and P.M. Shepard. 1960a. The evolution of mimicry in the butterfly *Papilio dardanus. Heredity* 14: 163-173.

Clarke, C.A., and P.M. Shepard. 1960b. Super-genes and mimicry. *Heredity* 14: 175-185.

Clarke, C.A., and P.M. Shepard. 1962. Disruptive selection and its effect on a metrical character in the butterfly *Papilio dardanus. Evolution* 16: 214-226.

Clarke, C.A., and P.M. Shepard. 1963. Interactions between major genes and polygenes in the determination of the mimetic patterns of *Papilio dardanus. Evolution* 17: 404-413.

Clarke, C.A., P.M. Shepard, and I.W.B. Thornton. 1968. The genetics of the mimetic butterfly *Papilio memnon* L. *Phil. Trans. Roy. Soc. Lond.* B 254: 39-89.

Clarkson, E.N.K. 1966. Schizochroal eyes and vision of some Silurian acastid trilobites. *Palaeontology* 9: 1-29.

Clarkson, E.N.K. 1979. The visual system of trilobites. *Palaeontology* 22: 1-22.

Clarkson, E.N.K., and R. Levi-Setti. 1975. Trilobite eyes and the optics of Descartes and Huygens. *Nature* 254: 663-667.

Cloud, P.E. 1948. Some problems and patterns of evolution exemplified by fossil invertebrates. *Evolution* 2: 322-350.

Cloud, P.E. 1968. Pre-metazoan evolution and the origin of the metazoa. In *Evolution and Environment*, ed. E.T. Drake, pp. 1-72. New Haven: Yale Univ. Press.

Clutton-Brock, T.H., S.D. Albon, and P.H. Harvey. 1980. Antlers, body size and breeding group size in the Cervidae. *Nature* 565-567.

Clutton-Brock, T.H., and P.H.Harvey. 1979. Comparison and adaptation. *Proc. Roy. Soc. Lond.* B205: 547-565.

Cock, A.G. 1966. Genetical aspects of metrical growth and form in animals. *Quart. Rev. Biol.* 41: 131-190.

Cock, A.G. 1969. Genetical studies on growth and form in the fowl. *Genet. Res. Camb.* 14: 237-247.

Cody, M.L. 1974. *Competition and the Structure of Bird Communities.* Princeton Univ. Press.

Coen, E., and G. Dover. 1983. Unequal exchanges and the coevolution of X and Y rDNA arrays in *Drosophila melanogaster*. *Cell* 33: 849-855.

Coen, E., T. Strachan, and G. Dover. 1982. Dynamics of concerted evolution of ribosomal DNA and histone gene families in the *melanogaster* species subgroup of *Drosophila*. *J. Mol. Biol.* 158: 17-35.

Cole, R.K. 1967. Ametapodia, a dominant mutation in the fowl. *J. Hered.* 58: 141-146.

Coleman, W. 1976. Morphology between type concept and descent theory. *J. Hist. Med.* 31: 149-175.

Collins, D., D. Briggs, and S.C. Morris. 1983. New Burgess shale fossil sites reveal Middle Cambrian faunal complex. *Science* 222: 163-167. Cambridge, Massachusetts.

Conway Morris, S. 1979. The Burgess Shale (Middle Cambrian) fauna. *Ann. Rev. Ecol. Syst.* 10: 327-349.

Conway Morris, S. 1985a. The Middle Cambrian metazoan *Wiwaxia corrugata* (Matthew) from the Burgess Shale and *Ogygopsis* Shale, British Columbia, Canada. *Phil. Trans. Roy. Soc. London* 307: 507-586.

Conway Morris, S. 1985b. The Ediacaran biota and early metazoan evolution. *Geol. Mag.* 122: 77-81.

Cooper, G.A. 1944. Phylum Brachiopoda. In *Index Fossils of North America*, ed. H.W. Shimer and R.R. Shrock, pp. 277-365. New York: John Wiley and Sons.

Cooper, G.A., and A. Williams. 1952. Significance of the stratigraphic distribution of brachiopods. *J. Paleontol.* 26: 326-337.

Corbet, G.B. 1964. Regional variation in the bank vole *Clethrionomys glareolus* in the British Isles. *Proc. Zool. Soc. London* 143: 191-219.

Corliss, B.H., M.-P. Aubry, W.A. Berggren, J.M. Fenner, L.D. Keigwin, Jr., and G. Keller. 1984. The Eocene/Oligocene boundary in the deep sea. *Science* 226: 806-810.

Cowen, R. 1981. Crinoid arms and banana plantations: an economic harvesting analogy. *Paleobiology* 7: 332-343.

Cowen, R. 1983. Algal symbiosis and its recognition in the fossil record. In *Biotic Interactions in Recent and Fossil Benthic Communities*, ed. M.J.S. Tevesz and P.L. McCall, pp. 431-478. London: Plenum.

Coyne, J.A. 1983. Genetic basis of differences in genital morphology in three sibling species of *Drosophila Evolution* 37: 1101-1118.

Coyne, J.A. 1984. Genetic basis of sterility in hybrids between two closely related species of *Drosophila Proc. Nat. Acad. Sci. U.S.A.* 81: 4444-7.

Cracraft, J. 1981. The use of functional and adaptive criteria in phylogenetic systematics. *Amer. Zool.* 21: 21-36.

Crane, J. 1975. *Fiddler Crabs of the World*. Princeton Univ. Press.

Crick, F.H.C. 1970. Diffusion in embryogenesis. *Nature* 225: 420-422.

Crick, F.H.C., and P.A. Lawrence. 1975. Compartments and polyclones in insect development. *Science* 189: 340-347.

Crompton, A.W. 1963. On the lower jaw of *Diarthrognathus* and the origin of the mammalian lower jaw. *Proc. Zool. Soc. Lond.* 140: 697-753.

Crompton, A.W., and F.A. Jenkins. 1968. Molar occlusion in Late Triassic mammals. *Biol. Rev.* 43: 427-458.

Crompton, A.W., and F.A. Jenkins. 1973. Mammals from reptiles: a review of mammalian origins. *Ann. Rev. Earth Planet. Sci.* 1: 131-153.

Crompton, A.W., and F.A. Jenkins. 1979. Origin of mammals. In *Mesozoic Mammals*, ed. J.A. Lillegraven, Z. Kielan-Jaworska, and W.A. Clemens, pp. 59-71. Berkeley: Univ. of California Press.

Crow, J.F. 1957. *Genetics* of insect resistance to chemicals. *Ann. Rev. Entomol.* 2: 227-246.

Crowder, L.B. 1980. Ecological convergence of community structure: a neutral model analysis. *Ecology* 61: 194-204.

Crumpacker, D.W., and J.S. Williams. 1974. Rigid and flexible chromosomal polymorphisms in neighboring populations of *Drosophila pseudoobscura. Evolution* 28: 57-66.

Darwin, C. 1859. *On The Origin of Species By Means of Natural Selection.* London: John Murray.

Darwin, C. 1874. *The Descent of Man,* 2nd Ed. New York: Hurst and Co.

Darwin, C. 1876. *On The Origin of Species By Means of Natural Selection.* 6th ed. New York: D. Appleton.

Davenport, R. 1979. *An Outline of Animal Development.* Reading, MA: Addison-Wesley.

Davidson, E.H. 1976. *Gene Activity in Early Development.* New York: Academic Press. 452 pp.

Davidson, E.H. 1982. Evolutionary change in genomic regulatory organization: speculations of the origins of novel biological structure. In *Evolution and Development,* ed. J.T. Bonner, pp. 65-84. Berlin: Springer-Verlag.

Davidson, E.H., and R.J. Britten. 1971. Note on the control of gene expression during development. *J. Theor. Biol.* 32: 123-130.

Davidson, E.H., B.R. Hough-Evans, and R.J. Britten. 1982. Molecular biology of the sea urchin embryo. *Science* 217: 17-26.

Davis, M., P. Hut, and R.A. Muller. 1984. Extinction of species by periodic comet showers. *Nature* 308: 715-717.

Dawkins, R. 1983. Universal Darwinism. In *Evolution From Molecules to Men*, ed. D.S. Bendall, pp. 403-428. Cambridge Univ. Press.

de Beer, G. 1958. *Embryos and Ancestors.* Oxford: Clarendon Press, 3rd edition.

del Pino, E.M., and R.P. Elinson. 1983. A novel development pattern for frogs: gastrulation produces an embryonic disk. *Nature* 306: 589-591.

Desmond, A. 1982. *Archetypes and Ancestors.* London: Blond and Briggs.

Desplan, C., J. Theis, and P.H. O'Farrell. 1985. the *Drosophila* developmental gene, *engrailed*, encodes a sequence-specific DNA binding activity. *Nature* 318:630-635.

Diamond, J.M. 1983. "Normal" extinctions of isolated populations. In *Extinctions*, ed. M. Nitecki, pp. 191-246. Field Museum Nat. Hist. Symp. 1982. Univ. Chicago Press.

DiMichelle, L., and D.A. Powers. 1982. Physiological basis for swimming endurance differences between LDH-B genotypes of *Fundulus heteroclitus. Science* 216: 1014-1016.

Dingus, L. 1984. Effects of stratigraphic completeness on interpretations of extinction rates across the Cretaceous-Tertiary boundary. *Paleobiology* 10: 420-438.

Doane, W.W. 1969. *Drosophila* amylases and problems in cellular differentiation. In *Problems in Biology: RNA In Development*, ed. E.W. Hanly, pp. 75-108. Salt Lake City: Univ. of Utah.

Doane, W.W., I. Abraham, M.M. Kolar, R.E. Marenson, and G.E. Deibler. 1975. Purified *Drosophila* alpha-amylase enzymes: genetical biochemical and molecular characterization. In *Isozymes: Genetics and Evolution*, vol. 4, ed. C.L. Markert, pp. 585-607. New York: Academic Press.

Dobzhansky, T. 1935. A critique of the species concept in biology. *Philos. Sci.* 2: 344-355.

Dobzhansky, T. 1937. *Genetics and the Origin of Species.* New York: Columbia University Press.

Dobzhansky, T. 1943. Genetics of natural populations IX. Temporal changes in the composition of populations of *Drosophila pseudoobscura. Genetics* 28: 162-186.

Dobzhansky, T. 1947. Adaptive changes induced by natural selection in wild populations of *Drosophila. Evolution* 1: 1-16.

Dobzhansky, T. 1948a. *Genetics* of natural populations. XVI. Altitudinal and seasonal changes produced by natural selection in certain populations of *Drosophila pseudoobscura* and *Drosophila persimilis. Genetics* 33: 158-176.

Dobzhansky, T. 1948b. Genetic structure of natural populations. *Yearb.* Carnegie Inst. Wash. 47: 193-203.

Dobzhansky, T. 1951. *Genetics* and the Origin of Species. 3rd. ed. New York: Columbia Univ. Press.

Dobzhansky, T. 1955. A review of some fundamental concepts and problems in population genetics. *Cold Spring Harbor Symp. Quant. Biol.* 20: 1-15.

Dobzhansky, T. 1970. *Genetics of the Evolutionary Process.* New York: Columbia Univ. Press.

Dobzhansky, T., and O. Pavlosky. 1957. An experimental study of interaction between genetic drift and natural selection. *Evolution* 11: 311-319.

Dobzhansky, T., and B. Spassky. 1954. *Genetics* of natural populations. XXII. A comparison of the concealed variability in *Drosophila prosaltans* with that in other species. *Genetics* 39: 472-487.

Dobzhansky, T., and B. Spassky. 1962. Genetic drift and natural selection in experimental populations of *Drosophila pseudoobscura. Proc. Nat. Acad. Sci. USA* 48: 148-156.

Dobzhansky, T., and S. Wright. 1941. *Genetics* of natural populations. V. Relations between mutation rate and accumulation of lethals in populations of *Drosophila pseudoobscura*. *Genetics* 26: 23-51.

Doolittle, W.F., and C. Sapienza. 1980. Selfish genes, the phenotype paradigm and genome evolution. *Nature* 284: 601-603.

Douthart, R.J., and F.H. Norris. 1982. Events in the evolution of pre-proinsulin. *Science* 217: 729-732.

Dover, G. 1982. Molecular drive: a cohesive mode of species evolution. *Nature* 299. 111-117.

Dover, G., S. Brown, E. Coen, J. Dallas, T. Strachan, and M. Trick. 1982. The dynamics of genome evolution and species differentiation. In *Genome Evolution*, ed. G.A. Dover and R.B. Flavell, pp. 343-372. London: Academic press.

Dover, G.A., and R.B. Flavell, eds. 1982. *Genome Evolution.* London: Academic Press.

Dungan, M.L. 1985. Competition and the morphology, ecology, and evolution of acorn barnacles: an experimental test. *Paleobiology* 11:165-173.

Dunn, E.R. 1942. Survival value of varietal characters in snakes. *Am. Nat.* 75: 104-109.

Dunnill, R.M., and D.V. Ellis. 1969. The distribution and ecology of sub-littoral species of *Macoma* off Moresby Island and in Satellite Channel. *Veliger* 12: 201-206.

Dykhuizen, D., and D.L. Hartl. 1983. Functional effects of PGI allozymes in *Escherichi coli*. *Genetics* 105: 1-18.

Dykhuizen, D., J. de Framond, and D.L. Hartl. 1984. Selective neutrality of glucose-6-phosphate dehydrogenase allozymes in *Escherichia coli*. *Mol. Biol. Evol.* 1: 162-170.

Eanes, W.F. 1984. Viability interactions, *in vivo* activity, and G6PD polymorphism in *Drosophila melanogaster Genetics* 106:95-107.

Eanes, W.F. 1987. Allozymes and fitness: Evolution of a problem. *Trends in Ecology and Evolution* 2: 44-48.

Eanes, W.F., B. Bingham, J. Hey, and D. Houle. 1985. Targeted selection experiments and enzyme polymorphism: negative evidence for octanoate selection at the G6PD locus in *Drosophila melanogaster*. *Genetics* 109: 379-391.

Easton, W.H. 1960. *Invertebrate Paleontology.* New York: Harper and Row.

Eck, R.V., and M.O. Dayhoff. 1966. *Atlas of Protein Sequence and Structure 1966.* Silver Spring, Maryland: National Biomedical Research Foundation.

Edelman, G. 1986. Evolution and morphogenesis: the regulator hypothesis. In *Genetics, Development, and Evolution,* ed. J.P. Gustafson, G.L. Stebbins, F.J. Ayala. New York: Plenum Publ. Co., pp. 1-27.

Edgell, M.H., S.G. Hardies, B. Brown, C. Voliva, A. Hill, S. Phillips, M. Comer, F. Burton, S. Weaver, and C.A. Hutchison III. 1983. Evolution of the mouse beta globin complex locus. In *Evolution of Genes and Proteins,* ed. M. Nei and R.K. Koehn, pp. 1-13. Sunderland, MA: Sinauer Associates,

Ehrlich, P.R., and P.H. Raven. 1969. Differentiation of populations. *Science* 165: 1228-1232.

Ehrman, L. 1967. Further studies on genotype frequency and mating success in *Drosophila. Am. Nat.* 101: 415-424.

Eldredge, N. 1971. The allopatric model and phylogeny in Paleozoic invertebrates. *Evolution* 25: 156-167.

Eldredge, N. 1974. Testing evolutionary hypotheses in paleontology: a comment on Makurath and Anderson (1973). *Evolution* 28: 479-481.

Eldredge, N. 1982. Phenomenological levels and evolutionary rates. *Syst. Zool.* 31: 338-347.

Eldredge, N. 1985. *Unfinished Synthesis: Biological Hierarchies and Modern Evolutionary Thought.* New York: Oxford Univ. Press.

Eldredge, N., and J. Cracraft. 1980. *Phylogenetic Patterns and the Evolutionary Process.* New York: Columbia University Press.

Eldredge, N., and S.J. Gould. 1972. Punctuated equilibria: an alternative to phyletic gradualism. In *Models In Paleobiology,* ed. T.J.M. Schopf, pp. 82-115. San Francisco: Freeman, Cooper, & Co.

Eldredge, N., and I. Tattersall. 1982. *The Myths of Human Evolution.* New York: Columbia Univ. Press.

Ellstrand, N.C., and J. Antonovics. 1985. Experimental studies of the evolutionary significance of sexual reproduction. II. A test of the density-dependent hypothesis. *Evolution* 39: 657-666.

Emiliani, C. 1982. Extinctive evolution. *J. Theor. Biol.* 97: 13-33.

Endler, J.A. 1977. *Geographic Variation, Speciation, and Clines.* Princeton Univ. Press.

Endler, J.A. 1978. A predator's views of animal color patterns. *Evol. Biol.* 11. 319-304.

Endler, J.A. 1986. *Natural Selection in the Wild.* Princeton Univ. Press.

Erwin, D.H., and J.W. Valentine. 1984. "Hopeful monsters," transposons, and Metazoan radiation. *Proc. Nat. Acad. Sci. USA* 81:5482-5483.

Estabrook, G.F. 1968. A general solution in partial orders for the Camin-Sokal model in phylogeny. *J. Theoret. Biol.* 21: 421-438.

Estabrook, G.F. 1972. Cladistic methodology: a discussion of the theoretical basis for the induction of evolutionary history. *Ann. Rev. Ecol. Syst.* 3: 427-456.

Estabrook, G.F. 1980. The compatibility of occurrence patterns of chemicals in plants. In *Chemosystematics: Principles and Practice*, ed. F.A. Bisby, J.G. Vaughan and C.A. Wright, pp. 379-397. Systematics Association Special Volume No. 16. London: Academic Press.

Estabrook, G.F., and W.R. Anderson. 1978. An estimate of phylogenetic relationships within the genus *Crusea* (Rubiaceae) using character compatibility analysis. *Syst. Bot.* 3:179-196.

Etkin, W. 1970. The endocrine mechanism of amphibian metamorphosis, an evolutionary achievement. *Mem. Soc. Endocrinol.* 18: 137-155.

Ewens, W.J. 1969. *Population Genetics.* London: Methuen.

Falconer, D.S. 1960. *Introduction to Quantitative Genetics.* New York: Ronald Press.

Falconer, D.S. 1981. *Introduction to Quantitative Genetics.* 2nd ed. New York and London: Longman.

Falkner, F.G., and H.G. Zachau. 1984. Correct transcription of an immunoglobulin κ gene requires an upstream fragment containing conserved sequence elements. *Nature* 310:71-74.

Fallon, J.F., and J. Cameron. 1977. Interdigital cell death during limb development of the turtle with an interpretation of evolutionary significance. *J. Embryol. Exp. Morph.* 40: 285-289.

Farris, J.S. 1970. Methods for computing Wagner trees. *Syst. Zool.* 19: 83-92.

Farris, J.S. 1971. The hypothesis of nonspecificity and taxonomic congruence. *Ann. Rev. Ecol. Syst.* 2: 277-302.

Farris, J.S. 1972. Estimating phylogenetic trees from distance matrices. *Am. Nat.* 106: 645-668.

Farris, J.S. 1975. Formal definitions of paraphyly and polyphyly. *Syst. Zool.* 23: 548-554.

Farris, J.S. 1976. Phylogenetic classification of fossils with Recent species. *Syst. Zool.* 25: 271-282.

Farris, J.S. 1979. The information content of the phylogenetic system. *Syst. Zool.* 28: 483-519.

Farris, J.S. 1983. The logical basis of phylogenetic analysis. In *Advances in Cladistics*, vol. 2, ed. N.I. Platnick and V.A. Funk, pp. 7-36. New York: Columbia Univ. Press.

Felsenstein, J. 1978. Cases in which parsimony or compatibility methods will be positively misleading. *Syst. Zool.* 27: 401-410.

Felsenstein, J. 1979. Alternative methods of phylogenetic inference and their interrelationship. *Syst. Zool.* 28:49-62.

Felsenstein, J. 1981. Evolutionary trees from DNA sequences: a maximum likelihood approach. *J. Mol. Evol.* 17: 368-376.

Felsenstein, J. 1982. Numerical methods for inferring evolutionary trees. *Quart. Rev. Biol.* 57: 379-404.

Felsenstein, J. 1983. Parsimony in systematics: biological and statistical issues. *Ann. Rev. Ecol. Syst.* 14: 313-333.

Fenchel, T. 1975. Character displacement and coexistence in mud snails. *Oecologia* 20: 19-32.

Ferris, S.D., W.M. Brown, W.S. Davidson, and A.C. Wilson. 1981. Extensive polymorphism in the mitochondrial DNA of apes. *Proc. Nat. Acad. Sci. USA* 78: 6319-6323.

Fink, W.L. 1982. The conceptual relationship between ontogeny and phylogeny. *Paleobiology* 8: 254-264.

Fischer, A.G. 1984. The two Phanerozoic supercycles. In *Catastrophes and Earth History, The New Uniformitarianism,* ed. W.A. Berggren, J.A. Van Couvering, pp. 129-150. Princeton Univ. Press.

Fischer, A.G., and M.A. Arthur. 1977. Secular variations in the pelagic realm. Soc. Econ. Paleontol. Mineral. Spec. Publ. 25: 19-50.

Fisher, R.A. 1930. *The Genetical Theory of Natural Selection.* Oxford: Oxford Univ. Press. 1st ed.

Fisher, W.L., P.V. Rodda and J.W. Dietrich. 1964. Evolution of *Athleta* petrosa stock (Eocene, Gastropoda) of Texas. Univ. Texas Publ. 6413, pp. 1-117.

Fitch, W.M. 1971. Toward defining the course of evolution: minimum change for a specific tree topology. *Syst. Zool.* 20: 406-416.

Fitch, W.M., and J.S. Farris. 1974. Evolutionary trees with minimum nucleotide replacements from amino acid sequences. *J. Mol. Evol.* 3: 263-278.

Fitch, W.M., and E. Margoliash. 1967. Construction of phylogenetic trees. *Science* 155: 279-284.

Flavell, R. 1982. Sequence amplification, deletion and rearrangement: major sources of variation during species divergence. In *Genome Evolution,* ed. G. Dover and R. Flavell, pp. 301-323. London: Academic Press.

Flessa, K.W. 1981. The regulation of mammalian faunal similarity among the continents. *J. Biogeogr.* 8: 427-437.

Flessa, K.W., and J. Imbrie. 1973. Evolutionary pulsations: evidence from Phanerozoic diversity patterns. In *Implications of Continental Drift to the Earth Sciences,* ed. D.H. Tarling and S.K. Runcorn, pp. 247-285. New York: Academic Press.

Flessa, K.W., and D. Jablonski. 1985. Declining Phanerozoic background extinction rates: effect of taxonomic structure? *Nature* 313:216-218.

Flessa, K.W., and J.S. Levinton. 1975. Phanerozoic diversity patterns: tests for randomness. *J. Geol.* 83: 239-248.

Flessa, K.W., K.V. Powers and J.L. Cisne. 1975. Specialization and evolutionary longevity in the Arthropoda. *Paleobiology* 1: 71-81.

Ford, E. B. 1975. *Ecological Genetics.* 4th ed. London: Chapman and Hall.

Ford, E.B., and J.S. Huxley. 1927. Mendelian genes and rates of development in *Gammarus chevreuxi. Brit. J. Exp. Biol.* 5: 112-134.

Ford, E.B., and J.S. Huxley. 1929. Genetic rate factors in *Gammarus. Arch. Entwicklgmech* 117: 67-79.

Forey, P.L. 1982. Neontological analysis versus paleontological stories. In *Problems of Phylogenetic Reconstruction,* ed. K.A. Joysey and A.E. Friday, pp. 119-157. London: Academic Press.

Fortey, R.A., and R.P.S. Jefferies. 1982. Fossils and phylogeny - a compromise approach. In *Problems of Phylogenetic Reconstruction,* ed. K.A. Joysey and A.E. Friday, pp. 197-234. London: Academic Press,

Franklin, I., and R.C. Lewontin. 1970. Is the gene the unit of selection? *Genetics* 65: 707-734.

Frazzetta, T.H. 1969. Adaptive problems and possibilities in the temporal fenestration of tetrapod skulls. *J. Morph.* 125: 145-158.

Frazzetta, T.H. 1970. From hopeful monsters to Bolyerine snakes. *Am. Nat.* 104: 55-70.

Frazzetta, T.H. 1975. *Complex Adaptations in Evolving Populations.* Sunderland, MA: Sinauer.

Fuentes, E.R. 1976. Ecological convergence of lizard communities in Chile and California. *Ecology* 57: 3-17.

Fuerst, P.A., R. Chakraborty, and M. Nei. 1977. Statistical studies on protein polymorphism in natural populations. I. Distribution of single locus heterogeneity. *Genetics* 86: 455-483.

Fürsich, F.T., and D. Jablonski. 1984. Late Triassic naticid drillholes: carnivorous gastropods gain a major adaptation but fail to radiate. *Science* 224: 78-80.

Futuyma, D.J. 1987. On the role of species in anagenesis. *Am. Nat.*, in press.

Futuyma, D.J., and G.C. Mayer. 1980. Non-allopatric speciation in animals. *Syst. Zool.* 29: 254-271.

Futuyma, D.J., R.C. Lewontin, G.C. Mayer, J. Seger, and J.W. Stubblefield III. 1981. Macroevolution conference (letter). *Science* 211:770.

Ganapathy, K. 1982. Evidence for a major meteorite impact on the earth 34 million years ago: implications for Eocene extinctions. *Science* 216: 885-886.

Gans, M., C. Audit, and M. Masson. 1975. Isolation and characterization of sex-linked female-sterile mutants in *Drosophila melanogaster. Genetics* 81: 683-704.

Garabedian, M.J., B.M. Shepherd, and P.C. Wensink. 1986. A tissue-specific transcription enhancer from the Drosophila yolk protein 1 gene. *Cell* 45: 859-867.

Garcia-Bellido, A. 1975. Genetic control of wing disc development in *Drosophila.* In *Cell Patterning.* CIBA Foundation Symp. 29, pp. 161-182.

Garcia-Bellido, A., P. Ripoll and G. Morata. 1973. Developmental compartmentalization of the wing disc of *Drosophila. Nature* 245: 251-253.

Garn, S.M., and A.B. Lewis. 1963. Third molar polymorphism and its significance to dental genetics. *J. Dent. Res.* 42 suppl.: 1334-1363.

Garstang, W. 1922. The theory of recapitulation. A critical restatement of the biogenetic law. *J. Linn. Soc. London, Zoology* 35: 81-101.

Gecow, A., and A. Hoffman. 1983. Self-improvement in a complex cybernetic system and its implications for biology. *Acta Biotheoretica* 32: 61-71.

Gehring, W.J. 1987. Homeo boxes in the study of development. *Science* 236: 1245-1252.

Gerhart, D. 1984. Prostaglandin A_2: an agent of chemical defense in the Caribbean gorgonian *Plexaura homomalla*. *Marine Ecology-Progress Series* 19: 181-187.

Gerhart, J.S., S. Black, R.A. Gimlich, and S. Scharf. 1983. Control of polarity in the amphibian egg. In *Time, Space, and Pattern in Embryonic Development*, ed. W.R. Jefferey and W.H. Raff, pp. 261-286. New York: Alan R. Liss.

Ghiselin, M.T. 1975. A radical solution to the species problem. *Syst. Zool.* 23: 536-544.

Gibson, J.B., N. Lewis, M.A. Adena, and S.R. Wilson. 1979. Selection for ethanol tolerance in two populations of *Drosophila melanogaster* segregating alcohol dehydrogenase allozymes. *Austr. J. Biol. Sci.* 32: 387-398.

Gilbert, F.S. 1981. Foraging ecology of hoverflies: morphology of the mouth parts in relation to feeding on nectar and pollen in some common urban species. *Ecological Entomology* 6: 245-262.

Gilinsky, N. 1981. Stabilizing species selection in the Archaeogastropoda. *Paleobiology* 7: 316-331.

Gilinsky, N., and R.K. Bambach. 1986. The evolutionary bootstrap: a new approach to the study of taxonomic diversity. *Paleobiology* 12: 251-268.

Gillespie, J. 1974. Polymorphism in patchy environments. *Am. Nat.* 108: 145-151.

Gillespie, J. 1986. Rates of molecular evolution. *Ann. Rev. Ecol. Syst.* 17: 637-665.

Gillespie, J.H., and C.H. Langley. 1974. A general model to account for enzyme variation in natural populations. *Genetics* 76: 837-848.

Gilmour, J.S.L. 1961. Taxonomy. In *Contemporary Botanical Thought*, ed. A.M. McLeod and L.S. Cobley, pp. 27-45. Edinburgh: Oliver and Boyd.

Gingerich, P.D. 1976. Paleontology and phylogeny: patterns of evolution at the species level in Early Tertiary mammals. *Amer.*

J. Sci. 276: 1-28.

Gingerich, P. D. 1979. Stratophenetic approach to phylogeny reconstruction in vertebrate paleontology. In *Phylogenetic Analysis and Paleontology.*, ed. J. Cracraft and N. Eldredge, pp. 41-79. New York: Columbia Univ. Press.

Gingerich, P.D. 1983. Rates of evolution: effect of time and temporal scaling. *Science* 222: 159-161.

Gingerich, P.D. 1001. Smooth curve of evolutionary rate: a psychological and mathematical artifact (Reply to Gould). *Science* 226: 995-996.

Gingerich, P.D., N.A. Wells, D.E. Russell, and S.M. Ibrahim Shah. 1983. Origin of whales in epicontinental remnant seas: new evidence from the early Eocene of Pakistan. *Science* 220: 403-406.

Ginzburg, L., P.M. Bingham, and S. Yoo. 1984. On the theory of speciation induced by transposable elements. *Genetics* 107: 331-341.

Glaessner, M.F. 1984. *The Dawn of Animal Life. A BioHistorical Study.* Cambridge Univ. Press, Cambridge.

Gō, M. 1981. Correlation of DNA exonic regions with protein structural units in haemoglobin. *Nature* 291: 90-92.

Goldschmidt, R. 1933. Some aspects of evolution. *Science* 78: 539-547.

Goldschmidt, R. 1938. *Physiological Genetics.* New York: McGraw-Hill.

Goldschmidt, R.B. 1940. *The Material Basis of Evolution.* New Haven: Yale Univ. Press.

Goldschmidt, R.B. 1945. Mimetic polymorphism, a controversial chapter of Darwinism. *Quart. Rev. Biol.* 20: 147-164, 205-330.

Goldsmith, M.R., and F.C. Kafatos. 1984. Developmentally regulated genes in silkmoths. *Ann. Rev. Genet.* 18:443-487.

Gooch, J.L., and T.J.M. Schopf. 1973. Genetic variability in the deep sea: relation to environmental variability. *Evolution* 26: 545-552.

Goodman, M.J. Czelusniak, G.W. Moore, A.E. Romero-Herrera, and G. Matsuda. 1979. Fitting the gene lineage into its species lineage, a parsimony strategy illustrated by cladograms constructed from globin sequences. *Syst. Zool.* 28: 132-163.

Goodman, M., M.L. Weiss, and J. Czelusniak. 1982. Molecular evolution above the species level: branching pattern, rates, and mechanisms. *Syst. Zool.* 31: 376-399.

Gottlieb, L.D. 1976. Biochemical consequences of speciation in plants. In *Molecular Evolution*, ed. F.J. Ayala, pp. 123-140. Sunderland, MA: Sinauer Assoc.

Gould, S.J. 1966. Allometry and size in ontogeny and phylogeny. *Biol. Rev.* 41: 587-640.

Gould, S.J. 1969a. Character variation in two land snails from the Dutch Leeward islands: geography, environment, and evolution. *Syst. Zool.* 18: 185-200.

Gould, S.J. 1969b. Ecology and functional significance of uncoiling in *Vermicularia spirata*: an essay on gastropod form. *Bull. Mar. Sci.* 19: 432-445.

Gould, S.J. 1972. Allometric fallacies and the evolution of Gryphaea: A new interpretation based on White's criterion of geometric similarity. *Evol. Biol.* 6: 91-119.

Gould, S.J. 1974. The evolutionary significance of 'bizarre' structures: antler size and skull size in the 'Irish Elk', *Megalocerus giganteus*. *Evolution* 28: 191-220.

Gould, S.J. 1975. Allometry in primates with an emphasis on scaling and the evolution of the brain. *Contr. Primatol.* 5: 244-292.

Gould, S.J. 1976. The genomic metronome as a null hypothesis. *Paleobiology* 2: 177-179.

Gould, S.J. 1977. *Ontogeny and Phylogeny.* Cambridge, MA: Harvard Univ. Press.

Gould, S.J. 1980a. Is a new and general theory of evolution emerging? *Paleobiology* 6: 119-130.

Gould, S.J. 1980b. G.G. Simpson, Paleontology and the Modern Synthesis. In *The Evolutionary Synthesis*, ed. E. Mayr and W.B. Provine, pp. 153-172. Cambridge: Harvard Univ. Press.

Gould, S.J. 1982a. Darwinism and the expansion of evolutionary theory. *Science* 216: 380-387.

Gould, S.J. 1982b. Change in developmental timing as a mechanism of macroevolution. In *Evolution and Development*, ed. J.T. Bonner, pp. 333-346. Berlin: Springer-Verlag.

Gould, S.J. 1982c. The meaning of punctuated equilibrium and its role in validating a hierarchical approach to macroevolution. In *Perspectives on Evolution*, ed. R. Milkman, pp. 83-104. Sunderland MA: Sinauer Associates.

Gould, S.J. 1983a. The hardening of the Modern Synthesis. In *Dimensions of Darwinism*, ed. M. Grene, pp. 71-93. Cambridge Univ. Press.

Gould, S.J. 1983b. Irrelevance, submission and partnership: the changing role of palaeontology in Darwin's three centennials and a modest proposal for macroevolution. pp. 347-366, In *Evolution From Molecules to Men*, ed. D.S. Bendall, pp. 347-366. Cambridge Univ. Press.

Gould, S.J. 1984a. Smooth curve of evolutionary rate: a psychological and mathematical artifact. *Science* 226: 994-995.

Gould, S.J. 1984b. Toward the vindication of punctuational change. In *Catastrophes and Earth History, The New Uniformitarianism*, ed. W.A. Berggren and J.A. Van Couvering, pp. 9-34. Princeton Univ. Press.

Gould, S.J. 1984c. *Hen's Teeth and Horse's Toes*. New York: W.W.Norton (paper).

Gould, S.J. 1985a. The paradox of the first tier: an agenda for paleobiology. Paleobiology 11: 2-12.

Gould, S.J. 1985b. Not necessarily a wing. *Natural History*. Oct. 1985, pp. 12-25.

Gould S.J., and C.B. Calloway. 1980. Clams and brachiopods - ships that pass in the night. *Paleobiology* 6: 383-396.

Gould, S.J., and N. Eldredge. 1977. Punctuated equilibria: the tempo and mode of evolution revisited. *Paleobiology* 3: 115-151.

Gould, S.J., and R.C. Lewontin. 1979. The spandrels of San Marco and the Panglossian paradigm: A critique of the adaptationist programme. *Proc. R. Soc. Lond.* 205B: 581-598.

Gould, S.J., D.M. Raup, J.J. Sepkoski, Jr., T.J.M. Schopf and D.S. Simberloff. 1977. The shape of evolution: a comparison of real and random clades. *Paleobiology* 3: 23-40.

Gould, S.J., and E.S. Vrba. 1982. Exaptation - a missing term in the science of form. *Paleobiology* 8: 4-15.

Grabert, B. 1959. Phylogenetische untersuchungen an *Gaudyrina* und *Spiroplectinata* (Foram.) besonders aus dem nordwestdeutschen Apt und Alb. *Abh. senckenb. naturf. Ges.* 498: 1-71.

Grant, P.R. 1985. Selection on bill characters in a population of Darwin's finches: *Geospiza conirostris* on Isla genovesa, Galapagos. *Evolution* 39: 523-531.

Grant, R. 1972. The lophophore and feeding mechanism of the Productidina (Brachiopoda). *J. Paleontol.* 46: 213-248.

Grassle, J.F., and J.P. Grassle. 1974. Sibling species in the marine pollution indicator, *Capitella capitata* (Polychaeta). *Science* 192: 567-569.

Green, E.L. 1962. Quantitative genetics of skeletal variations in the mouse. II. Crosses between four inbred strains. (C3H, DBA, C57BL, BALB/c). *Genetics* 47: 1085-1096.

Greenbaum, I.F. 1981. Genetic interactions between hybridizing cytotypes of the tent-making bat (*Uroderma bilobatum*). *Evolution* 35: 306-321.

Gregory, W.C. 1966. Mutation breeding. In *Plant Breeding*, ed. K. J. Frey, pp. 189-202. Ames: Iowa State Univ. Press.

Grewal, M.S. 1962. The development of an inherited tooth defect in the mouse. *J. Embryol. Exp. Morph.* 10: 202-211.

Gross, P. R. 1985. Laying the ghost: embryonic development in plain words. *Biol. Bull.* 168 (suppl.): 62-79.

Gruneberg, H. 1965. Genes and genotypes affecting the teeth of the mouse. *J. Embryol. Exp. Morph.* 14: 137-159.

Gupta, A.P., and R.C. Lewontin. 1983. A study of reaction norms in natural populations of *Drosophila pseudoobscura*. *Evolution* 36:934-948.

Gurdon, J.B. 1974. *The Control of Gene Expression in Animal Development.* London: Oxford Univ. Press.

Gustafson, T., and L. Wolpert. 1961. Studies on the cellular basis of morphogenesis in the sea urchin embryo. *Expl. Cell. Res.* 24: 64-79.

Hadorn, E. 1961. *Developmental Genetics and Lethal Factors.* (transl. from German) London: Methuen & Co.

Hadorn, E. 1967. Dynamics of determination. In *Major Problems in Developmental Biology,* ed. M. Locke, pp. 85-104. New York: Academic Press.

Haeckel, E. 1866. *Generelle Morphologie der Organismen, Allgemeine Grundzüge der Organischen formen - Wissenschaft, mechanisch begründet durch die von Charles Darwin reformerte Decendenz - Theorie,* vol. 1 and 2. Berlin: Georg. Reiner.

Hafner, J.C., D.J. Hafner, J.L. Patton, and M.F. Smith. 1983. Contact zones and genetics of differentiation in the pocket gophers *Thomomys bottae* (Rodentia: Geomyidae). *Syst. Zool.* 32:1-20.

Haldane, J.B.S. 1927. A mathematical theory of natural and artificial selection. Part V. Selection and mutation. *Proc. Camb. Phil. Soc.* 23:838-844.

Haldane, J.B.S. 1930. A note on Fisher's theory of the origin of dominance and on a correlation between dominance and linkage. *Am. Nat.* 64:87-90.

Haldane, J.B.S. 1932a. *The Forces of Evolution.* London and New York: Longmans Green and Co.

Haldane, J.B.S. 1932b. The time of action of genes, and its bearing on some evolutionary problems. *Am. Nat.* 66: 5-24.

Haldane, J.B.S. 1949. Suggestions as to quantitative measurement of rates of evolution. *Evolution* 3: 51-56.

Haldane, J.B.S. 1954. *The Biochemistry of Genetics.* Woking and London: Allen and Unwin.

Haldane, J.B.S., and S.D. Jayakar. 1963. Polymorphism due to selection of varying direction. *J. Genet.* 58: 237-242.

Hall, B.K. 1984. Developmental processes underlying heterochrony as an evolutionary mechanism. *Can. J. Zool.* 62: 1-7.

Hallam, A. 1968. Morphology, palaeoecology and evolution of the genus *Gryphaea* in the British Lias. *Phil. Trans. Roy. Soc. Lond.* B 254: 91-128.

Hallam, A. 1975. Evolutionary size increase and longevity in Jurassic bivalves and ammonites. *Nature* 258: 493-496.

Hallam, A. 1977a. Jurassic bivalve biogeography. *Paleobiology* 3: 58-73.

Hallam, A. 1977b. Secular changes in marine inundation of USSR and North America through the Phanerozoic. *Nature* 269: 769-772.

Hallam, A. 1978. How rare is phyletic gradualism and what is its evolutionary significance? *Paleobiology* 4: 16-25.

Hallam, A. 1982. Patterns of speciation in Jurassic *Gryphaea*. *Paleobiology* 8: 354-366.

Hallam, A. 1984a. The causes of mass extinctions. *Nature* 308: 686-687.

Hallam, A. 1984b. Pre-Quaternary sea-level changes. *Ann. Rev. Earth Planet. Sci.* 12: 205-243.

Hamer, D.H., and P. Leder. 1979. Splicing and the formation of stable RNA. *Cell* 18: 1299-1302.

Hampé, A. 1959. Contribution a l'etude du developement et de la regulation des deficiencies et des excedents dans la patte de l'embryon de poulet. *Arch. Anat. Microscop. Morphol. Exp.* 48: 347-478.

Hansen, T.A. 1980. Influence of larval dispersal and geographic distribution on species longevity in neogastropods. *Paleobiology* 6: 193-207.

Hansen, T.A. 1982. Modes of larval development in Early Tertiary neogastropods. *Paleobiology* 8: 367-377.

Harcourt, A.H., P.H. Harvey, S.G. Larson, and R.V. Short. 1981. Testis weight, body weight and breeding system in primates. *Nature* 293: 55-57.

Hardy, M.C. 1985. Testing for adaptive radiation: the ptychasipid (Trilobita) biomere of the Late Cambrian. In *Phanerozoic Diversity Patterns: Profiles in Macroevolution*, ed. J.W. Valentine, pp. 379-397. Princeton Univ. Press.

Harper, C.W., Jr. 1976. Phylogenetic inference in paleontology. *J. Paleontol.* 50: 180-193.

Harris, J.M., and T.D.White. 1979. *Evolution* of the Plio-Pleistocene African Suidae. *Trans. Amer. Phil. Soc.* 60(2): 1-128.

Harris, H. 1966. Enzyme polymorphism in man. *Proc. Roy. Soc. Lond.* Ser. B 164: 298-310.

Harvey, P.H., and P. M. Bennett. 1983. Brain size, energetics, ecology and life history patterns. *Nature* 306: 314-315.

Hatfield, C.B., and M.J. Camp. 1970. *Geol. Soc. Amer. Bull.* 81: 911-914.

Hayami, I. 1973. Discontinuous variations in an evolutionary species, *Cryptopecten vesiculosus*, from Japan. *J. Paleontology* 47: 401-420.

Hayes, J.D., J. Imbrie, and N.J. Shackleton. 1976. Variations in the earth's orbit: pacemaker of the ice ages. *Science* 194: 1121-1132.

Hecht, M.K. 1952. Natural selection in the lizard genus *Aristelliger*. *Evolution* 6: 112-134.

Hedrick, P.W., M.E. Ginevan, and E.P. Ewing. 1976. Genetic polymorphisms in heterogeneous environments. *Ann. Rev. Ecol. Syst.* 7: 1-32.

Hennig, W. 1966. *Phylogenetic Systematics* (transl. by D.D. Davis and R. Zangerl). Urbana, Illinois; University of Illinois Press.

Herbert, T.D., and A.G. Fischer. 1986. Milankovitch climatic origin of mid-Cretaceous black shale rhythms in central Italy. *Nature* 321: 739-743.

Hersh, A.H. 1934. Evolutionary relative growth in the Titanotheres. *Am. Nat.* 68: 537-561.

Hey, J. 1987. Speciation via hybrid dysgenesis: negative evidence from the *Drosophila affinis* subgroup; with a critique of proposed models. *Evolution*, submitted.

Hickey, L.J. 1981. Land plant evidence compatible with gradual, not catastrophic, change at the end of the Cretaceous. *Nature* 292: 529-531.

Hickey, L.J. 1984. Changes in the angiosperm flora across the Cretaceous-Tertiary boundary. In *Catastrophes and Earth History, The New Uniformitarianism*, ed. W.A. Berggren and J.A. Van Couvering, pp. 279-313. Princeton Univ. Press.

Hickey, L.J., R.N. West, M.R. Dawson, D.K. Choi. 1983. Arctic terrestrial biota: paleomagnetic evidence of age disparity with mid-northern latitudes during the Late Cretaceous and Early Tertiary. *Science* 221: 1153-1158.

Hilgendorf, F. 1879. Zur Streitfrage des *Planorbis multiformis*. *Kosmos* (Leipzig) 5: 10-22, 90-99.

Hill, W.G. 1982. Predictions to response to artificial selection for new mutations. *Genet. Res.* 40: 255-278.

Hinchliffe, J.R., and M. Gumpel-Pinot. 1981. Control of maintenance and anteroposterior skeletal differentiation of the anterior mesenchyme of the chick wing bud by its posterior margin (the ZPA). *J. Embryol. Exp. Morph.* 62: 63-82.

Hoffman, A. 1985. Biotic diversification in the Phanerozoic: diversity independence. *Palaontology*, 28:387-391.

Hoffman, A., and J. Ghiold. 1985. Randomness in the pattern of 'mass extinctions', and 'waves of origination.' *Geol. Mag.* 122: 1-4.

Holman, E.W. 1983. Time scales and taxonomic survivorship. *Paleobiology* 9: 20-25.

Hood, L., J.H. Campbell, and S.C.R. Elgin. 1975. The organization, expression, and evolution of antibody genes and other multigene families. *Ann. Rev. Gen.* 9: 305-352.

Hopson, J.A. 1966. The origin of the mammalian middle ear. *Am. Zool.* 6: 437-450.

Hubby, J.L., and R.C. Lewontin. 1966. A molecular approach to the study of genic heterozygosity in natural populations. I. The number of alleles at different loci of *Drosophila pseudoobscura*. *Genetics* 54:577-594.

Hull, D.L. 1973. *Darwin and His Critics*. Cambridge, MA: Harvard Univ. Press.

Hull, D.L. 1974. *Philosophy of Biological Science*. Englewood Cliffs NJ: Prentice-Hall.

Hull, D.L. 1976. Are species really individuals? *Syst. Zool.* 25:174-191.

Hull, D.L. 1980. Individuality and selection. *Ann. Rev. Ecol. Syst.* 11:311-332.

Huxley, J.S. 1931. The relative size of antlers in deer. *Proc. Zool. Soc. Lond.* 19: 819-864.

Huxley, J.S. 1932. *Problems of Relative Growth*. London: MacVeagh.

Huxley, J.S. 1939. Clines: an auxiliary method in taxonomy. *Bijdr. Dierk.* 27: 491-520.

Huxley, J.S. 1940. *Evolution: The Modern Synthesis*. London: Allen and Unwin.

Huxley, J.S. 1960. The emergence of Darwinism. In *Evolution After Darwin*, vol. 1., ed. S. Tax, pp. 1-21. University of Chicago Press.

Iljin, N.A. 1927. Studies in morphogenetics of animal pigmentation. IV. Analysis of pigment formation by low temperature. (In Russian, with English summary.) *Trans. Lab. Exptl. Biol. Zool. Park Moscow* 3: 183-200.

Ilmensee, K. 1976. Nuclear and cytoplasmic transplantation in *Drosophila*. In *Insect Development*, ed. P.A. Lawrence, pp. 76-96. Oxford: Blackwell.

Ilmensee, K., and P.C. Hoppe. 1981. Nuclear transplantation in *Mus musculus*: developmental potential of nuclei from preimplantation embryos. *Cell* 23: 9-18.

Imbrie, J. 1957. The species problem with fossil animals. In *The Species Problem*, ed. E. Mayr, pp. 125-153. Am. Assoc. Adv. Sci. Publ. 50.

Imbrie, J., and K.P. Imbrie. 1979. *Ice Ages: Solving the Mystery*. Short Hills, NJ: Enslow Publishers.

Imbrie, J., and N.G. Kipp. 1969. A new micropaleontological method for quantitative paleoclimatology: application to a late Pleistocene Caribbean core. In *The Late Cenozoic Glacial Ages*, ed. K.K. Turekian, pp. 71-181. New Haven: Yale Univ. Press.

Ives, P.T. 1954. Genetic changes in American populations of *Drosophila melanogaster*. *Proc. Nat. Acad. Sci. USA* 40:87-92.

Jaanusson, V. 1973. Morphological discontinuities in the evolution of graptolite colonies. In *Animal Colonies*, ed. R. Boardman, A. Cheethan, and R. Oliver, pp. 515-521. Stroudsburg, PA: Dowden, Hutchinson and Ross.

Jaanusson, V. 1981. Functional thresholds in evolutionary progress. *Lethaia*. 14: 251-260.

Jaanusson, V. 1985. Functional morphology of the shell in platycope ostracodes - a study of arrested evolution. *Lethaia* 18: 73-84.

Jablonski, D. 1979. *Paleoecology, paleobiogeography, and evolutionary patterns of Late Cretaceous Gulf and Atlantic Coastal Plain mollusks*. Ph.D. thesis, Yale Univ.

Jablonski, D. 1980. Apparent versus real biotic effects of transgression and regression. *Paleobiology* 6: 397-407.

Jablonski, D. 1985. Marine regressions and mass extinctions: A test using the modern biota. In *Phanerozoic Diversity Patterns: Profiles in Macroevolution*, ed. J.W. Valentine, pp. 335-354. Princeton Univ. Press.

Jablonski, D. 1986. Background and mass extinctions: the alternation of macroevolutionary regimes. *Science* 231:129-133.

Jablonski, D., J.J. Sepkoski, Jr., D.J. Bottjer, and P.M. Sheehan. 1983. Onshore-offshore patterns in the evolution of Phanerozoic shelf communities. *Science* 222: 1123-1124.

Jablonski, D., and J.W. Valentine. 1981. Adaptive strategies in Recent Pacific Rim benthos and implications for Cenozoic Paleobiogeography. In *Evolution Today, Proc. 2nd Intl. Congr. Syst. Evol. Biol.*, ed. J.L. Scudder and J.L. Reveal, pp. 441-453.

Jackson, J.B.C. 1974. Biogeographic consequences of eurytopy and stenotopy among marine bivalves and their evolutionary significance. *Am. Nat.* 108: 541-560.

Jacob, F. 1977. Evolution and tinkering. *Science* 196: 1161-1166.

Jacob, F. 1983. Molecular tinkering in evolution. pp. 131-144. In *Evolution from Molecules to Men*, ed. D.S. Bendall, pp. 131-144. Cambridge Univ. Press.

Jaffe, L.F., and C.D. Stern. 1979. Strong electrical currents leave the primitive streak of chick embryos. *Science* 206: 569-571.

Janzen, D.H. 1967. Why mountain passes are higher in the tropics. *Am. Nat.* 101: 233-249.

Järvinen, O. 1979. Geographical gradients of stability in European land bird communities. *Oecologia* 38: 51-69.

Jeffreys, A.J. 1982. *Evolution* of globin genes. In *Genome Evolution*, ed. G.A. Dover and R. B. Flavell, pp. 157-176. London: Academic Press.

Jeletzky, J.A. 1955. *Evolution* of Santonian and Campanian *Belemnitella* and paleontological systematics: exemplified by *Belemnitella praecursor* Stolley. *J. Paleontol.* 29: 478-509.

Jeletzky, J.A. 1965. Late Upper Jurassic and early Lower Cretaceous fossil zones of the Canadian western Cordillera, British Columbia. *Geol. Surv. Canada Bull.* 103: 1-70, pl. I-XXII.

Jell, P.A. 1980. Earliest known pelecypods on earth- a new Early Cambrian genus from South Australia. *Alcheringa* 4: 233-239.

Jenkins, F.A. Jr., A.W. Crompton, and W.R. Downs. 1983. Mesozoic mammals from Arizona: new evidence on mammalian evolution. *Science* 222: 1233-1235.

Jensen, E.O., K. Paludan, J.J. Hyldig-Nielsen, P. Jorgensen, and K.A. Marcker. 1981. The structure of a chromosomal leghaemoglobin gene from soybean. *Nature* 291: 677-679.

Jepsen, G.L., E. Mayr, and G.G. Simpson, eds. 1949. *Genetics, Paleontology, and Evolution*. Princeton Univ. Press.

Jinks, J.L. 1983. Biometrical genetics of heterosis. In *Heterosis: Reappraisal of Theory and Practice*, ed. R. Frankel, pp. 1-46. Berlin: Springer-Verlag.

Johnson, J.G. 1982. Occurrence of phyletic gradualism and punctuated equilibria through time. *J. Paleontol.* 56: 1329-1331.

Jones, J.S. 1981. An uncensored page of fossil history. *Nature* 293: 427-428.

Jones, J.S., R.K. Selander and G.D. Schnell. 1980. Patterns of morphological and molecular polymorphism in the land snail *Cepaea nemoralis*. *Biol. Jour. Linn. Soc.* 14: 359-387.

Jones, W.C., and F.C. Kafatos. 1982. Accepted mutations in a gene family: evolutionary diversification of duplicated DNA. *J. Mol. Evol.* 19: 87-103.

Jurgens, G. 1985. A group of genes controlling the spatial expression of the bithorax complex in *Drosophila*. *Nature* 316: 153-155.

Kaesler, R.L., and J.A. Waters. 1972. Fourier analysis of the ostracode margin. *Bull. Geol. Soc. America* 83:1169-1178.

Kafatos, F.C. 1983. Structure, evolution, and developmental expression of the chorion multigene families in silkmoths and *Drosophila*. In *Gene Structure in Regulation and in Development*, ed. S. Subtelny and F.C. Kafatos, pp. 33-61. New York: Alan R. Liss.

Karn, M.N., and L.S. Penrose. 1951. Birth weight and gestation time in relation to maternal age, parity, and infant survival. *Ann. Eugen.* 15: 206-233.

Kauffman, E.G. 1973. Cretaceous bivalvia. In *Atlas of Paleobiogeography*, ed. A. Hallam, pp. 353-384. Amsterdam: Elsevier Sci. Publ. Co.

Kauffman, E.G. 1978. Evolutionary rates and patterns among Cretaceous Bivalvia. *Phil. Trans. Roy. Soc. Lond.* 284B: 277-304.

Kauffman, E.G. 1984. The fabric of Cretaceous marine extinctions. In *Catastrophes and Earth History, The New Uniformitarianism*, ed. W.A. Berggren and J.A. Van Couvering, pp. 151-246. Princeton Univ. Press.

Kaufmann, R. 1933. Variationstatistische untersuchungen uber die "Artabwandlung" und "Artumbildung" an der Oberkambrischen Trilobitengattung *Olenus* Dalm. *Abhandl. Geol.-Pal. Institut Greifswald* 10: 1-54.

Kelley, P.C. 1983a. The role of within-species differentiation in macroevolution of Chesapeake Group bivalves. *Paleobiology* 9: 261-268.

Kelley, P.C 1983b. Evolutionary patterns of eight Chesapeake group molluscs: evidence for the model of punctuated equilbria. *J. Paleontol.* 57: 581-598.

Kellogg, D.E. 1975. The role of phyletic change in the evolution of *Pseudocubus vema* (Radiolaria). *Paleobiology* 1: 359-370.

Kemp, T.S. 1982. *Mammal-like Reptiles and the Origin of Mammals.* London: Academic Press.

Kennett, J.P., and M.S. Srinivasan. 1983. *Neogene Planktonic Foraminifera: A Phylogenetic Atlas.* Stroudsburg, PA: Hutchinson and Ross Co.

Kermack, K.A. 1954. A biometrical study of *Micraster coranguinum* and *M. (Isomicraster) senonensis. Phil. Trans. Roy. Soc. Lond.* 237B: 375-428, pl. 24-26.

Kettlewell, H.B.D. 1955. Selection experiments on industrial melanism in the Lepidoptera. *Heredity* 9: 323-342.

Key, K.H.L. 1974. Speciation in the Australian Morabine grasshoppers—taxonomy and ecology. In *Genetic Mechanisms of Speciation in the Insects,* ed. M.J.D. White, pp. 43-56. Sydney: Australian and New Zealand Book Co.

Kidwell, M.G., J.F. Kidwell, and J.A. Sved. 1977. Hybrid dysgenesis in *Drosophila melanogaster*: a syndrom of aberrant traits including mutation, sterility and male recombination. *Genetics* 86: 813-833.

Kidwell, M.G., J.B. Novy, and S.M. Feely. 1981. Rapid unidirectional change of hybrid dysgenesis potential in *Drosophila. J. Heredity* 72: 32-38.

Kidwell, S.M. 1986. Models for fossil concentrations: paleobiologic implications. *Paleobiology* 12: 6-24.

Killingley, J.S., and M.A. Rex. 1985. Mode of larval development in some deep-sea gastropods indicated by oxygen-18 values of their carbonate shells. *Deep-Sea Res.* 32:809-818.

Kimura, M. 1983. *The Neutral Theory of Evolution*. Cambridge Univ. Press.

Kimura, M., and T. Ohta. 1971. *Theoretical Aspects of Population Genetics*. Princeton Univ. Press.

King, J.L. 1967. Continuously distributed factors affecting fitness. *Genetics* 55: 483-492.

King, J.L., and T.H. Jukes. 1969. Non-Darwinian evolution. *Science* 164: 788-798.

King, M.C., and A.C. Wilson. 1975. *Evolution* at two levels in humans and chimpanzees. *Science* 188: 107-116.

Kirkpatrick, M. 1982. Quantum evolution and punctuated equilibria in continuous genetic characters. *Am. Nat.* 119: 833-848.

Kitchell, J.A., and T.R. Carr. 1985. Nonequilibrium model of diversification: faunal turnover dynamics. In *Phanerozoic Diversity Patterns: Profiles in Macroevolution*, ed. J.W. Valentine, pp. 277-309. Princeton Univ. Press.

Kitchell, J.A., D.L. Clark, and A.M. Gombos, Jr., 1986. The biological selectivity of extinction: a link between background and mass extinction. *Palaios* 1: 504-511.

Kitchell, J., and D. Pena. 1984. Periodicity of extinctions in the geologic past: deterministic versus stochastic explanations. *Science* 226: 689-692.

Kluge, A.G., and J.S. Farris. 1969. Quantitative phyletics and the evolution of anurans. *Syst. Zool.* 18: 1-32.

Koch, C.F. 1980. Bivalve species duration, areal extent and population size in a Cretaceous sea. *Paleobiology* 6: 189-192.

Koch, C.F., and N.F. Sohl. 1982. Preservational effects in paleoecological studies: Cretaceous mollusc examples. *Paleobiology* 9:26-34.

Koehl, M.A.R. 1976. Mechanical design in sea anemones. In *Coelenterate Ecology and Behavior*, ed. G.O. Mackie, pp. 23-31. New York: Plenum.

Koehl, M.A.R., and S.A. Wainwright. 1977. Mechanical adaptation of a giant kelp. *Limnol. Oceanogr.* 22: 1067-1071.

Koehn, R.K., and W.F. Eanes. 1977. Subunit size and genetic variation of enzymes in natural populations of *Drosophila*. *Theor. Pop. Biol.* 11: 330-341.

Koehn, R.K., J.G. Hall, D.J. Innes, and A.J. Zera. 1984. Genetic differentiation of *Mytilus edulis* in eastern North America. *Mar. Biol.* 79:117-126.

Koehn, R.K., R. Milkman, and J.B. Mitton. 1976. Population genetics of marine polecypods. IV. Selection, migration, and genetic differentiation in the blue mussel *Mytilus edulis*. *Evolution* 39: 2-32.

Koehn, R.K., A.J. Zera, and J.G. Hall. 1983. Enzyme polymorphism and natural selection. In *Evolution of Genes and Proteins*, ed. M. Nei and R.K. Koehn, pp. 115-136. Sunderland MA: Sinauer Assoc.

Kohn, A.J. 1959. The ecology of *Conus* in Hawaii. *Ecol. Monogr.* 29: 47-90.

Kollar, E.J., and C. Fisher. 1980. Tooth induction in chick epithelium: expression of quiescent genes for enamel synthesis. *Science* 207: 993-995.

Kozhov, M. 1963. *Lake Baikal and Its Life.* Monographiae Biologicae. The Hague: Dr. W. Junk. vol. XI.45.

Kreauter, J.N. 1974. Offshore currents, larval transport, and establishment of southern populations of *Littorina littorea* Linne along the U.S. Atlantic coast. *Thalassia Jugo.* 10: 159-170.

Kreiber, M., and M.R. Rose. 1986. Molecular aspects of the species barrier. *Ann. Rev. Ecol. Syst.* 17: 465-485.

Kurtén, B. 1953. On the variation and population dynamics of fossil and Recent mammal populations. *Acta Zool. Fenn.* 76: 1-122.

Kurtén, B. 1954. Observations of allometry in mammalian dentitions: its interpretation and evolutionary significance. *Acta Zool. Fenn.* 85: 1-13.

Kurtén, B. 1959a. Rates of evolution in fossil mammals. *Cold Spring Harbor Symp. Quant. Biol.* 24: 205-215.

Kurtén, B. 1959b. On the longevity of mammalian species in the Tertiary. *Soc. Scient. Fenn. Comm. Biol.* 21: 1-14.

Kurtén, B. 1960. Faunal turnover dates for the Pleistocene and late Pliocene. *Soc. Scient. Fenn. Comm. Biol.* 22: 1-14.

Kurtén, B. 1963. Return of a lost structure in the evolution of the felid dentition. *Soc. Scient. Fenn. Comm. Biol.* 26(14): 1-12.

Kurtén, B. 1981. The "gestalt" of hominid evolution. In *Les Processus de L'Hominisation. Coll. Internat. C.N.R.S.* no. 599, pp. 61-65.

Lande, R. 1976. The maintenance of genetic variability by mutation in a polygenic character with linked loci. *Genet. Res. Camb.* 26: 221-235.

Lande, R. 1978. Evolutionary mechanisms of limb loss in Tetrapods. *Evolution* 32: 73-92.

Lande, R. 1979a. Effective deme sizes during long-term evolution estimated from rates of chromosomal rearrangement. *Evolution* 33: 234-251.

Lande, R. 1979b. Quantitative genetic analysis of multivariate evolution, applied to brain: body size allometry. *Evolution* 33: 402-416.

Lande, R. 1980a. Genetic variation and phenotypic evolution during allopatric speciation. *Am. Nat.* 116: 463-479.

Lande, R. 1980b. Microevolution in relation to macroevolution. *Paleobiology* 6: 233-238.

Lande, R. 1983. The minimum number of genes contributing to quantitative variation between and within populations. *Genetics* 99: 541-553.

Lande, R. 1985. Expected time for random genetic drift of a population between stable phenotypic states. *Proc. Nat. Acad. Sci. USA* 82:7641-7645.

Lang, W.D. 1919. The Pelmatoporinae, an essay on the evolution of a group of Cretaceous Polyzoa. *Phil. Trans. Roy. Soc. Lond.* B 209: 191-228.

Langley, C.H., and W.M. Fitch. 1974. An examination of the constancy of the rate of molecular evolution. *J. Mol. Evol.* 3: 161-177.

Langley, C.H., R.A. Voelker, A.J. Leigh Brown, S. Ohnishi, B. Dickson, and E. Montgomery. 1981. Null allele frequencies at allozyme loci in natural populations of *Drosophila melanogaster*. *Genetics* 99: 151-156.

Laporte, L.F. 1983. Simpson's *Tempo and Mode in Evolution* revisited. *Proc. Amer. Phil. Soc.* 127: 365-417.

Larson, A. 1980. Paedomorphosis in relation to rates of morphological and molecular evolution in the salamander *Aneides flavipunctatus* (Amphibia, Plethodontidae). *Evolution* 34: 1-17.

Larson, A. 1984. Neontological inferences of evolutionary pattern and process in the salamander family Plethodontidae. Evolutionary Biology 17: 119-217.

Larson, A., D.B. Wake, L.R. Maxson, and R. Highton. 1981. A molecular phylogenetic perspective on the origins of morphological novelties in the salamanders of the tribe Plethodontini (Amphibia, Plethodontidae). *Evolution* 35: 405-422.

Larson, A., D.R. Wake, and K.P. Yanev. 1984. Measuring gene flow among populations having high levels of genetic fragmentation. *Genetics* 106: 293-308.

Larson, G.L. 1976. Social behavior and feeding ability of two phenotypes of *Gasterosteus aculeatus* in relation to their spatial and trophic segregation in a temperate lake. *Can. J. Zool.* 54: 107-121.

Lauder, G. 1981. Form and function in structural analysis in evolutionary paleontology. *Paleobiology* 7: 430-442.

Lawrence, P.A. 1981. The cellular basis of segmentation in insects. *Cell* 26: 3-10.

Lazarus, D. 1986. Tempo and mode of morphologic evolution near the origin of the radiolarian lineage *Pterocanium prismatium*. *Paleobiology* 12:175-189.

Le Douarin, N.M. 1980. The ontogeny of the neural crest in avian embryo chimaeras. *Nature* 286: 663-669.

Leopoldt, M., and J. Schmidtke. 1982. Gene expression in phylogenetically polyploid organisms. In *Genome Evolution*, ed. G. Dover, pp. 219-236. Cambridge Univ. Press.

LeQuesne, W.J. 1969. A method of selection of characters in numerical taxonomy. *Syst. Zool.* 18: 201-205.

Lerner, I.M. 1954. *Genetic Homeostasis*. Edinburgh: Oliver and Boyd.

Levene, H. 1953. Genetic equilibrium when more than one niche is available. *Am. Nat.* 87:331-333.

Leversee, G.J. 1976. Flow and feeding in fan-shaped colonies of the gorgonian coral, *Leptogorgia. Biol. Bull.* 151: 344-356.

Levin, D.A. 1982. Polyploidy and novelty in flowering plants. *Am. Nat.* 22:1-25.

Levins, R. 1970. Extinction. In *Some Mathematical Questions in Biology*, pp. 77-107. Providence, RI: The American Mathematical Society.

Levinton, J.S. 1970. The paleoecological significance of opportunistic species. *Lethaia* 3: 69-78.

Levinton, J.S. 1973. Genetic variation in a gradient of environmental variability. *Science* 180:75-76.

Levinton, J.S. 1974. Trophic group and evolution in bivalve molluscs. *Palaeontology* 17: 579-585.

Levinton, J.S. 1975. Levels of genetic polymorphism at two enzyme encoding loci in eight species of the genus *Macoma* (Mollusca:Bivalvia). *Mar. Biol.* 33: 41-47.

Levinton, J.S. 1977. The ecology of deposit-feeding communities: Quisset Harbor, Massachusetts. In *Ecology of Marine Benthos*, ed. B.C. Coull, pp. 191-228. Columbia: Univ. South Carolina Press.

Levinton, J.S. 1979. A theory of diversity equilibrium and morphological evolution. *Science* 204: 335-336.

Levinton, J.S. 1982a. *Marine Ecology*. Englewood Cliffs NJ: Prentice-Hall.

Levinton, J.S. 1982b. Estimating stasis: can a null hypothesis be too null? *Paleobiology* 8: 307.

Levinton, J.S. 1982c. Charles Darwin and Darwinism. *BioScience* 32: 495-500.

Levinton, J.S. 1983. Stasis in progress: the empirical basis of macroevolution. *Ann. Rev. Ecol. Syst.* 14: 103-137.

Levinton, J.S. 1984. Evolutionary biology today: historians as biologists; biologists as historians. *Paleobiology* 10: 377-383.

Levinton, J.S. 1986. Developmental constraints and evolutionary saltations: a discussion and critique. In *Genetics, Development, and Evolution*, ed. J.P. Gustafson, G.L. Stebbins, and F.J. Ayala, pp. 253-288. New York: Plenum.

Levinton, J.S., and R.K. Bambach. 1975. A comparative study of Silurian and Recent deposit-feeding bivalve communities. *Paleobiology* 1: 97-124.

Levinton, J.S., K. Bandel, B. Charlesworth, G. Müller, W.R. Nagl, B. Runnegar, R.K. Selander, S.C. Stearns, J.R.G. Turner, A.J. Urbanek, and J.W. Valentine. 1986. Genomic versus organismic evolution. In *Patterns and Processes in the History of Life*, ed. D.M. Raup and D. Jablonski, pp. 167-182. Berlin: Springer-Verlag.

Levinton, J.S., and D.J. Futuyma. 1982. Macroevolution: Pattern and Process. *Evolution* 36: 425-426.

Levinton, J.S., and L.R. Ginzburg. 1984. Repeatability of taxon longevity in successive foraminifera radiations and a theory of random appearance and extinction. *Proc. Nat. Acad. Sci. USA* 81: 5478-5481.

Levinton, J.S., and R.K. Koehn. 1976. Population genetics. In *Marine Mussels*, ed. B.L. Bayne, pp. 357-384. Cambridge Univ. Press.

Levinton, J.S., and H.H. Lassen. 1978. Experimental mortality studies and adaptation at the Lap locus in *Mytilus edulis*. In *Marine Organisms: Genetics, Ecology, and Evolution*. ed. B. Battaglia and J.L. Beardmore, pp. 229-254. New York: Plenum.

Levinton, J.S., and R. Monahan. 1983. The latitudinal compensation hypothesis: growth data and a model of latitudinal growth differentiation based upon energy budgets. II. Intraspecific comparisons between subspecies of *Ophryotrocha puerilis* (Polychaeta: Dorvilleidae). *Biol. Bull.* 165: 699-707.

Levinton, J.S., and C. Simon. 1980. A critique of the punctuated equilibria model and implications for the detection of speciation in the fossil record. *Syst. Zool.* 29: 130-142.

Levinton, J.S., and T.H. Suchanek. 1978. Geographic variation, niche breadth, and genetic differentiation at different geographic scales in the mussels, *Mytilus californianus* and *M. edulis. Mar. Biol.* 49: 363-375.

Levitan, P.J., and A.J. Kohn. 1980. Microhabitat resource use, activity patterns, and episodic catastrophe: *Conus* on tropical intertidal reef rock benches. *Ecol. Monogr.* 50: 55-75.

Lewin, R. 1980. Evolutionary theory under fire. *Science* 210:883-887.

Lewis, E.B. 1963. Genes and developmental pathways. *Am. Zool.* 3: 33-56.

Lewis, E.B. 1978. A gene complex controlling segmentation in *Drosophila. Nature* 276: 565-570.

Lewis, H. 1973. The origin of diploid neospecies in *Clarkia. Am. Nat.* 107: 161-170.

Lewontin, R.C. 1970. The units of selection. *Ann. Rev. Ecol. Syst.* 1: 1-18.

Lewontin,R.C. 1974. *The Genetic Basis of Evolutionary Change.* New York: Columbia Univ. Press.

Lewontin, R.C. 1978. Adaptation. *Scient. Amer.* 239(3): 156-169.

Lewontin, R.C. 1983a. Detecting population differences in quantitative characters as opposed to gene frequencies. *Am. Nat.* 123: 115-124.

Lewontin, R.C. 1983b. Gene, organism, and environment. In *Evolution from Molecules to Men,* ed. D.S. Bendall, pp. 273-285. Cambridge Univ. Press.

Lewontin, R.C., and J.L. Hubby. 1966. A molecular approach to the study of genic heterozygosity in natural populations. II.

Amount of variation and degree of heterozygosity in natural populations of *Drosophila pseudoobscura*. *Genetics* 54: 595-609.

Li, W.-H. 1983. *Evolution* of duplicate genes and pseudogenes. In *Evolution of Genes and Proteins*, ed. M. Nei and R.K. Koehn, pp. 14-37. Sunderland, MA: Sinauer Assoc.

Liem, K.F. 1973. Evolutionary strategies and morphological innovations: cichlid pharyngeal jaws. *Syst. Zool.* 22: 425-441.

Liem, K.F. 1980. Adaptive significance of intra- and interspecific differences in the feeding repertoires of chiclid fishes. *Amer. Zool.* 20: 295-314.

Lillegraven, J.A. 1972. Ordinal and familial diversity of Cenozoic mammals. *Taxon* 21: 261-274.

Lillie, F.R. 1895. The embryology of the Unionidae. *J. Morphology* 10: 1-100.

Lindenberg, H.G., and II. Mensink. 1979. Multivariate Gruppierungsmethoden in phylogenetisch orientierter Palaontologie (am Beispielder Gastropodenaus dem Steinheimer Becken). *Berliner geowiss. Abh.* A 15: 30-51.

Lindsley, D.L., and E.H. Grell. 1968. Genetic variations of *Drosophila melanogaster*. *Carnegie Inst. Wash. Publ.* 627.

Linsley, R.M. 1977. Some "laws" of gastropod shell form. *Paleobiology* 3: 196-206.

Lipps, J. 1970. Plankton evolution. *Evolution* 24: 1-22.

Locke, M. 1959. The cuticular pattern in an insect, *Rhodnius prolixus Stal. J. Exp. Biol.* 36: 459-477.

Locke, M., and P. Huie. 1981. Epidermal feet in insect morphogenesis. *Nature* 293: 733-735.

Lohman, G.P. 1983. Eigenshape analysis of microfossils: a general morphometric procedure for describing changes in shape. *J. Int. Assoc. Math. Geol.* 15: 659-672.

Long, E.O., and I.B. Dawid. 1979. Expression of ribosomal DNA insertions in *Drosophila melanogaster*. *Cell* 18: 1185-1196.

Lonsdale, D.J., and J.S. Levinton. 1985. Latitudinal differentiation in copepod growth: an adaptation to temperature. *Ecology* 66:1397-1407.

Lovtrup, S. 1974. *Epigenetics*. New York: John Wiley and Sons.

Lull, H.S. 1929. *Organic Evolution*. New York and London: Macmillan.

Lutz, R.A., D. Jablonski, and R.D. Turner. 1984. Larval development and dispersal at deep-sea hydrothermal vents. *Science* 226: 1451-1454.

Lynch, J.D. 1982. Relationships of the frogs of the genus *Ceratophrys* (Leptodactylidae) and their bearing on hypotheses of Pleistocene forest refugia in South America and punctuated equilibria. *Syst. Zool.* 31: 166-179.

Lynch, J.F. 1981. Patterns of ontogenetic and geographic variation in the Black Salamander, *Aneides flavipunctatus* (Caudata:Plethodontidae) *Smithsonian Contr. Zool.*, no. 324.

MacArthur, R.H., and E.O. Wilson. 1967. *The Theory of Island Biogeography*. Princeton Univ. Press.

MacFadden, B. J. 1985. Patterns of phylogeny and rates of evolution in fossil horses: Hipparions from the Miocene and Pliocene of North America. *Paleobiology* 11: 245-257.

Maderson, P.F.A. 1975. Embryonic tissue interactions as the basis for morphological change in evolution. *Amer. Zool.* 15: 315-327.

Maderson, P.F.A., P. Alberch, B.C. Goodwin, S.J. Gould, A. Hoffman, J.D. Murray, D.M. Raup, A. de Ricqles, A. Seilacher, G.P. Wagner, and D.B. Wake. 1982. The role of development in macroevolutionary change. In *Evolution and Development*, ed. J.T. Bonner, pp. 279-312. Berlin: Springer-Verlag.

Mahowald, A.P. 1968. Polar granules of Drosophila II. Ultrastructural changes during early embryogenesis. *J.Exp. Zool.* 167: 237-262.

Malmgren, B.A., W.A. Berggren and G.P. Lohman. 1983. Evidence for punctuated gradualism in the Late Neogene *Globorotalia tumida* lineage of planktonic foraminifera. *Paleobiology* 9: 377-389.

Malmgren, B.A., and J.P. Kennett. 1981. Phyletic gradualism in a Late Cenozoic planktonic foraminiferal lineage; DSDP Site 284, southwest Pacific. *Paleobiology* 7: 230-240.

Markert, C.L., J.B. Shaklee, and G.S. Whitt. 1975. *Evolution* of a gene. *Science* 189: 102-114.

Marks, J. 1983. Rates of karyotype evolution. *Syst. Zool.* 32: 207-209.

Marsh, O. C. 1892. Recent polydactyle horses. *Am. J. Sci.* 43: 339-355.

Marshall, L.G., S.D. Webb, J.J. Sepkoski, Jr., D.M. Raup. 1982. Mammalian evolution and the great American exchange. *Science* 215: 1351-1357.

Martin, G., D. Wiernasz, and P. Schedl. 1983. *Evolution* of *Drosophila* repetitive-dispersed DNA. *J. Mol. Evol.* 19: 203-213.

Martin, R.D. 1981. Relative brain size and basal metabolic rate in terrestrial vertebrates. *Nature* 293: 57-60.

Mather, K., and B.S. Harrison. 1949. The manifold effect of selection. *Heredity* 3: 1-52.

Maynard Smith, J. 1960. Continuous, quantized and modal variation. *Proc. Roy. Soc. London* 152B: 397-409.

Maynard Smith, J. 1976. What determines the rate of evolution. *Am. Nat.* 110:331-338.

Maynard Smith, J. 1979. Optimization theory in evolution. *Ann. Rev. Ecol. Syst.* 9: 31-56.

Maynard Smith, J. 1982. Overview–Unsolved evolutionary problems. In *Genome Evolution*, ed. G.A. Dover and R.B. Flavell, pp. 375-382. London: Academic Press.

Maynard Smith, J. 1983. Current controversies in evolutionary biology. In *Dimensions of Darwinism*, ed. M. Grene, pp. 273-286. Cambridge Univ. Press.

Maynard Smith, J., R. Burian, S. Kauffman, P. Alberch, J. Campbell, B. Goodwin, R. Lande, D. Raup, and L. Wolpert. 1985. Developmental constraints and evolution. *Qu. Rev. Biol.* 60:265-287.

Maynard Smith, J., and J. Haigh. 1974. The hitch-hiking effect of a favourable gene. *Genet. Res. Camb.* 23: 23-35.

Mayr, E. 1942. *Systematics and the Origin of Species.* New York: Columbia Univ. Press.

Mayr, E. 1963. *Animal Species and Evolution.* Cambridge MA: Belknap Press.

Mayr, E. 1969. *Principles of Systematic Zoology.* New York: McGraw Hill.

Mayr, E. 1974. Cladistic analysis or cladistic classification. *Z. Zool. Syst. Evol.-forsch.* 12: 94-128.

Mayr, E. 1976. *Evolution and the Diversity of Life.* Cambridge MA: Belknap Press.

Mayr, E. 1982a. *The Growth of Biological Thought.* Cambridge MA: Belknap Press.

Mayr, E. 1982b. Processes of speciation in animals. In *Mechanisms of Speciation,* ed. C. Barigozzi, pp. 1-19. New York: Alan R. Liss.

Mayr, E. 1983. How to carry out the adaptationist program. *Am. Nat.* 121:324-334.

Mayr, E., E.G. Linsley, and R.L. Usinger. 1953. *Methods and Principles of Systematic Zoology.* New York: McGraw Hill.

Mayr, E., and W.B. Provine. 1980. *The Evolutionary Synthesis: Perspectives on the Unification of Biology.* Cambridge, MA: Harvard Univ. Press.

McAlester, A.L. 1965. Systematics, affinities, and life habits of *Babinka,* a transitional Ordovician lucinoid bivalve. *Palaeontology* 8: 231-246.

McCarthy, B.J., and B.H. Hoyer. 1964. Identity of DNA and diversity of messenger RNA molecules in normal mouse. *Proc. Nat. Acad. Sci. USA* 52: 915-92

McCune, A.R. 1982. On the fallacy of constant extinction rates. *Evolution* 36: 610-614.

McDonald, J.F., S.M. Anderson and M. Santos. 1980. Biochemical differences between products of the *Adh* locus in *Drosophila. Genetics* 95: 1013-1022.

McGhee, G.R., Jr. 1980. Shell form in the biconvex articulate Brachiopoda: a geometric analysis. *Paleobiology* 6: 57-76.

McGhee, G.R., Jr. 1981. Evolutionary replacement of ecological equivalents in Late Devonian benthic marine communities. *Palaeogeogr. Palaeoclimatol. Palaeoecol.* 34: 267-283.

McGhee, G.R. Jr. 1982. The Frasnian-Famennian extinction event: a preliminary analysis of Appalachian marine ecosystems. *Geol. Soc. America Sp. Paper* 100, pp. 101 500.

McGhee, G.R. Jr., J.S. Gilmore, C.J. Orth, and E. Olsen. 1984. No geochemical evidence for an asteroidal impact at late Devonian mass extinction horizon. *Nature* 308: 629-631.

McGhee, G.R. Jr., C.J. Orth, L.R. Quintana, J.S. Gilmore, and E.J. Olsen. 1986. Late Devonian "Kellwasser Event" mass-extinction horizon in Germany: No geochemical evidence for a large-body impact. *Geology* 14: 776-779.

McGinnis, W., R.L. Garber, J. Wirz, A. Kuroiwa, and W.J. Gehring. 1984. A homologous protein-coding sequence in *Drosophila* homoeotic genes and its conservation in other metazoans. *Cell* 37: 403-408.

McGinnis, W., M.S. Levine, E. Haten, A. Kuroiwa, and W.J. Gehring. 1984. A conserved DNA sequence in homoeotic genes of the Drosophila Antennapedia and bithorax complexes. *Nature* 308: 428-433.

McGowan, C. 1984. Evolutionary relationships of ratites and carinates: evidence from ontogeny of the tarsus. *Nature* 307: 733-735.

McKenzie, J., and S. McKechnie. 1978. Ethanol tolerance and the Adh polymorphism in a natural population of *D. melanogaster*. *Nature* 272: 75-76.

McKinney, M.L., and R. M. Schoch. 1985. Titanothere allometry, heterochrony, and biomechanics: revising an evolutionary classic. *Evolution* 39:1352-1363.

McKnight, S.L., and R. Kingsbury. 1982. Transcriptional control of a eukaryotic protein-coding gene. *Science* 217: 316-324.

McLaren, D.J. 1970. Time, life and boundaries. *J. Paleontol.* 44: 801-815.

McMahon, T. 1973. Size and shape in biology. *Science* 179: 1201-1204.

McNab, B.K. 1978. The evolution of endothermy in the phylogeny of mammals. *Am. Nat.* 112: 1-21.

McNamara, K.J. 1982. Heterochrony and phylogenetic trends. *Paleobiology* 8: 130-142.

McPhail, J.D. 1984. Ecology and evolution of sympatric sticklebacks (*Gasterosteus*): morphological and genetic evidence for a species pair in Enos Lake, British Columbia. *Canad. J. Zool.* 62: 1402-1408.

Medawar, P. 1974. A geometric model of reduction and emergence. In *Studies in the Philosophy of Biology*, ed. F.J. Ayala and T. Dobzhansky, pp. 57-63. Berkeley: Univ. of California Press.

Meinhardt. H. 1983. A boundary model for pattern formation in vertebrate limbs. *J. Embryol. Exp. Morph.* 76: 115-137.

Meinhardt, H., and A. Gierer. 1974. Applications of a theory of biological pattern formation based on lateral inhibition. *J. Cell. Sci.* 15: 321-346.

Meyer, C.J.A. 1878. Micrasters in the English chalk - two or more species? *Geol. Mag. new ser.* 5: 115-117.

Michener, C.D. 1978. Dr. Nelson on taxonomic methods. *Syst. Zool.* 27: 112-128.

Mickevich, M.F. 1978. Taxonomic congruence. *Syst. Zool.* 27: 143-158.

Middleton, R.J., and H. Kaczer. 1983. Enzyme variation, metabolic flux and fitness: alcohol dehydrogenase in *Drosophila melanogaster*. Genetics 105: 633-650.

Milkman, R. 1967. Heterosis as a major cause of heterozygosity in nature. *Genetics* 55: 493-495.

Milkman, R. 1973. Electrophoretic variability in *E. coli* from natural sources. *Science* 182: 1024-1026.

Mitton, J.B., and M.C. Grant. 1984. Associations among protein heterozygosity, growth rate, and developmental homeostasis. *Ann. Rev. Ecol. Syst.* 15: 479-499.

Mivart, St. G. 1871. *The Genesis of Species.* New York: D. Appleton and Co.

Miyata, T., and T. Yasunaga. 1981. Rapidly evolving mouse alpha-globin-related pseudogene and its evolutionary history. *Proc. Nat. Acad. Sci. USA* 78: 450-453.

Miyazaki, J.M., and M.F. Mickevich. 1982. *Evolution* of *Chesapecten* (Mollusca:Bivalvia, Miocene-Pliocene) and the Biogenetic Law. *Evol. Biol.* 15: 369-409.

Moore, G.W., J. Barnabas, and M. Goodman. 1976. A method for constructing maximum parsimony ancestral amino acid sequences on a given network. *J. Theor. Biol.* 38: 459-485.

Moore, R.C. 1952. Evolutionary rates among crinoids. *J. Paleontol.* 26: 338-352.

Moore, R.C., C.G. Lalicker, and A.G. Fischer. 1952. *Invertebrate Fossils.* New York: McGraw-Hill.

Moos, J.R. 1955. Comparative physiology of some chromosomal types in *Drosophila pseudoobscura. Evolution* 9: 141-151.

Morgan, E. 1982. *The Aquatic Ape.* London: Souvenir Press.

Motro, U., and G. Thomson. 1982. On heterozygosity and the effective size of populations subject to size changes. *Evolution* 36: 1059-1066.

Mount, J.F., and P.W. Signor. 1985. Early Cambrian innovation in shallow subtidal environments: Paleoenvironments of Early Cambrian shelly fossils. *Geology* 13: 730-733.

Mukai, T., and O. Yamaguchi. 1974. The genetic structure of natural populations of *Drosophila melanogaster* XI. Genetic variability in a local population. *Genetics* 76: 339-376.

Müller, G. 1985. Experimentelle untersuchungen zur theorie des epigenetischen systems. In *Evolution, Ordnung und Erkenntnis.* ed. Ott, J.A., G.P. Wagner, and F.M. Wuketits, pp. 82-96. Berlin: Verlag Paul Parey.

Muller, H.J. 1939. Reversibility in evolution considered from the standpoint of genetics. *Biol. Rev.* 14: 261-280.

Muller, H.J. 1949. Redintegration of the symposium on genetics, paleontology, and evolution. In *Genetics, Paleontology and Evolution.* ed. G.L.Jepsen, G.G. Simpson, and E. Mayr, pp. 421-445. Princeton Univ. Press (reprinted 1963, New York: Atheneum).

Muller, M.M., A.E. Carrasco, and E.M. Derobertis. 1984. A homoeobox containing gene expressed during oogenesis in *Xenopus. Cell* 39: 157-162.

Murray, J.D. 1981. A pre-pattern formation mechanism for animal coat markings. *J. Theor. Biol.* 88: 161-199.

Nagel, E. 1961. *The Structure of Science.* New York: Hackett.

Neff, N., and L. Marcus. 1980. *A Survey of Multivariate Methods for Systematics.* New York: Privately Published (from L.F. Marcus, American Museum of Natural History, New York).

Nei, M. 1972. Genetic distance between populations. *Am. Nat.* 106:283-292.

Nei, M. 1975. *Molecular Population Genetics and Evolution.* Amsterdam: North-Holland.

Nei, M. 1980. Stochastic theory of population genetics and evolution. *Lecture Notes in Biomathematics,* ed. C. Bariguzzi, pp. 17-47. Berlin: Springer-Verlag.

Nei, M. 1983. Genetic polymorphism and the role of mutation in evolution. In *Evolution of Genes and Proteins,* ed. M. Nei and R.K. Koehn, pp. 165-190. Sunderland MA: Sinauer Associates.

Nei, M., and R.K. Koehn, eds. 1983. *Evolution of Genes and Proteins.* Sunderland MA: Sinauer Assoc.

Nelson, G.A. 1978. Ontogeny, phylogeny, paleontology and the biogenetic law. *Syst. Zool.* 27: 324-345.

Nelson, G.A., and N. Platnick. 1981. *Systematics and Biogeography: Cladistics and Vicariance.* New York: Columbia Univ. Press.

Nelson, K., and D. Hedgecock. 1980. Enzyme polymorphism and adaptive strategy in the Decapod crustacea. *Am. Nat.* 116: 238-280.

Nevo, E. 1982. Speciation in subterranean mammals. In *Mechanisms of Speciation*, ed. C. Barigozzi, pp. 191-218. New York: Alan R. Liss.

Nevo, E., and C.R. Shaw. 1972. Genetic variation in a subterranean mammal, *Spalax ehrenbergi. Biochem. Genet.* 7: 235-241.

Newell, N. D. 1937. Late Paleozoic Pelecypods: Pectinacea. *Kansas Geol. Surv. Bull.* 10 (1).

Newell, N.D. 1947. Intraspecific categories in invertebrate paleontology. *Evolution* 1: 163-171.

Newell, N.D. 1952. Periodicity in invertebrate evolution. *J. Paleontol.* 26: 371-385.

Newell, N.D. 1956. Fossil populations. In *The Species Concept in Paleontology*, Syst. Assoc. Spec. Publ. No. 2. pp. 63-82.

Newell, N.D. 1965. Classification of the Bivalvia. *Novitates* Amer. Mus. Nat. Hist. New York 2206: 1-25.

Newell, N.D. 1967. Revolutions in the history of life. In *Uniformity and Simplicity: A Symposium on the Principle of the Uniformity of Nature*, ed. C.C. Albritton, Jr., Geol. Soc. Amer. Sp. Pap. 89: 63-91.

Newell, N.D. 1969. Classification of Bivalvia. In *Treatise of Invertebrate Paleontology*, (N) Mollusca, vol. 1, ed. R.C. Moore, pp. N205-N224. Boulder, CO: Geological Society of America and Lawrence, KS: Univ. of Kansas.

Newell, N.D. 1971. An outline history of tropical organic reefs. *Novitates* Amer. Mus. Nat. Hist. New York 2465: 1-37.

Newell, N.D., and D.W. Boyd. 1975. Parallel evolution in early Trigoniacean Bivalves. *Bull. Amer. Mus. Nat. Hist.* N.Y. 154: 55-162.

Newman, C.M., J.E. Cohen, and C. Kipnis. 1985. Neo-Darwinian evolution implies punctuated equilibria. *Nature* 315:400-401.

Nichols, D. 1959. Changes in the chalk heart-urchin Micraster interpreted in relation to living forms. *Philos. Trans. Roy. Soc. Lond.* B242: 347-437, pl. 1-9.

Niklas, K.J., B.H. Tiffney, and A.H. Knoll. 1979. Apparent changes in the diversity of fossil plants. *Evol. Biol.* 12: 1-89.

Norris, G., and A.D. Miall. 1984. Arctic biostratigraphic heterogeneity. *Science* 224: 173-174.

Nuccitelli, R. 1983. Transcellular ion currents: signals and effectors of cell polarity. In *Modern Cell Biology*, vol. 2, ed. J.R. McIntosh, pp. 451-481. New York: Alan R. Liss.

Nüsslein-Volhard, C., and E. Wieschaus. 1980. Mutations affecting segment number and polarity in *Drosophila*. *Nature* 287: 795-801.

O'Brien, S.J., and W.G. Nash. 1982. Genetic mapping in mammals: chromosome map of domestic cat. *Science* 216: 257-265.

Odell, G.M., G. Oster, P. Alberch, and B. Burnside. 1981. The mechanical basis of morphogenesis. I. Epithelial folding and invagination. *Dev. Biol.* 85: 446-462.

Ohno, S. 1970. *Evolution By Gene Duplication*. Berlin: Springer-Verlag.

Ohno, S. 1973. Ancient linkage groups and frozen accidents. *Nature* 244: 259-262.

Ohta, T., and G. Dover. 1983. Population genetics of multigene families that are dispersed into two or more chromosomes. *Proc. Natl. Acad. Sci. USA* 80: 4079-4083.

Ohta, T., and G. Dover. 1984. The cohesive population genetics of molecular drive. *Genetics* 108: 501-521.

Ohta, T., and M. Kimura. 1971. On the constancy of the evolutionary rates of cistrons. *J. Mol. Evol.* 1: 18-25.

Okada, Y. 1981. Development of cell arrangement in ostracod carapaces. *Paleobiology* 7: 276-280.

Olsen, P.E. 1986. A 40-million-year lake record of Early Mesozoic orbital climatic forcing. *Science* 234: 842-848.

Olson, E.C. 1966. Community evolution and the origin of mammals. *Ecology* 47:291-302

Olson, E., and R. Miller. 1958. *Morphological Integration*. Univ. of Chicago Press.

Olvera, O., J.R. Powell, M.E. de la Rosa, V.M. Salceda, M.I. Gaso, J. Guzman, W.W. Anderson, and L. Levine. 1979. Population genetics of Mexican *Drosophila* VI. Cytogenetic aspects of the inversion polymorphism in *Drosophila pseudoobscura*. *Evolution* 33: 381-395.

Orgel, L.E., and F.H.C. Crick. 1980. Selfish DNA: the ultimate parasite. *Nature* 284: 604-607.

Orth, C.J., J.S. Gilmore, L.R. Quintana, and P.M. Sheehan. 1986. Terminal Ordovician extinction: Geochemical analysis of the Ordovician/Silurian boundary, Anticosti Island, Quebec. *Geology* 14: 433-436.

Orzack, S. 1981. The Modern Synthesis is partly Wright. *Paleobiology* 7:128-134.

Osborn, H.F. 1929. The Titanotheres of ancient Wyoming, Dakota and Nebraska. *U.S. Dept. Int., Geol. Survey. Monogr.* 55, 2 vols., Washington, D.C.

Osborn, H.F. 1934. Aristogenesis, the creative principle in the origin of species. *Am. Nat.* 68: 193-235.

Ospovat, D. 1981. *The Development of Darwin's Theory.* Cambridge Univ. Press.

Oster, G.F., J.D. Murray, and A.K. Harris. 1983. Mechanical aspects of mesenchymal morphogenesis. *J. Embryol. Exp. Morph.* 78: 83-125.

Ostrom, J.H. 1974. *Archaeopteryx* and the origin of flight. *Qu. Rev. Biol.* 49: 27-47.

Ouweneel, W. 1976. Developmental genetics of homoeosis. *Adv. Genet.* 16: 179-248.

Owen, R. 1859. On the orders of fossil and Recent Reptilia, and their distribution in time. *Rept. Brit. Assoc. Adv. Sci.*, Aberdeen, pp. 153-166.

Ozawa, T. 1975. *Evolution* of *Lepidolina multiseptata* (Permian foraminifer) in east Asia. *Mem. Fac. Sci. Kyushu Univ.* Ser. D (Geol.) 23: 117-164.

Palmer, A.R. 1982. Predation and parallel evolution: recurrent parietal plate reduction in balanomorph barnacles. *Paleobiology* 8: 31-44.

Palmer, A.R. 1984. The biomere problem: evolution of an idea. *J. Paleontology* 58: 599-611.

Palmer, A.R. 1985. Quantum changes in gastropod shell morphology need not reflect speciation. *Evolution* 39: 699-705.

Paquin, O.E., and J. Adams. 1983. Relative fitness can decrease in evolving asexual populations of *S. cerevisiae*. *Nature* 306: 368-371.

Parrington, F.R. 1971. On the upper Triassic mammals. *Phil. Trans. Roy. Soc. Lond.* 261B: 231-272.

Patterson, C. 1981. Significance of fossils in determining evolutionary relationships. *Ann. Rev. Ecol. Syst.* 12: 195-223.

Patterson, C. 1982. Morphological characters and homology. In *Problems of Phylogenetic Reconstruction*, ed. K.A. Joysey and A.E. Friday, pp. 21-74. Systematics Assoc. Sp. Publ. No. 21. London: Academic Press.

Patton, J.L. 1972. Patterns of geographic variation in karyotype in the pocket gopher, *Thomomys bottae* (Eydoux and Gervais). *Evolution* 26: 574-586.

Paul, C.R.C. 1982. The adequacy of the fossil record. In *Problems of Phylogenetic Reconstruction*, ed. K.A. Joysey and A.E. Friday, pp. 119-157. London: Academic Press.

Penney, D.F., and E.G. Zimmerman. 1976. Genic divergence and local population differentiation by random drift in the pocket gopher genus *Geomys*. *Evolution* 30: 473-483.

Penny, D. 1983. Charles Darwin, gradualism and punctuated equilibria. *Syst. Zool.* 32: 72-74.

Peterson, C.H. 1977. Competitive organization of the soft-bottom macrobenthic communities of southern California lagoons. *Mar. Biol.* 43: 343-359.

Petry, D. 1982. The pattern of phyletic speciation. *Paleobiology* 8: 56-66.

Phillips, D.C., M.J.E. Sternberg, and B.J. Sutton. 1983. Intimation of evolution from the three-dimensional structures of proteins. In *Evolution from Molecules to Men*, ed. D.S. Bendall, pp. 145-173. Cambridge Univ. Press.

Pianka, E.R. 1966. Latitudinal gradients in species diversity: a review of concepts. *Am. Nat.* 100: 33-46.

Place, A.R., and D.A. Powers. 1979. Genetic variation and relative catalytic efficiencies: Lactate dehydrogenase B allozymes of *Fundulus heteroclitus. Proc. Natl. Acad. Sci. USA* 76: 2354-2358.

Platnick, N.I. 1977a. Cladograms, phylogenetic trees, and hypothesis testing. *Syst. Zool.* 26: 438-442.

Playford, P.E., D.J. McLaren, C.J. Orth, J.S. Gilmore, W.D. Goodfellow. 1984. Iridium anomaly in the Upper Devonian of the Canning Basin, Western Australia. *Science* 226: 437-439.

Pojeta, J., Jr., and B. Runnegar. 1974. *Fordilla troyensis* and the early history of pelecypod mollusks. *Amer. Sci.* 62: 706-711.

Pojeta, J., Jr., B. Runnegar, and B. Kriz. 1973. *Fordilla troyensis* Barrande: the oldest known pelecypod. *Science* 180: 866-868.

Popper, K. 1974. Scientific reduction and the essential incompleteness of all science. In *Studies in the Philosophy of Biology,* ed. F.J. Ayala and T. Dobzhansky, pp. 259-284. Berkeley: Univ. California Press.

Powell, J.R. 1971. Genetic polymorphisms in varied environments. *Science* 174: 1035-1036.

Powell, J.R. 1983. Interspecific cytoplasmic gene flow in the absence of nuclear gene flow: evidence from *Drosophila. Proc. Nat. Acad. Sci. USA* 80: 492-495.

Powers, D.A., and A.R. Place. 1978. Biochemical genetics of *Fundulus heteroclitus* (L.). I. Temporal and spatial variation in gene frequencies of *Ldh-B, Mdh-A, Gpi-B, and Pgm-A. Biochem. Genet.* 16: 593-607.

Prakash, S., R.C. Lewontin, and J.L. Hubby. 1969. A molecular approach to the study of genic heterozygosity in natural populations. IV. Patterns of genic variation in central, marginal and isolated populations of *Drosophila psuedoobscura. Genetics* 61: 841-858.

Prothero, D.R., and D.B. Lazarus. 1980. Planktonic microfossils and the recognition of ancestors. *Syst. Zool.* 29: 119-129.

Provine, W.B. 1971. *The Origins of Theoretical Population Genetics.* Univ. of Chicago Press.

Provine, W.B. 1983. The development of Wright's theory of evolution: systematics, adaptation, and drift. In *Dimensions of Darwinism*, ed. M. Grene, pp. 43-70. Cambridge Univ. Press.

Punnett, R.C. 1915. *Mimicry in Butterflies.* Cambridge, U.K.: Cambridge Univ. Press.

Pyke, G.H. 1984. Optimal foraging theory: a critical review. *Ann. Rev. Ecol. Syst.* 15:523-575.

Rachootin, S., and K.S. Thomson. 1981. Epigenetics, paleontology, and evolution. In *Evolution Today*, Proc. Intl. Congr. Syst. Evol. Biol., ed. G.G.E. Scudder and J.L. Raveal, pp. 181-194. Pittsburgh: Hunt Inst. for Botanical Documentation, Carnegie Mellon Univ.

Radinsky, L. 1978. Evolution of brain size in carnivores and ungulates. *Am. Nat.* 112: 815-831.

Radinsky, L.B. 1982. Evolution of skull shape in carnivores. 3.The origin and early radiation of the modern carnivore families. *Paleobiology* 8: 177-195.

Raff, R., and T.H. Kaufman. 1983. *Embryos, Genes, and Evolution.* Macmillan: New York.

Ramirez, W.B. 1970. Host specificity of fig wasps (Agaonidae). *Evolution* 24: 680-691.

Rampino, M.R., and R.B. Stothers. 1984. Terrestrial mass extinctions, cometary impacts and the sun's motion perpendicular to the galactic plane. *Nature* 308: 709-711.

Raup, D.M. 1966. Geometric analysis of shell coiling: general problems. *J. Paleontol.* 40: 1178-1190.

Raup, D.M. 1967. Geometric analysis of shell coiling: coiling in ammonoids. *J. Paleontol.* 41: 43-65.

Raup, D.M. 1972a. Approaches to morphologic analysis. In *Models In Paleobiology*, ed. T.J.M. Schopf, pp. 28-44. San Francisco: Freeman, Cooper, and Co.

Raup, D.M. 1972b. Taxonomic diversity during the Phanerozoic. *Science* 177: 1065-1071.

Raup, D.M. 1975. Taxonomic survivorship curves and Van Valen's law. *Paleobiology* 1: 82-96.

Raup, D.M. 1976a. Species diversity in the Phanerozoic: a tabulation. *Paleobiology* 2: 279-288.

Raup, D.M. 1976b. Species diversity in the Phanerozoic: an interpretation. *Paleobiology* 2: 288-297.

Raup, D.M. 1978. Cohort analysis of generic survivorship. *Paleobiology* 4: 1-15.

Raup, D.M. 1983. On the early origins of major biologic groups. *Paleobiology* 9: 107-115.

Raup, D.M., and R.E. Crick. 1981. Evolution of single characters in the Jurassic ammonite Kosmoceras. *Paleobiology* 7: 200-215.

Raup, D.M., and R.E. Crick. 1982. *Kosmoceras*: evolutionary jumps and sedimentary breaks. *Paleobiology* 8: 90-100.

Raup, D.M., S.J. Gould, T.J.M. Schopf, and D.J. Simberloff. 1973. Stochastic models of phylogeny and the evolution of diversity. *J. Geol.* 81: 525-542.

Raup, D.M., and A. Michelson. 1965. Theoretical morphology of the coiled shell. *Science* 147: 1294-1295.

Raup, D.M., and J.J. Sepkoski, Jr.. 1982. Mass extinctions in the marine fossil record. *Science* 215: 1501-1503.

Raup, D.M., and J.J. Sepkoski, Jr. 1984. Periodicity of extinctions in the geologic past. *Proc. Nat. Acad. Sci. USA* 81: 801-805.

Raup, D.M., and J.J. Sepkoski, Jr. 1986. Periodic extinction of families and genera. *Science* 231:833-836.

Raup, D.M., and S.M. Stanley. 1971. *Principles of Paleontology*. San Francisco: W.H. Freeman and Co.

Rees, H., G. Jenkins, A.G. Seal, and J. Hutchinson. 1982. Assays of the phenotypic effects of changes in DNA amounts. In *Genome Evolution*, ed. G.A. Dover, and R.B. Flavell, pp. 287-297. London: Academic Press.

Reeve, E.C.R., and P.D.F. Murray. 1942. Evolution in the horse's skull. *Nature* 150: 402-403.

Reif, W.-E. 1983a. Evolutionary theory in German paleontology. In *Dimensions of Darwinism*, ed. M. Grene, pp. 173-203. Cambridge Univ. Press.

Reif, W. 1983b. Hilgendorf's (1863) dissertation on the Steinheim planorbids (Gastropoda: Miocene): the development of a phylogenetic research program for paleontology. *Paläont. Z.* 57: 7-20.

Reif, W.-E. 1983c. The Steinheim snails (Miocene:Schwabische Alb.) from a Neo-Darwinian point of view: a discussion. *Paläont. Z.* 57: 21-26.

Reimchen, T.E. 1983. Structural relationship between spines and lateral plates in threespine sticklebacks (*Gastrerosteus aculeatus*). *Evolution* 37: 931-946.

Rendel, J.M. 1959. Canalization of the scute phenotype. *Evolution* 13: 425-439.

Rensch, B. 1959. *Evolution Above the Species Level.* London: Methuen.

Reyment, R.A. 1974. Analysis of a generic level transition in Cretaceous-Ammonites. *Evolution* 28: 665-676.

Reyment, R.A. 1982a. Analysis of trans-specific evolution in Cretaceous ostracodes. *Paleobiology* 8: 293-306.

Reyment, R.A. 1982b. Phenotypic evolution in a Cretaceous foraminifer. *Evolution* 36: 1182-1199.

Reyment, R.A. 1982c. Threshold characters in a Cretaceous foraminifer. *Paleoclimatol. Paleogeogr. Paleoecol.* 38: 1-7.

Reyment, R.A. 1985. Phenotypic evolution in a lineage of the Eocene ostracod *Echinocythereis. Paleobiology* 10: 174-194.

Rhoads, D.C., and J.W. Morse. 1971. Evolutionary and ecological significance of oxygen-deficient marine basins. *Lethaia* 4:413-428.

Rhodes, F.H.T. 1983. Gradualism, punctuated equilibrium and the *Origin of Species. Nature* 305: 269-272.

Riddle, D.L., M.M. Swanson, and P.S. Albert. 1981. Interacting genes in a nematode dauer larva formation. *Nature* 290: 668-671.

Riedl, R. 1978. *Order in Living Organisms* (transl. from German by R.P.S. Jefferies). Chichester, U.K.: John Wiley and Sons.

Rightmire, G.P., 1981. Patterns in the evolution of *Homo erectus*. *Paleobiology* 7: 241-246.

Riska, B., and W.R. Atchley. 1985. Genetics of growth predict patterns of brain-size evolution. *Science* 229: 668-671.

Robertson, F.W. 1962. Changing the relative size of body parts of Drosophila by selection. *Genet. Res. Cambr.* 3: 169-180.

Robertson, F.W., and E.C.R. Reeve. 1952. Studies in quantitative inheritance. I. The effects of selection of wind and thorax in length in Drosophila. *J. Genet.* 50: 414-448.

Rockwood, E.S., C.G. Kanapi, M.R. Wheeler, and W.S. Stone. 1971. X. Allozyme changes during the evolution of Hawaiian *Drosophila*. Studies in Genetics VI. *Univ. Texas Publ.* 7103: 193-212.

Rodakis, G.C., and F.C. Kafatos. 1982. Origin of evolutionary novelty: how a high cystine protein has evolved. *Proc. Nat. Acad. Sci. USA* 79: 3551-3555.

Rohlf, F.J. 1965. A randomization test of the nonspecificity hypothesis in numerical taxonomy. *Taxon* 14:262-267.

Rohlf, F.J., D.H. Colless, and G. Hart. 1983. Taxonomic congruence reexamined. *Syst. Zool.* 32: 144-158.

Rohlf, F.J., and R.R. Sokal. 1981. Comparing numerical taxonomic studies. *Syst. Zool.* 30: 459-490.

Rose, K.D., and T.M. Bown. 1984. Gradual phyletic evolution at the generic level in early Eocene omomyid primates. *Nature* 309: 250-252.

Rosen, D.E. 1978. Vicariant patterns and historical explanations in biogeography. *Syst. Zool.* 27: 159-188.

Rosen, D.E., P.L. Forey, B.G. Gardiner, and C. Patterson. 1981. Lungfishes, tetrapods, paleontology, and plesiomorphy. *Bull. Amer. Mus. Nat. Hist.* 167: 163-275.

Rosenberg, J.B., C. Schröeder, A. Preiss, A. Kienlin, S. Coté, I. Riede, and H. Jäckle. 1986. Structural homology of the product of the *Drosophila* Krüppel gene with *Xenopus* transcription factor. *Nature* 319:336-339.

Rosenzweig, M.L. 1966. Community structure in sympatric Carnivora. *J. Mammal.* 47: 602-612.

Rosenzweig, M.L. 1975. On continental steady states of species diversity. In *Ecology and Evolution of Communities,* ed. M.L. Cody and J.M. Diamond, pp. 121-140. Cambridge MA: Belknap Press.

Rosenzweig, M.L. 1978. Competitive speciation. *Biol J. Linn. Soc.* 10: 275-289.

Rosenzweig, M.L., and J.L. Duek. 1979. Species diversity and turnover in an Ordovician marine invertebrate assemblage. In *Contemporary Quantitative Ecology and Related Econometrics,* ed. G.P. Patil and M.L. Rosenzweig, pp. 109-119. Fairfield, MD: International Cooperative Publ. House.

Rosenzweig, M.L., and J.A. Taylor. 1980. Speciation and diversity in Ordovician invertebrates: filling niches quickly and carefully. *Oikos* 35: 236-243.

Rowe, A.W. 1899. An analysis of the genus *Micraster* as determined by rigid zonal collection from the zone of *Rhynchonella cuvieri* to that of *Micraster coranguinum*. *Quart. J. Geol. Soc. Lond.* 55: 494-546.

Rowell, A.J., and M.J. Brady. 1976. Brachiopods and biomeres. *Brigham Young Univ. Geol. Ser.* 23(2): 165-180.

Rubin, G.M., M.G. Kidwell, and P.M. Bingham. 1982. The molecular basis of P-M hybrid dysgenesis: the nature of induced mutations. *Cell* 29: 987-994.

Rubin, G.M., and A.C. Spradling. 1980. Genetic transformation of *Drosophila* with transposable element vectors. *Science* 218: 348-353.

Rudwick, M.J.S. 1961. The feeding mechanism of the Permian brachiopod *Prorichtofenia*. *Palaeontology* 3: 450-471.

Rudwick, M.J.S. 1964. The inference of function from structure in fossils. *Brit. J. Phil. Sci.* 15: 27-40.

Rudwick, M.J.S. 1970. *Living and Fossil Brachiopods.* London: Hutchinson Univ. Library.

Rudwick, M.J.S. 1972. *The Meaning of Fossils.* London: Macdonald.

Runnegar, B. 1982. A molecular clock date for the origin of the animal phyla. *Lethaia* 15: 199-205.

Runnegar, B. 1986. Molecular palaeontology. *Palaeontology* 29: 1-24.

Runnegar, B., and C. Bentley. 1983. Anatomy, ecology, and affinities of the Australian early Cambrian bivalve *Pojetaia runnegari* Jell. *J. Paleontol.* 57: 73-92.

Runnegar, B., and J. Pojeta, Jr. 1974. Molluscan phylogeny: the paleontological viewpoint. *Science* 186: 311-317.

Sadler, P.H. 1981. Sediment accumulation rates and the completeness of stratigraphic sections. *J. Geol.* 89: 569-584.

Sadler, P.M., and L.W. Dingus. 1982. Expected completeness of sedimentary sections: estimating a time-scale dependent, limiting factor in the resolution of the fossil record. *Third N. Amer. Paleontol. Conv. Proc.* 2: 461-464.

Sage, R.D., and R.K. Selander. 1979. Hybridization between species of the *Rana pipiens* complex in central Texas. *Evolution* 33: 1069-1088.

Salmon, M., S.D. Ferris, D. Johnston, G. Hyatt, and G.S. Whitt. 1974. Behavioral and biochemical evidence for species distinctiveness in the fiddler crabs *Uca speciosa* and *U. spinicarpa*. *Evolution* 33: 82-91.

Salthe, S.N. 1985. *Evolving Hierarchical Systems: Their Structure and Representation.* New York: Columbia Univ. Press.

Salthe, S.N., and N.O. Kaplan 1966. Immunology and rates of enzyme evolution in the Amphibia in relation to the origins of certain taxa. *Evolution* 20: 603-616.

Sambol, M., and R.M. Finks. 1977. Natural selection in a Cretaceous oyster. *Paleobiology* 3: 1-16.

Sanders, H.L. 1955. The Cephalocarida, a new subclass of Crustacea from Long Island Sound. *Proc. Nat. Acad. Sci. USA* 41:61-66.

Sanders, H.L., and R.R. Hessler. 1969. Ecology of the deep-sea benthos. *Science* 163: 1419-1424.

Santibanez, S.K., and C.H. Waddington. 1958. The origin of sexual isolation between different lines within a species. *Evolution* 12:485-493.

Sarich, V.M., and A.C. Wilson 1967. Immunological time scale for hominid evolution. *Science* 158: 1200-1202.

Sarich, V.M., and A.C. Wilson. 1973. Generation time and genomic evolution in Primates. *Science* 179: 1144-1147.

Schaeffer, B., M.K. Hecht, and N. Eldredge. 1973. Phylogeny and paleontology. *Evol. Biol.* 6:31-46.

Schaffner, K.F. 1974. The peripherality of reductionism in the development of molecular biology. *J. Hist. Biol.* 7: 111-139.

Schaffner, K.F. 1984. Reduction in biology: prospects and problems. In *Conceptual Issues in Evolutionary Biology: An Anthology*, ed. E. Sober, pp. 428-445. Cambridge, MA: MIT Press.

Scharloo, W. 1964. The effect of disruptive and stabilizing selection on the expression of a *cubitus interruptus* mutant in Drosophila. *Genetics* 50: 553-562.

Scheltema, R.S. 1971. Larval dispersal as a means ofegenetic exchange between geographically separated populations of shallow- water benthic marine gastropods. *Biol. Bull.* 140: 284-322.

Schindel, D.E. 1980. Microstratigraphic sampling and the limits of paleontological resolution. *Paleobiology* 6: 408-426.

Schindel, D.E. 1982a. Resolution analysis: a new approach to the gaps in the fossil record. *Paleobiology* 8: 340-353.

Schindel, D.E. 1982b. The gaps in the fossil record. *Nature* 297:282-284.

Schindel, D.E. 1982c. Punctuations in the Pennsylvanian evolutionary history of *Glabrocingulum* (Mollusca: Arachaeogastropoda). *Geol. Soc. America Bull.* 93: 400-408.

Schindewolf, O.H. 1936. *Palaeontologie, Entwickslungslehre un Genetik.* Berlin: Bornträger.

Schindewolf, O.H. 1950. *Grundfragen der Palaontologie.* Stuttgart: Schweizerbart.

Schmalhausen, I.I. 1949. *Factors of Evolution, The Theory of Stabilizing Selection.* Philadelphia: Blakiston.

Schoener, T.W. 1983. Field experiments in interspecific competition. *Am. Nat.* 122: 240-285.

Schopf, T.J.M. 1974. Permo-Triassic extinctions: relation to sea-floor spreading. *J. Geol.* 82: 129-143.

Schopf, T.J.M. 1981. Evidence from findings of molecular biology with regard to the rapidity of genomic change: implications for species durations. In *Paleobotany, Paleoecology and Evolution: Festschrift for Harlan P. Banks,* ed. K.J. Niklas. New York: Praeger.

Schopf, T.J.M. 1982. Extinction of the Dinosaurs: a 1982 understanding. In *Geological Implications of Impacts of Large Asteroids and Comets on the Earth,* ed. L.T. Silver and P.H. Schultz, *Geol. Soc. America, Spec. Pap.* 190, pp. 415-422.

Schopf, T.J.M., J.B. Fisher, and C.A.F. Smith. 1978. Is the marine latitudinal diversity gradient merely another example of the species area curve? In *Marine Organisms: Genetics, Ecology and Evolution.* ed. B. Battaglia and J.L. Beardmore, pp. 365-386. New York: Plenum Press.

Schopf, T.J.M., and J.L. Gooch. 1971. Gene frequencies in a marine ectoproct: a cline in natural populations related to sea temperature. *Evolution* 25: 286-289.

Schopf, T.J.M., D.M. Raup, S.J. Gould, and D.S. Simberloff. 1975. Genomic versus morphologic rates of evolution: influence of morphologic complexity. *Paleobiology* 1: 63-70.

Schuchert, C. 1893. A classification of the brachiopoda. *Am. Geologist* 11: 141-167.

Schuh, R.T. 1976. Pretarsal structure in the Miridae (Hemiptera) with a cladistic analysis of the relationships of the family. *Amer. Mus. Novitates* 2601: 1-39.

Schuh, R.T., and J.S. Farris. 1981. Methods for investigating taxonomic congruence and their application to the

Leptopodomorpha. *Syst. Zool.* 30: 331-351

Schull, W.J., and J.V. Neel. 1965. *The Effects of Inbreeding on Japanese Children.* New York: Harper and Row.

Schwartz, R.D., and P.B. James. 1984. Periodic mass extinctions and the sun's oscillation about the galactic plane. *Nature* 308: 712-713.

Scotese, C.R., R.K. Bambach, C. Barton, R. Van der Voo, and A.M. Ziegler. 1979. Paleozoic base maps. *J. Geol.* 87: 217-268.

Scott, J.P. 1937. The embryology of the guinea pig. III. The development of the polydactyous monster. A case of growth accelerated at a particular period by a semi-dominant lethal gene. *J. Exp. Zool.* 77: 123-156.

Seed, R. 1978. Systematics and evolution of Mytilus galloprovincialis LmK. In *Marine Organisms: Genetics, Ecology and Evolution,* ed. B. Battaglia, and J.L. Beardmore, pp. 447-468. New York: Plenum Press.

Seeley, R.H. 1986. Intense natural selection caused a rapid morphological transition in a living marine snail. Proc. Nat. Acad. Sci. 83: 6897-6901.

Seilacher, A. 1970. Arbeitskonzept zur construktionsmorphologie. *Lethaia* 3: 393-396.

Seilacher, A. 1973. Fabricational noise in adaptive morphology. *Syst. Zool.* 22: 451-465.

Seilacher, A. 1979. Constructional morphology of sand dollars. *Paleobiology* 5: 191-221.

Seilacher, A. 1983. Precambrian metazoan extinctions. *Geol. Soc. America, Abstr. with Progr.* 15: 683.

Selander, R.K., D.W. Kaufman, R.J. Baker, and S.L. Williams. 1974. Genic and chromosomal differentiation in pocket gophers of the *Geomys bursarius* group. *Evolution* 28: 557-564.

Selander, R.K., and B.R. Levin. 1980. Genetic diversity and structure in *Escherichia coli* populations. *Science* 210: 545-547.

Selander, R.K., and T.S. Whittam. 1983. Protein polymorphism and the genetic structure of populations. In *Evolution of Genes and*

Proteins, ed. M. Nei and R.K. Koehn, pp. 89-114. Sunderland MA: Sinauer Assoc.

Sepkoski, J.J., Jr. 1975. Stratigraphic biases in the analysis of taxonomic survivorship. *Paleobiology* 1: 343-355.

Sepkoski, J.J., Jr. 1976. Species diversity in the Phanerozoic: species-area effects. *Paleobiology* 2: 298-303.

Sepkoski, J.J., Jr. 1978. A kinetic model of Phanerozoic taxonomic diversity. I. Analysis of marine orders. *Paleobiology* 4: 223-251.

Sepkoski, J.J., Jr. 1981. A factor analytic description of the Phanerozoic marine fossil record. Paleobiology 7: 36-53.

Sepkoski, J.J., Jr. 1982. A compilation of fossil marine families. Milwaukee Publ. Mus. Contr. Biol. Geol. 51: 1-125.

Sepkoski, J.J., Jr. 1984. A kinetic model of Phanerozoic taxonomic diversity. III. Post-Paleozoic families and mass extinctions. *Paleobiology* 10: 246-267.

Sepkoski, J.J., Jr. 1987. Environmental trends in extinction during the Paleozoic. *Science* 235: 64-66.

Sepkoski, J.J., Jr., R.K. Bambach, D.M. Raup, and J.W. Valentine. 1981. Phanerozoic marine diversity and the fossil record. *Nature* 293: 435-437.

Sepkoski, J.J., Jr., and P.M. Sheehan. 1983. Diversification, faunal change, and community replacement during the Ordovician radiations. *Biotic Interactions in Recent and Fossil Communities,* ed. M.J.S. Tevesz and P.L. McCall, pp. 673-717. New York: Plenum.

Shackleton, N.J., and J.P. Kennett. 1975. Paleotemperature history of the Cenozoic and the initiation of Antarctic glaciation: oxygen and carbon isotope analyses in DSDP sites 277, 279, and 281. *Init. Rep. Deep Sea Drilling Proj.* 24: 743-755.

Shaw, A.B. 1964. *Time in Stratigraphy.* New York: McGraw-Hill.

Shaw, A.B. 1969. Adam and Eve, paleontology, and the non-objective arts. *J. Paleontol.* 43: 1085-1098.

Shea, B.T. 1983. Paedomorphosis and neoteny in the Pygmy chimpanzee. *Science* 222: 521-522.

Sheehan, P.M., and T.A. Hansen. 1986. Detritus feeding as a buffer to extinction at the end of the Cretaceous. *Geology* 14: 868-870.

Sheehan, P.M., and C.L. Morse. 1986. Cretaceous-Tertiary dinosaur extinction. *Science* 234: 1171-1172.

Sherrington, C. 1949. *Goethe on Nature and on Science.* Cambridge Univ. Press.

Sibley, C.G., and J.E. Ahlquist. 1983. The phylogeny and classification of birds based upon the data of DNA-DNA hybridization. *Current Ornithology* 1:245-292.

Siegel, A.F., and R.H. Benson. 1982. A robust comparison of biological shapes. *Biometrics* 38: 341-350.

Signor, P.W., and J.H. Lipps. 1982. Sampling bias, gradual extinction patterns and catastrophes in the fossil record. In *Geological implications of impacts of large asteroids and comets on the Earth,* ed. Silver, L.T., and P.H. Schultz, *Geol. Soc. America Sp. Pap.* 190: 291-296.

Signor, P.W., and J.F. Mount. 1986. Paleoenvironmental gradients in adaptive innovation: nearshore innovations or evolutionary persistence? *Soc. Econ. Pal. Min. Annual Midyr. Mtg. Raleigh, NC, Program with Abstr.,* p. 103.

Silver, L.T., and P.H. Schultz, eds. 1982. *Geological implications of impacts of large asteroids and comets on the Earth. Geol. Soc. Am. Sp. Paper* 190: 1-528.

Simberloff, D. 1986. The proximate causes of extinction. In *Patterns and Processes in the History of Life,* ed. D.M. Raup and D. Jablonski, pp. 259-276. Berlin: Springer-Verlag.

Simon, H.A. 1962. The architecture of complexity. *Proc. Am. Phil. Soc.* 106: 467-482.

Simpson, G.G. 1944. *Tempo and Mode in Evolution.* New York: Columbia Univ. Press.

Simpson, G.G. 1950. *The Meaning of Evolution.* London: Oxford Univ. Press.

Simpson, G. G. 1951. *Horses.* Oxford: Oxford Univ. Press.

Simpson, G.G. 1952. Periodicity in vertebrate evolution. *J. Paleontol.* 26: 359-370.

Simpson, G.G. 1953. *The Major Features of Evolution.* New York: Columbia Univ. Press.

Simpson, G.G. 1960. Diagnosis of the classes Reptilia and Mammalia. *Evolution* 14: 388-392.

Simpson, G.G. 1961. *The Principles of Animal Taxonomy.* New York: Columbia Univ. Press.

Simpson, G.G. 1975. Recent advances in methods of phylogenetic inference. In *Phylogeny of the Primates, a Multidisciplinary Approach,* ed. W.P. Luckett and F.S. Szalay, pp. 3-19. New York: Plenum Press.

Sinnott, E.W., and L.C. Dunn. 1935. The effect of genes on the development of size and form. *Biol. Rev. Cambr.* 10: 123-151.

Skevington, D. 1967. Probable instance of genetic polymorphism in the graptolites. *Nature* 213: 810-812.

Slatkin, M. 1973. Gene flow and selection in a cline. *Genetics* 75: 733-756.

Slatkin, M. 1981. A diffusion model of species selection. *Paleobiology* 7: 421-425.

Slatkin, M. 1985. Rare alleles as indicators of gene flow. *Evolution* 39: 53-65.

Slatkin, M. 1987. Gene flow and the geographic structure of natural populations. *Science* 236: 787-792.

Sloss, L.L. 1950. Rates of evolution. *J. Paleontol.* 24: 131-139.

Smith, A.B. 1984. Classification of the echinodermata. *Palaeontology* 27: 431-459.

Smith, A.G., and J.C. Briden 1977. *Mesozoic and Cenozoic Paleocontinental Maps.* Cambridge Univ. Press.

Smith, D.B., and R.B. Flavell. 1974. The relatedness and evolution of repeated nucleotide sequences in the genomes of some gramineae species. *Biochem. Genet.* 12: 243-256.

Smith, C.T. 1945. The biostratigraphy of *Glycymeris ventchii* in California. *J. Paleontol.* 19: 35-44.

Smith, G.R. 1981. Late Cenozoic freshwater fishes of North America. *Ann. Rev. Ecol. Syst.* 12: 163-193.

Smith, J.D. 1976. Comments on flight and the evolution of bats. In *Major Patterns in Vertebrate Evolution*, ed. M.K. Hecht, P.C. Goody, and B.M. Hecht, pp. 427-437. New York: Plenum Press.

Sneath, P.H.A., and R.R. Sokal. 1973. *Numerical Taxonomy*. San Francisco: W.H. Freeman and Co.

Snyder, T.P., and J.L. Gooch. 1973. Genetic differentiation of *Littorina saxatilis* (Gastropoda). *Mar. Biol.* 22: 177-182.

Sober, E. 1983. Parsimony in systematics: philosophical issues. *Ann. Rev. Ecol. Syst.* 14: 335-357.

Sober, E., ed. 1984a. *Conceptual Issues in Evolutionary Biology: An Anthology*. Cambridge, MA: MIT Press.

Sober E. 1984b. *The Nature of Selection: Evolutionary Theory and Philosophical Focus*. Cambridge, MA: Bradford Books of MIT Press.

Sohndi, K.C. 1962. The evolution of a pattern. *Evolution* 16: 186-191.

Sokal, R.R. 1962. Variation and covariation of characters of Alate *Pemphigus populi-transversus* in eastern North America. *Evolution* 16: 227-245.

Sokal, R.R. 1983a. A phylogenetic analysis of the caminalcules. II. Estimating the true cladogram. *Syst. Zool.* 32: 185-201.

Sokal, R.R. 1983b. A phylogenetic analysis of the caminalcules. III. Fossils and classification. *Syst. Zool.* 32: 248-258.

Sokal, R.R. 1983c. A phylogenetic analysis of the caminalcules. IV. Congruence and character stability. *Syst. Zool.* 32: 259-275.

Sokal, R.R. 1986. Phenetic taxonomy: Theory and methods. *Ann. Rev. Ecol. Syst.* 17: 423-442.

Sokal, R.R., and J.H. Camin. 1965. The two taxonomies: areas of agreement and conflict. *Syst. Zool.* 14: 176-195.

Sokal, R.R., and T.J. Crovello. 1970. The biological species: a critical evaluation. *Am. Nat.* 104: 127-153.

Sokal, R.R., K.L. Fiala, and G. Hart. 1984. OTU stability and factors determining taxonomic stability: Examples from the Caminalcules and the Leptopodomorpha. *Syst. Zool.* 33: 387-407.

Sokal, R.R., and K. Shao. 1985. Character stability in 39 data sets. *Syst. Zool.* 34: 83-89.

Sokal, R.R., and P.H.A. Sneath. 1963. *Principles of Numerical Taxonomy*. San Francisco: W.H. Freeman and Co.

Spemann, H. 1938. *Embryonic Development and Induction*. New York: Hafner (1967 reprint).

Spieth, H.T., and W.R. Heed. 1972. Experimental systematics and ecology of *Drosophila*. *Ann. Rev. Ecol. Syst.* 3. 209-288.

Spiller, J. 1977. *Evolution of Turritellid gastropods from the Miocene and Pliocene of the Atlantic coastal plain*. Ph.D. dissertation, State Univ. of New York at Stony Brook.

Spjeldnaes, N. 1964. Climatically induced faunal migrations: examples from the littoral fauna of the Late Pleistocene of Norway. *Problems in Palaeoclimatology,* ed. A.E.M. Nairn, pp. 353-356.

Spradling, A.C., and G.M. Rubin. 1981. *Drosophila* genome organization: conserved and dynamic aspects. *Ann. Rev. Gen.* 15: 219-264.

Stanley, S.M. 1968. Post-Paleozoic adaptive radiation of infaunal bivalve molluscs - a consequence of mantle fusion and siphon formation. *J. Paleontol.* 42: 214-229.

Stanley, S.M. 1970. Relation of Shell Form to Life Habits in the Bivalvia (Mollusca). *Geol. Soc. America Mem.* 125.

Stanley, S.M. 1972. Functional morphology and evolution of byssally attached bivalve mollusks. *J. Paleontol.* 46: 165-212.

Stanley, S.M. 1973a. Effect of competition on rates of evolution with special reference to bivalve mollusks and mammals. *Syst. Zool.* 22: 486-506.

Stanley, S.M. 1973b. An ecological theory for the sudden origin of multicellular life in the late Precambrian. *Proc. Nat. Acad. Sci. USA* 72:646-650.

Stanley, S.M. 1973c. An explanation for Cope's rule. *Evolution* 27: 1-26.

Stanley, S.M. 1975. A theory of evolution above the species level. *Proc. Nat. Acad. Sci. USA* 72: 646-650.

Stanley, S.M. 1979. *Macroevolution: Pattern and Process*. San Francisco: W.H. Freeman.

Stanley, S.M. 1981. *The New Evolutionary Timetable*. New York: Basic.

Stanley, S.M. 1982. Macroevolution and the fossil record. *Evolution* 36: 460-473.

Stanley, S.M. 1983. Marine mass extinctions: a dominant role for temperature. In *Extinctions*, ed. M. Nitecki, pp. 69-118. Univ. of Chicago Press.

Stanley, S.M. 1984. Temperature and biotic crises in the marine realm. *Geology* 12: 205-208.

Stanley, S.M. 1985. Rates of evolution. *Paleobiology* 11:13-26.

Stanley, S.M. 1986a. Population size, extinction, and speciation: the fission effect in Neogene Bivalvia. *Paleobiology* 12: 89-110.

Stanley, S.M. 1986b. Anatomy of a regional mass extinction: Plio-Pleistocene decimation of the western Atlantic bivalve fauna. *Palaios* 1: 17-36.

Stanley, S.M., and L.D. Campbell. 1981. Neogene mass extinction of western Atlantic molluscs. *Nature* 293: 457-459.

Stanley, S.M., and W.A. Newman. 1980. Competitive exclusion in evolutionary time: the case of the acorn barnacles. *Paleobiology* 6: 173-183.

Stanley, S.M., P.W. Signor, S. Lidgard, and A.F. Karr. 1981. Natural clades differ from "random" clades: simulations and analyses. *Paleobiology* 7: 115-127.

Stanley, S.M., and X. Yang. 1987. Approximate evolutionary stasis for bivalve morphology over millions of years: a multivariate, multilineage study. *Paleobiology*, In press.

Stearns, S.C. 1976. Life-history tactics: a review of the ideas. *Qu. Rev. Biol.* 51: 3-47.

Stebbins, G.L. 1950. *Variation and Evolution in Plants*. New York: Columbia Univ. Press.

Stebbins, G.L. 1971. *Chromosomal Evolution in Higher Plants*. New York: Addison-Wesley.

Stebbins, G.L. 1974. Adaptive shifts and evolutionary novelty: a compositionist approach. In *Studies in the Philosophy of Biology*, ed. F.J. Ayala and T. Dobzhansky, pp. 285-338. Berkeley: Univ. of California Press.

Stebbins, G.L. 1983a. Plant speciation. In *Mechanisms of Speciation*, ed. C. Barigozzi, pp. 21-39. New York: Alan R. Liss.

Stebbins, G.L. 1983b. Mosaic evolution: an integrating principle for the modern synthesis. *Experientia* 09. 020 004.

Stebbins, G.L., and F.J. Ayala. 1981. Is a new evolutionary synthesis necessary? *Science* 213: 967-971.

Stebbins, G.L., and L. Ferlan. 1956. Population variability, hybridization and introgression in some species of *Ophrys*. *Evolution* 10: 32-46.

Stehli, F.G., R. Douglas, and I. Kafescegliou. 1972. Models for the evolution of planktonic foraminifera. In *Models in Paleobiology*, ed. T.J.M. Schopf, pp. 116-128. San Francisco: Freeman Cooper and Co.

Stehli, F.G., and J.W. Wells. 1971. Diversity and age patterns in hermatypic corals. *Syst. Zool.* 20: 115-126.

Stenseth, N. C., and J. Maynard Smith. 1984. Coevolution in ecosystems: Red Queen evolution or stasis. *Evolution* 38: 870-880.

Stenzel, H.B. 1949. Successional speciation in paleontology: the case of the oysters of the *sellaeformis* stock. *Evolution* 3: 34-50.

Stern, C. 1968. *Mosaics and Other Essays*. Cambridge MA: Harvard Univ. Press.

Stevens, G.R. 1973. Jurassic belemnites. In *Atlas of Palaeobiogeography*, ed. A. Hallam, pp. 259-274. Amsterdam: Elsevier Sci.

Stewart, A.D., and D.M. Hunt. 1982. *The Genetic Basis of Development*. Glasgow and London: Blackie.

Stitt, J.H. 1971. Repeating evolutionary pattern in Late Cambrian trilobite biomeres. *J. Paleontol.* 45: 178-181.

Stock, G.B., and S.V. Bryant. 1981. Studies of digit regeneration and their implications for theories of development and evolution of vertebrate limbs. *J. Exp. Zool.* 216: 423-433.

Stoecker, D. 1978. Resistance of a tunicate to fouling. *Biol. Bull.* 155: 615-626.

Strachan, T., D.A. Webb, and G.A. Dover. 1985. Transition stages of molecular drive in multiple-copy DNA families of *Drosophila.* *EMBO Journal* 4:1701-1708.

Strathman, R.R., and M. Slatkin. 1983. The improbability of animal phyla with few species. *Paleobiology* 9:97-106.

Struhl, G. 1984. Splitting the bithorax complex of *Drosophila.* *Nature* 308: 454-457.

Sturtevant, A.H. 1913. The Himalayan rabbit case, with some considerations of multiple allelomorphs. *Am. Nat.* 47: 234-238.

Summerbell, D. 1981. The control of growth and the development of pattern across the anteroposterior axis of the chick limb bud. *J. Embryol. Exp. Morph.* 63: 161-180.

Surlyk, F., and M. B. Johansen. 1984. End-Cretaceous brachiopod extinctions in the chalk of Denmark. *Science* 223: 1174-1177.

Swinnerton, H.H. 1940. The study of variation in fossils. *Qu. J. Geol. Soc. London* 96: 87-118.

Tan Sin Hok. 1939. The results of phylomorphogenetic studies of some larger Foraminifera (a review). *De Ing. in Ned.-Indie. IV. Mijnb. en Geol.* 6(7): 93-97.

Teichert, C. 1946. Obituary. Rudolph Kaufmann. *Am. J. Sci.* 244: 808-810.

Teichert, C. 1949. Permian Crinoid Calceolispongia. *Geol. Soc. America Mem.* 34.

Templeton, A.R. 1977. Analysis of head shape differences between two interfertile species of Hawaiian *Drosophila. Evolution* 31: 630-641.

Templeton, A.R. 1979. The unit of selection in *Drosophila mercatorium.* II. Genetic revolution and the origin of coadapted genomes in parthogenetic strains. *Genetics* 92: 1265-1282.

Templeton, A.R. 1981. Mechanisms of speciation-a population genetic approach. *Ann. Rev. Ecol. Syst.* 12: 23-48.

Templeton, A.R. 1982. Genetic architecture of speciation. In *Mechanisms of Speciation*, ed. C. Barigozzi, pp. 105-121. New York: Alan R. Liss.

Templeton, A., and L. Val Giddings. 1981. Macroevolution conference (letter). *Science* 211:770-771.

Thayer, C.W. 1983. Sediment-mediated biological disturbance and the evolution of marine benthos. In *Biotic Interactions in Recent and Fossil Benthic Communities*, ed. M.J.S. Tevesz and P.L. McCall, pp. 480-625. New York: Plenum Press.

Thoday, J.M. 1959. Effects of disruptive selection. 1. Genetic flexibility. *Heredity* 13: 187-203.

Thoday, J.M., and J.B. Gibson. 1962. Isolation by disruptive selection. *Nature* 193: 1164-1166.

Thompson, D.W. 1915. *On Growth and Form.* Cambridge Univ. Press.

Thompson, D.W. 1952. *On Growth and Form.* 2nd ed. Cambridge Univ. Press.

Thorson, G. 1950. Reproductive and larval ecology of marine bottom invertebrates. *Biol. Rev.* 25: 1-45.

Tiffney, B.H. 1981. Diversity and major events in the evolution of land plants. In *Paleobotany, Paleoecology, and Evolution*, ed. K.V. Niklas, pp. 193-230. New York: Praeger.

Tompkins, R. 1978. Genic control of axolotl metamorphosis. *Am. Zool.* 18: 313-319.

Turelli, M. 1984. Heritable genetic variation via mutation-selection balance: Lerch's zeta meets the abdominal bristle. *Theor. Pop. Biol.* 25: 138-193.

Turing, A.M. 1952. The chemical basis of morphogenesis. *Phil. Trans. Roy. Soc. Lond.* 237B: 37-72.

Turner, B.J. 1983. Genic variation and differentiation of remnant natural populations of the desert pupfish, *Cyprinidon macularius. Evolution* 37: 690-700.

Turner, B.J., T.A. Grudzien, K.P. Adkisson, and R.A. Worrell. 1985. Extensive chromosomal divergence within a single river basin in the goodeid fish, *Ilyodon furcidens*. *Evolution* 39: 122-134.

Turner, J.R.G. 1967. On supergenes. I. The evolution of supergenes. *Am. Nat.* 101: 195-221.

Turner, J.R.G. 1977. Butterfly mimicry: the genetical evolution of adaptation. *Evol. Biol.* 10: 163-206.

Turner, J.R.G. 1981. Adaptation and evolution in *Heliconius*: a defense of neoDarwinism. *Ann. Rev. Ecol. Syst.* 12: 99-121.

Turner, J.R.G. 1983. "The hypothesis that explains mimetic resemblance explains evolution": the gradualist-saltationist schism. In *Dimensions of Darwinism*, ed. M. Grene, pp. 129-169, Cambridge Univ. Press.

Vail, P.R., R.M. Mitchum, Jr., and S. Thompson. 1977. Seismic stratigraphy and global changes of sea level, part 4. In *Seismic Stratigraphy*, ed. C.E. Peyton, *Amer. Assoc. Petr. Geol. Mem.* 26: 83-97.

Val, F.C. 1977. Genetic analysis of the morphological differences between two interfertile species of Hawaiian *Drosophila*. *Evolution* 31: 611-629.

Valentine, J.W. 1968. The evolution of ecological units above the population level. *J. Paleontol.* 42: 253-267.

Valentine, J.W. 1969. Patterns of taxonomic and ecological structure of the shelf benthos during Phanerozoic time. *Palaeontology* 12: 684-709.

Valentine, J.W. 1971a. Plate tectonics and shallow marine diversity and endemism, an actualistic model. *Syst. Zool.* 20: 253-264.

Valentine, J.W. 1971b. Resource supply and species diversity patterns. *Lethaia* 4: 51-61.

Valentine, J.W. 1986. Fossil record of the origin of Baupläne and its implications. In Patterns and Processes in the History of Life, D.M. Raup and D. Jablonski, eds., pp. 209-222. Berlin: Springer-Verlag.

Valentine, J.W., and C.A. Campbell. 1975. Genetic regulation and the fossil record. *Am. Sci.* 63:673-680.

Valentine, J.W., T.C. Foin, and D. Peart. 1978. A provincial model of Phanerozoic marine diversity. *Paleobiology* 4: 55-66.

Valentine, J.W., and E.M. Moores. 1971. Global tectonics and the fossil record. *J. Geol.* 80: 167-184.

Valentine, J.W., and T.D. Walker. 1987. Extinctions in a model taxonomic hierarchy. *Paleobiology*, in press.

van Delden, W. 1982. The alcohol dehydrogenase problem in *Drosophila melanogaster*: selection at an enzyme locus. *Evol. Biol.* 15: 187-222.

Van Valen, L. 1973a. Are categories in different phyla comparable? *Taxon* 22: 333-373.

Van Valen, L. 1973b. A new evolutionary law. *Evol. Theory* 1: 1-30.

Van Valen, L. 1974. Two modes of evolution. *Nature* 252: 298-300.

Van Valen, L. 1984a. A resetting of Phanerozoic community evolution. *Nature* 307: 50-52.

Van Valen, L. 1984b. Catastrophes, expectations, and the evidence. *Paleobiology* 10: 121-137.

Van Valen, L. 1985. How constant is extinction? *Evol. Theory* 7: 93-106.

Vavilov, N. 1922. The law of homologous series in variation. *J. Genetics* 12: 47-69.

Vawtor, L., and W.M. Brown. 1986. Nuclear and mitochondrial DNA comparisons reveal extreme rate variation in the molecular clock. *Science* 234: 194-196.

Vermeij, G.J. 1977. The Mesozoic marine revolution: evidence from snails, predators and grazers. *Paleobiology* 3: 245-258.

Vermeij, G.J. 1978. *Adaptation and Biogeography.* Cambridge, MA: Harvard Univ. Press.

Vermeij, G.J. 1982. Phenotypic evolution in a poorly dispersing snail after arrival of a predator. *Nature* 299: 349-350.

Vermeij, G.J. 1983. Shell-breaking predation through time. In *Biotic Interactions in Recent and Fossil Benthic Communities*, ed. M.J.S. Tevesz and P.L. McCall, pp. 649-669. New York: Plenum Press.

Vermeij, G.J., D.E. Schindel, and E. Zipser. 1981. Predation through geological time: evidence from gastropod shell repair. *Science* 214: 1024-1026.

Vogel, S. 1981. *Life In Moving Fluids.* Boston: Willard Grant Press.

Vrba, E.S. 1980. Evolution, species, and fossils: How does life evolve? *S. Afr. J. Sci.* 76: 61-84.

Vrba, E.S. 1983. Macroevolutionary trends: new perspectives on the roles of adaptation and incidental effect. *Science* 221: 387-389.

Vrba, E.S., and N. Eldredge. 1984. Individuals, hierarchies and processes: towards a more complete evolutionary theory. *Paleobiology* 10: 146-171.

Vrba, E.S., and S.J. Gould. 1986. The hierarchical expansion of sorting and selection: sorting and selection cannot be equated. *Paleobiology* 12:217-228.

Vrijenhoek, R.C. 1978. Coexistence of clones in a heterogeneous environment. *Science* 199:549-552.

Waagen, W. 1869. Die Formenreihe des Ammonites subradiatus. Versuch einer palaeontologischen Monographic. *Geognostische und Palaontologische Beitrage* ser. 2, 2: 181-256.

Waddington, C.H. 1940. *Organizers and Genes.* Cambridge Univ. Press.

Waddington, C.H. 1942. Canalization of development and the inheritance of acquired characters. *Nature* 150: 563-565.

Waddington, C.H. 1956. Genetic assimilation of the *bithorax* phenotype. *Evolution* 10: 1-13.

Waddington, C.H. 1957. *The Strategy of the Genes.* London: Allen and Unwin.

Waddington, C.H. 1962. *New Patterns in Genetics and Development.* New York: Columbia Univ. Press.

Wake, D.B., and A. Larson. 1987. Multidimensional analysis of an evolving lineage. *Science* 238: 42-48.

Wake, D.B., and A.H. Brame. 1969. Systematics and evolution of neotropical salamanders of the *Bolitoglossa helmrichi* group. *Los Angeles Co. Mus. Contr. Sci.* no. 175: 1-40.

Wake, D.B., G. Roth, and M.H.Wake. 1983. On the problem of stasis in organismal evolution. *J. Theor. Biol.* 101: 211-224.

Walker, K.R., and L.F. Laporte. 1970. Congruent fossil communities from Ordovician and Devonian carbonates of New York. *J. Paleontology* 44: 928-944.

Ward, B.L., W.T. Starmer, J.S. Russell, and W.B. Heed. 1974. The correlation of climate and host plant morphology with a geographic gradient of an inversion polymorphism in *Drosophila pachea*. *Evolution* 28: 565-575.

Ward, I.W., and B.W. Blackwelder. 1975. *Chesapecten*, a new genus of Pectinidae (Mollusca:Bivalvia) from the Miocene and Pliocene of eastern North America. *U.S. Geol. Surv. Prof. Paper* 861: 1-21.

Ward, P.D. 1981. Shell sculpture as a defensive adaptation in ammonoids. *Paleobiology* 7:96-100.

Ward, P.D., and P.W. Signor III. 1983. Evolutionary tempo in Jurassic and Cretaceous ammonites. *Paleobiology* 9: 183-198.

Ward, P.D., J. Wiedmann, and J.F. Mount. 1986. Maastrichtian molluscan biostratigraphy and extinction patterns in a Cretaceous/Tertiary boundary section at Zumaya, Spain. *Geology* 14: 899-903.

Webb, S.D. 1969. Extinction-origination equilibria in Late Cenozoic land mammals of North America. *Evolution* 23: 688-702.

Wei. K.Y., and J.P. Kennett. 1983. Nonconstant extinction rates of Neogene planktonic foraminifera. *Nature* 305: 218-220.

Weller, J.M. 1960. *Stratigraphic Principles and Practice*. New York: Harper and Row.

Wells, G.P. 1957. Variation in *Arenicola marina* (L.) and the status of *Arenicola glacialis* Murdoch. Polychaeta. *Proc. Zool. Soc. Lond.* 129: 397-419.

Werdelin, L. 1981. The evolution of lynxes. *Ann. Zool. Fennici* 18: 37-71.

Wessells, N.K. 1977. *Tissue Interactions and Development*. Menlo Park, CA: W.A. Benjamin.

West, R.M. 1979. Apparent prolonged evolutionary stasis in the
    middle Eocene hoofed mammal *Hyopsodus*. *Paleobiology* 5:
    252-260.

Westoll, T.S. 1949. On the evolution of the Dipnoi. In *Genetics,
    Paleontology, and Evolution*, ed. G.L. Jepsen, E. Mayr, and
    G.G. Simpson, pp. 121-184. Princeton Univ. Press.

White, J.F., and S.J. Gould. 1965. Interpretation of the coefficient in
    the allometric equation. *Am. Nat.* 99: 5-18.

White, M.J.D. 1968. *Modes of Speciation*. San Francisco: W.H.
    Freeman.

White, M.J.D. 1973. *Animal Cytology and Evolution*. London:
    William Clowes and Sons.

White, M.J.D. 1974. Speciation in Australian Morabine
    grasshoppers: the cytogenetic evidence. In *Genetic Mechanisms
    of Speciation in Insects*, ed. M.J.D. White, pp. 57-68. Sydney:
    Australia and New Zealand Book Co.

White, M.J.D. 1982. Rectangularity, speciation, and chromosome
    architecture. In *Mechanisms of Speciation*, ed. C. Barigozzi, pp.
    75-103. New York: Alan R. Liss.

White, M.J.D., R.C. Lewontin, and L.E. Andrew. 1963. Cytogenetics
    of the grasshopper *Moraba scurra*. VII. Geographic variation of
    adaptive properties of inversions. *Evolution* 17: 147-162.

White, T.D., and J.M. Harris. 1977. Suid evolution and correlation
    of African Hominid localities. *Science* 198: 13-21.

Whitmire, D.P., and A.A. Jackson IV. 1984. Are periodic mass
    extinctions driven by a distant solar companion? *Nature* 308:
    713-715.

Whitt, G.S., W.F. Childers, P.L. Cho. 1973. Allelic expression at
    enzymic loci in an intertribal hybrid sunfish. *J. Heredity* 64:
    55-61.

Wiley, E.O. 1981. *Phylogenetics*. New York: John Wiley and Sons.

Wilkins, H. 1971. Genetic interpretation of regressive evolutionary
    processes: studies on hybrid eyes of two *Astyanax* cave
    populations (Characidae, Pisces). *Evolution* 25: 530-544.

Williams, A., and A.J. Rowell. 1965. Classification. In *Treatise of Invertebrate Paleontology* (H) *Brachiopoda*, vol. 1, ed. R.C. Moore, pp. H214-H237. Boulder, CO: Geological Society of America, Lawrence KS: Univ. of Kansas. University of Kansas.

Williams, E.E. 1950. Variation and selection in the cervical central articulations of living turtles. *Bull. Amer. Mus. Nat. Hist.* 94: 509-561.

Williams, E.E. 1972. Origin of faunas. Evolution of lizard congeners in a complex island fauna - a trial analysis. *Evol. Biol.* 6: 47-89.

Williams, G.C. 1966. *Adaptation and Natural Selection: A Critique of Some Contemporary Thought.* Princeton Univ. Press.

Williams, G.C. 1985. A defense of reductionism in evolutionary biology. In *Oxford Surveys in Evolutionary Biology*, ed. R. Dawkins and M. Ridley, vol. 2, pp. 1-27.

Williams, G.C., R.K. Koehn, and J.B. Mitton. 1973. Genetic differentiation without isolation in the American eel, *Anguilla rostrata. Evolution* 27: 192-204.

Williams, N.E. 1984. An apparent disjunction between the evolution of form and substance in the genus *Tetrahymena. Evolution* 38: 25-33.

Williamson, P.G. 1981. Palaeontological documentation of speciation in Cenozoic molluscs from Turkana basis. *Nature* 293: 437-443.

Williston, S.W. 1879. Are birds derived from dinosaurs? *Kansas City Rev. Sci.* 3: 457-460.

Wills, C. 1981. *Genetic Variability.* Oxford: Clarendon.

Wilson, A.C., R.L. Cann, S.M. Carr, M. George, U.B. Gyllensten, K.M. Helm-Bychowski, R.G. Higuchi, S.R. Palumbi, E.M. Prager, R.D. Sage, and M. Stoneking. 1985. Mitochondrial DNA and two perspectives on evolutionary genetics. *Biol. J. Linn. Soc.* 26: 375-400.

Wilson, A.C., S.S. Carlson, and T.J. White 1977. Biochemical Evolution. *Ann. Rev. Biochem.* 46: 573-639.

Wilson, A.C., L.R. Maxson, and V.M. Sarich. 1974a. *Proc. Nat. Acad. Sci. USA* 71: 2843-2847.

Wilson, A.C., V.M. Sarich, and L.R. Maxson. 1974b. The importance of gene rearrangement in evolution: evidence from studies on rates of chromosomal, protein, and anatomic evolution. *Proc. Nat. Acad. Sci. USA* 71: 3028-3030.

Wilson, E.O. 1975. *Sociobiology: The New Synthesis.* Cambridge, MA: Belknap Press of Harvard Univ. Press.

Wilson, M.V.H. 1983. Is there a characteristic rate of radiation for the insects? *Paleobiology* 9: 79-85.

Wilson, T.G. 1981a. Expression of phenotypes in a temperature-sensitive allele of the apterous mutation in *Drosophila melanogaster. Dev. Biol.* 85: 425-433.

Wilson, T.G. 1981b. A mosaic analysis of the Apterous mutation in *Drosophila melanogaster. Dev. Biol.* 85: 434-435.

Wimsatt. W.C. 1980. Reductionistic research strategies and their biases in the units of selection controversy. In *Scientific Discovery, vol. 2, Case Studies.* ed. T. Nickles, pp. 213-259. Dordrecht: D. Reidel and Co.

Wise, K.P., and T.J.M. Schopf. 1981. Was marine faunal diversity in the Pleistocene affected by changes in sea level? *Paleobiology* 7: 394-399.

Wolfe, J.A., and D.M. Hopkins. 1967. Climatic changes recorded by Tertiary land floras in northwestern North America. In *Tertiary Correlations and Climatic Changes in the Pacific.* pp. 67-76.

Wolpert, L. 1969. Positional information and the spatial pattern of cellular differentiation. *J. Theoret. Biol.* 25: 1-47.

Wolpoff, M.H. 1984. *Evolution* of *Homo erectus*: the question of stasis. *Paleobiology* 10: 389-406.

Woodburne, M. G., and B. J. MacFadden. 1982. A reappraisal of the systematics, biogeography, and evolution of fossil horses. *Paleobiology* 8: 315-327.

Woodring, W. 1952. Discussion. *J. Paleontol.* 26: 386-394.

Woodruff, D.S., and S.J. Gould. 1980. Geographic differentiation and speciation in *Cerion* - a preliminary discussion of patterns and processes. *Biol. J. Linn. Soc.* 14: 389-416.

Woodruff, D.S., and J.N. Thompson. 1980. Hybrid release of mutator activity and the genetic structure of natural populations. *Evol. Biol.* 12:129-162.

Wright, S. 1932. The roles of mutation, inbreeding, crossbreeding and selection in evolution. *Proc. 6th Int. Congr. Genet.* 1: 356-366.

Wright, S. 1934. An analysis of variability in number of digits in an inbred strain of guinea pigs. *Genetics* 19: 506-551.

Wright, S. 1935a. Polydactylous guinea pigs. *J. Hered.* 25: 359-362.

Wright, S. 1935b. A mutation of the Guinea pig, tending to restore the pentadactyl foot when heterozygous, producing a monstrosity when homozygous. *Genetics* 20: 84-107.

Wright, S. 1938. Size of population and breeding structure in relation to evolution. *Science* 87: 430-431.

Wright, S. 1940. Breeding structure of populations in relation to speciation. *Am. Nat.* 74: 232-248.

Wright, S. 1977. *Evolution and the Genetics of Populations.* vol. 3. University of Chicago Press.

Wright. S. 1978. *Evolution and the Genetics of Populations. vol. 4, Experimental results and evolutionary deduction.* Univ. of Chicago Press.

Wu, C.-I., and W.-H. Li. 1985. Evidence for higher rates of nucleotide substitutions in rodents than in man. *Proc. Nat. Acad. Sci. USA* 82: 1741-1745.

Wynne-Edwards, V.C. 1962. *Animal Dispersion in Relation to Social Behaviour.* Edinburgh and London: Oliver and Boyd.

Yamazaki, T., U. Tachida, M. Ichinose, H. Yoshimaru, Y. Matsuo, and T. Mukai. 1983. Reexamination of diversifying selection by using population-cages in *Drosophila melanogaster. Proc. Nat. Acad. Sci. USA* 80: 5789-5792.

Yourno, J., T. Kohno, and J.R. Roth. 1970. Enzyme evolution: generation of a bifunctional enzyme by fusion of adjacent genes. *Nature* 228: 820-824.

Zachar, Z., and P.M. Bingham. 1982. Regulation of *white* locus expression: the structure of mutant alleles at the *white* locus of *Drosophila melanogaster*. *Cell* 30:529-541.

Zinsmeister, W.J., and R.M. Feldman. 1984. Cenozoic high latitude heterochroneity of southern hemisphere marine faunas. *Science* 224: 281-283.

Zuckerkandl, E. 1975. The appearance of new structures and functions in proteins during evolution. *J. Mol. Evol.* 7:1-57.

Zuckerkandl, E., and L. Pauling. 1965. Evolutionary divergence and convergence in proteins. In *Evolving Genes and Proteins*, ed. V. Bryson and H.J. Vogel, pp. 97-166. New York: Academic Press.

Glossary

Accelerating Differentiation Increasing geographic differentiation over many loci, following isolation established by differences at only a few loci. [3]

Adaptation An historical process, where natural selection leads to an increase in performance of an organism. Performance is judged against a model of optimality, given a series of boundary conditions of evolutionary history and environmental constraints. [3]

Adaptedness A static description of the current functional superiority of one phenotype (genotype) over another. [3]

Adaptive Landscape A model describing a series of fitness peaks and valleys, as a function of genotype. The highest peak connotes the genotype with the highest fitness, though small peaks may be present. [3]

Allometry The relationship between change in shape and overall size. [6]

Allopatric When two populations or species are isolated from each other.

Allopatric speciation The origin of two or more species by means of one or more physical (geographic) separations. [4]

Allozyme An enzyme which is one of a series of alternative gene products of a given locus. [3]

Analogy The possession of a similar character state by two taxa, on the same location of the body, but owing to a factor other than common evolutionary origin of the taxa.

Background extinction The average extinction rate during a period of geological time, excluding times of extraordinary and sudden high extinction. [8]

Balance School School of population genetics that argued for the importance of maintenance of polymorphism chiefly by heterozygote advantage. [3]

Biological species concept View of species as a series of groups that are not interfertile. [4]

Canalization The case where the same phenotype is expressed, despite some variation in the genotype. The more constancy of expression, the stronger the degree of canalization. [3]

Character stasis Long-term constancy in a phenotypic character within a lineage. [7]

Clade A group of species that have descended from the same species. [2]

Cladistics The study of genealogy, principally using the method of grouping taxa by their shared derived character states (synapomorphies). [2]

Cladogram A diagram, in the form of a tree, grouping taxa by their synapomorphies. [2]

Classical School A school of population genetics that argued for the importance of directional selection, and, by extension, the likelihood of reduced polymorphism in natural populations. [3]

Cline A regular geographic change in average phenotype or genotype in species. [3]

Coding regions Those parts of the DNA used in the transcription process that leads to production of a protein. [3]

Codon A triplet nucleotide sequence that codes for a specific amino acid, or for termination of transcription. [3]

Concerted Evolution The presence of families of identical genes. [3]

Constructional morphology The field of morphology that argues that an organism's morphology can be explained as a combination of phylogenetic history, functional considerations, and boundary conditions imposed by the architecture, chemistry, etc., of the organism. [6]

Conversion See Gene conversion.

Correlated progression A hypothesis of multicharacter evolution which argues that character complexes evolve as an interacting process, whereby some evolutionary changes in the phenotype cannot be made until others occur. [7]

Darwin A rate of phenotypic evolution corresponding to a change by a factor of e per million years. [7]

Developmental constraint A factor in development, such as an obligatory tissue interaction, that might prevent or channel an evolutionary change. [5]

Directional Selection Natural selection, where a monotonic shift in phenotypic value is favored. [3]

Disruptive Selection Natural selection, where an intermediate phenotypic value is in disfavor, causing increases in frequency of divergent phenotypes. [3]

Downward causation In hierarchies, the case where a change in a hierarchical level causes a concomitant change at a lower hierarchical level. [1]

Duplicative transposition See Transposable element.

Effective Population Size A complex estimate of the number of individuals in a population that contribute genes to subsequent generations. Effective population size increases with an increase in the number of breeding individuals, an equalization of the sex ratio, and an increase of the geometric mean of population size of successive generations. [3]

Enhancer A nucleotide sequence at some distance from the structural gene that influences expression of the gene. [3]

Epigenetic pleiotropy Condition when a given developmental event affects widely disparate parts of the overall phenotype. [5]

Evolutionary Systematics A school of classification that groups taxa based upon the two criteria of evolutionary relationship and the degree of phenotypic difference. [2]

Exon The nucleotide sequence of the structural gene that codes for the protein. It may be interrupted by introns. [3]

Fitness The relative ability of different genotypes to leave offspring. [3]

Frameshift Mutation A mutation owing to a loss or gain of a small segment of DNA, and a subsequent shift of the adjacent DNA sequence. [3]

Frequency-dependent selection Natural selection whose strength and direction depends upon the frequency of alleles of a locus in the population. [3]

Gene conversion The process by which a specific sequence of DNA is converted to resemble another sequence. [3]

Gene flow The sum of successful dispersal and breeding of individuals originating in one population but arriving in another. [3,4]

Genealogy A set of relationships among taxa, based upon their evolutionary branching sequence. [2]

Genetic Drift The change in gene frequency, owing to random processes of gains and loss in a population. [3]

Genetic pleiotropy The effect of a gene on more than one aspect of the overall phenotype. [3,5]

Geographic stratigraphic completeness Proportion of the geographic range represented by the rocks containing a given fossil species. [7]

Heritability (broad-sense heritability) The proportion of phenotypic variance that can be explained by genetic variance. see Narrow-Sense heritability. [3]

Heterochrony A change in order of appearance of a trait during development between an ancestor and its evolutionary descendant. [5]

Hitchhiking (species level) The process by which a character state increases in frequency among species, by virtue of its coincidental association with a group that is increasing or decreasing in species richness. [4]

Homology Possession of a trait by two or more taxa because of a common evolutionary origin. [2]

Homoplasy Possession of two or more taxa of a character state not because of a common evolutionary origin, but because of convergence, parallel evolution, or evolutionary reversal. [2]

Hopeful monsters Hypothetical new mutants of major phenotypic difference that might give rise to strong evolutionary change in a population. [1]

Intron A nucleotide sequence that is part of a structural gene, but is not part of the DNA message that codes for the amino acid sequence of the protein. [3]

Inversion A reversal of some part of a chromosome, relative to a reference chromosome. [4]

Iterative evolution Repeated origin of evolutionary stocks of similar morphology.

Key innovation A new genetically-determined phenotype with strong potential for further evolutionary change. [6]

Lethal Genes A gene whose presence (usually in double dose in a homozygote) is lethal to the organism. [3]

Linnaean concept of species A species concept which idealizes the species by an essence, distinguishing it immutably from all other species. [4]

Lyellian curves Plots of the percentage of a fossil fauna that is still extant. Variants of this are common, such as: percentage of a fossil fauna that is present some number of years after a reference time datum. [8]

Macroevolution The sum of those processes that explain the character-state transitions that diagnose evolutionary differences of major taxonomic rank. [1]

Macromutations Mutations of strong phenotypic effect. [1,3]

Mass extinction An extinction that is characterized by loss of many taxa in a geologically brief time period. Also thought of as a statistically significant and higher rate of extinction, relative to the background extinction over a longer time period. [8]

Microevolution As commonly used, evolutionary change occurring within a species. [1]

Microstratigraphic acuity Amount of time represented by a given sediment sample. [7]

Modern Synthesis A movement of evolutionary biology, arguing that variation within natural populations must be studied to understand the overall evolutionary process, species arise from changes in within-population variation, and that natural selection is preeminent in the formation of the phenotype. [1]

Molecular clock Hypothesis that the rate of evolutionary change in DNA is constant over geologically long periods of time. [3]

Molecular drive Theory that argues that gene conversion, unequal crossing over, and duplicative transposition all contribute towards the construction of gene families. [3]

Monophyletic group A group of species that includes a species (or an hypothesized ancestral species defined by a set of character states) and all of its descendants. [2]

Morphological species concept Division into species mainly on the basis of morphological differences.

Mosaic evolution Case where different sets of traits in an organismal lineage evolve at different rates and apparently independently of each other. [7]

Mutation The set of processes that causes a change in a nucleotide sequence in an organism. [3]

Narrow-sense heritability The proportion of phenotypic population variance that can be explained by additive genetic variation (variation that can be ascribed to among-allele differences). [3]

Natural Selection The process of change of gene frequencies in a population, owing to fitness differences among genotypes. [3]

Neutral Theory of Molecular Evolution Theory arguing that DNA sequence evolution is governed by stochastic processes and that molecular polymorphism is neutral with regard to fitness. [3]

Paracentric inversion Inversion not including the centromere. [4]

Paraphyletic group A group of species that includes the ancestral species, but not all of the known descendant species.

Parsimony The use of the shortest number of evolutionary steps as a criterion for constructing a cladogram. [2]

Peripatric speciation A hypothesis of speciation by budding off of small founder populations into geographically marginal, and usually ecologically marginal, areas. Such budding is accompanied by strong natural selection, genetic drift, and rearrangement of genetic determination of phenotypic traits. [4]

Phenetics A theory and practice of classification involving grouping of taxa by virtue of their overall similarity. [2]

Phyletic evolution Evolutionary change within a lineage, without cladogenesis. [7]

Phylogenetic systematics The area of study where genealogies are established using cladistic methods. [2]

Phylogeny A hypothesis of the evolutionary history of a group, including hypothesizing ancestors, and exact order and geographic locus of cladogenesis. [2]

Pleiotropy See Genetic Pleiotropy, Epigenetic Pleiotropy

Plesiomorphy The state where two taxa share ancestral character states for lack of any divergent evolutionary change. [2]

Point mutation A DNA mutation where one nucleotide has been changed into another, with no change in overall number of sites.

Polymorphism The presence of variants, genetic or phenotypic, *within* a population. [3]

Polytypism The situation where a species consists of a series of geographically and phenotypically (or genotypically) distinct subpopulations. [4]

Primary Intergradation Case where a cline is due to processes excluding a history of isolation and secondary contact. [3,4]

Promoter A nucleotide sequence immediately "upstream" of the structural gene that is necessary for initiation of transcription. [3]

Protein Polymorphism A polymorphism in a population of different protein products of a locus. See allozyme. [3]

Pseudoextinction The apparent extinction of a taxon, because it has evolved phyletically into another taxon that is sufficiently distinctive to been given a new taxonomic name. [7]

Pseudogenes Non-functioning (and probably former) structural genes. [3]

Punctuated equilibrium theory A theory that asserts that species do not change during most of their history, but change is mostly concentrated at speciation events. [1,7]

Quantum evolution A greatly increased rate of phyletic evolution over a relatively short period of time. [7]

Reference Time Datum (RTD) A time horizon in the geological record, which is used as a starting point for the study of taxonomic survivorship, either forward or backward in time. [8]

Regression In geological parlance, a lowering of sea level, explained either by a global fall of sea level, or a local rise of land elevation. [7,8]

Regulatory Gene A gene whose product is a polypeptide that affects the degree of expression of a structural gene. [3]

Robertsonian fusion Fusion of two acrocentric chromosomes. [4]

RTD See Reference Time Datum.

Secondary contact After allopatric speciation, the reintroduction of contact between the offspring species by elimination of the geographic barrier. [4]

Selection Coefficient A fraction representing the proportional difference in fitness of one genotype relative to others. [3]

Shifting Balance Theory Theory of interaction of gene flow, local population differentiation, and genetic drift. Genetic drift or intermittent gene flow permits a population to leave a local adaptive peak and climb a higher one. [3]

Speciation The formation of two or more new species from one former species. [4]

Species drift The increase or decrease of number of species in a taxonomic group owing to random speciation and extinction processes. [4]

Species selection A change in the number of species in a taxonomic group owing to its deterministically different speciation or extinction rate. [4]

Species stasis Lack of change in morphological characters of a species throughout its history. [7]

Stabilizing Selection Natural selection that favors the same genotype or phenotype throughout successive generations. [3]

Step Cline A cline with a relatively sharp change over a small amount of geographic distance, relative to very little change on either side of the step. [3]

Stratigraphic completeness See Temporal stratigraphic completeness, Geographic stratigraphic completeness.

Structural gene A gene whose product is a protein involved in cellular metabolism, or at least does not act by binding to DNA and thereby directly regulate expression of other genes. [3]

Sympatric When two populations or species are in sufficient contact for an opportunity for interbreeding to occur.

Sympatric speciation Division of a species into two or more offspring species in the absence of geographic isolation. [4]

Synapomorphy A character state shared among two or more taxa, but to no others outside of this group. [2]

Taxonomic Longevity The time over which a given taxon lives. [8]

Temporal scope The period of time represented by a given column of rock. [7]

Temporal stratigraphic completeness The proportion of geological time of the total temporal scope represented by the preserved rock column. [7]

Transgression A deepening of water at a site, owing either to a rise in sea level, or a local fall of the sea bed. [7,8]

Transilience mechanisms of speciation Speciation mechanisms that involve changes in genetic organization, such as polyploidy. [4]

Transposable Elements DNA sequences that are capable of moving from one part of the nuclear DNA to another, or are capable of producing a replicate that can move to another site (duplicative transposition). [3]

Truncation Selection Natural selection that eliminates all individuals beyond a given phenotypic value. [3]

Upward causation In hierarchies, the case where a change in a lower hierarchical level has an effect on higher levels. [1]

Author Index

Subject Index

accelerating differentiation, 126
acceleration, *see* heterochrony
adaptation, 90, 114, 282-284,
   303-304;
   in mammals 393-400
Aegilops, 195
Aegilops speltoides, 195
Aegilops squarrosa, 195
Afrobolivina afra, 360, 363, 364
adaptedness, 114
adaptive landscape, 130
advancement index, 63
allometry, 309-320;
   allometric coefficient, 309;
   and developmental constraint,
312-320;
   exponent, 309
allopatric speciation, *see* speciation
allopatric-dumbbell model, *see*
   speciation
allozymes, *see* molecular
polymorphism
Ambystoma mexicanum, 257
ametapodia mutant in chicken, 236
ammonoids, 297, 330
ammonites, 347
analogous structures, 60
Aneides flavipunctatus, 264, 265
angiosperms, 401
Anguilla rostrata, 126
Anolis, 437
antennapedia, 155
apical epidermal ridge (AER), 242
apomorphous, 49
apterous mutation, 252
Arachnoides, 295
Archaeopteryx, 248, 400, 414

architectural constraints, 290
Arisaigia postornata, 360
Aristelliger praesignis, 120
Artabwandlung, 406-407
Artemia salina, 215, 216, 236, 250
ascite caudal mutant in *Pleurodeles
   waltl*, 232
atavism, 247-279
Athleta petrosa, 267, 268
atomism, 11
autapomorphy, 52
Aysheaia pedunculata, 415
Babinka, 82
Bachia, 245
background extinction, 450
Balaenoptera borealis, 169
balance school, 110
bauplan, 6;
   gradual assembly, 391-407
bears, fossil, 348
biogenetic law, 58
biogeography, 437-450
biometricians, 21
biorthogonal analysis, 322-324
Biston betularia, 118, 363
bithorax complex of *Drosophila*, 98,
   103, 155, 252
Bivalvia, 48, 441
Bolitoglossa, 59, 254, 255, 263, 264,
   265
brachiopod, 79, 297, 308
bryozoa, and punctuated
equilibrium, 377
building, in development, 220
burden, 16, 217, 292
Caenorhabditis elegans, 101, 221,
   230, 250, 491

627